PERGAMON INTERNATIONAL LIBRARY
of Science, Technology, Engineering and Social Studies

The 1000-volume original paperback library in aid of education, industrial training and the enjoyment of leisure

Publisher: Robert Maxwell, M.C.

Biomagnetism: Applications & Theory

BIOMAGNETISM:
Applications & Theory

Proceedings of the Fifth World Conference on Biomagnetism
Vancouver, Canada, August 1984

Edited by
Harold Weinberg
Simon Fraser University
Burnaby, British Columbia, Canada

Gerhard Stroink
Dalhousie University
Halifax, Nova Scotia, Canada

Toivo Katila
Helsinki University of Technology
Espoo, Finland

Pergamon Press
New York • Oxford • Toronto • Sydney • Paris • Frankfurt

Pergamon Press Offices:

U.S.A.	Pergamon Press Inc., Maxwell House, Fairview Park, Elmsford, New York 10523, U.S.A.
U.K.	Pergamon Press Ltd., Headington Hill Hall, Oxford OX3 0BW, England
CANADA	Pergamon Press Canada Ltd., Suite 104, 150 Consumers Road, Willowdale, Ontario M2J 1P9, Canada
AUSTRALIA	Pergamon Press (Aust.) Pty. Ltd., P.O. Box 544, Potts Point, NSW 2011, Australia
FRANCE	Pergamon Press SARL, 24 rue des Ecoles, 75240 Paris, Cedex 05, France
FEDERAL REPUBLIC OF GERMANY	Pergamon Press GmbH, Hammerweg 6, D-6242 Kronberg-Taunus, Federal Republic of Germany

Library of Congress Cataloging in Publication Data

World Conference on Biomagnetism (5th : 1984 :
Vancouver, B. C.)
Biomagnetism : applications & theory.
Includes index.
1. Biomagnetism—Congresses. I. Weinberg, Harold.
II. Stroink, Gerhard. III. Katila, T. E. (Toivo E.) IV. Title.
QH504.W67 1984 574.19'12 85–3723
ISBN 0-08-031971-8 (pbk.)

Printed in Great Britain by A. Wheaton & Co. Ltd, Exeter

CONTENTS

CONTENTS

CONTENTS

CONTENTS

OTHER STUDIES

PUBLIC LECTURE

Biomagnetism: Applications & Theory

PREFACE

Watson: "This is indeed a mystery, What do you imagine it means?"

Holmes: "I have no data yet. It is a capital mistake to theorize before one has data. Insensibly one begins to twist facts to suit theories, instead of theories to suit facts."

Sir Arthur Conan Doyle
<u>A Scandal in Bohemia</u>

As in any science, and especially in a relatively new science, there is and must be a continuous and imaginative modification of theory relative to data. This book reflects the most current and the most sophisticated data available. The book also gives the reader an understanding of which aspects of theory are actively under review.

The science of Biomagnetism applies a technology which was originally developed for the measurement of extremely small magnetic fields in physics. The measurement of small fields must include the means through which they can be discriminated from the surrounding magnetic noise which is of both biological and non-biological origin. Consequently, the problem of distilling small signals from noise, especially at very low signal frequencies, is the focus of much contemporary research in the field. Methods of signal extraction have been improved with advances in instrumentation and also with innovative statistical treatments. The present technology is capable of measuring biomagnetic fields in the very noisy environments of urban settings and hospital laboratories, without expensive shielding. These developments have led to the increased proficiency of laboratories in the study of many biological systems including the function of heart, lung, brain, muscle, liver and others.

In this book are the first comprehensive reports of the measurement of biomagnetic fields using dc-SQUIDs (Superconducting Quantum Interference Device). The high sensitivity achieved with this technology will open new areas in Biomagnetism and will be an important new tool of future research. Several groups describe new multiprobe systems, or active plans to acquire them. Although this is not the first time such systems have been described, this book includes the most extensive discussion to date of their advantages and possible capabilities.

Many of the studies reported are concerned with modeling sources or generators of magnetic fields. Such models can be found in this book under the section dealing with heart, lung, brain and nerve function. With respect to lung function, the relaxation technique for the study of lung clearance and microphage activity reported here has been a very important development. Nerve studies reported here, which combine measurements of electrical and magnetic fields, have accurately and reliably derived some of the physical parameters important in nerve conduction.

An understanding of the complexity of biomagnetic sources and their location is necessary to analyse functions of biological systems, and to utilise Biomagnetism for the diagnosis and assessment of pathological states. Great progress has been made in these directions. For the study of electrophysiological processes of the heart and brain, many research groups have developed advanced mapping techniques which emphasize the complimentary use of both magnetic and electrical data. This has allowed the identification of new features in the magnetocardiograph (MCG), in both normal and abnormal heart function. Progress in the use of the MCG for diagnosis will undoubtedly increase when more data on abnormal heart function becomes available in the future.

An important maturation of the field has been the development of models for the localisation of simple and complex sources in the brain. Although the emphasis in this book has been on the study of magnetic fields, many of the experimental methodologies and techniques of electrophysiology have been applied in the studies of Biomagnetism, primarily for the purpose of identifying the current sources of Event Related Electrical Potentials associated with complex information processing and sensory function in humans. Another body of research which is very promising is the measurement of magnetic fields to localise sources which may be active in focal epilepsy. The combination of electrical and magnetic data has proved to be very useful in clarifying the degree to which electrical potentials originate in different brain systems.

The book reflects a growing interest in imaging techniques as they are applied to Biomagnetism. Imaging of sources may in the future be able to combine better spatial and temporal resolution for the purpose of differentiating function within well defined physiological structures.

Contemporary issues, new data and advances in theory are evaluated in the Panel Discussions. The format of the Panel Discussion was used for the first time in Biomagnetism conferences and these discussions have helped, together with the invited talks, to bring into focus the future directions of research.

Acknowledgments: At the 5th World Conference on Biomagnetism there were 152 participants from laboratories in Australia, Brazil, Canada, Czechoslovakia, Denmark, Federal Republic of Germany, Finland, France, Great Britain, Italy, Japan, Poland, Spain, Switzerland, The Netherlands, U.S.A., U.S.S.R., and Yugoslavia. The research reports, poster presentations and other papers were reviewed by recognised experts in the field. We would like to acknowledge their valuable contribution to the conference and to the substance of this book.

The Programme Committees of the Conference were:

Chairman of the Conference - Harold Weinberg
Departments of Psychology and Biological Sciences
Simon Fraser University, Burnaby, B.C., Canada

North American Committee - Gerhard Stroink (Chairman)
Department of Physics
Dalhousie University, Halifax, N.S., Canada

Samuel J. Williamson
Department of Physics
New York University, New York City, N.Y., U.S.A.

John P. Wikswo, Jr.
Department of Physics and Astronomy
Vanderbilt University, Nashville, Tennessee, U.S.A.

The Conference and this Book were funded in part by the following:

Natural Sciences and Engineering Research Council of Canada
Ministry of Industry and Small Business of British Columbia, Canada
Science Council of British Columbia, Canada
Simon Fraser University, British Columbia, Canada

Technical assistance in the preparation of this book was provided by:

Dalhousie University, Nova Scotia, Canada

H. Weinberg, G. Stroink, and T. Katila
(Editors)

INTRODUCTION: A 1984 PERSPECTIVE ON BIOMAGNETISM

David Cohen

Francis Bitter National Magnet Laboratory
Massachusetts Institute of Technology
Cambridge, Massachusetts 02139 USA

ABSTRACT

A brief overview is given of biomagnetism, including a definition, a brief history, and some comments about the present status and future prospects.

KEYWORDS

Biomagnetism definition, biomagnetism milestones, external magnetic field, applied magnetic field, natural ion currents, Fe_3O_4 particles.

INTRODUCTION

I believe I have been asked to give this talk because I've been working in biomagnetism for about 20 years now, essentially from the beginning. Therefore, perhaps I can give some perspective to biomagnetism, especially to the newcomers at this conference. To do this, I will first state the definition of biomagnetism, then give a brief history, then make some comments about where we are now, and finally make some remarks about the future. In a sense, this is a brief update or continuation of a perspective I gave several years ago (Williamson and colleagues, 1983).

DEFINITION OF BIOMAGNETISM

Most of us accept and use this definition: Biomagnetism is the phenomenon where biological material produces an external magnetic field.

We explain this definition by stating that the biological material can produce the external magnetic field in three ways. First, it can produce a field in a natural way; muscles or nerves naturally generate electrical ion currents in the body, and these currents produce the external field. Examples of these are the currents generated by the heart muscle or the brain; these produce a magnetic field over the chest (the magnetocardiogram or MCG), or a magnetic field around the head (the magnetoencephalogram or MEG). In the second way, biological material can produce a magnetic field due to iron-like magnetic particles within the material, usually Fe_3O_4 particles. These can be contaminating particles which enter the body in various ways, such as being inhaled into the lung; these produce a remanent magnetic field, say over the lung region, after they have been aligned by an applied magnetic field which is then removed. These Fe_3O_4 particles can also be produced naturally in the biological material, as in some bacteria and in chitons and other creatures; here, in some cases the particles may already be naturally aligned. In the third way, biological material itself (without particles) can produce a magnetic field in response to an applied field; this induced field exists only during the time the applied field is maintained, so it must be measured in the presence of the much stronger applied field. The induced field depends on the magnetic susceptibility of the biological material, which in turn depends on the composition of the biological tissue. At this conference, reports are presented which involve all three ways of producing a magnetic field.

We note that this definition of biomagnetism does not include the effects of an external magnetic field on living biological material. That is another phenomenon, not directly related to our definition, and is often called Magnetobiology. Nevertheless, various members of this conference maintain an interest in this phenomenon, and several papers of this type are presented here.

A BRIEF HISTORY

I here list only the major milestones of biomagnetism. More complete information along these lines can be found in several reviews (Cohen, 1975; Williamson and Kaufman, 1981; Williamson and colleagues, 1983). I will first list the major experimental milestones, then the major theoretical milestones.

The first reliable biomagnetic measurements were made by Baule and McFee in 1963. They used two side-by-side large coils (a gradiometer) to record the first MCG; the gradiometer greatly reduced the magnetic background, and for further reduction they measured their subject out in a rural field, away from urban magnetic disturbances. Next, a different approach in coping with the magnetic background was used by me, that of the shielded room. This allowed a clearer measurement with a smaller coil, and in 1967 I was easily able to confirm the Baule-McFee heart measurement; also I was able to detect the magnetic field of the brain and other organs, where signal averaging was necessary because of the large intrinsic coil noise relative to the weak signals.

Later I built a better room, and in 1969 Jim Zimmerman brought me one of his first reliable point-contact SQUIDs to use in that room. This combination of room and SQUID allowed, for the first time, high quality (low noise) MCGs; it also allowed direct measurements (without signal averaging) of the brain and muscle, and the first good dc measurements from various organs in the body, including measurements of particles in the lung. These new measurements with the SQUID began to interest both physicists and physiologists, and started the ball rolling in biomagnetism. Next, in 1971 Zimmerman showed that a first-order SQUID gradiometer could be used without shielding to measure the MCG, and several years later Sam Williamson and his group showed that a second-order gradiometer could be used to measure the MEG in an urban environment without shielding; these demonstrations of SQUID use without shielding gave further impetus to biomagnetism. Then, in 1980 David Farrell and his group showed that the magnetic susceptibility of the liver could be measured and iron-overload disease detected in this way. Also, at that time, John Wikswo showed that the magnetic field of a single nerve bundle could be measured, using a toroidal coil around the nerve bundle of a frog.

Concerning the major theoretical milestones, first in 1965 Baule and McFee showed that the concept of the lead field could be used in biomagnetic theory; this allowed rapid estimation of the field expected from a dipole source. They also showed that a radial dipole in a spherical conductor produces no external field, so important in understanding the MEG. Then, in 1970 David Gezelowitz provided the mathematical basis for calculating magnetic fields due to biomagnetic sources generally, for any arrangement of sources and volume conductors. Next, in 1972 Robert Plonsey showed that the magnetic and electric fields can, in principle, contain different information about the electrical sources, so that magnetic and electrical biomagnetic measurements are complementary at the most fundamental level. Then, in 1976 our MIT group showed that the magnetic field component normal to the surface of special conductors receives no contribution from the volume current.

So much for the major milestones. Next, I will give a brief history of these biomagnetic conferences. The first conference was held in Cambridge, Mass. (USA) in 1976. Thereafter, they were held at two-year intervals. The 1978 conference was held in Grenoble (France); the 1980 conference in Berlin (Germany); and the 1982 conference was held in Rome (Italy). At the first conference at MIT there were 22 official attendees, while here in Vancouver, there are about 150 official attendees, so we have certainly grown during these eight years. As a further indication of our growth, a parsimonious colleague pointed out that the registration at the first conference was free, while now the registration fee has risen to $200!

Finally, from the historical point of view, I'll show a biomagnetic measurement made in the "early days" and, for comparison, I'll also show some corresponding measurements made today. The early biomagnetic measurement is one of the first recorded MEGs, measured with a room-temperature coil; the corresponding measurements made today are MEGs measured with the latest type of SQUID. These are shown in Fig. 1. The point is that, in 1968, eight minutes of signal averaging was necessary to see only one cycle of the alpha rhythm; the method was cumbersome, where an elaborate trigger for the averaging was derived from the EEG. In contrast, today, direct (no averaging) MEG recordings can be made of the alpha rhythm and other spontaneous signals. Not only are these recordings as clear as with the EEG, but in one sense are better than the EEG, in that the MEG bandwidth can be extended down to dc; this is difficult or impossible to do with the EEG. Thus, in 16 years there has been dramatic progress in the MEG specifically, and in biomagnetism technology generally.

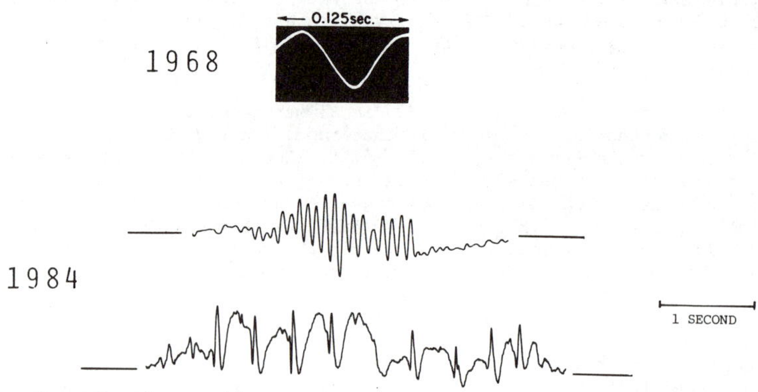

Fig. 1. An early biomagnetic measurement compared with corresponding measurements made today. The 1968 trace is the MEG of the alpha rhythm, due to eight minutes of signal averaging (Cohen, 1968); only one cycle could be averaged. In contrast, the two 1984 traces are single-trace MEGs (no averaging). The upper 1984 trace is a burst of alpha rhythm, while the lower trace is a burst of spike-and-wave from an epileptic patient. In these two traces the low frequency limit is at dc, therefore they have a true baseline (horiz. end-lines), and the low-frequency aspects of the signals can be seen with respect to this baseline. These dc measurements cannot reliably be made with the EEG.

PRESENT STATUS

Perhaps one way to see the status of biomagnetism today is to look at some numbers, shown in Tables 1 and 2. Table 1 outlines the status in a general, practical way; the numbers are only approximate, and represent an informal estimate. The number of groups around the world includes many new groups, some of whom are not yet represented at this conference. The number of professional researchers includes graduate students, and the number of refereed papers excludes those presented at these conferences. The funding awards are meant to be a worldwide estimate, as are the other numbers. We bear in mind that the total number of proposals actually written every year is much greater than those funded; I estimate 200 of these, which is a substantial activity.

TABLE 1 Status of Biomagnetism in 1984

No. of groups around the world:	40 to 60
No. of professional researchers (incl. grad. students):	200 to 250
No. of refereed papers each year (excl. biomagn. conf.):	50 to 60
No. of active funding awards:	40 to 60

Table 2 indicates the relative amount of biomagnetic work being done today (via the papers presented here), and compares this with the effort of two years ago. Again, the numbers are only approximate. It is seen that the instrumentation activity has remained fairly constant. However, the brain papers have increased substantially. They are the largest block of work presented at this conference, and include an increase in theoretical papers, which mostly deal with solutions of the inverse problem, that is, determining the internal generators from MEG measurements. The number of heart papers has remained fairly constant, although these include an increase in theoretical papers, mostly Wikswo's work on fibers. The susceptibility work has decreased (no liver work is presented). However, there is a significant increase in work involving Fe_3O_4 in the lungs, and then a surprising increase in the papers involving relaxation; these are a spin-off from the lung work, where the phenomenon of magnetic relaxation had been shown to be due to the macrophage cells, and relaxation in these cells is now investigated separately. This is a new area within biomagnetism which can be called "cytomagnetism."

TABLE 2 Change in No. of Conference Papers Presented

Subdivision	Rome	Vancouver
Instrumentation	10	13
Brain (incl. theory)	15	25
Heart (incl. theory)	13	15
Susceptibility	3	1
Lung	2	7
Relaxation in cells	1	6
Other	2	2
Total	46	69

Not shown in the above numbers are the problems in biomagnetism today. Two of these will be mentioned here. Certainly, in the U.S. the most serious problem is funding; this problem is apparently not as serious in Europe where, for example, the German and Italian groups appear to be well funded. In the U.S., biomagnetism proposals receive a low priority, partly because biomagnetism has not shown direct practical application to date. However, the situation seems to be improving somewhat, and government agencies are taking notice of the increasing effort in biomagnetism with its prospects for future application. Above, I estimated a present ratio of proposals written/funded of about four; this is too large a ratio for a new and promising technology, and we hope to see an improvement in this ratio.

The second problem was not foreseen until recently. Although we can now readily map the magnetic field around the heart and lung, and especially the brain, how do we solve inversely to determine the sources? After all, that is the main purpose of these measurements, to learn the distribution and strength of the sources. Unfortunately, solving the inverse problem is still in a primitive state. Work in this area began years ago for the ECG, was extended into the EEG, then recently was further extended into biomagnetism, especially for the brain. The advances have been impressive; however, there is a long way to go in order to extract all possible information contained in the magnetic field. To see what is needed, as an example, we can consider the technology of weather forecasting; in some ways the problem of determining the electrical sources in, say, the brain from MEG maps is similar to predicting the weather from meteorological measurements. However, in the latter case, an elaborate computing effort is used to extract much of the information contained in the measurements and is becoming quite successful. We don't do this in the MEG inverse problem because we have not learned how to connect the different instantaneous states, that is, we don't use data recorded at one instant to enhance the data at the next instant. Yet the information is there; for example, the somatosensory evoked map at 22 msec certainly "knows about the map" at 20 msec, etc., but we don't feed this information back and forth. Therefore, what is needed in the case of the MEG is a large effort to perform iterative procedures.

FUTURE PROSPECTS

Based on Table 2, it appears from this 1984 perspective that the largest application of biomagnetism will be in brain measurement. This is the area growing most rapidly, and where a new technology is badly needed by the research and medical communities. Because the EEG has various limitations, and the combination of MEG and EEG is much better than EEG alone, the MEG has a good future. However, this combination would be better exploited if we could move ahead with a larger effort for the inverse problem.

Another application which is especially promising, based on Table 2, is the use of magnetic particles as a probe of cell function (cytomagnetism). The cyto-researchers at this conference indicate that this is promising because there is no competing way of measuring the motions within macrophage cells, and these cells are involved in organs other than the lung, for example, in the liver. Further, magnetometry in this area is far less expensive than in the brain, because only the inexpensive fluxgate need be used here instead of the SQUID.

A third application that is promising (not apparent in Table 2) is the magnetic measurement of nerves and muscle fibers, using a toroidal coil around the fiber. This is because the magnetic technique has some advantages over the potential-measuring technique; for example, the fiber is not impaled, and the current is directly determined. Further the number of neuroscience users could be very large. Of course, there are other promising areas, such as the lung measurements, and certainly there will be more surprises as the instrumentation advances and biomagnetism develops and moves along. Thus, viewed from 1984, biomagnetism is a growing and promising area of research.

REFERENCES

Cohen, D. (1968). Magnetoencephalography: evidence of magnetic fields produced by alpha-rhythm currents. Science, 161, 784-786.

Cohen, D. (1975). Magnetic fields of the human body. Physics Today, August, 34-43.

Williamson, S., and L. Kaufman. (1981). Biomagnetism. J. Magn. Magn. Mat., 22, 129-201.

Williamson, S., G. Romani, L. Kaufman, and I. Modena (Eds.). (1983). Biomagnetism: An Interdisciplinary Approach. Plenum Press, New York. Chap. 1 by D. Cohen, 5-16.

THEORETICAL MODELS FOR SOURCE LOCALIZATION

B.N. Cuffin[1], T.E. Katila[2], M. Pelizzone[3], and J.P. Wikswo, Jr.[4]

[1] Nat. Magnet Lab., MIT, Cambridge, MA, USA

[2] Dept. of Tech. Physics, Helsinki Univ. of Technology, Espoo, Finland

[3] Dept. of Physics, NYU, NY, NY, USA

[4] Dept. of Physics and Astronomy, Vanderbilt Univ., Nashville, TN, USA

ABSTRACT

The purpose of this panel is to discuss the present state of source localization methods and models. The panel concludes that the choice of the models for the source and the body has a great effect on the accuracy of source localization. The single current dipole seems to be an adequate model for the source in many, but not all, situations. The adequacy of the presently used models of the body is largely unknown. The panel also concludes that source localization acccuracy for simple, dipolar sources in the heart may be in the range of several centimeters while the accuracy for such sources in the brain may be in the range of several millimeters. The accuracy for sources in an isolated nerve axon is in the range of a fraction of a millimeter. Finally, the panel concludes that the help that magnetic measurements provide in source localization as compared to electric measurements depends on the experimental situation.

KEYWORDS

Source localization; modeling; inverse solutions; neuromagnetism; MEG; MCG; biomagnetism.

INTRODUCTION

Once measurements have been made of the magnetic fields produced by an electrical source in the human body or other biological organism, the next step is to use these measurements to obtain information about the source. The information which is of greatest clinical and research use is the location of the source. In order to obtain this information, it is necessary to assume some models of the source and the body. The parameters of the source model are then adjusted to give the best fit between the measured data and that produced by the source model in the body model. This source model is then taken to represent the actual source in the body.

The purpose of this panel is to discuss, in a general way, the present state of source localization methods and models. To this end, each panel member will give his views on the following three questions.

1) What is the importance of the modeling assumptions? How useful is the dipole as a source model? How adequate are present models of the body?

2) What is the present accuracy of source localization?

3) How can magnetic measurements help in source localization?

PANEL MEMBER RESPONSES

1) What is the importance of the modeling assumptions?

KATILA: We have investigated the use of the magnetic multipole expansion (MME) for magnetocardiographic studies (Karp and co-workers, 1980). The MME takes into account all currents flowing in the body, both the primary current sources in the heart and the secondary volume currents that are generated in the body by the primary sources. There is a theoretical problem with the MME which is that the MME strictly represents the fields only outside of the sphere containing all the currents and measurements on the thorax fall well inside of this sphere. Nevertheless, the lowest order term (magnetic dipole) in this expansion was able to account for up to 80% of the measured field. The disadvantage of the MME is that there is no clear connection between the physiological sources and the equivalent sources of the MME. This is also true of the vector magnetocardiogram. On the other hand, several medical uses of bioelectric measurements are very empirical, e.g., the 12-lead ECG.

The current multipole expansion (CME) can be used for measurements made inside the sphere containing the volume currents (Katila and Karp, 1983). However, the influence of the volume currents on the measured field component presents a problem. One possible way to avoid this problem is to model the body by such special geometric shapes as a sphere or semi-infinite slab and use the radial or perpendicular component of the magnetic field only. The CME introduces the current dipole as the lowest order source term. The next order terms (current quadrupoles) then contain the magnetic dipole term of the MME. Using an idealized semi-infinite half-space model for the thorax, the current dipole term alone was to represent up to 80% of the measured field during the QRS of the normal MCG.

We have also compared the dipole terms of the MME and CME as equivalent sources for the signals caused by eye blinks (Antervo and co-workers, 1984). Both dipoles (magnetic and current) are able to explain the measured magnetic field distributions almost equally well. However, the current dipole model explains the electric potential distribution as well and it may be more easily understood as a physiological source.

If the dimensions of the source volume are small as compared to the distance between the source and the measurement point, the dipole term alone will often explain the measured field. In some cases, such as the eyes, the use of two or even more dipoles may be physiologically postulated. For an extended source, the CME offers a unique presentation. We have found it useful even in calculating components of the electric heart vector from magnetic measurements. The MME has been practical in studies of the magnetic contamination of the lungs (Brauer and Stroink, 1984; Stroink and co-workers, 1984).

The computer models available today for studies of the MCG have helped much in understanding the spatial and time dependencies of cardiac signals. Problems still remain, e.g., during the maximum QRS the models suggest approximately zero signal in the region above the heart while large signals are measured experimentally in this region. For studies of the brain, such sophisticated computer model do not yet exist.

PELIZZONE: Neuromagnetic studies during the past five years have established that the single current dipole in a homogeneous conducting sphere are the most appropriate models for many types of brain activity. In order to use a single dipole, it must be assumed that only one discrete region of the brain is active at a given instant in time, and that it is relatively far from the point where the magnetic field is observed. The law of Biot and Savart specifies then how the current density within each infinitesimal volume in space contributes to the magnetic field at a given point. This assumption may not be generally valid, but in fact, it turns out that in many cases the field pattern of a single current dipole is not statistically distinguishable from that due to a variety of spatially extended sources simultaneously active because of the inherent averaging by the pick-up coil and the experimental variablity in the data. (Okada, 1984) Thus, except for a few special cases, the quality of the actual measurements justify the use of the single dipole model.

The use of a homogeneous sphere as the model for the head is based in part on the fact that concentric layers of different conductivity and the numerical values of the conductivities have no effect on the normal component of the magentic field outside the sphere. (Grynszpan and Geselowitz, 1973) The only important parameter is the sphere radius. This can easily be understood if one remembers that the secondary source produced at a boundary produces no magnetic field along its own normal and that the normal component of the magnetic field outside the sphere originates from the primary source only, i.e., the dipole. Small departures from the spherical symmetry are believed to produce only negligeable corrections. (Tripp, 1977) However, by mapping the magnetic field at constant distance from the surface of the head, one may deviate from a spherical surface by as much as 1 cm in some cases. This can lead to an asymmetry in the measured field pattern because the distance between one extremum and the active region is different from the distance between the other extremum and the active region. On that particular point one has yet to define the best mapping strategy: Is it better to measure the field patterns strictly normal to an imaginary spherical surface, or is it better to map normal to the head and try to correct for the nonsphericity of the head by adjusting the distance between the sensor the skull? Some numerical and experimental estimates of these effects are needed for interpreting high quality data and will help in the design of the future large arrays of sensors.

The use of the homogeneous sphere as the head model also assumes only concentric variations of the conductivity, thus ignoring the distortions induced by non-concentric regions of inhomogeneous electrical conductivity. The field pattern measured by Okada (1983a) for electrical stimulation of the ankle is highly asymmetric. The evoked source lies in the somatosensory cortex at the upper end of the Rolandic fissure near the midline. It would be interesting to have practical estimates of the effects of the inhomogeneous conductivity in this region due to the falx cerebri. Recent theoretical calculations (Cuffin, 1982) show that an internal sphere whose conductivity is only five times that of the rest of the volume conductor significantly affects the symmetry of the field pattern. Thus, this mechanism could be important in such special situations.

Within these assumptions, the current dipole in a homogeneous sphere gives excellent results when applied to fit the dominant features of the neuromagnetic fields. In the somatic, visual, and auditory modalities, the locations and orientations of the generators have been shown to be consistent with the neuroanatomy (Okada, 1983a, 1983b; Tuomisto, 1983) in that they agree fairly well with the dominant landmarks of the brain determined by procedures described as in Gray (1977). However, systematic comparisons between the neuroanatomy of individual subjects as revealed by Nuclear Magnetic Resonance (NMR) scanners and the location(s) of the generator(s) deduced from the neuromagnetic data should be done in the future. The UCLA group has shown promising results of that kind in their study of focal epilepsy by the neuromagnetic technique (Barth and co-workers, 1984).

WIKSWO: The detailed theoretical study of the biomagnetic fields produced by the heart and brain is greatly complicated by the fact that the actual biological current sources that produce the fields can seldom be accurately specified because of the difficulty in performing the invasive measurements necessry to determine the source distribution. Thus when measured magnetic fields are studied in terms of models, the inability to make an accurate comparison of the modeled, or equivalent, sources with the actual ones is a major limitation in verifying the model. If both electric potential and magnetic field data are available for the same source distribution, the problem is alleviated somewhat, since one set of data can be used in an inverse calculation to generate the model and the modeled equivalent sources can then be used in a forward calculation to generate the complementary field.

The theoretical and experimental analysis of the magnetic field of an isolated nerve axon (Swinney and Wikswo, 1980; Barach, Roth, and Wikswo, 1984; Woosley, Roth,and Wikswo, 1984; Roth and Wikswo, 1984) is unique in the field of biomagnetism in that the actual sources of the fields are determined by the highly invasive penetration of the nerve cell with a glass microelectrode. This provides us with the ultimate test of a model, since sufficent data are available to perform forward and inverse calculations using each field or potential to predict the other measured fields and potentials.

We model the single axon as a thin, infinitely long cylindrical membrane that contains the active source. The currents inside and outside the membrane are Ohmic and obey Laplace's equation. From the point of view of this discussion, the goal of a magnetic measurment is to determine the transmembrane potential, which is the true source of the measured signals. The model used for this analysis presents a set of very restrictive but most realistic assumptions: the location of the axon is known and the constraint of cylindrical symmetry means that our source can only exhibit axial variation. These modeling assumptions are important, since they allow the forward and inverse problems to be solved analytically.

CUFFIN: The choice of appropriate source and body models is of great importance for accurate source localization since improper choices can produce inaccurate and misleading results. For example, computer model studies (Cuffin, 1981a) show that except for MCG measurements directly over the heart, modeling the external surface of the torso as the surface of a semi-infinite volume can lead to significant errors in source localization.

The use of the current dipole as a source model is adequate for many situations. Computer simulation studies (Cuffin, 1984) indicate that the source must be large as compared to the area over which the measurements are made before the single dipole source model no longer gives a reasonable estimate of the location of the approximate center of the source as well as of its net dipolar electrical activity. Previous work (Cuffin, 1981b) has shown that using a source model with more than two dipoles quickly leads to problems in obtaining physiologically meaningful solutions, i.e., the dipoles do not represent the electrical sources in the region near them. Hence, except for special cases where additional information about the source configuration is available, the single dipole model is the only choice. The accuracy of the dipole in representing the net dipolar activity of the actual source can be improved somewhat by using a multipole expansion source model so that any non-dipolar terms associated with the actual source do not produce errors in the solution for the dipolar term. (Cuffin, 1981b) Unfortunately, this seems to be the only use for multipole expansion models since there is little physiologically useful information that can yet be obtained from the higher order terms.

The adequacy of the simple body models presently being used is largely unknown. While a large number of studies of the effects of various body parameters on source localization have been performed, the effects of realistic variations in these parameters in a realistically shaped body have not been determined. For example, the effects of a local variation in the thickness or conductivity the skull are unknown; such non-concentric inhomogeneous regions might well have significant effects on the magnetic field produce by a source in the brain. Much more research must be done in order to determine the effects of the various parameters in the body and in developing methods of tailoring a model to a particular situation or subject.

2) What is the present accuracy of source localization?

KATILA: For an extended source, such as occurs in the human heart during QRS, the localization error using the current dipole approximation can be several centimeters. We have studied in vivo the effect of the volume currents on the localization problem (Varpula and co-workers, 1984). The volume current distribution was changed by immersing the subject in a conducting bath and by varying the stage of respiration. There was remarkable change in the magnetic field measured, in agreement with computer calculations. Nevertheless, the change in the estimated location of the equivalent current dipole was small. Thus, at least for the heart, a straightforward interpretation of the field with the aid of the simple current dipole leads to inaccurate source localization. The error could be less for small localized sources. In the eye blink experiment, the localization error would be a few millimeters.

In several biomagnetic experiments, the reproducibility of the signals and thus the reproducibility of the localization results can be much higher than the real accuracy. For instance in the MCG studies mentioned above, the same location of the equivalent dipole source was found within a few millimeters in subsequent experiments while the physiology showed that the location error was an order of magnitude higher.

PELIZZONE: We have obtained excellent reproducibility of recorded field patterns during sessions extending on a time scale of many weeks. This is demonstrated by the 100 ms (N1) component of the transient response to tone bursts. For the same subject, the parameters of the best fit dipoles gave the same surface location of the source within ±2 mm and the same depth within ±1 mm. The 90% confidence intervals on the parameters of the fit for 10% noise in the data (i.e., standard errors of 20 fT. for a response of 200 fT. at the extrema) are: ±4 mm along the axis joining the extrema, ±9 mm along the axis perpendicular to this direction, ±3 mm for the depth, ±18% for the strength, and ±15° for the orientation.

WIKSWO: Using the cylindrical axon model, we have calculated potential and field waveforms which have the general shape of the measures waveforms, but there are errors in amplitude and detailed shape. Errors in amplitude result primarily from uncertainty in the axon diameter and the intracellular conductivity. Errors in shape are more fundamental and have a variety of causes including uncertainities in conduction velocity, frequency compensation, and the effect of the averaging of the field over the toroid cross-section. However, we believe that magnetic measurements on a single nerve axon can be used to determine the spatial distribution of the sources to within a fraction of a millimeter.

CUFFIN: With present measurement and modeling methods, the best accuracy for source localization is probably in the range of ±1 cm for simple, dipolar sources in the heart and somewhat better than that for such sources in the brain where measurements are closer to the source. This is only an estimate since little if any experimental data exist to verify the accuracy of present source localizations. While computer model studies have provided some information on localization accuracy, care must be taken in applying these results to the much more complex case that exists for the actual experimental situation.

The method of making the localization calculation can also be important in source localization accuracy. In the simpliest method, i.e., the feature-matching method, a best fit match is obtained between only particular features such as peaks or nulls in the map produced by the source model and the measured map. While this method is adequate for a measured map with low noise and a dipolar source, it is not so adequate for situations where significant noise and other measurement errors exist. A more accurate method is to get a best fit between all the measured and calculated data. This method uses all the available data rather than just several points as in the feature-matching method. This method also has the advantage of providing information about the dipolar nature of the source; the closeness of the fit can indicate how dipolar the actual source is.

3) How can magnetic measurements help in source localization?

KATILA: Within the current dipole approximation, the electric and magnetic field patterns are orthogonal so that the magnetic measurements provide another orthogonal direction for localizing the source. It appears that this is often the direction of shifts of interest, e.g., in studies of cortical bioelectric activity. The magnetic measurements reflect essentially only the tangential part of the current sources. Thus the "center of gravity" revealed by electric and magnetic measurements could be different. On the other hand, the fact that there is one less component to deal with than in source localization using electric measurements can facilitate the solution.

PELIZZONE: The amount of new information brought by the magnetic technique to the previous data obtained by electrical recordings varies strongly from one experimental situation to another. If one looks at the 30 ms component of the transient somatosensory response following an electrical stiumation of the median nerve, the magnetic field and the electric potential patterns are very similar. They both show a clear dipolar nature and, as expected, are perpendicular to each other. The current dipole model gives essentially the same results when applied to both measurements: the best fitting dipoles lie near the approximate location of the Rolandic fissure and are oriented tangentially to the surface of the scalp (Wood, 1982). Even if the later components of the somatic response show a less perfect correlation, the magnetic and electric data provide the same information about the 30 ms component. However, you need to know the conductivities for analysing the electric data, whereas you only need to know the head radius for analysing the magnetic data.

On the other hand, comparisons of similar results for the magnetic and electric auditory evoked responses demonstrate some advantages of magnetic measurements. Wood and Wolpaw (1982) measured the electric potential pattern evoked by click stimuli. For five different latencies, the best fitting dipole was located near the midline. However, one has to conclude that in this particular case, this only demonstrates the approximate dipolar shape of the electric data, as there is no physiological basis for a single dipole in the center of the head. Instead, equivalent dipoles located deep near the midline is what is expected for either the bilateral auditory cortex sources or a diffuse cortical source and, in this case, the model does not help to discriminate between alternate hypotheses. This kind of ambiguity is less likely to occur with the magnetic data because the field patterns remain clearly bilaterally localized whereas the potential patterns merge into one unique, almost dipolar pattern.

An additional advantage of magnetic measurements is that by choosing the geometric dimensions of the gradiometer used for the measurements, one is free to adjust the spatial sensitivity or selectivity of the sensor. Thus, the contribution from far sources can be made smaller and less likely to disturb significantly the magnetic field patterns. Furthermore, there is no problem of the reference point as in electric measurements. The evoked magnetic fields depend on a smaller number of parameters than the evoked electric potentials. First, the pattern of the magnetic fields is due only to the tangential component of the generators whereas the pattern of the electric potentials is due to both tangential and radial components of the generators. Therefore, by not containing information about the radial component, the magnetic field pattern becomes in general simplier. Second, the only parameter of a spherical conducting medium that affects the magnetic field is the radius of the sphere, whereas the electrical potentials are affected by the degree of inhomogeneity in the conductivity and by the thickness of the different layers such as the skull overlying the brain. Since the magnetic data depend on a smaller number of parameters than the electrical data, the modeling is in general easier and at least in principle more accurate.

WIKSWO: The analysis of the cylindrical nerve model has contributed substantially to the understanding of the merits of magnetic measurements relative to electric ones. This analysis shows that the shape of the transmembrane potential may be determined more readily and more accurately using the magnetic measurement than by using transmembrane microelectrodes. Magnetic techniques provide a direct measurement of currents, whereas electric techniques provide a measurement of voltages. The combination of the two yields information regarding resistances. In the case of the one-dimensional nerve axon, both the intracellular and transmembrane resistances can be determined. Alternatively, if the intracellular resistance is known, the magnetic technique provides the additional advantage of allowing determination of the transmembrane action potential without the need to penetrate the membrane with a microelectrode.

CUFFIN: The help provided by magnetic measurements depends entirely on the experimental situation. For the case of a simple, dipolar source in a body with uniform conductivity, magnetic measurements provide little if any help and indeed can fail to detect unfavorably oriented sources. For more complex situations in which non-dipolar sources are present or where the body has regions of different conductivity, magnetic measurements can provide help. As an example, recently magnetic measurements have been used to show that the sources of the N20 and P30 components of the somatosensory response to median nerve stimulation is a single tangential dipole rather than a pair of opposed radial dipoles (Wood and co-workers, 1984). Magnetic measurements of the brain will be quite helpful since the problem of dealing with the effects of the skull, scalp, etc. are much reduced as compared with electric measurements.

CONCLUSIONS

1) What is the importance of the modeling assumptions?

The modeling assumptions greatly influence the accuracy of source localization. Overly simlified and/or unrealistic models can produce significant errors in source localization. Fortunately, it appears that a single current dipole is an adequate source model for a large number of situations. These situations are where the actual source is of a simple, dipolar nature. Little confidence can be placed in source localizations using a single dipole when the actual source is widespread. Except in those situations where additional information is available about the source configuration, such as for the isolated nerve axon, the use of a multi-dipolar source model tends to produce inaccurate solutions. Multipole expansion source models provide only limited useful physiological information. The adequacy of the body models presently used remains largely unknown. However, evidence is available which indicates that at least for the heart unrealistic body models can produce significant errors in source localization.

2) What is the present accuracy of source localization?

For heart measurements, it appears that source localization accuracy is in the range of several centimeters for simple, dipolar sources. For brain measurements, the accuracy may be somewhat better for such sources and may be in the range of several millimeters. It should be emphasized that none of these estimates has been verified experimentally. For non-dipolar sources and for situations where complex body geometry exists, the location accuracy may be considerably less than the above estimates. For isolated nerve axon measurements, source localization accuracy is probably in the range of a fraction of a millimeter. The evidence from the study of the isolated nerve indicates that if the source and body models can be more accurately defined, then source localization accuracy will be improved significantly. Such improvement will require the use of additional information about the body that can probably be provided by CAT and/or NMR scans.

3) How can magnetic measurements help in source localization?

The help provided by magnetic measurements depends on the experimental situation. For simple, dipolar sources in a body with uniform conductivity, they provide little if any help. However, there may be some technical measurement advantages to magnetic mesurements. In those situations where non-dipolar sources or complex body geometry exist, magnetic measurements can provide signficant help. For example, for brain measurements, magnetic measurements can help in source localization because such measurements are much less affected by the skull, scalp, etc. than electric measurements. Magnetic measurements can also provide significant help in those situations where accurate determination of a current source is required, e.g., the isolated nerve axon. The most powerful way to use magnetic measurements is in conjuction with electric measurements.

REFERENCES

Antervo, A., R. Hari, T. Katila, T. Ryhanen, and M. Seppanen (1984). Magnetic field pattern produced by eye blinking. Submitted for publication.

Barach, J.P., B.J. Roth, and J.P. Wikswo, Jr. (1984). Magnetic measurements of currents in a single nerve axon: a core-conductor model. Submitted to for publication.

Barth, D.S., W. Sutherling, J. Engle, and J. Beatty (1984). Neuromagnetic evidence of the spatially distributed sources underlying epileptiform spikes in the human brain. Science, 223, 293-296.

Brauer, F., and G. Stroink (1984). Submitted to IEEE Trans. Biomed. Engng. for publication.

Cuffin, B.N. (1981a). The effects of torso geometry on magnetocardiographic lead fields. IEEE Trans. bio-med. Engng., BME-28, 742-749.

Cuffin, B.N. (1981b). Effects of modeling errors on the least squares error solution to the inverse problem of electrocardiology. Ann. Biomed. Engr. (USA), 9, 369-382.

Cuffin, B.N. (1982). Effects of inhomogeneous regions on electric potentials and magnetic fields: two special cases. J. Appl. Physics, 53, 9192-9197.

Cuffin, B.N. (1984). A comparison of moving dipole inverse solutions using EEG's and MEG's. In preparation.

Gray, H. (1977). Anatomy: Descriptive and Surgical, T.P. Pick and R. Howden (Eds.), Boundy Books, New York.

Grynszpan, F., and D.B. Geselowitz (1973). Model studies of the magnetocardiogram. Biophys. J., 13, 911-925.

Karp, P.J., T.E. Katila, M. Saarinen, P. Siltanen, and T.T. Varpula (1980). The normal human magnetocardiogram. II. a multipole anaylsis. Circulation Res., 47, 117-130.

Katila, T.E., and P.J. Karp (1983). Magnetocardiography: morphology and multipole analysis. In S.J. Williamson, G.L. Romani, L. Kaufman, and I. Modena (Eds.), Biomagnetism, Plenum Press, New York, pp. 237-263.

Okada, Y.C. (1983a). Evoked magnetic fields. In S.J. Williamson, G.L. Romani, L. Kaufman, and I. Modena (Eds.), Biomagnetism, Plenum Press, New York, pp. 399-468.

Okada, Y.C. (1983b). Inferences concerning anatomy and physiology of the human brain based on its magnetic field. Il Nuovo Cimento, 2D, 379-409.

Okada, Y.C. (1984). In preparation.

Roth, B.J., and J.P. Wikswo, Jr. (1984). The magnetic field of a single axon: a comparison of theory and experiment. Submitted for publication.

Stroink, G., F. Brauer, C. Purcell and P. Krieger (1984). An inverse solutio solution in magnetopneumography. This volume.

Swinney, K.R., and J.P. Wikswo, Jr. (1980). A calculation of the magnetic field of a nerve action potential. Biophys. J., 32, 719-732.

Tripp, J.H. (1977). Effects of torso geometry on the magnetocardiogram. Biophys. J., 18, 269-273.

Tuomisto, T., R. Hari, T. Katila, T. Poutanen, and T. Varpula (1983). Studies of auditory evoked magnetic and electric respones: modality specifity and modelling. Il Nuovo Cimento, 2D, 471-483.

Varpula, T, T. Katila, T. Poutanen, and M. Seppanen (1984). In vivo study of the effect of the secondary currents on the MCG. Submitted for publication.

Wood, C.C. (1982). Application of dipole localization methods to source identification of the human evoked potentials. Ann. N.Y. Acad. Sci., 388, 139-155.

Wood, C.C. and J.R. Wolpaw (1982). Scalp distribution of human auditory evoked potentials. II. evidence for overlapping sources and involvement of auditory cortex. Electroenceph. clin. Neurophysiol., 54, 25-38.

Wood, C.C., D. Cohen, B.N. Cuffin, M. Yarita, and T. Allison (1984). Electrical sources in the human somatosensory cortex: identification by combined magnetic and potential field recordings. Submitted to Science.

Woosley, J.K., B.J. Roth, and J.P. Wikswo, Jr. (1984). The magnetic field of a single axon: a volume conductor model. Submitted for publication.

CRYOGENIC INSTRUMENTATION FOR BIOMAGNETIC MEASUREMENTS

Introduction by
James E. Zimmerman
National Bureau of Standards
Division 724.03
Boulder CO, 80303

ABSTRACT

This chapter, including the introduction, consists of five sections by different authors on new and projected developments in cryogenic support systems for SQUID magnetometers and gradiometers, and on the gradiometers themselves. It includes a review of the present technology of liquid-helium dewars with centralized liquefaction facilities, and possible alternative systems using small dedicated cryocoolers or liquefiers integrated with the SQUID instrumentation. The final section discusses high-order magnetic gradiometers for noisy environments.

KEYWORDS

Biomagnetism; cryocooler; cryogenics; dewar; Gifford-McMahon; gradiometer; Joule-Thomson; liquid helium; magnetic noise; magnetometer; refrigeration; safety; Stirling; SQUID

INTRODUCTION

In considering cryogenics for SQUIDs, we are returning to a problem that was addressed by Prof. Sam Collins of Mass. Institute of Technology forty years ago--how to provide an economical cryogenic facility for fundamental and applied research at very low temperatures. Collins designed the first commercially produced helium liquefier/cryostat. If I remember correctly, his machine was originally marketed as a "cryostat" in which one could install the measuring equipment and samples necessary to carry out research on the properties of materials and devices at temperatures down to about 2 K. This machine was a key factor in the explosion of low-temperature research that has occurred since about 1950, and it immediately developed that the demands of large numbers of students for cryogenic facilities required that the Collins machine be used primarily as a liquefier, rather than as a cryostat, so that each machine could provide liquid helium to many satellite cryostats, rather than just one. The satellite cryostats were, of course, simply vacuum-insulated vessels ("dewars") with appropriate instrumentation, in which the liquid helium slowly evaporated. In time, larger, centrally located helium liquefiers were built; and so it has developed that virtually all research and development in superconductivity and superconducting devices (such as SQUIDs for biomagnetic measurements) has been and still is dependent upon the immediate and continuing availability of liquid helium from some central source, special storage and transport vessels, reliable transportation facilities, some knowledge of the special handling requirements of the storage vessels, special equipment to transfer the helium from the storage vessel to the user's cryostat, and (in most countries) provisions to catch the evaporating helium, re-compress it, and get it back to the central facility for reuse.

The next section of this chapter is a short description of the state of the art of dewars for biomagnetic measurements. It deals with the special requirements for low vibration and low magnetic noise and interference, while at the same time attending to other desirable characteristics such as low evaporation rate, reasonable cost, convenience in handling, and safety.

Needless to say, the system described above, consisting of a central liquefier and remotely located helium dewars, can be highly inconvenient for instruments which must be operated in distant and inaccessible locations. Even in the best of circumstances the liquid-helium supply can be interrupted owing to communication, transportation, or equipment failures, not to mention mistakes by novice or expert operators. In recent years there is evidence of a certain yearning to be free of the cost and inconvenience of this rather complex set of facilities and services, to have a self-contained, reliable integrated system consisting of a low-temperature refrigerator and SQUID gradiometer, requiring only to be plugged into a source of electrical power. The third and fourth sections of this chapter deal with such futuristic systems, first a summary of a number of conceptual and practical possibilities, and second a particular experimental cryocooler which may indeed meet our requirements.

The final section describes the use of second- and third-order gradiometers for use in noisy environments, and the critical problems of building and balancing such instruments.

There is not space in this chapter for perhaps the most exciting development in instrumentation for biomagnetism. This is the development of multi-channel arrays of magnetic sensors. At this moment, to my knowledge, only two multi-channel arrays are in use, at New York University and at Helsinki Technical University. However, there seems to be general agreement that results obtained with such arrays will be a qualitative breakthrough in biomagnetism, and that their development, while relatively expensive, is both essential and inevitable if biomagnetism is to reach its full potential as a clinical and medical research tool.

Finally, a critical reader will note that the contributors to this chapter are all from the western US and Canada. This in no way represents the geographical distribution of significant work on SQUID instrumentation and cryogenics. Advanced development of compatible small cryocoolers is currently being carried on at several laboratories in Japan, in China, and in several other countries: the Netherlands, Germany, the Soviet Union, Brazil, and elsewhere in the US, to name a few.

THE DESIGN AND USE OF DEWARS FOR BIOMAGNETIC MEASUREMENTS

D. Crum
S.H.E. Corporation, San Diego, California

ABSTRACT

The fabrication of liquid helium dewars designed for use with superconducting biomagnetometers is reviewed, with emphasis on fabrication materials, thermal design, and internal sources of magnetic noise. Techniques for safe operation and efficient transferring of liquid helium are also discussed.

INTRODUCTION

The SQUIDs and superconductors typically used to fabricate biomagnetometers require a stable cryogenic environment that is usually obtained by direct immersion in a bath of liquid helium. Helium, which boils at about 4.2K, has a very small latent heat (30 mV evaporates 1 liter per day) and requires a well insulated storage vessel called a "dewar". Typical dewars used with biomagnetometers have a capacity of between 5 and 15 liters of liquid helium and evaporate between 1 and 3 liters per day.

A simplified dewar drawing is shown in Fig. 1. The dewar is largely fabricated from a fiber reinforced plastic such as an epoxy/fiberglass composite. The helium reservoir is suspended from the top plate by a "neck tube" which also provides access to the reservoir. This tube is thin walled to minimize heat conduction and is the weakest structural component. Although far stronger than necessary for normal operation, it is overdesigned to survive the torture of international shipment. At the bottom of the helium reservoir there is commonly a reduced diameter "tail section" in which the pickup coils are located.

The region between the helium reservoir and outer case is evacuated to prevent gaseous heat conduction. A "getter" comprising molecular sieve or activated charcoal is glued to the outside of the helium reservoir to cryopump any gas in this region. Thermal shielding, consisting of one or more vapor cooled shields and "superinsulation" (multiple layers of aluminized mylar) is mounted in this vacuum to intercept thermal radiation. Superinsulation works by reflecting thermal radiation. The vapor-cooled shields, on the otherhand, absorb thermal radiation and conduct the heat up to the neck tube where it is transferred to the cold stream of evaporating helium gas.

FABRICATION

Most biomagnetometer dewars are built from glass/epoxy composite tubes and plates. This material is available from a number of suppliers (e.g., Synthane Taylor, Westinghouse,

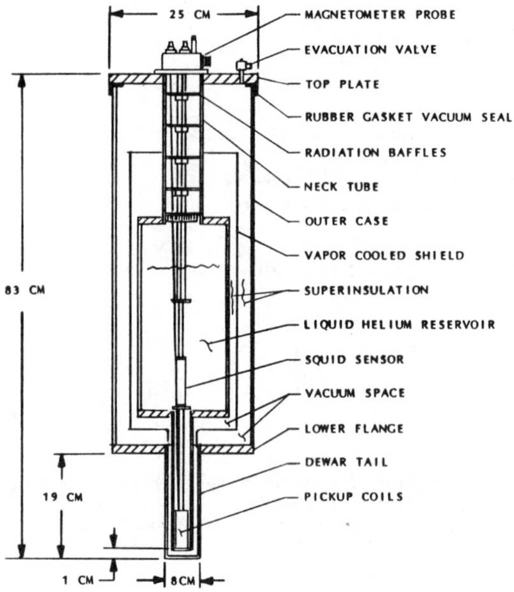

FIGURE 1
TYPICAL BIOMAGNETOMETER SYSTEM

Spaulding) and is commonly called G-10 or G-11 fiberglass. Unfortunately, its specifications pertain mainly to dielectric strength since its primary use is for printed circuit boards. There is no guarantee that the material will be vacuum leak tight at room temperature, much less at 4K. In our experience, a large percentage leaks at room tembuilt from glass/epoxy composite tubes and plates. This material is available from a number of suppliers (e.g., Synthane Taylor, Westinghouse, Spaulding) and is commonly called G-10 or G-11 fiberglass. Unfortunately, its specifications pertain mainly to dielectric strength since its primary use is for printed circuit boards. There is no guarantee that the material will be vacuum leak tight at room temperature, much less at 4K. In our experience, a large percentage leaks at room temperature and over 5% of the "good" material will develop leaks after several thermal cycles. SHE Corp. currently has 27 leaky helium reservoirs in inventory, most of which failed after final assembly. Some suppliers offer a "cryogenic" grade of G-10 but our experience with it has been about the same. Since most leaks result from a partial delamination of the material and have long, complicated leak paths, patching is not recommended.

Custom materials made from cloth-epoxy laminates are an alternative to standard G-10. We have found these to be more reliable, a fact which is most likely due to the care taken in their manufacture rather than any fundamental difference in material or fabrication. Hundreds of epoxies are available and several different reinforcing fibers can be used, including glass, carbon, quartz, and aramid (Kevlar) fibers. Filament wound parts are also available and can be made in a variety of shapes. The fiberglass pieces are easily joined using epoxy glue. Care should be taken to ensure that both surfaces are thoroughly wet with glue. Press fits are sometimes used for increased mechanical strength.

The thickness of fiberglass required for structural safety should be calculated using standard equations for pressure vessels. In these equations, appropriate values of Young's modulus and

Poisson's ratio are approximately 10^6 psi and 0.3 respectively for G-10 and most other epoxy composites (Kasen, 1975). A safety factor of from 3 to 5 should be used in the design.

Several other factors influence the material thickness chosen. Walls are often made thicker to reduce the probability of leaks, to improve the dewar's chances of surviving shipping, or to reduce flexing of plates where close spacing is required. Large plates are always much thicker than the calculations would indicate to reduce flexing which could crack the glue joints. The spacing between the pickup coil and the outside of the dewar is minimized to allow close access and is typically 1 cm at the bottom. Without special construction, it is not possible to reduce this spacing much. If the spacing were less than 7 mm when the dewar is cold, differential thermal expansion between the helium reservoir and the outer case would force the inner tail through the bottom of the outer case when the dewar was warmed to room temperature.

THERMAL DESIGN

Heat leaks into the helium occur via conduction, convection and radiation.

Conduction. The heat flow ($\dot{Q}$) along a uniformly shaped object is given by $\dot{Q} = \kappa A \, dT/d\ell$ where A is the object's cross sectional area, κ is its thermal conductivity, and $dT/d\ell$ is the temperature gradient. Integrating this equation is difficult since κ is often a strong function of temperature (Kasen, 1975; Childs, Eriks and Powell, 1973; Powell and Blanpied, 1954). Since many dewar components have their two ends at 4K and 300K, it is common to use a single value for κ , averaged over this temperature range, to perform calculations. Table 1A lists these κ values and Table 1B gives the corresponding helium consumption for objects commonly found in biomagnetometers. Values are given both for operation in a vacuum and for operation in the cold stream of evaporating helium gas. The latter value reflects the approximate 20-fold reduction obtainable when the enthalpy of the cold gas is used to cool objects running between 4K and room temperature. This enormous reduction results from the fact that 75 times as much heat is needed to warm helium gas from 4K to 300K as is needed to evaporate the liquid (Long and Loveday, 1968). To achieve this reduction, the object must be in good thermal contact with the evolving gas. If electrical leads are located in the vacuum, they must be thermally lagged to the tube which vents the helium gas.

TABLE 1A - Average Thermal Conductivity of Materials (300K-4K)

MATERIAL	k avg. (W/cm K)
Copper, 99.9%	6
Copper O.F.H.C.	5
CuNi (30% Ni)	0.3
Stainless Steel (316)	0.08
BeCu (2% Beryllium)	0.6
Aluminum (99.9%)	2
Aluminum (6061)	0.8
G-10 Fiberglass	0.005

TABLE 1B - Helium Evaporation Rate for Some Items Commonly Used in Biomagnetometers

	Helium Consumption (liters/day)	
ITEM	Not Gas Cooled	Gas Cooled
Two #36 Cu Wires	0.5	0.03
Two #36 Manganin Wires	0.07	0.004
CuNi Tube (3mm O.D. x .15mm Wall)	2	0.1
G-10 Neck Tube (5cm O.D.x 2mm Wall)	6	0.3
BeCu Coax (2mm O.D.)	5	0.3
316 Stainless Steel Tube (3mm O.D. x 15mm Wall)	0.8	0.04

Convection. Since most dewars have their low temperature region at the bottom, convection is rarely a design consideration. In fact, it is nearly impossible to develop a temperature inversion or horizontal temperature gradient in the lower portion of the neck tube. This follows from the very low viscosity of cold helium gas and its large heat capacity per unit volume. Also, as the gas warms from 4K to 8K, which may occur over a distance of only 1 cm, its weight per unit volume, which is the driving force for convection, changes by a factor of two. Together, these properties make it possible to convectively transfer large quantities of heat with small horizontal temperature gradients. As a result, the gas remains almost perfectly statified in temperature as it moves very slowly up the lower section of the neck tube (less than 0.01 cm/sec at 6K). Higher up the neck, the situation is quite different. The flow velocity of the gas rapidly increases (≃ 0.5 cm/sec at 300K) while both the heat capacity and weight per unit volume decrease. Convection may then no longer be adequate to keep everything isothermal at a given height. It is then desirable to force the gas to flow preferentially along items which represent the dominant heat leak.

If a dewar is tipped from vertical, horizontal thermal stratification begins to thermally short out part of the neck tube. The effect is usually small but it can be extremely large at large angles. When horizontal, the neck tube becomes an efficient heat pipe that convectively transfers heat from room temperature to the helium.

Radiation. If no insulation were placed in the vacuum, a typical dewar would evaporate 3000 liters/day due to thermal radiation. Fortuntely, this is easily reduced.

The heat radiated per unit area from an object at temperature T is given by $\dot{Q} = \varepsilon\sigma T^4$ where σ is the Stefan-Boltzmann constant and ε is the emissivity of its surface which can range from 0 to 1. For an ideal "black-body", $\varepsilon = 1$, but values for common materials rarely exceed 0.6. As shown in Table 2, most clean shiny surfaces have an effective ε of only about 0.02 (Chelton, 1956). However, values ten times worse can result if air or moisture is condensed on the surface.

TABLE 2 - Emissivity of Selected Materials

Material	Emissivity
Fiberglass	0.6
Aluminized Mylar (Superinsulation)	0.04
Condensed Air on Aluminized Mylar	0.1
Gold on Mylar	0.02
Polished Gold Surface	0.01
Polished Copper	0.02
Aluminum Kitchen Foil	0.02
Aluminum (Sanded surface)	0.06

Multiple layers of superinsulation are usually used in the vacuum to reduce radiation into the helium reservoir. The use of n layers of superinsulation reduces this heat leak (Scott and others, 1968) to

$$\dot{Q} = \varepsilon\sigma(T_h^4 - T_c^4)/2(n+1)$$

where ε is the emissivity of the superinsulation, and T_h and T_c are the temperatures of the surfaces being insulated (e.g., 300K and 4K respectvely for a dewar with no vapor-cooled shield). Typical dewars use numerous (≃ 50) layers of superinsulation along with one or more vapor-cooled shields, the coldest of which is about 60K. As a practical problem, the quality of superinsulation varies widely, even when purchased from the same supplier. We routinely measure infrared transmission and have found variations of a factor of 30.

Radiation also enters the helium reservoir down the neck tube or through other ports. The problem here is different than radiation through the vacuum since baffles in the neck are cooled by the evaporating helium. As a result, it is not important to use low emissivity baffles. If the last baffle is cold ($\lesssim$ 30K), its radiation into the helium reservoir is negligible regardless of its emissivity. It is only important is to minimize the cross-sectional area of any direct, unbaffled paths to the helium bath.

THE DEWAR VACUUM

Before filling a dewar with helium, the pressure in its vacuum space should be less than 100 millitorr, of which the partial pressure of helium must be much less than 1 millitorr. If the primary contaminant in the vacuum space is helium, it is helpful to dilute its concentration by evacuating and backfilling the vacuum space with nitrogen gas several times. Reevacuation can usually be accomplished in less than one hour using a mechanical pump, although the initial evacuation after fabrication can require days to remove adsorbed gasses. One must then be careful not to pump for long periods at low pressure to prevent oil from backstreaming out of the pump and spoiling the superinsulation. (If extensive pumping is necessary, a coldtrap should be used.) One must also be very careful to never evacuate or vent the vacuum space rapidly since abrupt pressure changes can collapse or rupture the thermal shields and superinsulation located therein.

The use of a "getter" is essential for long term operation since fiberglass/epoxy composites diffuse helium gas at a substantial rate. Fortunately, this diffusion decreases rapidly with temperature and is negligible everywhere except at the top of the neck. The getter allows continuous operation for over a year without reevacuation. Otherwise, the pressure would rise enough to significantly increase the helium boiloff in less than a month. When a dewar is warmed to room temperature, its getter releases the adsorbed helium gas. This makes it advisable to reevacuate a dewar if it was previously used for over a month. One should never leave a warm dewar filled with helium gas, since this exposes large areas of warm fiberglass to helium and rapidly degrades the vacuum.

MAGNETIC NOISE

Helium dewars can increase the noise of SQUID magnetometers in a variety of ways, including magnetic fields generated by Johnson noise currents in conductors, fields due to currents driven by thermal EMF's, and noise due to local eddy currents induced by ambient fields.

Nyquist showed that as a consequence of the 2nd law of thermodynamics, a random, fluctuating noise voltage having a mean square amplitude given by $\overline{V_N}^2 = 4kTRB$ present across any resistance R at a temperature T; k is Boltzmann's constant and B is the "noise bandwidth". Accordingly, the noise spectrum is independent of frequency, a property which is referred to as "white" noise. The noise voltage can result in magnetic field noise by causing currents to flow with mean square amplitude $T_N^2 = 4\ kTB/R$. The important properties of this noise source are that its noise power is proportional to temperature, inversely proportional to resistance, and is strongly dependent on the size of the conducting loop and its mutual inductance to the pickup coils. When designing dewars, one must minimize the size and temperature of conductors and must locate them away from the pickup coils.

Calculations of the magnetic noise coupled into a gradiometer are difficult since the noise results largely from small, unintentional assymetries in fabrication. For the simple case of a conducting cylinder, a good approximation of its low frequency magnetic noise can be made by treating the cylinder as a single turn (sheet) solenoid with a noise current appropriate to its resistivity. For a cylinder of OFHC Copper at 4K with resistivity $2 \times 10^{-10}\ \Omega$ m, 2 cm diameter x 10 cm long x 0.005 cm thick, the approximate axial noise is 1 pT/$\sqrt{Hz}$. Slotting the cylinder reduces the axial noise but has little effect on transverse noise.

The dewar components most likely to introduce noise are the superinsulation and vapor-cooled shields. Vapor-cooled shields are commonly made from narrow strips ($\approx$ 1 mm x 0.1 mm) of aluminum or copper that are connected together only where they fasten to the dewar neck. The superinsulation is not a serious problem unless hundreds of layers are used or if it has excessive metalization. The metal layer on the superinsulation can be slit or etched into strips or small squares to reduce conducting paths, if necessary. Regardless, it is important not to electrically short adjoining layers.

Scattering of ambient magnetic fields by conductors can introduce noise into a gradiometer. This scattering results from eddy currents generated in response to changing fields. Even minor field distortions introduce noise due to the high degree of uniform field rejection built into a gradiometer. Fortunately, most ambient magnetic noise, other than line frequency noise, is at very low frequencies and is ineffective at generating eddy currents. However, rotating machinery such as vacuum pumps can be a problem. Calculating the size of this effect is complicated by the geometries involved, but a good rule is to keep the total thickness of conductors near the coils less than 0.01 times the electrical skin depth at the frequencies of concern and to position them symmetrically with respect to the coils.

Magnetic fields are also generated by thermal EMF's which exist due to the large thermal gradients in a dewar. Fortunately, fluctuations in these voltages (and fields) are usually below the bandwidth of interest and rarely cause difficulties.

LIQUID HELIUM TRANSFER

Liquid helium is typically received in 30, 100, or 250 liter metallic storage dewars that evaporate only 1% of their total capacity per day. A typical biomagnetometer dewar is refilled twice a week by transferring from a storage dewar. Transferring requires about 15 minutes and is accomplished by inserting a vacuum insulated tube (shaped like an inverted "U") into the two dewars. Liquid helium is forced to flow into the biomagnetometer by raising the storage dewar's pressure about 100 torr (2 psi), usually using a cylinder of helium gas.

There are numerous sources of helium loss encountered when transferring. Helium is first boiled away by inserting a warm transfer tube into the storage dewar. The storage dewar is then pressurized which pumps heat (via hot gas) into the dewar and slowly warms the liquid as its vapor pressure comes into equilibrium. Cold liquid is forced through the warm transfer tube and is completely evaporated until the tube reaches 4K. Continued evaporation of helium occurs during the transfer due to heat leaks into the transfer tube. Also, as the helium passes through the transfer tube, its pressure drops and part of it evaporates, cooling the liquid. Finally, the storage dewar is depressurized and more liquid is vaporized as its contents cool back to equilibrium at atmospheric pressure. An efficient transfer wastes only about 1 liter or 15% of the amount transferred, whichever is greater. If the transfer tube is inserted slowly into the storage dewar (so as to use the enthalphy of the gas to cool the tube) only a few tenths of a liter are consumed. A similar quantity is required to cool the inside of the tube. The heat leak during transfer into a typical 2 meter long, flexible stainless steel transfer tube will evaporate less than 1 liter during a 10 minute transfer.

Losses from pressurization and depressurization increase with the time required by the transfer. Since liquid helium is a poor thermal conductor, it remains thermally stratified in the storage dewar so that during short transfers only its surface warms significantly when the pressure is increased. If a storage dewar is kept at a pressure significantly above atmospheric for long periods of time, the entire contents will come to equilibrium at an elevated temperature. Then, for a pressure 100 torr above atmospheric, about 5% of the contents will be lost when the dewar is vented to insert the transfer tube. It is only practical to keep a storage dewar at an elevated pressure if the transfer tube can be inserted without venting the dewar. If the dewar is pressurized only for short periods, these losses can be kept to less than 10% of the amount actually transferred. The loss due to flashoff at the exit of the transfer tube (due to isenthalphic expansion) is about 6% for a 100 torr pressure drop (Scott, and others, 1968).

To reduce helium loss, one should minimize both the time required to transfer the helium and the pressure used to drive the helium through the transfer tube. This requires a low impedance transfer tube whose inner diameter should be no smaller than 4 mm. The transfer tubes supplied with commercially available biomagnetic systems require a 100 torr (2 psi) pressure difference to transfer helium at a rate of about 1 liter per minute, which is fine for dewars up to 15 liters in capacity.

Slow transfers are advisable when <u>first</u> cooling the system and allow efficient utilization of the cold evaporated helium gas to cool the dewar to 4K. When refilling a dewar that contains liquid helium, it is important not to insert the transfer tube all the way until after liquid begins to exit the tube. Otherwise, the warm helium gas initially exiting the tube will evaporate the remaining liquid.

LIQUID HELIUM AVAILABILITY

Helium gas makes up only 5 ppm of the earth's atmosphere and cannot be economically separated at this time. Over ninety-five percent of the world's supply comes from a natural gas field in Kansas (United States) which contains 0.85% helium. Although it is estimated that this field will be empty in ten years, other fields are known to have reasonably high helium content which are not now economically viable for natural gas production. Enough helium gas is available to satisfy our current rate of demand (4 x 10^{10} liters/year of gas which is equivalent to 150,000 liquid liters per day) for several decades. The introduction of whole-body NMR scanners (now called Magnetic Resonance Imaging Systems, or MRI to avoid the use of "Nuclear") does not appear to be a serious problem. Their helium consumption has been reduced to less than 5 liters per day which, based on current estimates, would represent only 4% of the helium gas consumed in 1986 for all applications. Nevertheless, we are near the limit of the current supply rate; short term (seasonal) shortages have already occurred in some parts of the world and are predicted to be common as early as 1985. It seems that the use of open-cycle systems (freely vented dewars) is viable for many years, but consideration should be given to closed-cycle systems (reliquifaction of gas or closed-cycle refrigerators) for use in areas where helium supply is difficult. Small closed-cycle refrigerators would also be helpful in the fabrication of smaller biomagnetometers. Closed-cycle refrigerators are likely to become important for biomagnetic measurements during the next 10 years.

Liquid helium is readily available from a number of suppliers including, in the U.S., Linde, Airco, and Air Products, all of whom deliver the liquid in most metropolitan areas. In remote areas, it is necessary to arrange special shipping. Although this is not difficult, we know of delays caused by a pilot who refused to carry the shipment and of cases where the dewar arrived empty. The cost of liquid helium in the U.S. now ranges between $3.25 and $10 per liter. Assuming it is purchased in a 100 liter storage dewar, extra charges of 25¢ per mile may be necessary for shipping; demurrage charges may run an extra $75 per week. In other parts of the world, the delivered cost per liter can exceed $50. Still, it is impossible to economically justify the use of closed-cycle refrigerators on biomagnetometer systems today. I estimate that the extra price for an installed refrigerator would be about $50,000 and that it would cost about $5,000 per year for electricity to run its compressors (5kW at $.11 per kW hour). Even at $25 per day, liquid helium would cost only $9,000 per year. This is small compared to the total budget of a biomagnetism laboratory and cannot justify the cost of installing a refrigerator in a system whose useful life before obsolesence is probably less than 5 years.

SAFETY

As more researchers having little or no experinece in cryogenics enter the field of biomagnetism, it is important to review and eliminate all potential safety hazards. To our knowledge, in over ten years of biomagnetic research activity, there has never been an injury to a test subject, nor have there been any "close calls". Even so, the potential for incident exists and only one serious injury could severely impact research funding.

Although asphyxiation by helium gas is possible, it usually represents a negligible danger. If ten liters of helium were instantly vaporized, it would rise to the ceiling and displace 25 m^3 of air. Since most laboratories have volumes in excess of 100 m^3, asphyxiation is simply not possible. However, in small shielded rooms, essentially all the air could be displaced and there is potential for panic when one is completely engulfed in a cold cloud of water vapor. Such rooms require a carefully designed ventilation system.

A more likely danger is rupture of the helium reservoir due to plugging of the neck tube with solid air. If the helium transfer port is left open, air can be cryopumped back into the neck tube where it will freeze. This would be unlikely to completely plug a 5 cm diameter neck tube, even if the port were left open for a week (after which time one typically would have had to retransfer helium and would have noticed the open port). The frozen air would remain in the dewar neck, however. Since biomagnetometers may be operated continuously for a year or more, the possibility exists for repeated errors or small continuous leaks to result in eventual plugging. Once plugged, the pressure will continue to rise until the plug is blown out, or until the plug begins to soften and melt or, in the worst case, until the reservoir ruptures. It is difficult to plug a 5 cm diameter or larger neck since the pressure build up is likely to blow out the plug prior to rupturing the reservoir. With smaller neck tubes, the possibility of rupture is far greater. Softening of the neck plug will occur at about 60K which sets an upper limit in the reservoir of 150 atmospheres. If the reservoir ruptures, the vacuum will be lost and the outer case, which in our designs is held together only by vacuum, would simply fall apart, rather than rupture. The primary danger results from the close spacing normally used at the tail. In the event of an explosive rupture of the reservoir, its tail section could be driven against the outer case at a high velocity. This has happened only once at SHE (due to severe intentional mishandling) but there is little doubt that a person would have been injured if his head had been under the dewar.

This hazard can be minimized by supporting the dewar case at the tail flange plate (rather than around the large outer case tube as is commonly done) to keep the tail section from being driven downward in the event of a rupture. One should also monitor the flow of evaporated helium gas and stop measurements if the flow falls below normal. A separate safety-vent line connected to a high-pressure relief valve can also be used, but its design is complicated by its natural tendency to produce thermal (Taconis) oscillations (Bannister, 1966).

ACKNOWLEDGEMENTS

I would like to thank William Goree for useful discussions and his excellent review paper (Goree, 1981) which served as an outline for much of the material contained herein.

REFERENCES

Bannister, J.D. (1966). Spontaneous Pressure Oscillations in Tubes Connecting Liquid Helium Reservoirs to 300°K Environments, in Pure and Applied Cyrogenics Vol. 6, Liquid Helium Technology, Pergamon Press Oxford, New York.

Chelton, D. (1956). Emissivities of Materials, UCRL - 3421 Monograph, p-91.

Childs, G. E. Eriks, L. J. and Powell, R. L. (1977). "Thermal Conductivity of Solids at Low Temperatures. A Review of the Literature," N.B.S. Monograph 131.

Goree, W. S. (1981). Cryogenics for Field Applications, in SQUID Applications to Geophysics, ed. by Weinstock, Society of Exploration Geophysicists, 85–91.

Kasen, M.B. (1975). "Mechanical and Thermal Properties of Filamentary-Reinforced Structural Composites at Cryogenic Temperatures." Cryogenics 15, p 327–701.

Long, Hugh M. and Loveday, Paul E. (1968). Safe and Efficient Use of Liquid Helium, in Technology of Liquid Helium, ed. by Kropschot, Birmingham and Mann, U.S. National Bureau of Standards Monograph III, Chap. 6.

Powell, Robert L. and Blanpied, William A. (1954). Thermal Conductivity of Metals and Alloys at Low Temperatures, N.B.S. Circular 556.

Scott, L.E., VanMeerbake, R.C., Kirk, B.S., and Nubel, G.C. (1968) Storage, Distribution and Handling, in Technology of Liquid Helium, ed. by Kropschot, Birmingham and Mann, U.S. National Bureau of Standards Monograph III. Chap. 2.

NEW APPROACHES TO SMALL SCALE CRYOGENIC FACILITIES

Calvin Winter
Quantum Technology Corp., 6237 - 148 Street,
SURREY, B.C.; CANADA V3S 3C3

ABSTRACT

A survey of current approaches to providing cooling to superconducting temperatures, without the use of externally supplied liquid cryogens, is presented. The approaches covered include regenerative isothermal (Stirling) cryocoolers, isenthalpic (Joule-Thomson) cryocoolers, adiabatic expansion (Simon) cryocoolers, adiabatic demagnetization cryocoolers and combinations of the above. Emphasis is placed on the current state of development of each approach. Possible directions of future practical developments are suggested.

APPROACHES TO CRYOGENIC REFRIGERATION

There are a large number of different types of cryogenic refrigerators which are capable of attaining superconducting temperatures. From a purely practical standpoint they may be divided into four groups. The first group includes all the large scale helium liquifiers. These will not be discussed further here. Current users of liquid helium are well aware of the high cost of these liquifiers and of the costs and inconvenience in transporting volatile liquid helium over large distances. The second group includes all the machines based on regenerative cryocoolers of the types commonly used in cryopumps. The third group includes all the machines based on the small scale plastic cryocoolers developed by Zimmerman. The fourth group is for all other types of cryocoolers. An adiabatic demagnetization stage can achieve intermittent cooling to 1K with a warmup of 0.3K/hour (Valera and co-workers, 1984). This will not be discussed further here, as the large magnetic fields used (1.4T) make this approach unsuitable for biomagnetism measurements. Various approaches to cryogenic refrigeration are summarized in Table 1.

Regenerative Cryocoolers of the Cryopump Type

The recent explosion in use of cryopumps in high vacuum systems, especially in the semiconductor industry, has resulted in mass production of the associated cryocoolers. This gives these cryocoolers both the advantage of a relatively low price (6-20 k$) and the advantages of product standardization with both these cryocoolers and their parts available from a number of manufacturers.

A typical cryocooler of this type has a compressor unit which consumes 1.5kW of electricity and may be forced air or water cooled. The compressor itself is a sealed, oil lubricated, rotary or piston compressor of the type mass produced for consumer air conditioning or refrigeration systems. A disposable adsorption filter is used to remove the oil vapour from the helium gas before it is sent to the cold head or 'expander'. The adsorption filter requires replacement every 10,000 hours. The life expectancy of the compressors is about 30,000 hours with considerable variation from one unit to the next. The cold head or 'expander' is connected to the compressor unit by two flexible helium lines and an electrical connection. The operating cycle is the Gifford-Mahon or Split-Stirling cycle. The displacer in the cold head is moved either by an electic motor and cam, or by cyclic pneumatic control at a frequency near one hertz. A small model may produce 1 Watt of cooling at 15K and 10 Watts at 80K. The ultimate (no heat load) cold temperature of refrigerators of this design vary from 12 to 8K. This is cold enough to operate some superconducting devices, however it is not sufficient for the majority of applications. The ultimate (no heat load) temperature limit of 8K is due to the loss of efficiency in the lead shot regenerator below this temperature. In order to attain lower temperatures (ie. 4.2K) it is necessary to add a separate cooling stage to these cryocoolers.

Cryopump Type Cryocoolers with a J-T Stage Added

Liquid helium temperature, typically 4.5K with cooling powers of 1-3 Watts can be achieved by adding a Joule-Thomson stage and an additional compressor to basic 15K cryocooler described above. Initial systems had poor reliability due to contaminants freezing and plugging the tiny J-T orifice. Much better reliability can now be expected from these cryocoolers (Higa and Wiebe 1977). The electrical consumption is typically 5kW. Several firms, including our own, now have designs for smaller units, near 1/4 Watt of cooling. Many of the approximately 50 existing units (with prices ranging from 30-100 k$US) are installed on radio telescopes. We have developed a cryocooler with room-temperature access via a dip tube to a pool of liquid helium. Several designs related to cooling superconducting electronics (computers) have also been developed recently (Longsworth, 1984). From the point of view of many applications to biomagnetic sensing these cryocoolers will almost certainly produce too much vibration and magnetic interference. However, their ability to liquify helium suggests that they be used to liquify a quantity of helium (in overnight operation for example) and then be turned off for extended periods of low-level magnetic measurement. This mode of operation might be attractive wherever such intermittent measurements are appropriate. More work is needed in this area.

Cryopump Type Cryocoolers with a Simon Expansion Stage Added

An adiabatic (Simon) expansion stage added to a standard cryopump type cryocooler can provide cooling at liquid helium temperature. An inexpensive accessory is connected to a standard helium gas cylinder. When the pressure is released a pool of liquid helium is created which remains for up to 15 minutes. We have also developed a cryocooler which allows the Simon expansion stage to be recycled and the liquid helium thereby produced to be collected in a reservoir. This cryocooler provides continuous cooling at liquid helium temperature with a cooling capacity of three milliwatts. Temperatures below 4K can be achieved by pumping on the pool of liquid helium. Unlike the J-T design it does not have any small orifices to plug and therefore can be operated with impure helium gas (Winter, Gygax, Myrtle and Barton, 1984).

Cryopump Type Cryocoolers with an Independent Stirling-Cycle Stage Added

By adding an additional displacer with its own independent gas compressor it is also possible to reach liquid helium temperatures. An experimental cryocooler has been built which reaches 4.8K with no load and 5.8K with an 8mW heat load, all with a 12RPM operating speed (Matsubara and Yasukochi, 1983). Zimmerman and Sullivan (1979) also demonstrated a Stirling stage that gave 4mW at 4K, with the warm end at 12 to 15K.

Plastic Regenerative Cryocoolers of the Zimmerman Type

Plastic cryocoolers, capable of achieving temperatures as low as 7K directly from room temperature have been built in a number of laboratories (Zimmerman, Daney and Sullivan, 1983; Myrtle, Winter and Gygax 1982; Sager and Paulson, 1983). The non-magnetic and non-metallic construction makes cryocoolers of this type well suited to measurements of biomagnetism. Magnetocardiograms have been recorded by Sullivan, Zimmerman and Ives (1981) using a SQUID cooled by a plastic cryocooler. The theoretical design parameters are well understood (Zimmerman and Sullivan, 1982; Sullivan and co-workers, 1983; Myrtle, Gygax, Plateel and Winter, 1983). There remain, however, a number of practical problems which have prevented attaining a reliability sufficient for the commercialization of this design. The first problem is that of providing a contamination-free compressor. Many of our experiments have ended after operating for one week because of an increase in friction caused by contaminants in the helium gas freezing out in the critical clearance gap at the cold end of the displacer. The second problem is the fact that helium diffuses through plastic, eventually degrading the vacuum insulation of the dewar. The third problem is finding a reliable means of moving the displacer which will not produce an undesirable magnetic signal for the SQUID sensor. Zimmerman (1984) has solved the second problem by laminating a thin metal foil into the plastic to prevent diffusion. He has also built a low-vibration gas displacer drive which may largely solve the third problem, but measurements of the interference levels have not yet been made.

The problems with this type of cryocooler are difficult, but not insurmountable. As the market for cryocoolers for superconducting applications expands, it is quite likely that inexpensive cryocoolers of this type will be mass produced.

Cascaded Joule-Thomson cryocoolers

Joule-Thomson cryocoolers have no moving parts and therefore do not have the reliability problems associated with friction. Because of the properties of gasses it is not possible to produce cooling at superconducting temperatures directly. A cascade of three coolers would be required, successive stages using high pressure nitrogen, hydrogen and helium. Open cycle cascaded Joule-Thomson cryocoolers have been built which produce 500mW of refrigeration in the 4K range (Currie, 1966). The major difficulty is the rapid consumption of compressed gasses. A very high pressure, contamination free, compressor is needed. A number of groups are currently attempting to build the required compressors (Chan, Tward and Elleman, 1984). See also the article by Tward in this chapter. Much smaller J-T cryocoolers have been made using the new glass slide with etched channel technology (Duboc, 1983). This technology may soon be extended to cascaded cryocoolers, but again, contamination-free compressors will be needed for long term operation.

CONCLUSIONS

A number of small and relatively inexpensive cryocoolers which can reach liquid helium temperature are commercially available. These machines all utilize additional cooling stages added to standard cryopump type cryocoolers. Recent innovations, including providing for a pool of liquid helium at the bottom of an accessible dip dewar, make these cryocoolers well suited for many applications of superconductivity. The magnetic noise and vibration produced by these machines, while operating, present difficulties for biomagnetism measurements.

There are a number of other tested cryocooler designs which can provide cooling to superconducting temperatures. The major stumbling block in all of these designs is the room temperature compressor component. A reliable compressor which will not introduce impurities into the pure helium gas is needed. For the Zimmerman type cryocoolers the pressure requirements are a modest 5 to 10 atmospheres. The J-T cryocoolers operate better at pressures in excess of 100 atmospheres.

There are many exciting new developments in the field of cryogenic refrigeration. A refrigerator which is ideal for cooling superconducting biomagnetic sensors has not yet been built. But, with new developments in any of a number of approaches, one may be available soon.

TABLE 1 Summary of Approaches to Cryogenic Cooling

REGENERATIVE CRYOCOOLERS OF THE TYPE USED IN CRYOPUMPS

PRO: -Mass produced for the cryopump market.
-Priced within the range of researcher's budget ($6,000-$20,000 US).
-Reasonably reliable, 10,000 hours between scheduled maintenance.
-High power, 1W at 15K and 10W at 80K (or more, depending on the model).
-Typical no load temperature of 10K.

CON: -Cannot reach below 8K.
-Vibration and magnetic noise from an electric motor on expander.
-Continuous electical power consumption of 1 to 5kW.

NEW: -Active competition and mass production has resulted in price decreases.
-A wide range of models available.
-Over 10,000 units in existence.

REGENERATIVE CRYOCOOLERS OF THE TYPE USED IN CRYOPUMPS WITH A J-T STAGE

PRO: -Produce liquid helium temperatures, typically 4.5K at 1-3W.
-Can produce 3K with an additional Joule-Thomson stage.

CON: -History of poor reliability due to J-T orifice plugging.
-Reliability problems due to possibility of leakage into vacuum space.
-Vibration and magnetic noise as above.
-Requires an additional compressor.
-High continuous electrical consumption (5kW).
-Price affordable for some researchers (30-100,000 US$).

NEW: -Recent progress developing smaller (1/4 Watt) units, improving reliability and lowering price.
-Approximately 50 units in existence, many on radio telescopes.

REGENERATIVE CRYOCOOLERS OF THE TYPE USED IN CRYOPUMPS WITH A THIRD SIMON EXPANSION STAGE AND A FOURTH STAGE LIQUID HELIUM COLLECTOR.

PRO: -Produces a stable 4.2K and continuous cooling.
-Can produce lower temperatures by pumping on the liquid helium.
-No small orifices to plug - higher reliability.
-May be used with an additional compressor, or with compressed gas cylinders.
-Third and fourth stages may be operated only when 4.2K is required.

CON: -Low cooling power (3mW). Sufficient for many superconducting devices.
-Vibration and magnetic noise as above.

NEW: -This is a new design.

REGENERATIVE CRYOCOOLERS OF THE TYPE USED IN CRYOPUMPS WITH A THIRD STAGE USING A STIRLING-CYCLE DISPLACER IN AN INDEPENDENT LOW PRESSURE GAS SYSTEM.

PRO: -Can achieve temperatures of 4.8K or below 4K (with no heat load).
-Low speed operation of 12RPM.

CON: -Low cooling power 8mW at 5.8K, or 4mW at 4K.
-Requires an additional displacer drive and an additional compressor.

NEW: -This is an experimental design.

PLASTIC REGENERATIVE CRYOCOOLERS OF THE ZIMMERMAN TYPE.

PRO: -Multi-stage stepped cylinders or conical design.
-Can achieve temperatures as low as 7K (with no heat load).
-Have been demonstrated operating a SQUID for biomagnetism measurements.
-No magnetic parts, low vibration and magnetic noise.
-Low power consumption (hundreds of Watts, depending on compressor).

CON: -History of poor reliability due to contamination of helium.
-Helium diffuses through uncoated plastic ruining vacuum insulation.
-Room temperature parts remain the major stumbling block. (ie. Reliable, contamination free, low power helium compressor).
-Low cooling power (10mW range).
-No load temperatures range from 20K to 7K depending on design and quality (low friction) of construction.

NEW: -Metal foil laminated in plastic has eliminated helium diffusion.
-Gas displacer drive promises to have a low magnetic signature.

PLASTIC REGENERATIVE CRYOCOOLERS OF THE ZIMMERMAN TYPE WITH AN ADDITIONAL SMALL JOULE-THOMSON STAGE ADDED.

PRO: -Can achieve constant cooling near 4K with powers near 5mW.

CON: -Poor reliability due to plugging of J-T orifice after a few hours of operation.

NEW: -This is a new design.

SINGLE-STAGE JOULE-THOMSON CRYOCOOLERS.

PRO: -Can achieve temperatures near 70K in a few seconds.
-No moving parts.
-Low cost ($1,000 and up). Models from 1/10 Watt to 10 Watts.

CON: -Open cycle operation consumes compressed argon or nitrogen.
-Plugging of J-T orifice by contaminants in gas is a reliability problem.

NEW: -Recent progress has been made producing miniature J-T crycoolers by etching flow channels in glass slides.

MULTI-STAGE JOULE-THOMSON CRYOCOOLERS.

PRO: -Two stage design with nitrogen and hydrogen gas can achieve 23K.
-Three stage design with nitrogen, hydrogen and helium gasses to 4.4K.
-No moving parts. Low vibration and magnetic noise.

CON: -Reliability problem due to plugging of J-T orifices.
-Reliability problem due to leakage of gas lines into vacuum space.
-Open cycle operation feasible only for short interval due to contamination from impurities in compressed gas cylinders.
-Closed cycle operation requires, contamination-free, reliable, small size, very high pressure compressors which are yet to be developed.

NEW: -Open cycle coolers which reach liquid hydrogen or helium temperatures have been built.
-Current attempts to make multi-stage coolers with the new glass-slides
-Three stage closed-cycle coolers remain a theoretical concept.
-A satisfactory compressor has yet to be demonstrated but considerable research is currently underway to develop one.

REFERENCES

Chan, C. K., E. Tward and D. D. Elleman (1984). Kinetics of a gas adsorption compressor. Advances in cryogenic engineering, 29, 533-542.

Currie, R. B. (1966). A 4K Joule-Thomson laboratory refrigerator. Advances in cryogenic engineering, 12, 557-563.

Duboc, R. M. Jr. (1983). Low cost microminiature refrigerators for large unit volume applications. In Max Gasser (Ed.) Refrigeration for cryogenic sensors. NASA conference publication 2287, pp. 431-443.

Higa, W. and E. Wiebe (1977). One million hours at 4.5 Kelvin. In J. E. Zimmerman and T. M. Flynn (Eds.) Applications of closed-cycle cryocoolers to small superconducting devices. NBS special publication 508, pp. 99-108.

Longsworth, R. C. (1984). Interfacing small closed-cycle refrigerators to liquid helium cryostats. Cryogenics, 24, 175-178.

Matsubara, Y. and K. Yasukochi (1983). An application of gap regenerator/expander precooled by two stage G-M refrigerator. In Max Gasser (Ed.) Refrigeration for cryogenic sensors. NASA conference publication 2287, pp. 157-168.

Myrtle, K., C. Winter and S. Gygax (1982). A 9K conical Stirling-cycle cryocooler. Cryogenics, 22, 139-141.

Myrtle, K., S. Gygax, C. Plateel and C. Winter (1983). Regeneration efficiency, shuttle heat loss and thermal conductivity in epoxy-composite annular gap regenerators from 4K to 80K. In Max Gasser (Ed.) Refrigeration for cryogenic sensors. NASA conference publication 2287, pp. 131-140.

Sager, R. E. and D. N. Paulson (1983) Regeneration experiments below 10K in a regenerative-cycle cryocooler. In Max Gasser (Ed.) Refrigeration for cryogenic sensors. NASA conference publication 2287, pp. 81-94.

Sixsmith, H. (1984). Miniature cryogenic expansion turbines - a review. Advances in cryogenic engineering, 29, 511-523.

Sullivan, D. B., J. E. Zimmerman and J. T. Ives (1981). Operation of a practical SQUID gradiometer in a low-power stirling cryocooler. In J. E. Zimmerman, D. B. Sullivan and S. E. McCarthy (Eds.) Refrigeration for cryogenic sensors and electronic systems. NBS special publication 607, pp. 186-194.

Sullivan, D. B., R. Radebaugh, D. E. Daney and J. E. Zimmerman (1983). An approach to optimization of low-power Stirling cryocoolers. In Max Gasser (Ed.) Refrigeration for cryogenic sensors. NASA conference publication 2287, pp. 107-130.

Valera, A. M., J. Maia Alves and E. Ducla-Soares (1984). An adiabatic demagnetization technique to reach 1K with a closed-cycle refrigerator. Physics in Canada, 40, No. 3, page 30.

Winter, C., S. Gygax, K. Myrtle and R. Barton (1984). Stable 4K cryocooler using adiabatic (Simon) expansion. To be published in Cryogenics.

Zimmerman, J. E. and D. B. Sullivan (1979). A milliwatt Stirling cryocooler for temperatures below 4K. Cryogenics, 19, 170-171.

Zimmerman, J. E. and D. B. Sullivan (1982). A study of design principles for refrigerators for low-power cryoelectronic devices. NBS Technical note 1049.

Zimmerman, J. E., D. E. Daney and D. B. Sullivan (1983). A cryocooler for applications requiring low magnetic and mechanical interference. In Max Gasser (Ed.) Refrigeration for cryogenic sensors. NASA conference publication 2287, pp. 95-105.

Zimmerman, J. E. (1984). Recent developments in self-contained cryocoolers for SQUIDS and other low-power cryoelectronic devices. To be published in the proceedings of ICEC-10. Presented at Helsinki, Finland, 1984 July 31.

A SMALL CLOSED CYCLE COOLER FOR SQUIDS

E. Tward

Jet Propulsion Laboratory
California Institute of Technology
Pasadena, California, USA 91109

ABSTRACT

A small closed cycle cascade Joule Thompson (J-T) refrigerator designed to cool a biomagnetic probe has been constructed. Three of the four cooling stages have been operated and have demonstrated satisfactory performance.

KEYWORDS

Refrigerator, Joule Thompson, SQUID.

INTRODUCTION

The cooling of Josephson junction-based magnetic sensors (SQUIDS) to cryogenic temperatures is a challenging task. These SQUID-based sensors are used only because of their unparalleled sensitivity to magnetic field variations, a sensitivity that is essential for the detection of tiny biomagnetic fields. This sensitivity imposes a severe constraint on the coolers which are used to achieve the required operating temperatures and results in the almost exclusive use of open cycle liquid helium dewars. The dewars are very carefully constructed of non-magnetic materials with special attention paid to the thermally necessary conductive materials such as super-insulation. Active closed cycle coolers have been rejected for these applications because small commercial cryocoolers use thermodynamic cycles requiring expansion engines with moving pistons. These devices inherently are a source of vibration and time varying magnetic fields which must be decoupled from the magnetic pickup coils of the sensor. Attempts have been made to build low cooling power Stirling cycle cryocoolers (Zimmerman and colleagues) constructed of plastics in order to overcome the moving conductive material problem. However, vibration still is a problem and, in addition, a low temperature stage (such as a J-T cooler) is required to reduce the temperature below approximately 10K.

For these reasons, we have designed and built a closed cycle cascade J-T refrigerator which has no cold moving parts and has the additional virtue of flexible geometry and small size.

CASCADE J-T COOLERS

The J-T process is a widely used refrigeration process. It is conceptually very simple (Walker, 1983) and, for small working powers, is very easy to build (Figure 1).

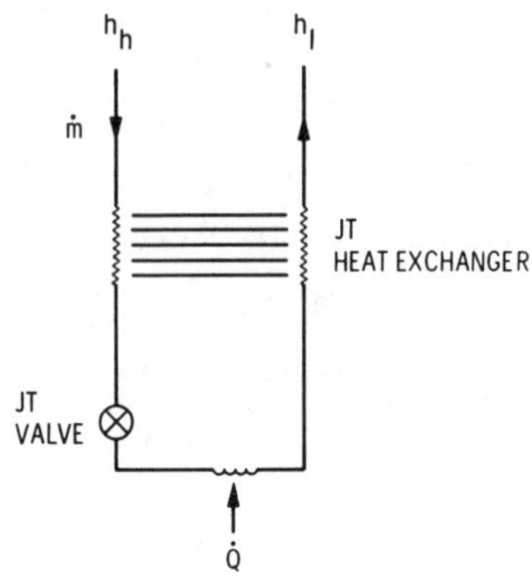

Fig. 1. Gas at high pressure (enthapy h_h) with flow rate, m, enters the heat exchanger, expands and cools at the J-T valve and exists through the low pressure side of the heat exchanger (enthapy H_L). The cooling power Q is given by $Q = m(h_h - h_L)$.

It has not been used for small closed cycle coolers chiefly because of the lack of suitable small compressors providing extremely clean high pressure gas. We have addressed this problem by designing and building a high compression ratio compressor which requires no lubricants in the compression chamber so that minimal filtering results in the required extremely pure gas. Since the J-T process produces cooling (rather than heating) only if the working gas enters the refrigerator below the gas inversion temperature, a minimum of three different gases are required to reach liquid helium temperatures. In the machine which is described here, four gases were used because of increased thermodynamic efficiency and because lower pressures are needed from the compressor. In the cascade J-T refrigerator, the four independent fluid loops provide four cooling stages between room temperature and liquid helium temperature. Each loop contains the components shown in Fig. 2.

The four loops are configured as shown in Fig. 3. The cold contaminant filters in each loop were provided as a precaution to assure that condensibles would not clog the very small J-T orifices.

Because of the low gas flows, the eight gas lines to the refrigerator (2 per stage) can be long enough for the compressor to be far from the refrigerator (we have used 3.2 meter lengths) and can be of small diameter (0.16 cm OD). This provides a very flexible interface to the sensor cryogenics. The refrigerator operates in any orientation which should make the device much easier to use than sensors using liquid helium dewars. The prototype refrigerator shown in Fig. 4 was specifically designed to cool a SHE biomagnetic probe (SHE Corp., San Diego) with a second-order gradiometer input coil. A schematic of the cryostat is shown in Fig. 5. To date, only three of the four stages have been operated. Figures 6 and 7 show cool down curves for the various stages. The longer than necessary cool down times are due to the fact that these measurements were taken with a leak in one of the loops which resulted in excessive heat leakage through the inadequately maintained dewar vacuum. More rapid cool down will be possible when thermal switches,

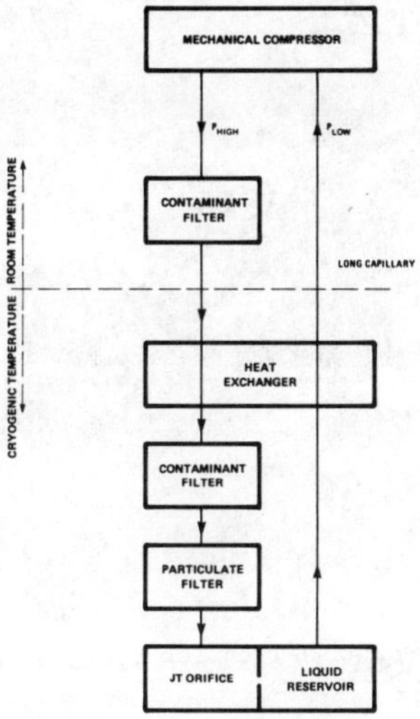

Fig. 2. Schematic of single stage J-T cooler.

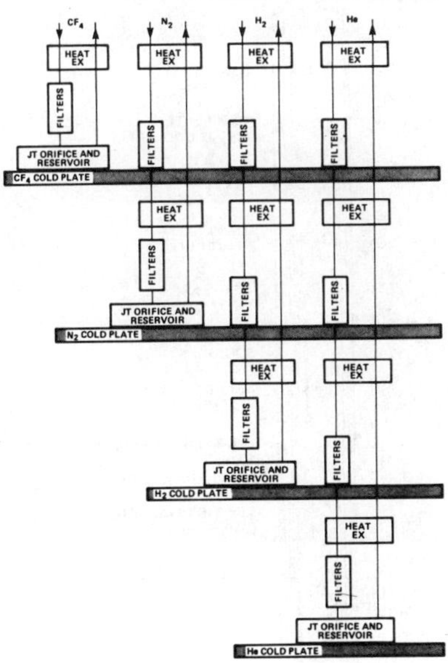

Fig. 3. Schematic of cascaded four-stage J-T cooler.

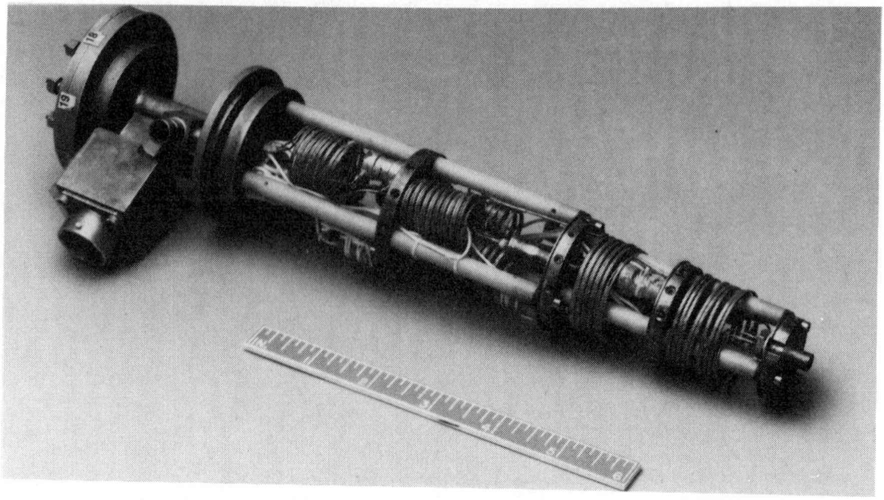

Fig. 4. Four stage cascaded J-T cooler designed to cool biomagnetic probe.

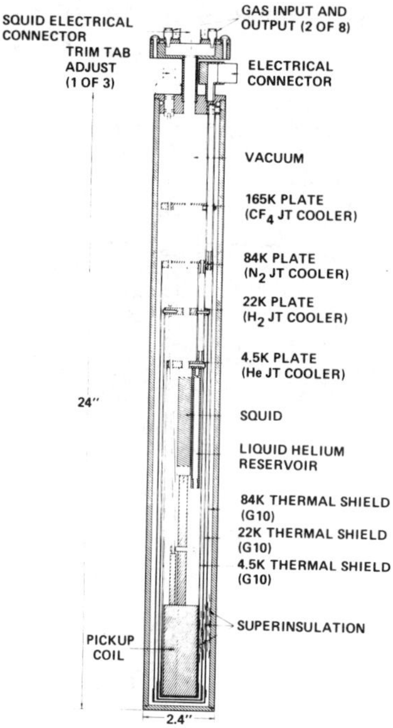

Fig. 5. Schematic of single sensor neuromagnetometer incorporating closed cycle J-T cooler.

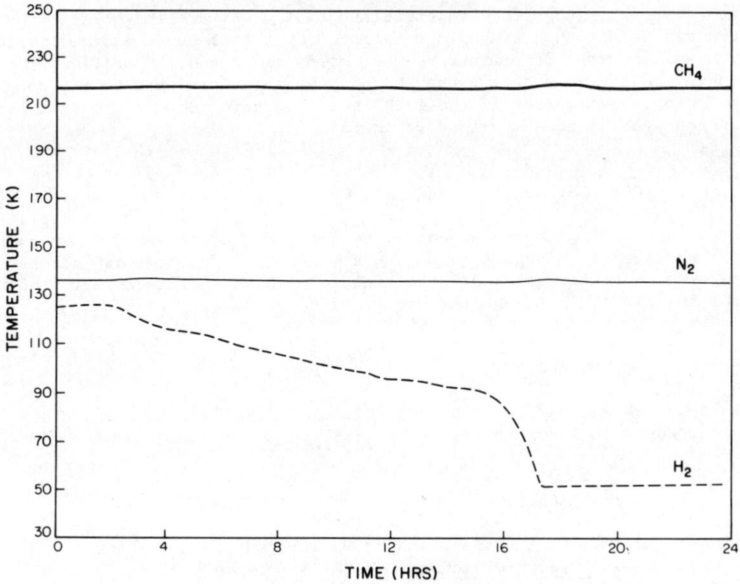

Fig. 6. Cool down of upper two stages of refrigerator. The H_2 is being cooled by conduction. The curve labelled CF_4 indicates the outlet temperature of the heat exchanger. The low temperature after cool down is due to excess cooling power.

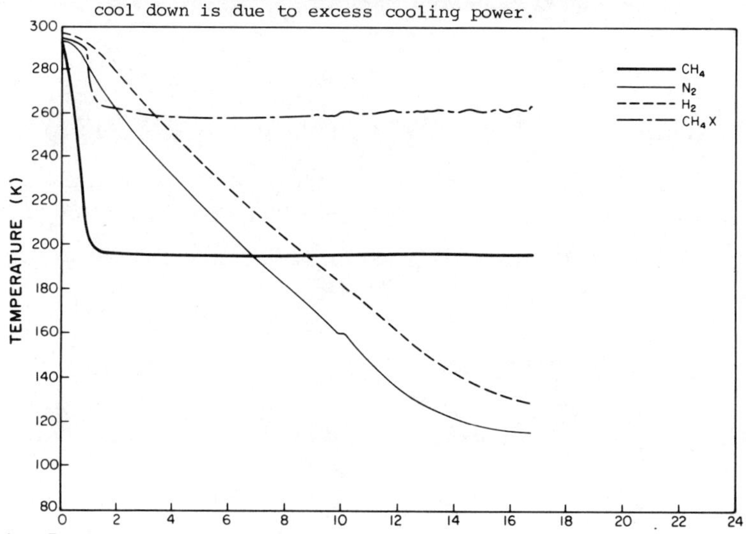

Fig. 7. H_2 stage coold down. Note temperature stability. The rapid cool down below 70K is due to increased gas flow and cooling power as the impedance of the refrigerator drops with temperature.

which thermally short the stages during cool down, are installed. The temperature reached at each stage is determined by the pressure at the liquid reservoir which is, in turn, determined by the impedance of the J-T orifice, the pressure drop in the low side heat exchanger and the inlet pressure to the compressor. Particularly noteworthy is the excellent long term temperature stability which has been achieved in these early experiments.

CONCLUSION

Cascade J-T refrigerators appear to be ideal for cooling biomagnetic sensors. Their small size, adequate temperature stability, orientation independent performance and low maintenance potential should make them attractive to the user who finds the use of liquid helium cryostats cumbersome.

ACKNOWLEDGMENTS

This research was supported by the Office of Naval Research and the Air Force Medical Research Laboratories.

REFERENCES

Zimmerman, J. E., D. E. Daney, and D. B. Sullivan. In M. Gasser (Ed.), Refrigeration for Cryogenic Sensors. NASA Conference Publication 2287. pp. 95-105.
Walker, G. (1983). Cryocoolers. Plenum Press, New York.
Tward, E., and W. A. Steyert. In M. Gasser (Ed.), Refrigeration for Cryogenic Sensors. NASA Conference Publication 2287. pp. 419-425.
SHE Corporation, San Diego, California, USA.

BIOGRADIOMETER DESIGN FOR NOISY ENVIRONMENT

A.A. Fife*, J. Vrba*, J. McCubbin*, H. Weinberg**, and P. Brickett**

* CTF Systems Inc., #15-1750 McLean Ave., Port Coq., B.C.
** Brain Behaviour Laboratory, Simon Fraser University, Burnaby, B.C.

ABSTRACT

The design features of unshielded biogradiometers operating in noisy environments are discussed. Topics include the compromise between good spatial discrimination and near field sensitivity, gradiometer errors and field and first gradient balancing.

KEYWORDS

SQUID, biogradiometer, flux transformer, common mode, first gradient balancing.

INTRODUCTION

There are three principal operational modes for SQUID biomagnetic systems:
a) in a magnetic or eddy-current shielded enclosure (e.g. Cohen, 1970; Zimmerman, 1977);
b) in an unshielded rural environment remote from urban noise (e.g. Farrell and co-workers, 1980);
c) in an unshielded urban environment where noise is rejected by use of the spatial discrimination properties of second or higher order gradiometers (e.g. Brenner and co-workers, 1974; Vrba and co workers, 1982).

In the basic configuration, the (RF or DC) SQUID sensor in its cryogenic enclosure is coupled both to gradiometer coils mounted on a rigid substrate, and also to suitable room temperature readout electronics. In optimising the readout sensitivity, one must pay attention to many system parameters which are often interactive:
a) signal and interference source parameters - distance, magnitude, and frequency spectrum;
b) SQUID energy sensitivity, white noise level, low frequency noise onset, slope;
c) gradiometer type, constructional accuracy, coil sizing, baseline;
d) Rf shield design;
e) magnetic noise levels contributed by <u>all</u> constructional materials;
f) temperature stability and heat load in the cryogenic environment (liquid helium dewar or cryocooler operation or a combination of both);
g) inter-component and system mechanical rigidity.

It is not the purpose of this paper to examine all of these features and their interactions in detail. Rather, the discussion will concentrate on gradiometer coil design, the effects of imperfections, and the importance of field and 1st gradient balancing.

GRADIOMETER FLUX TRANSFORMERS

Higher order gradiometer coil geometries are designed to spatially discriminate between near field biomagnetic signals of interest and far field sources. In near field, the system displays magnetometer response, and in far field, the designed sensitivity decay with distance r. Typical coil designs are:

i) second order with r^{-5} far field response (Fig. 1a and 1b);

ii) third order with r^{-6} far field response (Fig. 1c).

The parameters important for gradiometer selection include: i) desired near field sensitivity and coil baseline, ii) far field interference rejection and iii) achievable balance. The selection process requires careful consideration of the properties of the biomagnetic source of interest and the interference signals.

Sensitivity And Baseline Considerations

The near field resolution B_{min} of a gradiometer depends mainly on the sense coil area and the SQUID sensor input energy sensitivity (Romani and co-workers, 1982). By making the flux transformer design asymmetric (Figures 1b, 1c), one can approach magnetometer sensitivity in the near field region (Zimmerman, 1977; Duret, Karp, 1984). Typical projected (white noise) sensitivities are as follows: for a 1.5 cm diameter wire-wound symmetric 2nd order gradiometer matched to a SQUID of energy sensitivity $1\text{x}10^{-30}$J/Hz, $B_{min} \simeq 8$ fT rms/$\sqrt{\text{Hz}}$. If one uses a 3rd order asymmetric gradiometer with a 2.54 cm dia. sense coil, SQUID energy sensitivity $5\text{x}10^{-29}$J/Hz, $B_{min} \simeq 16$ fT rms/$\sqrt{\text{Hz}}$. The projected near field resolution is generally not attained in unshielded operation in noisy environments.

The selection of the optimum baseline unit is an involved problem, which depends on the distance between the sense coil and the source of interest, and the properties of the interference signals. Short baselines improve the far-field rejection but reduce near-field sensitivity. As a compromise, the baseline length D is typically chosen to be ≃ 1.5 - 2.5 times the sense coil diameter.

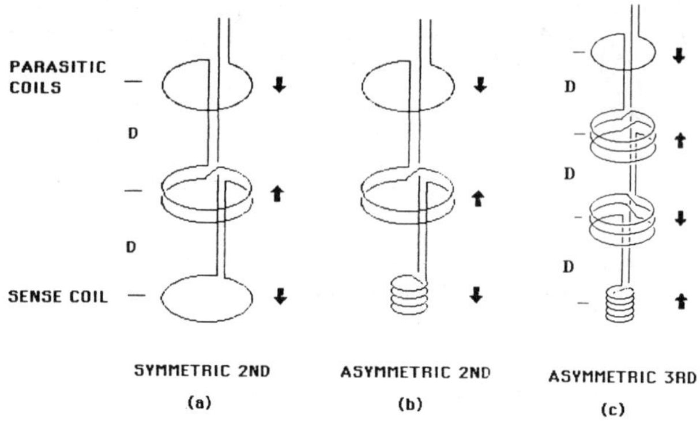

Fig. 1: 2nd and 3rd order gradiometer flux transformers.

Far Field Rejection

The selection of the gradiometer order depends on the character of the magnetic interference in the vicinity of the instrument. In quiet areas, the use of the 1st gradiometer may be sufficient. On the other hand, in noisy, unshielded urban areas, 2nd or 3rd gradiometers are mostly used. The far field noise rejection generally improves with the gradiometer order, with the 3rd order being found to exhibit superior interference rejection capabilities relative to 2nd order, while still maintaining good near field resolution. This is demonstrated in Fig. 2, where the spatial responses of ideal 2nd and 3rd gradiometers (solid lines) are compared. The noise rejection of the 3rd gradiometer is far superior, for larger noise source-gradiometer distances. However, in practice, these ideal responses at large distances are degraded by various common mode signals (discussed in the next section). The experimental confirmation of the predicted spatial behaviour for near sources is shown in Fig. 2 by solid circles. The implication of Fig. 2 is that for many confined urban environments, it may be preferable to aim at a compromise of good far field rejection and slightly reduced near field sensitivity for the best overall biogradiometer performance.

Gradiometer Errors And Balancing

Errors in the construction of gradiometer coils lead to additional (common mode) terms in the output. For example, 3rd gradiometer response is written as:

$$S_3 = C_B{*}B + C_{G1}{*}G_1{*}D + C_{G2}{*}G_2{*}D^2 + G_3{*}D^3$$

(Vrba and co-workers, 1982). B is the field vector, and G_1, G_2, G_3, are the 1st, 2nd, and 3rd gradient tensors. C_B, C_{G1}, C_{G2}, are the 3-, 5-, and 7- component common-mode vectors for field, 1st, and 2nd gradients. D is the gradiometer baseline unit (Fig. 1). For a 2nd gradiometer, results are similar. Thus the instrumental response is a complicated mix of unwanted common mode signals and the gradient signals of interest. In near field distance $r \simeq D$, S_2 and S_3 are designed to have magnetometer response, and in far field ($r \gg D$), they decay as r^{-5}, r^{-6}, respectively until limited by various common mode signals. In order to appreciate

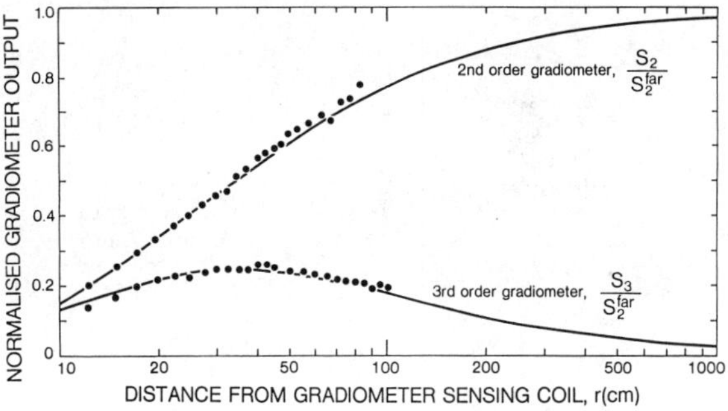

Fig. 2: Comparison of 2nd and 3rd order gradiometer spatial responses. Solid lines represent the calculated responses of 2nd and 3rd order gradiometers to a dipole located at distance r. Both responses are normalised by the far field approximation of the 2nd gradiometer. Circles represent experimental points fitted at one point.

the effects of the presence of common mode errors on signal detection, Fig. 3 compares the responses of 2nd and 3rd gradiometers to a dipole source, using realistic magnitudes of common mode vectors with $C_B = 10^{-5}$, $C_{G1} = 5 \times 10^{-3}$, $C_{G2} = 2 \times 10^{-2}$ (wire-wound coils). The outputs are normalised by the dipole field $B_0 = M/r^3$. It is clear that in the intermediate distances 2-20m, the 3rd gradiometer provides superior rejection of interference signals by factors of 2.5 - 15 over that of the 2nd order gradiometer. Fig. 3 complements the spatial transfer functions of Fig. 2.

Prior to signal measurement, the gradiometer must be balanced. This involves minimising the unwanted terms in equation 1. Balancing with respect to field has been the standard procedure in the past, either by means of movable superconducting vanes (Zimmerman, Frederick, 1971; Overweg, Walter-Peters, 1978), electronic subtraction (Jaworski, Crum, 1980) or by software methods (Vrba 1980). However it has become clear that balancing with respect to 1st gradient is also important (Zimmerman, 1977; Barbanera and co-workers, 1981; Vrba, McCubbin, 1983). Measurements of the 1st gradient coefficients C_{G1} for a 3rd gradiometer indicate that no 1st gradient term is dominant, and that all 5 have to balanced out. A demonstration of the effects of 1st gradient balancing on the frequency response of a field balanced 3rd gradiometer equipped with 3 first gradient sensors G_{13}, G_{23}, and G_{33}, is shown in Fig. 4. The (unshielded) instrument* was positioned ≃ 20m from a highway and periods of light and heavy traffic resulted in spectra A and B. Using correlation methods for the determination of the coefficients C_{G1}, the 1st gradient common mode contributions were subtracted from the 3rd gradiometer output. The resultant curve C indicates a considerable reduction in 1/f noise in the frequency band < 1Hz. The cancellation is not perfect due probably to the use of only 3 channels and to the neglect of field common mode effects in the 1st gradient sense channels.

In conclusion, it is clear that the design of unshielded biogradiometers operating in magnetically noisy environments depends on a judicious compromise between near field sensitivity and far field rejection capability. The importance of good common mode rejection ratio for both field and 1st gradient has also been emphasised.

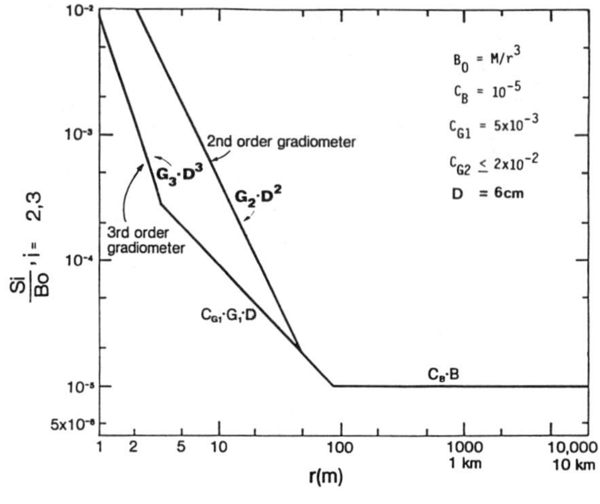

Fig. 3: Response of gradiometers to a dipole source.

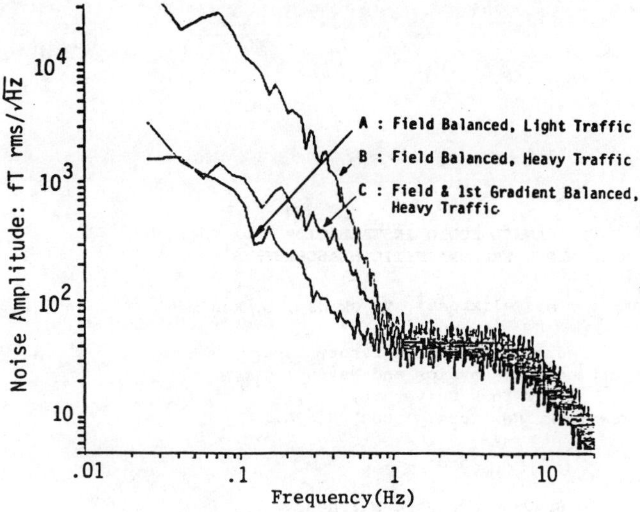

Fig. 4: The effects of 1st gradient balancing on the spectrum of a field balanced 3rd gradiometer.

ACKNOWLEDGEMENTS

This work was supported by grants from the Secretariat on Science, Research and Development of British Columbia, the National Research Council and the Federal Government Department of Supply and Services, and Transport.
* Equipment Manufactured by CTF Systems Inc.

REFERENCES

Barbanera, S., Carelli, P., Fenici, R., Leoni, R. Modena, I., Romani, G.L., (1981) IEEE Trans. Magn., Mag-17, 849 - 852.
Brenner, D., Williamson, S.J., Kaufman, L., (1975). Science, 190, 480 - 482.
Cohen, D., (1970). Rev. Phys. Appl., 5, 53 .
Duret, D., Karp, P., (1984) Figure of Merit and Spatial Resolution of Superconducting Flux Transformers. To be Publ. J. Appl. Phys.
Farrell, D.E., Tripp, J.H., Norgren, R., Teyler, T.J., (1980). Electroenceph. Clin. Neurophys., 49, 31 - 41.
Jaworski, F.B., Crum, D.B., (1981). In SQUID Application to Geophysics Workshop, edited by Weinstock, H., Overton Jr., W.C. Society of Exploration Geophysicists, Tulsa, Oklahoma, (Pub.). 19-25.
Overweg, J.A., Walter-Peters, M.J., (1976). Cryogenics 18, 529.
Romani, G.L., Williamson, S.J., Kaufman, L., (1982). Rev. Sci. Inst., 53, 1815-1845.
Vrba, J., (1980). Phys. Can 36, 3 - 9.
Vrba, J., Fife, A.A., Burbank, M.B., Weinberg, H., Brickett, P.A., (1982). Can. J. Phys. 60, 1060 - 1073.
Vrba, J., McCubbin, J., (1983). Il Nuo. Cim. 2D, 142 - 152.
Zimmerman, J.E., Frederick, N.V., (1971). National Bureau of Standards Report #10.
Zimmerman, J.E., (1977). J. Appl. Phys. 48, 702 - 710.

FIVE CHANNEL SQUID INSTALLATION FOR UNSHIELDED NEUROMAGNETIC MEASUREMENTS

S.J. Williamson*, M. Pelizzone+, Y. Okada*, L. Kaufman*

Neuromagnetism Laboratory
Departments of Physics and Psychology
New York University
New York, New York, 10003, U.S.A.

and

D.B. Crum and J.R. Marsden

S.H.E. Corporation
4174 Sorrento Valley Boulevard
San Diego, California, 92121, U.S.A.

ABSTRACT

A system containing nine SQUIDs has been introduced for neuromagnetic studies in an unshielded environment. Five dc SQUIDs monitor brain activity detected by an array of second-order gradiometers pointing toward the center of the subject's head. Four rf SQUIDs monitor three components of the ambient field and one component of the gradient so that by electronic cancellation these signals can be used for noise reduction in the five signal channels.

KEY WORDS

Neuromagnetism; magnetic sensor; magnetoencephalogram; evoked responses; SQUID

INTRODUCTION

A novel installation has been developed for unshielded neuromagnetic studies with a cryogenic system incorporating nine SQUIDs in a single dewar. Conventional measurements of the magnetic field pattern near the scalp with a single sensor are laborious because the field at only one point at a time can be measured. The single-sensor method is treated in detail in a recent multi-authored text (Williamson et al., 1983). Five dc-SQUIDs of this new system measure the magnetic field of interest at five positions over the scalp, and for electronic cancellation of background noise four rf-SQUIDs monitor the three components of the ambient field and the component of the field gradient along the dewar's axis. The effectiveness of electronic cancellation of low frequency ambient noise in an urban setting was previously demonstrated by Bastuscheck with two single sensor systems in the Neuromagnetism Laboratory at N.Y.U. (Romani, Williamson and Kaufman, 1982). Recently, a sensing system for use within a magnetically shielded room has been reported that contains four SQUIDs in a single dewar for simultaneous measurements at four positions over the scalp (R. Ilmoniemi, R. Hari, and K. Reinikainen, 1984). The system described here does not require magnetic shielding for carrying out a wide variety of neuromagnetic studies and is presently operating in a laboratory environment.

SENSING SYSTEM

The assembly of SQUIDs and detection coils is mounted on a probe that is suspended within a fiberglass dewar of 10 liter capacity (Fig. 1). The SQUIDs are nestled within a superconducting shield positioned at the bottom of the dewar's belly. Shielded superconducting leads connect the input coils of the SQUIDs to their respective detection coils which are supported by a frame braced securely against the bottom of the dewar's tail. The detection coils thus serve as primaries for superconducting flux transformers, whose secondaries are the input coils that are magnetically coupled to the SQUIDs. Each of the five detection coils for the signal channels has the geometry of a second-order gradiometer with a 1.5-cm diameter and 4.0-cm baseline separating adjacent coils. The array of these "signal coils" consists of one central coil aligned along the dewar's axis and four equally spaced surrounding coils whose bottom coils lie on a 4-cm diameter circle. The axes of the four outer coils are tipped outward at the top by 10 deg so that they point to a common position approximately 9 cm below the bottom of the dewar's tail, where the subject's head will be centered. The bottom of the dewar's tail has a concavity with a 9-cm radius so that it can be positioned close to the scalp, and the tilt of the outer detection coils places their axes on a 3.4-cm diameter circle on this surface. The sensor for each of the five signal channels is a dc DYNABIAS SQUID™, and despite the proximity of the detection coils, cross-talk between the channels is calculated to be less than 0.2%.

Just above these signal detection coils is an array of four detection coils for the reference channels, each of which serves as the primary for a flux transformer leading to an rf HYBRID SQUID™. Three magnetometer coils are orthogonally oriented to sense the axial and two transverse components of the ambient field. The fourth coil is a first-order gradiometer oriented in the axial direction.

The dewar with its sensors is mounted in a specially designed device (called Scanner) that moves the tail of the dewar over a spherical surface of selectable radius while keeping the dewar's axis pointing toward the center of the sphere, coinciding with the center of the subject's head. As shown in Fig. 2, the dewar is supported by a carriage that rides on two parallel, semicircular, fiberglass arches whose ends are supported by frames that enable the arches to be rotated about the horizontal diameter. This diameter serves as the polar axis of the dewar's spherical coordinate system: rotation of the arches corresponds to scanning the azimuthal angle, while movement of the carriage corresponds to scanning the polar angle. Angular indicators on Scanner give these angles so that the orientation and position of the detection coils over the subject's head are precisely known. The angular position can be established with a precision of better than 0.5 deg, corresponding to less than 1 mm at the scalp.

ELECTRONIC SYSTEM

The voltages in the five signal and four reference channels pass through individual bandpass filters with 48 db/octave rolloff to select the bandwidth of interest and then through comb filters to reduce the pickup of noise at the power line frequency and its harmonics. The time derivative of each reference is taken to provide greater flexibility in compensating for the effects of eddy currents

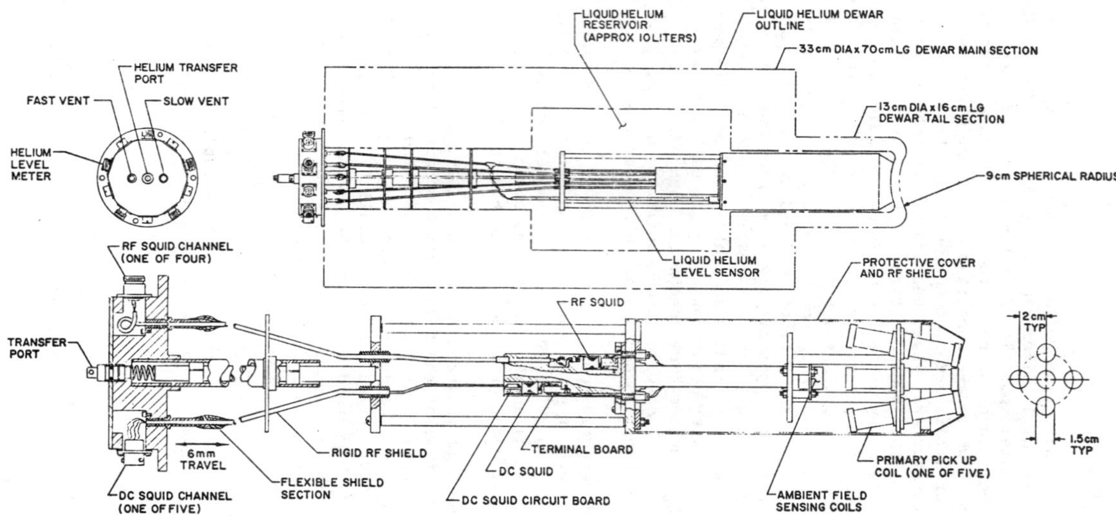

Fig. 1. Five-channel magnetic sensor, with a top view of the dewar at the upper right, a longitudinal section of the dewar at the lower right, and (not to scale) a longitudinal section of the probe at the left.

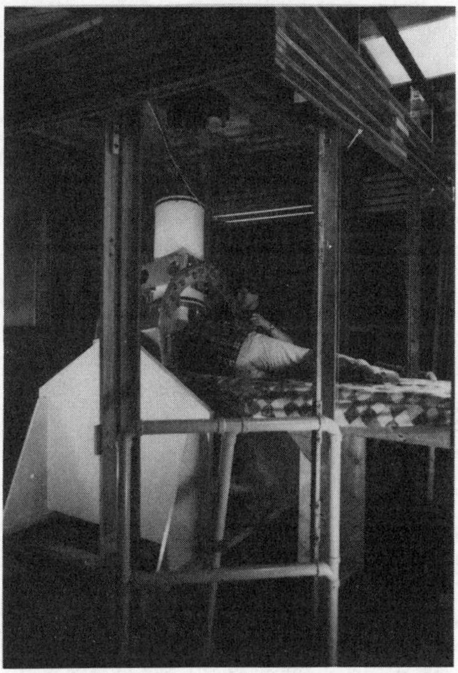

Fig. 2. Dewar supported by Scanner, with a subject on a bed under the sensor being positioned for measurements. A wood framework surrounding Scanner supports the coils that apply magnetic fields when optimizing reference weights.

and hysteritic magnetic materials in the environment. These derivatives together with the original references are then appropriately scaled by manually- or computer-controlled attenuators and subtracted from each of the five signal channels in a summation circuit, as illustrated in the block diagram of Fig. 3. Thus a total of 40 "weights" for the signal channels must be selected. The five summed signals coming from the summation circuits then pass through anti-aliasing, low-pass filters before entering a PDP 11/34 computer.

Optimization of the reference weights can be accomplished in less than one hour by using the ac technique of applying a uniform field in a given direction and, with the aid of a lock-in amplifier as readout, adjusting the weights until the summed signal in each channel is minimized. This procedure is carried out sequentially for the two transverse orthogonal field directions, as well as for the vertical

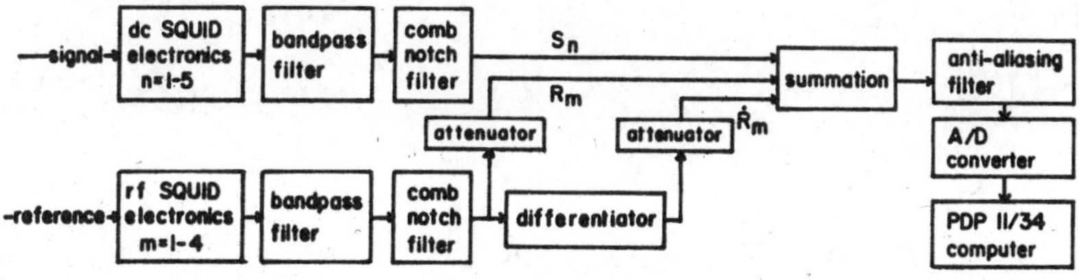

Fig. 3. Block diagram of the electronic system.

components of field and gradient, with the axis of the dewar oriented vertically. The procedure is easily accomplished with the aid of permanently installed sets of Helmholtz coils for the transverse field and axial gradient, and a four-coil set (Merritt, Purcell, and Stroink, 1983) for the axial field. The arrangement is illustrated in Fig. 2 with the coils for the vertical field and gradient raised, to provide convenient access to the subject and Scanner for neuromagnetic measurements.

PERFORMANCE

Measurements of the noise spectra exhibited by the five signal channels at S.H.E. Corporation in a light industry area showed intrinsic system noise of about 20 $fT/Hz^{1/2}$ extending from high frequencies down to approximately 0.5 Hz without use of electronic cancellation with the reference channels. The ambient noise level in the Neuromagnetism Laboratory at N.Y.U. is several orders of magnitude greater at frequencies below 1 Hz, as indicated by the magnetometer outputs. At 1 Hz it is even greater than the noise level measured in a hospital in Helsinki (Jarvinen et al. 1980) and in a laboratory in Berlin (Erne et al. 1981) whose spectra are displayed by Romani, Williamson, and Kaufman (1982).

As illustrated in Fig. 4, without electronic cancellation, the signal channels showed ambient noise rising above the intrinsic noise in the frequency range 15-20 Hz. By use of all the references for cancellation, the intrinsic noise level could be extended to below about 5-8 Hz before the ambient noise became apparent in a linear plot. The effect of including the axial gradient reference was comparatively small, amounting to an improvement of less than 20%. The achieved noise level is better than that of a single-channel DYNABIAS SQUID™ system operating in the same laboratory that relies on moveable superconducting tabs for affecting fine balance of the area-turns products of the detection coil. Furthermore the noise level is not senstivie to the orientation of the dewar. This five-channel system is thus adequate for a wide variety of neuromagnetic studies without shielding. The present system brings several added advantages, including: (1) five simultaneous measurements; (2) verification of signal stability when the

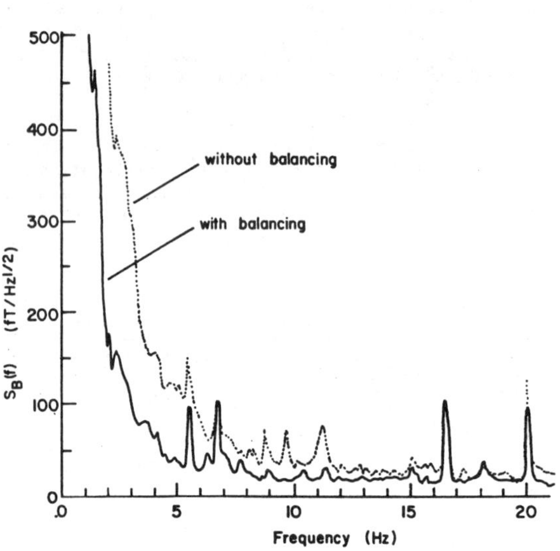

Fig. 4. Noise spectra for a signal channel with and without electronic cancellation of weighted references.

dewar is moved to a new position by placing one of the signal sensors over a position previously measured; (3) enhanced accuracy of positioning of the sensor; and (4) availablility of reference signals for subsequent computer processing to improve electronic cancellation. Measurements in progress demonstrate that a multiple SQUID system can be operated successfully in an unshielded urban environment using electronic cancellation techniques for studies of event-related cortical and subcortical activity of the human brain.

ACKNOWLEDGEMENTS

We thank Dr. Donald Woodward of the Office of Naval Research for his interest and support of this development and express appreciation to Sarah Curtiss and Dr. Ken Schafer for helping construct the Helmholtz coils. We also thank Prof. J. Anthony Movshon for kindly providing much of the computer software used in data analysis.

REFERENCES

*Supported in part by Office of Naval Research Contract N00014-76-C-0568, with ancillary support provided by Air Force Office of Scientific Research Contract F49620-82-K-0014.

+Supported by the Swiss National Science Foundation.

Erne, S.N., H.-D. Hahlbohm, and Z. Trontelj (1981). The Berlin magnetically shielded room (BMSR) Section B - Performance. In S.N. Erne, H.-D. Hahlbohm, and H. Lubbig, (Eds.), Biomagnetism, Walter de Gruyter, Berlin. pp. 79-87.

Ilmoniemi, R., R. Hari, and K. Reinikainen (1984). A four-channel SQUID magnetometer for brain research. Electroenceph. Clin. Neurophysiol., in press.

Jarvinen, M.L., T. Katila, R. Maniewski, P. Siltanen, and T. Varpula (1980). Measurement of biomagnetic fields in a hospital environment. Postepy Fiz. Med., 15, 201-210.

Merritt, R., C. Purcell, and G. Stroink. (1983). Uniform magnetic field produced by three, four, and five square coils. Rev. Sci. Instrum., 54, 879-882.

Romani, G.L., S.J. Williamson, and L. Kaufman (1982). Biomagnetic instrumentation. Rev. Sci. Instrum., 53, 1815-1845.

Williamson, S.J., G.L. Romani, L. Kaufman, and I. Modena, Eds. (1983). Biomagnetism: An Interdisciplinary Approach. Plenum Press, New York.

INTEGRATED BIOMAGNETIC ROBOTIC SYSTEM

J. Vrba*, M. Burbank*, H. Ensing*, A. Fife*, E. Heijster*, C. Marshall*, J. McCubbin*, D. McKenzie*, M. Tillotson*, K. Watkinson*, H. Weinberg**, and P. Brickett**

* CTF Systems Inc., 15-1750 McLean Ave., Port Coquitlam, B.C., Canada, V3C 1M9

** Simon Fraser University, Burnaby, B.C., Canada, V5A 1S6.

ABSTRACT

This paper describes the development of an integrated robotic scanning and mapping system designed for the manipulation of a SQUID biomagnetometer around a subjects head, or other anatomical region of interest. The system integrates a mechanical scanning mechanism, head shape modelling, magnetic and electric data acquisition and storage, stimulus presentation and data processing and display.

KEYWORDS

Biomagnetism, Biogradiometers, Automation, Software processing, Robotics.

INTRODUCTION

In recent years, SQUID biomagnetometers have been the subject of considerable development and innovation. Their designs have been sufficiently refined, such that high sensitivity measurements can be successfully performed even in the magnetically harsh environment of a conventional laboratory or hospital. This good performance in noisy environments has been made possible by configuring the biomagnetometers as 2nd or 3rd order gradiometers (Vrba and McCubbin, 1983; Deecke and co-workers, 1982) and through careful balancing of their field and first gradient common mode errors.

One of our principal interests has been in the development of biomagnetic instrumentation as it applies to brain studies. In this case, the localization of magnetic sources is very important and has been achieved through the mapping of the magnetic signals. The magnetic maps are compiled from magnetic measurements performed at various positions over the head surface and configured into isocontour overlays.

In order to facilitate the measurement and interpretation of biomagnetic signals, we have been developing an integrated biomagnetic measuring and mapping system for rapid scanning of anatomical surfaces (such as the head) and for fast accumulation, recording, and display of magnetic data. The main emphasis during the above development has been placed on system integration, i.e. unification of all tasks into a single instrument under computer control. This will permit fast and flexible coordination of head scanning (at predetermined points), stimulus presentation, and data manipulation. In this paper we would like to report on the progress of the above development.

DESCRIPTION OF THE SYSTEM

The system consists of the CONTROLLER, the DATA COLLECTION SYSTEM, the MECHANICAL GANTRY for manipulation of the biomagnetometer around the head, the HEAD MODELLING SYSTEM, and the STIMULUS DISPLAY SYSTEM. The various system elements are integrated through distributed computer units as illustrated in Fig. 1.

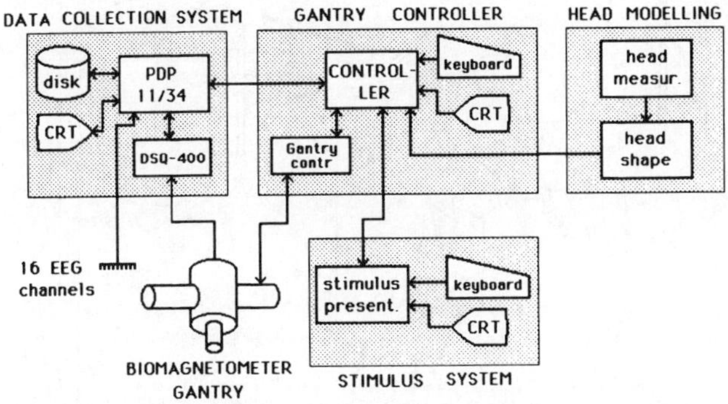

Fig. 1: Block diagram of the Automated Biomagnetic System.

The key element is the controller, which communicates with all computers in the system, and it directly controls the motion of the dewar around the head. It also provides for the operator's interaction with the system (e.g. the head scanning schedule is entered or modified through this computer), and coordinates the stimulus presentation and data accumulation. The controller is also connected to a 3-dimensional head measuring device, which is used to acquire the data for the determination of the head shape model and also during the "zeroing" phase to determine the appropriate referencing between the subject's head and gantry position.

A drawing of the mechanical gantry for the manipulation of the biomagnetometer is shown in Fig. 2. The gantry has five degrees of freedom (three translational and two rotational) and it allows for biomagnetometer tilts up to 45^0 in any direction anywhere in the working volume of 46 x 46 x 76 cm^3. The design of the mechanical gantry was constrained by a variety of boundary conditions. First, the gantry must be as safe as possible and the safety system must provide for various system failures and operator errors. Second, the gantry construction must be such that its interference with magnetic measurements is negligible. Finally the gantry mechanical response must be sufficiently fast and precise for automation of biomagnetic measurements on any head shape.

The safety of the system has been assured by the provision of three levels of positions control. First, the gantry is set into motion according to a predetermined motion schedule. The actual position of the gantry is also monitored by independent position detectors which are periodically compared to the scheduled position. If a disagreement is found, all system motions are stopped and can be restarted only through operator intervention. For further safety enhancement, the tip of the biomagnetometer dewar tail is also equipped with a proximity detector. If for some unforeseen reasons, the above dual safety procedure should fail and the dewar tip contacts the subject's head, the proximity detector will immediately stop all system motions. Finally, in the event of a power failure, all motions stop; the system is designed such that any gravity driven motions or tilts are not possible.

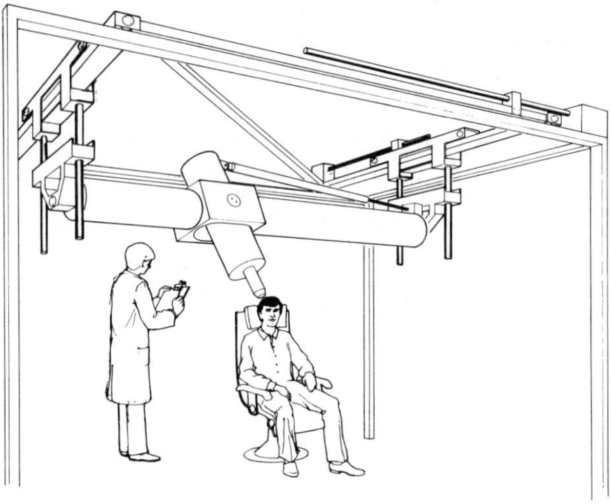

Fig. 2: Artist's rendition of the mechanical gantry for the manipulation of the biomagnetic dewar around the subject's head.

The gantry is driven by stepping motors. Its dimensions have been designed such that the magnetic interference produced by any of the motors, position encoders, and other material components is negligible. All mechanical structures are designed for maximum rigidity to assure not only optimum positional accuracy but also to minimize any magnetic interference through structural vibrations.

The motion of the dewar tip over the head surface is generated by the gantry according to a head model resident in the CONTROLLER computer. The head model for each subject is constructed from measurement of the head shape prior to a magnetic mapping session. The head shape is measured using a 3-dimensional digitizer. The data are then fitted by overlapping second order patches which represent the head shape model. An example of a surface fitted to the measurements taken from a human skull is shown in Fig. 3.

The data acquisition software is resident in a PDP 11/34 computer (see Fig. 1). The computer is connected to four channels of SQUID electronics (DSQ 400 units*), and a 16 channel Analogue to Digital acquisition system for a simultaneous measurement of EEG and MEG. The software configuration allows for a maximum sampling rate of 1024 samples/second and the simultaneous recording of all 16 analogue EEG channels and the 4 digital SQUID channels. The data collection is correlated with the mechanical manipulation of the biomagnetometer dewar and the software also provides for a variety of collection modes, including data averaging, reviewing, storage and recording in addition to maintaining a history of the experiment.

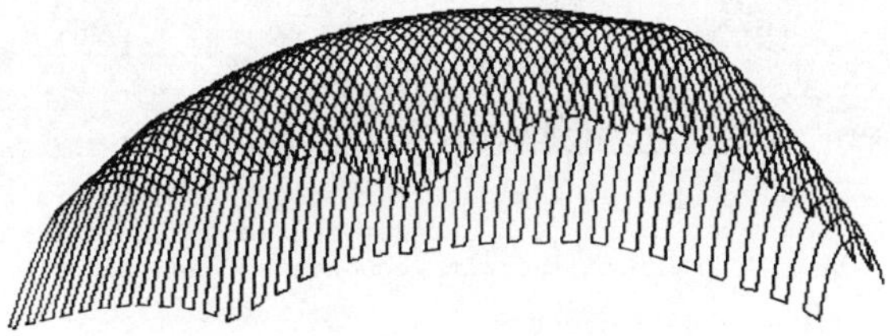

Fig. 3: An example of a head model fitted to the surface of a human skull. The data for the model are obtained from a 3-dimensional head measurements.

The biomagnetometer measuring system consists of a 3rd order gradiometer* with hardware field balancing provision and with three additional first gradient sensors for first gradient software balancing. The 3rd gradiometer white noise level is approximately 40 ft rms/$\sqrt{}$Hz and low frequency noise rises below 0.3 to 0.9 Hz (Vrba and co-workers, 1982; Fife and co-workers, 1984).

The data can be transmitted between various computers in the system and various software processing and mapping provisions have been developed. An example of some of the data processing and reduction features is shown in Fig. 4. This figure represents a magnetic map of a current dipole inserted into a skull filled with a conducting liquid. The map can be displayed either in colour mode on a computer screen or in different shades of black and white on a computer printer.

The system under development will be initially installed at Simon Fraser University, Brain Behaviour Laboratory where it will be used for the study of human performance evaluation and epileptic studies. The latter will be undertaken in collaboration with Dr. M. Low of the Vancouver General Hospital.

* Manufactured by CTF Systems Inc., Port Coquitlam, B.C.

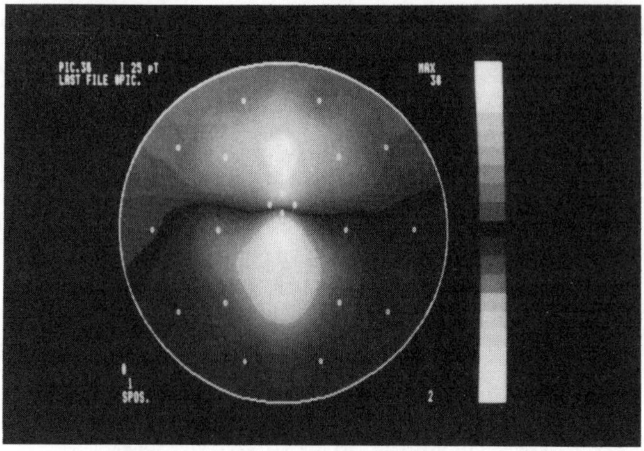

Fig. 4: An Example of a magnetic map of the human brain corresponding to a current dipole inserted inside of a skull filled with conducting liquid.

ACKNOWLEDGEMENTS

The authors wish to acknowledge the support of the British Columbia Science Council, the Department of Supply and Services, National Research Council, Transport Development Center, Simon Fraser University, and CTF Systems Inc.

REFERENCES

Deecke, L., Weinberg, H., Brickett, P.A., (1982). Magnetic Fields of the Human Brain Accompanying Voluntary Movement: Bereitshaftsmagnetfield. Exp. Brain Res. 48, 144 - 148.

Fife, A.A., Vrba, J., McCubbin, J., Weinberg, H., Brickett, P.A., (1984). Biogradiometer Design for Noisy Environments. This conference.

Vrba, J., Fife, A.A., Burbank, M., Weinberg, H., Brickett, P.A., (1982). Spatial Discrimination in SOUID gradiometers and 3rd order gradiometer Performance. Can J. Phys. 60, 1060 - 1073.

Vrba, J., McCubbin, J., (1983). First-Gradient Balancing of Higher-Order Gradiometers. Il. Nuo. Cim. 2D, 142 - 152.

MEASUREMENT OF THE TANGENTIAL COMPONENT OF THE MAGNETIC FIELD ASSOCIATED WITH RHYTHMIC ALPHA ACTIVITY IN THE HUMAN BRAIN

V. L. Vvedensky, S. P. Naurzakov, V. I. Ozhogin and S. Yu. Shabanov

Kurachatov Institute of Atomic Energy, 123182, Moscow, USSR

ABSTRACT

Instrumentation convenient for neuromagnetic research is described. Non-diagonal SQUID-gradiometer measures component of magnetic field parallel to the scalp with sensitivity 35 ft/$\sqrt{Hz}$.

KEYWORDS

Biomagnetism; cryostat; portable system; gradiometer; neuromagnetic fields.

INTRODUCTION

The progress in biomagnetic research and study of neuromagnetic phenomena require the permanent improvement of the equipment and techniques for magnetic measurements. We modified our instrumentation, described earlier (Vvedensky and others, 1983), keeping it small-size and introducing changes that make it convenient for measurements of the neuromagnetic fields of the human brain.

INSTRUMENTATION

The small change in the design of our first SQUID-gradiometer increased the time between the consequent refills of the dewar from 6 to 48 hours, without a considerable increase in its size. The outer diameter of the dewar is 105 mm, the height is 500 mm, the mass is 2.4 kg (with electronics), and the volume of liquid helium inside is 1 litre. The distance between the bottom of the liquid helium vessel and the outer wall in the lower part of the dewar is 8 mm. The diameter of the bottom is 80 mm. This is large enough to install several gradiometers in case we want to make a many channel system, as is expected in the future. Due to the long time between refills, the device can be kept permanently cold.

The small mass of the gradiometer allows easy manipulation when complex distributions of neuromagnetic fields are measured. A rather light (13 kg) tripod is used for reliable fixation of the gradiometer in respect to the subject's head. This fixation technique proved reliable in the previous experiments (Vvedensky and others, 1983). The new tripod is provided with an upper part, which can rotate around the vertical axis and the dewar fixed to it can be tilted at an angle up to 40°, see Fig. 1. This support structure allows easy displacement of the gradiometer in the natural spherical coordinates, with the centre of rotation placed in the centre of the head. The support gives the possibility to determine the local curvature of the scull in the area of measurment. This is important for precise localization of a neuromagnetic source in the head. The tripod and the nonmagnetic armchair can be disassembled when transportation to another experimental place is needed. The overall mass of the system with the monitoring electronics and battery power supply is no more than 70 kg.

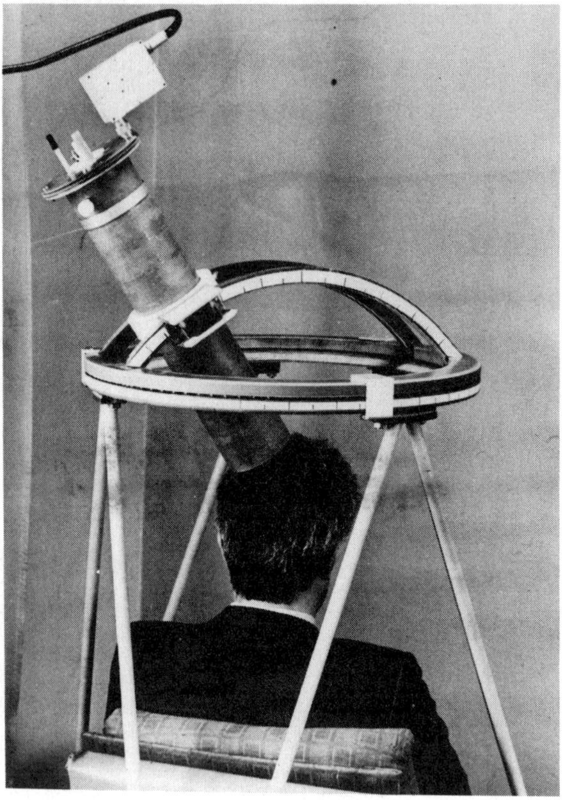

Fig. 1. The equipment for neuromagnetic research.

Our experiments have been performed inside a thick-walled (50 mm) room made of pure aluminium (99.8%) with internal dimensions 2.0 x 2.3 x 2.0 m^3, which gives protection from magnetic noise at frequencies 10 Hz and higher. The entrance (corridor) of the room is provided with a flexible door made of brass net to shield against rf interference. The small size of our device is important when performing experiments in the restricted volume of the room. Manipulation with more bulky instrumentation would be difficult.

We used a single non-diagonal first-order gradiometer in our experiments (Fig. 2). One can use a second-order gradiometer in our dewar when working in an unshielded environment, or several first-order gradiometers in a many channel system. The choice of the present design of our gradiometer was based upon the assumption that the magnetoencephalographic (MEG) technique has an advantage when one needs to localize a source of certain electrical activity in the brain (Vvedensky, Ozhogin, 1981; Proceedings 1983). This technique would be useful also when one needs to discriminate signals from several spacially separated sources in the cortex.

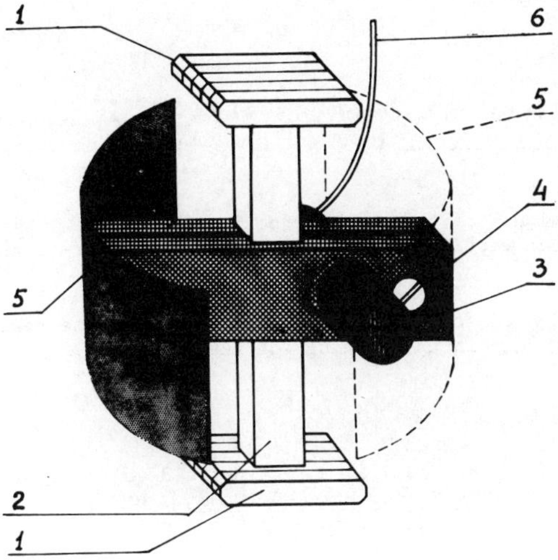

Fig. 2. The SQUID-gradiometer: 1 - quartz slabs with the coils of flux transformer; 2 - quartz pillar with parallel butt-ends; 3 - SQUID in ampule; 4 - glass fibre support; 5 - glass fibre springs; 6 - coaxial rf cable.

It can be shown that the measurement of the tangential component of the neuromagnetic field is preferable for certain types of sources than the usual measurement of radial component. For example, the source of the current dipole type often recorded in neuromagnetic experiments (Proceedings 1983), produces the maximum of the tangential component just above the source when the detection coil is perpendicular to the current dipole. The depth of the source is determined through the width of the maximum and the rate of the signal fall-off when the detector is moved away from the head (Cuffin and Cohen 1977).

The distribution of the characteristic neuromagnetic fields determined the shape of our detection coil. It is long and flattened, 15 x 30 x 5 mm^3. The coil (9 turns) lies on the bottom of the dewar, its axis parallel to the bottom. A similar coil serves as a compensating part of the flux transformer. The distance between the coils is 68 mm (Fig. 2). The coils are wound with 0,1 mm niobium wire on precisely ground quartz slabs glued to parallel butt-ends of a quartz pillar. This design allows an easy change to an asymmetric gradiometer (to increase the sensitivity) through the construction of another coil-form for the upper coil.

MEASUREMENTS

The measured sensitivity is 35 ft/$\sqrt{Hz}$. This allows a study of the distribution around the human head of the tangential component of the magnetic field associated with rhythmic alpha activity. A change in the position of the gradiometer relative to the head can be made very smoothly by the operator without disturbing the subject. This secures the relative stability of the subject's physiological state during an experiment. An example of the recorded signal is presented in Fig. 3. The maximum of amplitude is detected near the middle line of the head between vertex and inion. The direction of the neuromagnetic field is perpendicular to the middle line.

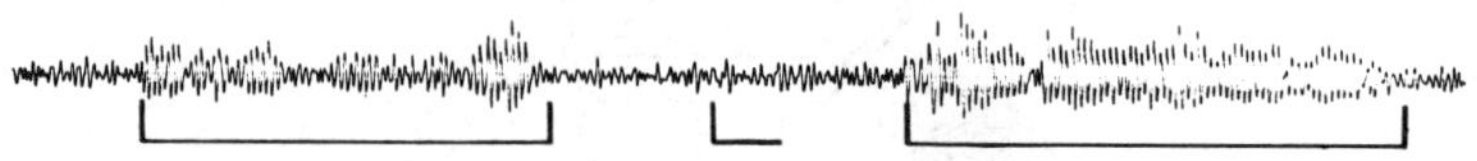

Fig. 3. Magnetic signal of the rhythmic occipital alpha activity. Brackets - eyes closed; vertical bar - 1 pT; horizontal bar - 1 sec.; bandwidth 7-13 Hz.

The authors are indebted to I. K. Kikoin for supporting the research, and to N. M. Barabanov for help in the assembly of the equipment.

REFERENCES

Cuffin, B. N., and D. Cohen (1977). Magnetic fields of a dipole in special volume conductor shapes. IEEE Trans. bio-med. Engng., BME-24, 372-381.

Proceedings of IV International Workshop on Biomagnetism, 1982, Rome (1983). Nuovo Cimento, 2D, N.2.

Vvedensky, V. L., S. P. Naurzakov, V. I. Ozhogin, and S. Yu. Shabanov (1983). A portable biomagnetic measuring system. Nuovo Cimento, 2D, N2, 224-230.

Vvedensky, V. L., and V. I. Ozhogin (1981). Supersensitive magnetometry and biomagnetism. Kurchatov Institute of Atomic Energy, Moscow (in Russian).

THE FLUX TRANSFORMER FOR MAGNETIC DIPOLE LOCALIZATION

Z. Dunajski, Z. Tomasiński, M. Nałęcz

Institute of Biocybernetics and Biomedical Engineering,
Warszawa, ul. KRN 55, Poland

ABSTRACT

A special construction of the flux transformer is described. This flux transformer enables one to find the position, direction and magnitude of the magnetic dipole located in biological objects.

KEYWORDS

Biomagnetism (including magnetocardiography); superconducting devices, flux transformers.

INTRODUCTION

Magnetic sources in biomagnetic research can be ascribed to current or magnetic dipoles. Primary sources such as ionic currents on the cellular level are described in terms of current dipoles (Williamson and Kaufman, 1981; Cohen and Hosaka, 1976). Magnetic dipoles are generated when the body of a chosen organ is magnetised by an ambient magnetic field. They may be created by clusters of paramagnetic or ferromagnetic particles which originate as a result of illness or contamination (for example, within lungs (Wikswo, 1980)). Diamagnetic organs such as a heart placed in a uniform magnetic field can be considered as a magnetic dipole and there is frequently a need to localize them. An example is the investigation of heart volume changes or heart movements. A magnetic dipole may also be introduced into the body to observe physiological phenomena. The flux transformer described in this work enables one to find the position, direction and magnetic moment of the dipole.

THE FLUX TRANSFORMER

The configuration of the flux transformer is shown in Fig. 1. The first and third coils are wound in the same sense, the middle coil is wound in the opposite sense as shown by arrows in Fig. 1. The diameters of the coils and the distances between them are such that the total flux threading the coils is equal to zero, when the dipole lies on the axis of the flux transformer at a defined distance $z = a$ from the pick up coil.

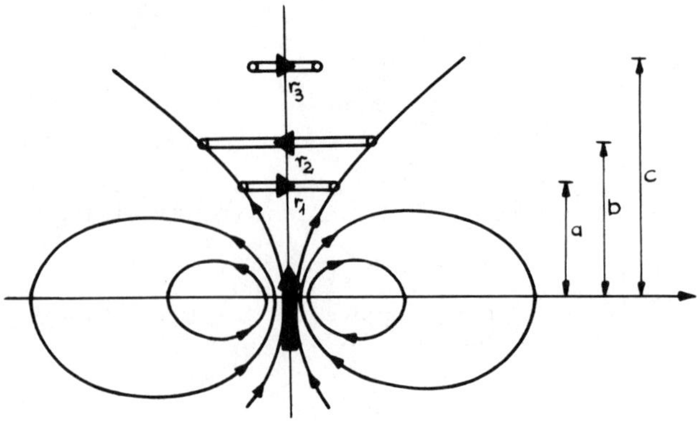

Fig. 1. The configuration of the flux transformer.

The radii of the coils r_1, r_2, r_3 and the distances a, b, c should fulfill the equations (1) and (2). From the condition that the total magnetic flux threading the coils $\Phi = \Phi_1 - \Phi_2 + \Phi_3$ is equal to zero, where $\Phi_i = m\mu/4\pi \cdot r_i^2(z_i^2 + r_i^2)^{-3/2}$ (i = 1,2,3) follows:

$$r_1^2(a^2+r_1^2)^{-3/2} - r_2^2(b^2+r_2^2)^{-3/2} + r_3^2(c^2+r_3^2)^{-3/2} = 0 \qquad (1)$$

The flux transformer will not be sensitive to the uniform magnetic field if:

$$r_1^2 - r_2^2 + r_3^2 = 0 \qquad (2)$$

Choosing particular values for a, b and r_2 the other parameters can be calculated.

Taking into account the dimensions of the cryostat in which the flux transformer was to be placed, the following values of parameters were taken: a = 40 mm, b = 50 mm, r_2 = 15 mm. The value of parameters c, r_1, r_3 were calculated from the equations (1) and (2), which were solved by the iteration method. The following values were obtained: c = 55 mm (for this value of c the equations (1) and (2) are noncontradictory), r_1 = 6 mm, r_3 = 13.75 mm.

PRINCIPLE OF OPERATION

The dependence of magnetic flux on the position and the angle of the magnetic dipole for this transformer was calculated and verified experimentally. In the calculations the following equation, defining the flux through a single coil as a function of the position and the angle of the dipole, was used:

$$\Phi = \int_S B_z dS = \int_0^{2\pi} \int_0^{\mathrm{artg}\frac{r_i}{z_i}} B_z \frac{z_i^2 \sin^2\Theta}{\cos^3\Theta} d\Theta dy \qquad (3)$$

The component of the magnetic flux density in the z direction B_z for the magnetic dipole $\bar{m} = m(\sin\alpha\cos\beta, \sin\alpha\sin\beta, \cos\alpha)$ is:

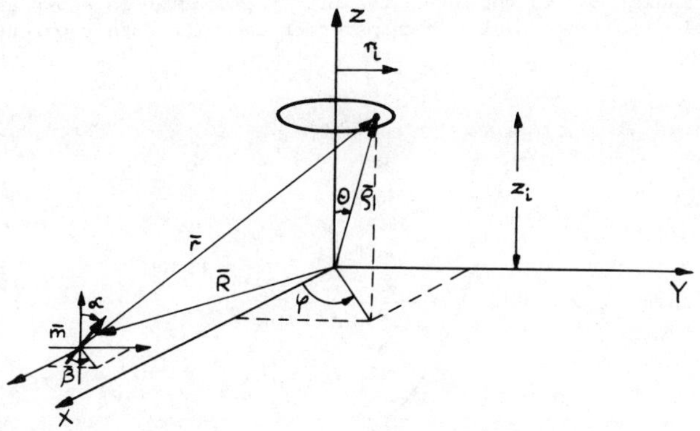

Fig. 2. A magnetic dipole for which the magnetic flux Φ through a coil is calculated.

$$B_z = \frac{m\mu}{r^3}\,[3(z_i tg\Theta \cos y - X)(\cos\alpha (z_i tg\Theta \cos y - X) - (z_i - Z)\sin\alpha \cos\beta + $$

$$-3(z_i tg\Theta \sin y - Y)(\sin\alpha \sin\beta (z_i - Z) - \cos\alpha (z_i tg\Theta \sin y - Y)) - 2\cos y] \qquad (4)$$

Figure 3 shows the flux Φ through the transformer as a function of the lateral displacement of the flux transformer for three different distances over the dipole oriented as in Fig. 1.

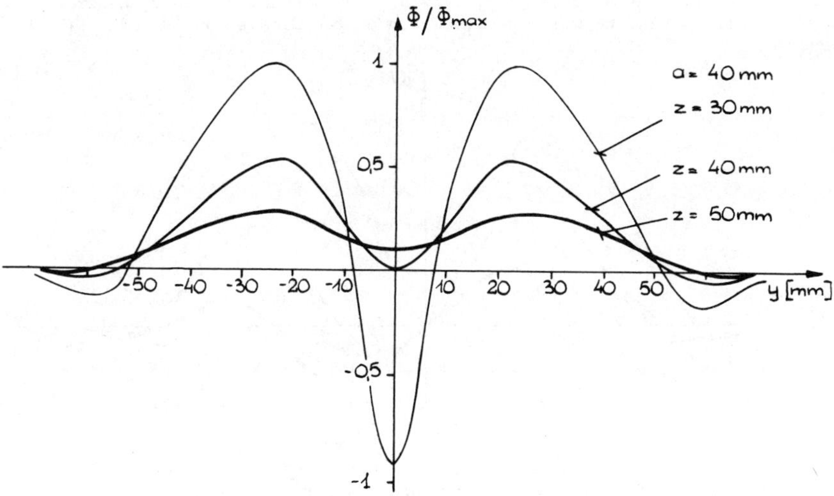

Fig. 3. The magnetic flux as a function of the lateral displacement of the transformer.

In Fig. 4 the magnetic flux threading the flux transformer is shown as a function of the vertical displacement of the transformer over the dipole oriented as in Fig. 1.

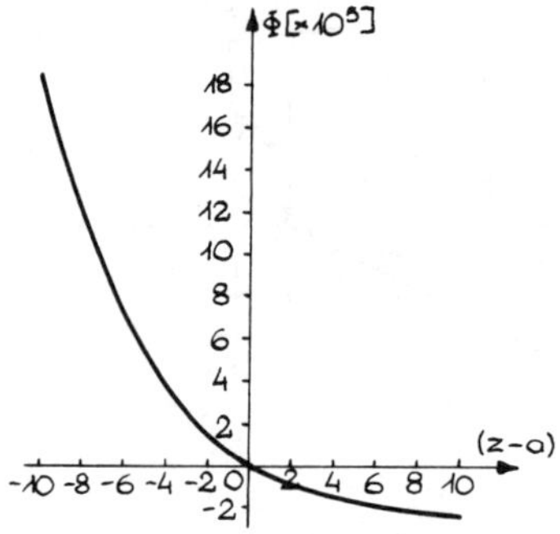

Fig. 4. The magnetic flux as a function of the vertical displacement of the transformer.

When the transformer passes z = a the output signal changes from positive to negative.

The localization of a dipole is carried out in the following way. We search for the minimum output signal over the dipole and then change the distance in such a way as to get a zero output signal (at y = 0 in Fig. 3). The direction of a thus localized dipole will not necessarily lie on the axis of the transformer, so the magnetic flux will not be described by curves shown in Fig. 3. The three curves in Fig. 5 present Φ as a function of lateral displacement of the transformer for three different angles between the direction of the dipole and the axis of the transformer.

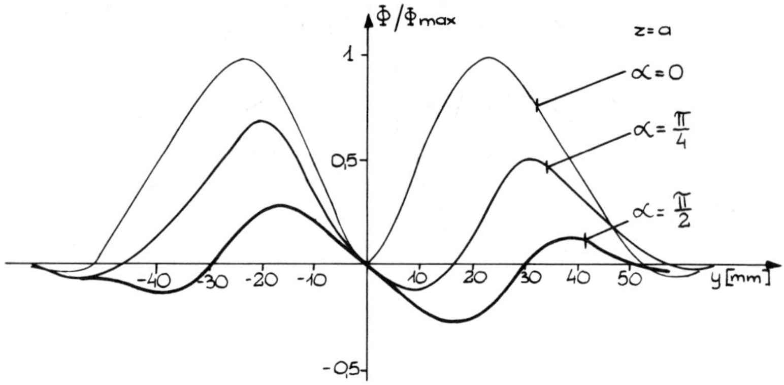

Fig. 5. The magnetic flux as a function of the lateral displacement of the transformer for various angles.

The ratio of the maxima on both sides of the point y = 0 is a single-valued function of the angle α as shown in Fig. 6.

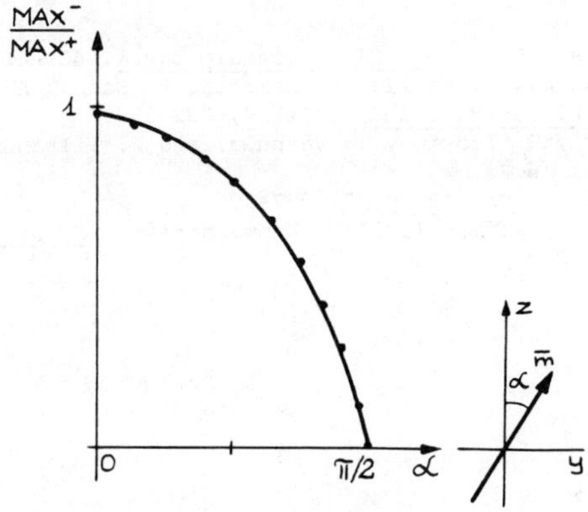

Fig. 6. The ratio of the maxima of the flux Φ as a function of the angle α.

This enables us to find the direction of the dipole. The values of the maxima are proportional to the magnetic moment m. A calibrated magnetometer enables us to measure the magnetic moment.

THE SECOND ORDER GRADIOMETER

This flux transformer may also be used as a second order gradiometer. The magnetic induction on the axis of the flux transformer in Taylor's series is:

$$B(z) = B(0) + \frac{\partial B(0)}{\partial z} \cdot z + \frac{\partial B(0)}{\partial z^2} \cdot \frac{z^2}{z} + \ldots \tag{5}$$

The magnetic flux through the coils is:

$$\Phi_1 = + \pi r_1^2 B(a) \tag{6}$$

$$\Phi_2 = - \pi r_2^2 [B(a) + \frac{\partial B(a)}{\partial z} (b-a) + \ldots] \tag{7}$$

$$\Phi_3 = + \pi r_3^2 [B(a) + \frac{\partial B(a)}{\partial z} (c-a) + \ldots] \tag{8}$$

The flux transformer will not be sensitive to a constant gradient when:

$$\pi r_2^2 \cdot \frac{\partial B(a)}{\partial z} (b-a) = \pi r_3^2 \cdot \frac{\partial B(a)}{\partial z} (c-a) \tag{9}$$

i.e., when

$$r_2^2 (b-a) = r_3^2 (c-a) \tag{10}$$

For the flux transformer to operate as a second order gradiometer, the parameters a, b, c, r_1, r_2, r_3 must fulfill equations (1), (2) and (10).

REFERENCES

Cohen, D., and H. Hosaka (1976). J. Electrocardiology, 4, 409-417.

Kalliomaki, P. L., P. Karp., T. Katila, P. Makipaa, P. Saar, and A. Tossavainen (1976). Scand. Work. Environ. and Health, 4, 232-239.

Katila, T., R. Maniewski, T. Tuosimo, T. Varpula, and P. Siltanen (1982). IEEE Trans. Biom. Eng., 1, 16-25.

Wikswo, J. P., Jr. (1980). Med. Phys., 7, 297-306.

Williamson, S. J., and L. Kaufman (1981). Biomagnetism. J. of Magnetism and Magn. Mat., 22, 129-202.

SPATIAL DISCRIMINATION: AN ALTERNATIVE APPROACH

A.C. Bruno, P. Costa Ribeiro, J.P. von der Weid and I.R. Eghrari

Departamento de Fisica, Pontificia Universidade Catolica
Cx.Postal 38071, Rio de Janeiro - RJ, BRASIL

ABSTRACT

The treatment of gradiometers as nonrecursive discrete spatial filters in a high-pass configuration is proposed. This approach makes possible an analytical description of the gradiometer's transfer function which is the major characterization of any instrument of measurement.

KEYWORDS

SQUID, Gradiometer, Digital Filtering.

INTRODUCTION

After the introduction of the SQUID, it was the use of gradiometers as sensing coils (Zimmerman, 1971), thereby avoiding the expensive shielded rooms (Cohen, 1975), that led to an increasing number of laboratories studying Biomagnetism. Different types of gradiometers (symmetric, assymmetric, with different orders, sizes and areas) have been built in many laboratories. The main goal in designing a gradiometer has always been to achieve the largest signal-to-noise ratio, and the fact that each group uses a particular type of gradiometer has not been a source of trouble. Nevertheless, with the need to compare data between groups, this lack of standardization must be overcome. In cardiomagnetism, where the magnetic field of the extended source falls off less rapidly than the field of a single current dipole, this problem is especially important. In fact, with such a slow fall off of the field, not only the lowest coil, but the upper coils in the gradiometer contribute to the signal. It would therefore, be interesting to have an analytical description of this device in order to characterize it, enabling one to be conscious of the amount of disturbance introduced in the signal.

GRADIOMETERS SENSING COILS

Traditionally, the equations that define a Nth order gradiometer can be found by considering the total flux, $\phi(t)$, induced in N+1 coils connected in series,

$$\phi(t) = A \left[\sum_{i=0}^{N} n_i \, B_z(z_i) \right] f(t) \tag{1}$$

where z_i is the distance from the ith-coil to the origin, $B_z(z)$ is the component of the magnetic field perpendicular to the coil's plane, and f(t) incorporates the time dependence of the flux. Each coil consists of n_i turns (positive or negative) taken by simplicity with the same area A. Expanding B_z in Taylor series about z_o and calling $b_i = z_i - z_0$ the distance between each coil and the lowest one, expression (1) can be given in matrix form:

$$\phi(t) = A\,[111\ldots]\begin{bmatrix} n_0+n_1+n_2+\ldots & & & 0 \\ & n_1b_1+n_2b_2+\ldots & & \\ & & \frac{n_1b_1^2}{2}+\frac{n_2b_2^2}{2}\ldots & \\ 0 & & & \ddots \end{bmatrix}\begin{bmatrix} B_z(z_0) \\ B_z^{(1)}(z_0) \\ B_z^{(2)}(z_0) \\ \vdots \end{bmatrix} f(t) \qquad (2)$$

The diagonal matrix represents the effect on the field, (B_z), and its derivatives, $(B_z^{(i)})$, at the lowest coil, (z_0), due to the gradiometer. If the gradiometer is designed so that the first diagonal terms in the square matrix are cancelled, then the contribution of a far source is diminished but the near source signal is also effected.
Such approach is similar to the classical electronic filtering of a time dependent signal. For instance, the matrix representation of an ideal high-pass filter acting on a function F(t) expanded in Fourier series,

$$F(t) = \sum_{n=0}^{\infty} C_n e^{j\omega_n t} \qquad (3)$$

would be:

$$\begin{bmatrix} 0 & & & & & 0 \\ & 0 & & & & \\ & & 1 & & & \\ & & & 1 & & \\ & & & & 1 & \\ 0 & & & & & \ddots \end{bmatrix}\begin{bmatrix} c_0 \\ c_1 \\ c_2 \\ \vdots \end{bmatrix} = \begin{bmatrix} 0 \\ 0 \\ c_2 \\ \vdots \end{bmatrix} \qquad (4)$$

This expression is similar to that of e.g. a second order gradiometer,

$$\begin{bmatrix} 0 & & & 0 \\ & 0 & & \\ & & \frac{n_1b_1^2}{2}+\frac{n_2b_2^2}{2}+\ldots & \\ 0 & & & \ddots \end{bmatrix}\begin{bmatrix} B_z(z_0) \\ B_z^{(1)}(z_0) \\ B_z^{(2)}(z_0) \\ \vdots \end{bmatrix} = \begin{bmatrix} 0 \\ 0 \\ \gamma B_z^{(2)}(z_0) \\ \vdots \end{bmatrix} \qquad (5)$$

where the γ's are the non zero diagonal terms of the square matrix. The similarity between a gradiometer and an electronic filter is stressed by the usual procedure to check gradiometer's unbalance. To measure the cancellation of the first two terms of the gradiometer matrix in a second order gradiometer, a constant field and a constant gradient field are usually applied to the system. That routine is equivalent to determining the transfer function of an electronic filter by connecting to it a sine wave generator and measuring the output intensity for each frequency. Nevertheless, up to this point, the analogy between a gradiometer and a filter is confined only to the matrix representation. Since the Taylor polynomials do not constitute an orthonormal basis, the gradiometer's transfer function cannot be derived by using the Taylor approach.

GRADIOMETER, A NONRECURSIVE DISCRETE FILTER

A gradiometer detects a signal that is a continuous function of time. The sampling takes place at discrete points in space corresponding to the position of the coils. The net signal measured is a weighted sum of these sampled values. Thus, it is simple to apply digital signal processing techniques (Lehmann, 1983) (Schwartz, 1975) to handle this problem. The output of a nonrecursive or moving-average filter can be defined by the well known expression

$$y_m = \sum_{i=-\infty}^{\infty} h_i \, x_{m-i} \tag{6}$$

This operation defines a new set of discrete numbers y_m from a set of numbers x_m corresponding to values of a variable sampled at constant intervals. Different filters correspond to different coeficients h_i. Although it is common to fit a monotonic function like the $1/r^3$ or $1/r^2$ magnetic field from a magnetic or electric current dipole by a polynomyal expansion, it is possible to describe the field by a Fourier expansion. The discrete Fourier transform of the signal at some instant of time will be

$$X(k) = \sum_{m=-\infty}^{\infty} x_m \, e^{-jmk\lambda_s} \tag{7}$$

where k is the wave number and λ_s is the distance between two successive sampling points.

Assuming that the distances bi can all be expressed as multiples of λ_s, that is

$$b_i = \lambda_s \beta_i \tag{8}$$

β_i being integers and λ_s being therefore just the maximum common divisor of all the b_i, the signal detected by the gradiometer will be given by:

$$Y = \sum_{i=0}^{N} n_i \, x_i \tag{9}$$

where N+1 is the number of coils, n_i is the number of turns of each coil and x_i is the magnetic flux at a given instant of time at the ith coil. The transfer function, defined by Y(k) = H(k) X(k), is obtained by taking the Fourier transform of the sequence of filter coefficients:

$$H(k) = \sum_{i=0}^{N} n_i \, e^{-jb_i k}$$

A far source of field (background noise) will mainly have low frequency components, (X(k) will be large near k=0) while a near source (signal) will have high frequency components.

The spatial discrimination will be achieved by making the moving average filter (gradiometer) into a high-pass filter. It is interesting to note that the design conditions for a Nth order gradiometer, obtained by the traditional Taylor approach (Karp, 1980),

$$\sum_{i=0}^{N} n_i \frac{b_i^{\alpha}}{\alpha!} = 0 \qquad \alpha=0,\ldots,N \tag{11}$$

are obtained in the Fourier approach by imposing:

$$\left.\frac{\partial^{\alpha}}{\partial k^{\alpha}} H(k)\right|_{k=0} = 0 \qquad \alpha=0,\ldots,N \tag{12}$$

CHOOSING GRADIOMETERS' DESIGNS

This alternative approach to the study of gradiometers provides a new tool for optimizing the choice of gradiometer parameters for each specific situation, as is illustrated by figures 1 and 2 and the brief outline presented below. Figure 1 shows the transfer function magnitude $H(k)$ as defined by (10) for first, second and third order symmetric gradiometers. The gradiometers' total length are the same (10cm) and therefore the sampling period λ_s are respectively 10, 5 and 2.5 cm. As is the case for the discrete Fourier transform, the highest meaningful frequency is π/λ_s or 0.31, 0.62 and 1.26cm^{-1} respectively.

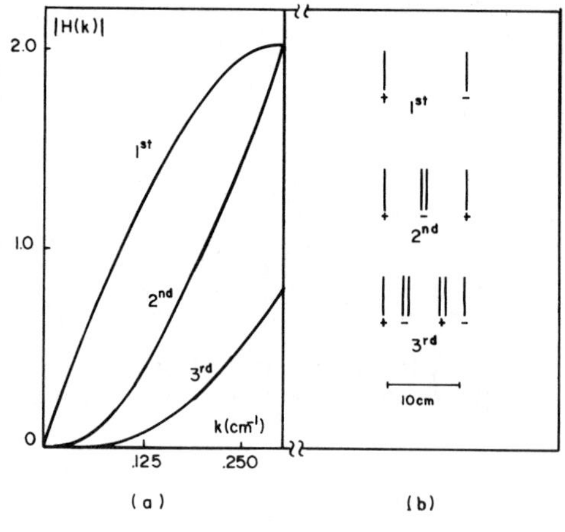

Figure 1(a). Magnitude of the gradiometers' transfer function and (b) their schematic geometries.

All the transfer functious are high-pass filters with the gradiometers of higher order more effectively attenuating low frequency background noise.
For frequencies larger than 0.3 cm^{-1}, the Fourier components of the signal are very small, representing only a few percent of the total power spectrum. This is particularly true for the cardiac magnetic signal. (Bruno, 1984).
The choice between gradiometers of different order will then be determined by the behaviour of the transfer functions for frequencies below 0.3 cm^{-1}. For a severe background noise environment due to not too distant sources, figure 1 indicates

the use of a third order gradiometer even if the signal is also drastically reduced.

The magnitude of the transfer function $|X(k)|$ for different gradiometers described in literature are shown in figure 2. Their areas have been normalized to unity for simplicity. The use of a first order gradiometer with a large base line (curve 1) is perfectly justified in a suburban area where the background noise is very low. The cardiac signal amplitude will therefore, be large since for intermediate frequencies the transfer function amplitude is large. The best choice of the gradiometer lengths can be suggested by comparing two second order gradiometers (curves 2a and 2b). For short lengths (large π/λ_s) (curve 2a) the filter more efficiently attenuates low frequency components, having a large background noise rejection, but also reduces the signal amplitude by attenuating the intermediate frequencies. This short gradiometer is indeed a good choice for studies of brain evoked fields when the gradiometer is near a $1/r^2$ source whose signal contains high frequency components in the signal.

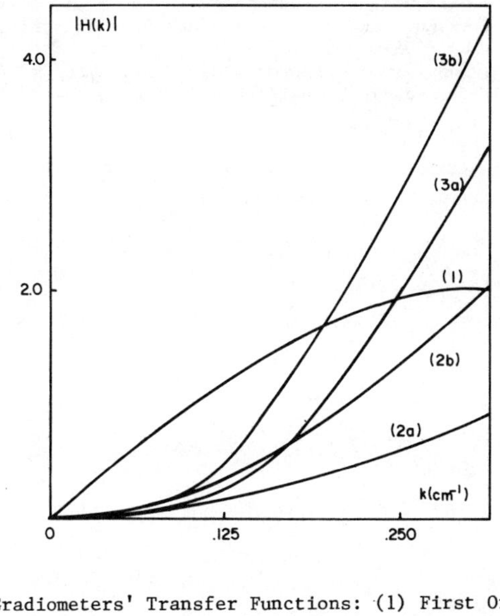

Figure 2. SQUID Gradiometers' Transfer Functions: (1) First Order Gradiometer (total length 10cm) (Saarinen, 1978); (2a) Second Order Gradiometer (total length 6.4cm) (Williamson, 1978); (2b) Second Order Gradiometer (total length 10cm) (Barbanera, 1981); (3a) Third Order Gradiometer (total length 16.5cm)(Vrba, 1982); (3b) Third Order Gradiometer (total length 20cm) (Bruno, 1983).

When measuring cardiac signals in a large background noise environment the choice of a third order gradiometer seems more reasonable to obtain a larger signal to noise ratio (curve 3a and 3b). A second order gradiometer with a larger baseline curve 2b, seems to be a good compromise to handle either cardiac and brain signals. The choice among different gradiometers designs will be imposed by the observation of their transfer functions and the Fourier power spectrum of the signal and the noise.

Finally, the usual measurement of the gradiometer's unbalance to check how far is the real gradiometer from its design, would be replaced by measuring its actual transfer function obtained by detecting the signal due to a magnet, displacing it along the axis of the gradiometer and comparing the Fourier transform of this signal with the Fourier transform of the dipolar source (Bruno, 1984).

REFERENCES

Barbanera, S., P. Carelli, R. Fenici, R. Leoni, I., Modena, and G.L. Romani, IEEE Trans. Magn. MAG-17 (1981), p. 849.

Bruno, A.C., P. Costa Ribeiro, J.P. von der Weid and O. Symko, (1984) (to be published).

Bruno, A.C. and P. Costa Ribeiro, Cryogenics, vol.23 (1983), pp. 324-326.

Cohen, D., Report of the Low Field Group: The Magnetocardiogram MIT Dec. (1975).

Karp, P., D. Duret, Journ. Appl. Phys. 51 (1980) pp. 1267-1272.

Lehmann, H.P., Signal Processing, Biomagnetism An Interdisciplinary Approach, S.J. G.L. Romani, L. Kaufman and I. Modena (Eds), New York, NATO ASI Series vol.66 (1983), pp. 591-624.

Saarinen, M., P. Siltanen, P.J. Karp and T.E. Katila, Ann. Clin. Res., vol 10, Suppl.21 (1978), pp. 1-42.

Schwartz, M. and S. Leonard, Signal Processing, McGraw-Hill, Tokyo (1975)

Vrba, J., A.A. Fife, M.B. Burbank, H. Weinberg and P.A. Brickett, Can. Journ. Phys., vol.60 (1982), pp. 1060-1071.

Williamson, S.I., D. Brenner and L. Kaufman, AIP Conf. Proc. 44 (1978), p.106.

Zimmerman, J.E. and N.V. Frederick, Appl. Phys. Letters, vol.19 (1971), pp.16-18.

$]U_2 = [S_2]^2[S_1]^{-2}$.

ght side of the equation is a rix, U_2 is simply the matrix ors that diagonalizes $Z_{12}^{t}Z_{12}$. e of U_2 produces an estimate , and S_2.

ion comparison of the sub- rs of leads produced the nal vectors graphed in Fig. four signals are those in l leads. The signal s the unique signal observed superposed graphs come from of different pairs of leads. " signal (CN) was calculated lowing observation. The erage vector is a linear of the basic signal vectors , plus any signals that are .e., with low variance) for Using the stable subspace age, the constant signal can ated, and estimates of these all leads except F3, and both ads) appear as the last set of graphs.

sents signal vectors derived techniques, for a second e first two signal vectors similar to vectors II and , because in each case the resemble damped sinusoids at quency, but different phase. the constant vectors, which kind of residual, are milar, both beginning at

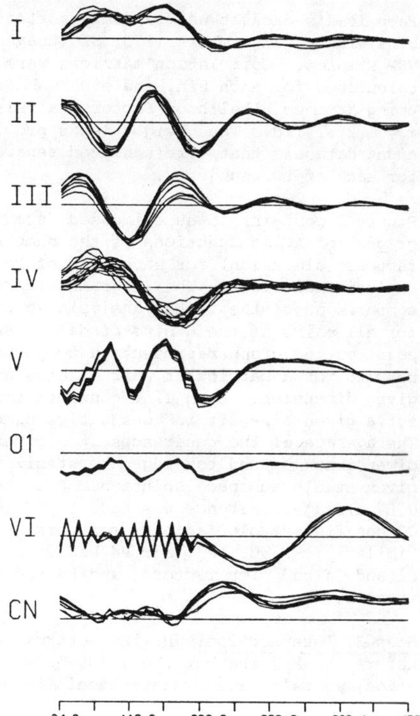

Fig. 2. Derived signals for first subject.

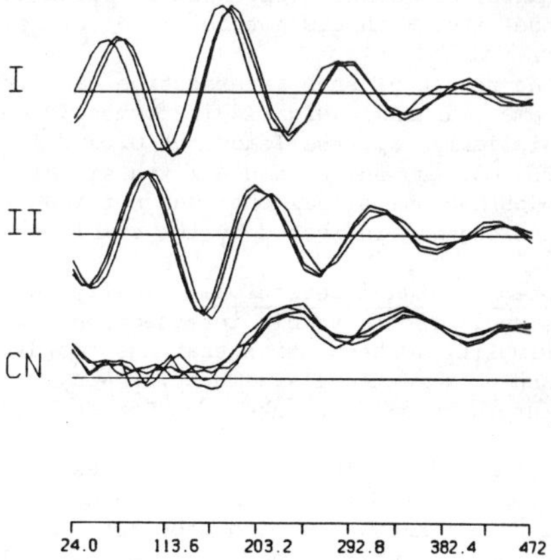

Fig. 3. Derived signals for second subject.

DISCUSSION

ectors shown here are not any other multivariate tech- to the authors), and it is they have a physical inter- It should be emphasized that mathematical restriction on the signal vectors. There- ncouraging that certain the observed signal vectors physiological ranges that ected for visual system For example, the two signal appear to be in common for s (II and III, in Fig. 2,

A NEW STATISTICAL METHOD FOR RESOLVING SUPERPOSED SIGNALS

L. J. Rogers* and R. R. Douglas**

*EEG Department, Vancouver General Hospital
**Mathematics Department, University of British Columbia
Vancouver, B.C., Canada

ABSTRACT

Magnetoencephalograms (and electroencephalograms) are known to be composed of several superposed signals and noise (in almost all paradigms). To date, no analytical tool has been developed to separate the superposed signals. The present paper provides a brief sketch of a statistical tool that can resolve superposed signals under certain very general assumptions that are likely to be met by the experimental system.

KEYWORDS

Magnetoencephalogram; event-related field; factor analysis; multivariate statistics; event-related potential.

INTRODUCTION

The purpose of the work described here is to discover and measure linearly independent "signals" produced by distinct generating processes in brains, under the hypothesis that the measured time series are linear combinations of these basic signals, together with additive noise. Thus the methods apply equally well to magnetically or electrically measured time series data, since both follow a superposition law. In fact, the methods apply to a wide variety of types of data in which several basic signals add linearly to produce the observed signal.

The mathematical model employed is (maximum likelihood) factor analysis (MLFA); however, the new methods to be described use factor analysis in a more strictly empirical context than has been achieved in previous applications of the power- ful MLFA algorithm. It is well known that factor analysis, as it is usually practiced, has two major indeterminacies: number of generators and rotation. This paper sketches resolutions of these two indeterminacies. Suppose that X is a (pxN) mean zero dataset, with p variables and N samples. We have the model equation: $X = LY + E$. L is the (pxm) "factor loading matrix" whose columns represent the m basic "signals". Y is the (mxN) matrix of coefficients represent- ing the "amplitude" of each signal for a given sample. E is the (pxN) matrix of uncorrelated noise.

These methods provide an empirical solution for the basic signal vectors which appear as the columns of L. A brief sketch of the three-step method, illustrated by examples of results from each step, will be presented here for one subject. Results from the last step for a second subject will be presented for comparison. Theory, simulations, and an EEG application of these methods have been published in a series of four papers (Douglas and Rogers, 1983, and preprint; Rogers and Douglas, 1984, 1984a).

METHODS

Visual evoked potentials (VEPs) were recorded from each of eight electrode placements (referenced to "linked ears") for data from two experimental subjects. The stimulus was a reversing full-field checkerboard, viewed binocularly; this is a standard stimulus used at the EEG Laboratory, Vancouver General Hospital (Morton Low, Director).

RESULTS

Step 1, Stable Subspace: The first step is to solve for what we term the "stable subspace" of the given dataset. The dataset is divided into (sufficiently large, random) sub-datasets, and the factor analyses of the sub-datasets are compared (pairwise) to find the largest dimension for which the factor analysis "results" are the same (for all sub-datasets). Realizing such comparisons effectively, depends critically on the observation that because of the factor analysis rotation indeterminacy, the actual "result" of a factor analysis computation is a subspace. In other words, each column of the factor loading matrix produced by standard factor analysis algorithms is well-known to be some linear combination of the columns of L (the basic signals) (Morrison, 1976). Thus the columns of this factor loading matrix and L both form bases of the same subspace of p-space (where p is the number of variables). Therefore, we required and defined a comparison measure for subspaces, which we have termed "ambient matrix coherence". Using this ambient coherence measure in cross-validation comparisons, yields an empirical technique for the determination of the number of independent generators in a given dataset, and the corresponding "stable subspace". Here, "stability" refers to the requirement that the factor analysis results are nearly the same for all sub-datasets.

The formula for computing the square of the (non-negative) "Ambient Coherence" (AC) of two (pxm) matrices, B_1 and B_2 (both of rank m), is:

$$[AC(B_1,B_2)]^2 = [\det(B_1^tQB_2)]^2/[\det(B_1^tQB_1)\det(B_2^tQB_2)]$$

where "det" is the (mxm) determinant function and Q is the (pxp) symmetric, positive definite matrix of the ambient inner product. For the remainder of this discussion, we simply set Q = I, the (pxp) unit matrix. Thus:

$$[AC(B_1,B_2)]^2 = [\det(B_1^tB_2)]^2/[\det(B_1^tB_1)\det(B_2^tB_2)] \quad .$$

Notice that AC satisfies precisely the correct (very strong) invariance property:

$$AC(B_1,B_2) = AC(B_1U,B_2V)$$

for arbitrary non-singular (mxm) real matrices, U and V. Thus, AC is invariant of all so-called "oblique rotations", separately in each of its two arguments. This allows resolution of one factor analysis indeterminacy (determination of the number of generators) in a way that is independent of the other factor analysis indeterminacy (rotation). Resolution of the second factor analysis indeterminacy can then follow, under certain very general assumptions.

Each lead's dataset was randomly sorted into 6 bins with equal numbers (870) of (unaveraged) VEP samples. Correlation matrices were then calculated for each bin, and each was analyzed using Maximum Likelihood Factor Analysis, for m = 1,2,...,10. Thus, eight leads provided eight datasets that were analyzed separately for each of two subjects.

Figure 1 contains 15 superimposed "stability graphs" of AC as functions of the number of factors (dimension) for each pair of factor analysis solutions. The 15 graphs represent comparisons of the factor analysis solutions for all pairs of the 6 bins of data. Each point on the graph represents a comparison of results for a particular pair of bins at the given dimension. Each line connects the points for a given bin-pair in consecutive dimensions. The average of the coherences at a given dimension was used to judge the stability of a given stable subspace solution. A criterion of 0.87 average coherence was used to decide that O2 had five stable factors, for this subject. Similarly, F3, C3, P3, and P4 had seven factors. O1 and C4 had five factors, and F4 had four factors.

Step 2, Intersection: Having determined the stab[...] natural to ask whether the leads have any signals [...] ation, we calculate a statistical "intersection".

Suppose that B_1 and B_2 are both rank m (pxm) matr[...] spaces calculated from two leads' datasets. It i[...] product of (all m of) the canonical correlations [...] $<B_2>$. If the datasets have k signals in common, [...] canonical correlations will be close to 1, and al[...] the first k should not be close to one.

One result of such intersection calculations show[...] same four basic signals in common (denoted I, II, [...] Similarly, all the leads but O1 and F4 had an add[...] F3, C3, P3, and P4 had a sixth signal (VI). Five [...] each had one signal that was not found at anothe[...] the signals indicated by the stability computati[...]

Step 3, Signal Detection: Finally, signal detec[...] signals within stable intersection subspaces. F[...] simplifying assumption that the two datasets hav[...] and no more signals. Thus, $L_1 = L_2$. The genera[...] intersection is treated in Douglas and Rogers ([...]

Let B_1 and B_2 (both (pxm) matrices) represent th[...] Step 1. $B_1 = L_1S_1U_1$ is a representation of the [...] indeterminacy, assuming stochastically independe[...] orthogonal matrix, and S_1 is a diagonal matrix. [...] vectors. Let $Z_{12} = (B_1^tB_1)^{-1}B_1^tB_2$. Beginning wi[...] straightforward linear algebra can be used to sl[...]

$U_2^t[Z$[...]

Since th[...] diagonal [...] of eigen[...] This esti[...] of L_2 (= [...]

Signal de[...] spaces of [...] estimated [...] 2. The f[...] common fo[...] labelled [...] at O1. T[...] compariso[...] The "cons[...] using the [...] dataset's [...] combinati[...] derived h[...] "constant[...] the datas[...] and the a[...] also be c[...] signals ([...] occipital [...] superimpos[...]

Figure 3 p[...] by the sam[...] subject. [...] appear to [...] III of Fig[...] two vector[...] the same f[...] Furthermor[...] represent [...] strikingly [...] about 180 [...]

The signal [...] derivable [...] nique (know[...] likely tha[...] pretation. [...] there was r[...] the forms c[...] fore, it is [...] features of [...] fall withir[...] might be ex[...] responses. [...] vectors tha[...] both subjec[...]

correspond to I and II, in Fig. 3, respectively) are both approximately 8 cps, which is the lower boundary of the alpha range. Furthermore, most of the signal vectors decay in the first two to three hundred msec. Finally, the frequencies of the signal vectors never increase during the 500 ms observation time. Of course, ascertaining any physiological significance of these signal vectors would involve exploring the effects of various experimental conditions.

Because of the considerable complexity of these signal vectors, they cannot be identified by their peaks from average evoked potentials. Once the signal vectors have been derived, the average event-related response can be used to estimate the amplitudes (and polarities) of the signals, for a given dataset.

REFERENCES

Douglas, R. R., and L. J. Rogers (1983). Comparative factor analysis models for an empirical study of EEG data. International Journal of Neuroscience, 18, 211-226.

Douglas, R. R., and L. J. Rogers. Stable factor analysis, stable intersections, and generalized coherence. Preprint, University of B.C.

Morrison, D. F. (1976). Multivariate Statistical Methods (2nd ed.). McGraw-Hill New York.

Rogers, L. J., and R. R. Douglas (1984). Comparative factor analysis models for an empirical study of EEG data, II: A data-guided resolution of the rotation indeterminacy. International Journal of Neuroscience, 22, 263-276.

Rogers, L. J., and R. R. Douglas (1984a). Comparative factor analysis models for an empirical study of EEG data, III: Resolution of stationary generating signals with confirmation by variable deletion. International Journal of Neuroscience, 24, 167-184.

AN EXPERIMENTAL AND THEORETICAL ANALYSIS OF THE MAGNETIC FIELD OF A SINGLE AXON

B.J. Roth, J.K. Woosley and J.P.Wikswo, Jr.

Department of Physics and Astronomy
Vanderbilt University
Nashville, TN 37235, U.S.A.

ABSTRACT

To obtain a detailed understanding of the magnetic field associated with a propagating nerve action potential, we have made quantitative theoretical and experimental studies of both the electric and magnetic fields of a single nerve axon in the isolated nerve cord of the crayfish.

KEYWORDS

Cellular magnetic field, nerve action potential, crayfish giant axon

INTRODUCTION

An active nerve axon has associated with it both a propagating action potential and a magnetic field. Traditionally only the electrical potential is measured; however, the magnetic field represents an additional source of information about the nerve. In this paper we discuss the behavior of the magnetic field, how the field can be measured, and how accurately these measurements can be made. The discussion is based on the predictions of the volume conductor model (Woosley, Roth and Wikswo, 1984) with which we can calculate the magnetic field from the transmembrane potential.

We performed an experiment in which we simultaneously measured the transmembrane potential and the magnetic field of a crayfish giant axon (Roth and Wikswo, 1984). The 2 μA action current produced a 200 pT magnetic field which was measured by threading the nerve through a wire-wound, ferrite-core toroidal pick-up coil (1.1 mm inner radius, 1.9 mm outer radius, and 1.2 mm width). As the action current propagated past the toroid, the changing magnetic field induced a current in the pick-up coil that we measured using a low-noise, room-temperature amplifier (Wikswo and colleagues, 1984). The action potential of approximately 100 mV was recorded using a microelectrode. Typical action potentials lasted 1 ms and had a rise time of about 0.2 ms. The conduction velocity was approximately 15 m/s, implying that d, the spatial extent of the depolarization, was 3 mm. This distance, rather than the axon radius, sets the natural length scale for the problem. The behavior of both the electric and magnetic fields will be different depending upon whether the fields are measured at distances much greater than or much less than d.

Neither microelectrodes nor toroids provide raw output proportional to the field they attempt to measure. Microelectrodes have a considerable capacitance to the bath which acts as a low pass filter. The output can be corrected using "negative capacitance," which amplifies the high frequency components of the signal to compensate for the capacitance of the electrode. Analogously, the toroid senses the time derivative of the magnetic field, so that the low frequency components of the magnetic signal are suppressed. These components of the toroid output must be amplified to produce a signal proportional to the magnetic field (Wikswo, 1982). The transmembrane potential and the magnetic field of a single axon after the frequency compensation, but with no averaging, are shown in Fig. 1.

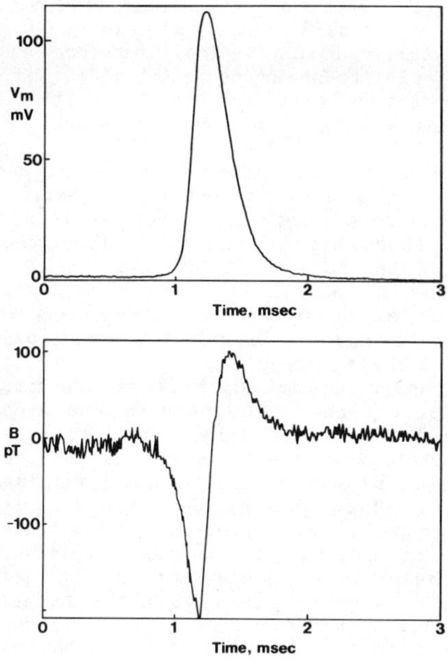

Fig. 1 The transmembrane potential and the magnetic field of a single axon after frequency compensation but with no averaging.

The magnetic field can be calculated from the transmembrane potential using the laws of electromagnetism. We model the axon as an infinitely long cylindrical membrane lying in an unbounded conducting bath. Solving the electromagnetic field equations in cylindrical coordinates leads naturally to mathematics characteristic of cylindrical geometry: Fourier transforms and Bessel functions. Although the results of the volume conductor model are most precisely expressed mathematically, we can explain many features of the magnetic field using physical arguments and experimental data. For instance, the cylindrical symmetry of the current pattern ensures that the magnetic field will be entirely azimuthal; the field lines will circle around the axon.

THE EFFECTS OF RETURN CURRENTS

To explain the spatial dependence of the magnetic field, we can take two points of view. If the measurement of the magnetic field is made at a distance large compared to d, then the magnetic field looks as if it were produced by two oppositely directed current dipoles placed along the axis of the nerve, representing the current pattern inside the axon (Swinney and Wikswo, 1980, Tripp, 1981). As the dipoles propagate past the toroid, the field as a function of time will be a smooth symmetrical biphasic signal, falling off in space as ρ^3, where ρ is the distance from the axon to the position where the field is measured. This point of view can be made precise by calculating the magnetic field from the law of Biot and Savart, in which the volume integral of the current density reduces to an integration of the axial current along the axon length. If the measurement of the magnetic field is made at a distance small compared to d, then the field is like that produced by a long straight wire. It is proportional to the current in the axon which is in turn proportional to the first derivative of the transmembrane potential (Barach, Roth and Wikswo, 1984). In this case, the field strength will fall off in space as $1/\rho$. This view can be made precise using Ampere's law, which states that the magnetic field is proportional to the current flowing through the toroid.

When the measuring distance, i.e. the size of the toroid, is comparable to d, neither of the limiting approximations are accurate. One can say that higher order multipoles of the current distribution begin to contribute to the dipole field, or alternatively that the currents outside the axon, termed return currents, begin to contribute to the net current passing through the toroid. In our experiment the toroids are small but their radius is not negligible compared to the spatial extent of depolarization. Therefore, considering the magnetic field as the derivative of the transmembrane potential is the most helpful view for getting an intuitive understanding of the signal, but to calculate quantitatively the magnetic field, return currents must be considered.

To see how much effect these return currents have on our measurements, we present the results of an unplanned experiment which occurred in our lab. When taking data from a crayfish nerve, the toroid and axon were lifted out of the bath for a short time. (In this experiment the toroid was larger than the one used to obtain the data in Fig. 1: 1.5 mm inner radius, 2.9 mm outer radius and 1.3 mm width). When again placed in the bath an air bubble was trapped in the center of the toroid, filling the space between the axon and the toroid inner surface. The bubble was able to block most of the return current while not damaging the nerve. Taking advantage of this fortunate occurrence, data were taken with and without the bubble present. The results are shown in Fig. 2. The magnetic field with the bubble present (no return currents) is narrower and larger than the field with the toroid filled with saline (including return currents).

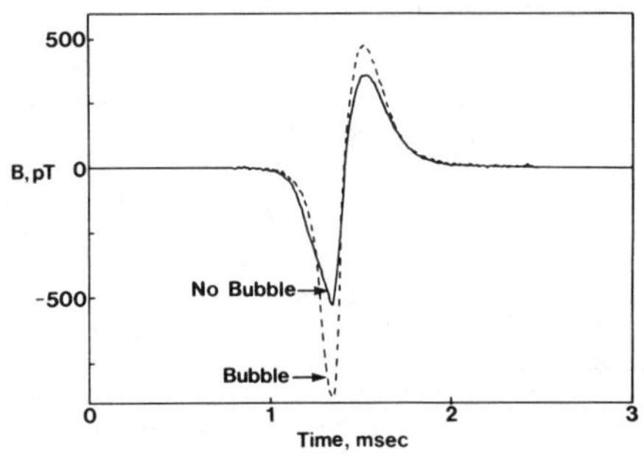

Fig. 2 The measured magnetic field of an axon with (dashed) and without (solid) a bubble in the toroid blocking the return current (64 averages).

The bubble experiment shows how large a contribution the return current makes to the magnetic signal. To see how accurately the volume conductor model accounts for this return current, let us use data from our first experiment, with the smaller toroid, and compare in Fig. 3 our measured magnetic field (512 averages, no bubble) to the volume conductor model prediction. If our model was unable to account for return currents, we would expect a discrepancy between theory and experiment on the order of the difference between the two traces in Fig. 2. Note that in Fig. 3 the two fields are plotted on different scales so that we can compare the shapes of the signals more easily. The difference between the signals is quite small. In particular, the width and the asymmetry of the two peaks is predicted almost exactly. We conclude that the volume conductor model can accurately predict the shape of the magnetic field, even for distances on the order of d, where consideration of the return currents is essential for quantitative comparisons between theory and experiment.

DISCUSSION

The difference between the amplitudes of the measured and calculated fields can be traced to an uncertainty in our knowledge of the axial resistance of the nerve. The current inside the axon is related to the transmembrane potential through Ohm's law, so it depends on the resistance of the axon. We can write r_i, the axon axial resistance per unit length, as $1/(\sigma_i \pi a^2)$, where σ_i is the conductivity of the fluid in the axon, approximately 1 to 2 $(\Omega m)^{-1}$, and a is the axon radius,

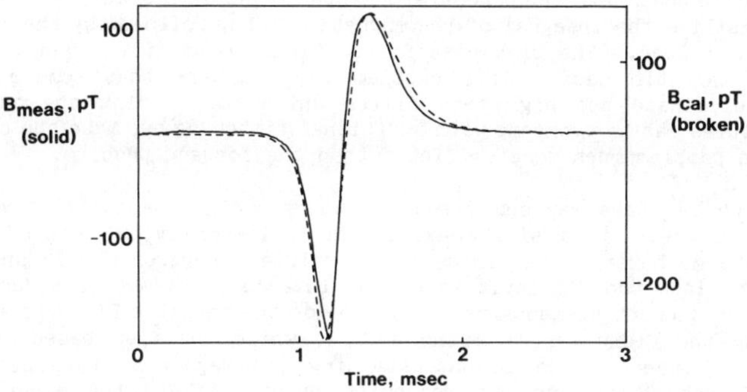

Fig. 3. The magnetic field of a single crayfish axon, comparing the measured and predicted magnetic fields.

typically 100 μm. Viewed from the external bath, the axon acts like a current source so that the return current is independent of the bath resistance, but does depend on r_i. Therefore the magnetic field, when calculated from the transmembrane potential, is proportional to $1/r_i$. It was difficult to determine the values of a or σ_i with high accuracy, producing a large uncertainty in r_i. This accounts for the difference in the size of the measured and calculated fields in Fig. 3. Note that the axon resistance is an important factor in calculating the current from the potential, but plays no role in calculating the magnetic field from the current. Thus magnetic measurements allow us to determine action currents without knowing the size and the internal conductivity of the axon.

When comparing theory and data in Fig. 3, we must consider several sources of systematic error in the experiment. For instance, the toroid does not measure the magnetic field at a point, but averages the field over its entire cross-section. For toroids of the size we used in this experiment we can assign an "effective radius" which, when used to calculate the magnetic field with our model, gives the field with an error of less than one percent. Because our toroids are cylindrically symmetric, the magnetic signal is fairly insensitive to the position of the axon in the toroid. In our experiment, errors due to the axon being off-axis in the toroid are less than two percent. Also, the toroid presents an obstacle to the current lines in the bath. However, calculations show that the net return current is changed by less than one percent due to the presence of the toroid. The bath boundaries have even a smaller effect on the magnetic field. In summary, no systematic errors were found which would significantly change the results in Fig. 3. The uncertainty in the axon resistance represented the only important source of error in our experiment, and affects only the size of the field, not its shape.

We can picture the calculation of the magnetic field from the transmembrane potential as a filtering problem. The magnetic signal close to the nerve is proportional to the derivative of the transmembrane potential, but as we move away from the axon, more and more of the high frequency information of the signal is attenuated. It is as if a low pass filter were being applied, the cutoff frequency of the filter depending on the distance from the axon. This interpretation of the relationship between the magnetic field and the transmembrane potential as a filtering problem is not merely a convenient device for understanding the field behavior, but is implied by the underlying mathematics of Fourier transforms.

The filtering interpretation is also helpful in understanding the inverse problem of calculating the transmembrane potential from the magnetic field. The transmembrane potential looks like the integral of the magnetic field, followed by the application of a high pass filter. The high pass filter tends to amplify any high frequency noise in the magnetic data. This problem is so severe that some method must be found to eliminate the high pass filter while not ruining the calculation. These difficulties can be overcome with additional filtering but make the calculation of the inverse problem much more difficult than the forward problem.

We find that we can take various views in understanding the qualitative features of the magnetic field of a single axon. The filtering view is useful in that it reflects the mathematics required by the volume conductor model, and it gives us some insight into the difficulties of the inverse problem. Considering dipole and higher order current elements as the source of the magnetic field is appropriate when measuring the field far from the source, while the view based on Ampere's law and return currents aids in understanding the magnetic field close to the axon. In all cases, however, to determine quantitatively the magnetic field, we must use the mathematics of the volume conductor model. Figure 3 suggests that the precise calculation accurately predicts the measured magnetic field. Magnetic measurements therefore provide a new technique for the study of nerve axons.

ACKNOWLEDGEMENTS

This research has been supported by ONR Contract N00014-82-K-0107, NIH Grant 1-R01 NS 19794-01, and the Vanderbilt University Research Council.

REFERENCES

Barach, J. P., B. J. Roth, and J. P. Wikswo, Jr. (1984). Magnetic measurements of action currents in a single nerve axon: a core-conductor model. (submitted for publication)

Roth, B. J., and J. P. Wikswo, Jr. (1984). The magnetic field of a single axon: comparison between theory and experiment. (in preparation).

Swinney, K. and J. P. Wikswo, Jr. (1980). A calculation of the magnetic field of a nerve action potential. Biophys. J. 32, 719-732.

Tripp, J. (1981). Biomagnetic fields and cellular action currents. In Ernè, S. N., H. D. Hahlbohm, and H. Lübbig, (Eds.) Biomagnetism, Walter de Gruyter, New York.

Wikswo, J. (1982). Improved instrumentation for measuring the magnetic field of cellular action currents. Rev. Sci. Instrum., 53, 1846-1850.

Woosley, J. K., B. J. Roth and J. P. Wikswo, Jr. (1984). The magnetic field of a single axon: a volume conductor model. (submitted for publication)

Wikswo, J.P., Jr., W.P. Henry, P.C. Samson, and R.P. Giffard (1984). A current probe system for measuring intracellular action currents. These proceedings.

CURRENT PROBE SYSTEM FOR MEASURING CELLULAR ACTION CURRENTS

J.P. Wikswo, Jr., W.P. Henry, and P.C. Samson

Department of Physics and Astronomy, Vanderbilt University
Nashville, Tennessee, 37235 U.S.A.

R.P. Giffard

Hewlett-Packard Laboratories, Palo Alto, California 94304, U.S.A.

ABSTRACT

We report an improved current probe system that is optimized for the magnetic measurement of cellular action currents in nerve and cardiac muscle tissue.

KEYWORDS

Neuromagnetic Field, Current Probe, Low-Noise Amplifier

INTRODUCTION

The Neuromagnetic Current Probe is an electronic amplifier system optimized for measuring the magnetic field from nerve impulses. Because the magnetic field associated with the electrical impulses in nerves is quite weak compared to the sensitivity of conventional magnetometers and magnetic noise in the environment, it has previously been quite difficult to use magnetic measurements to study nerve conduction. SQUID magnetometers used for most biomagnetic measurements have 1 cm diameter pickup coils that can be placed no closer than 0.5 to 1 cm from the biological sample. While these instruments are ideal for studying the magnetic fields produced by the intact brain or heart, they lack the spatial resolution required for measurements on small electrophysiological preparations such as nerves. The magnetic field of an isolated nerve was first measured using a SQUID magnetometer with a room-temperature toroidal pickup (Wikswo, Barach and Freeman, 1980; Barach, Freeman and Wikswo, 1980; Wikswo, 1982). To overcome the limitations of SQUID magnetometers, we developed a unique room-temperature amplifier (Wikswo, Samson and Giffard, 1983) that allows the magnetic measurement of the electrical signal that propagates along an exposed nerve bundle. In this paper, we report an improved amplifier integrated into a measurement system that we have termed the Neuromagnetic Current Probe and which was developed over the past 3 years. This complete system has greatly simplified neuromagnetic measurements and has improved the signal-to-noise ratio by several orders of magnitude over that described in our first SQUID measurements. In addition, the Neuromagnetic Current Probe provides the needed sensitivity and spatial resolution at one tenth the cost of a SQUID magnetometer and without liquid helium. The most important feature of the Neuromagnetic Current Probe is its capability of directly measuring cellular action currents by means of their magnetic field, without assumptions regarding cellular resistivities. The refined probe described in this paper has a maximum sensitivity of 120 pA/$\sqrt{\text{Hz}}$ at 20 KHz, and can measure a 4 μA nerve impulse with a 100 to 1 signal-to-noise ratio in a 20 kHz bandwidth.

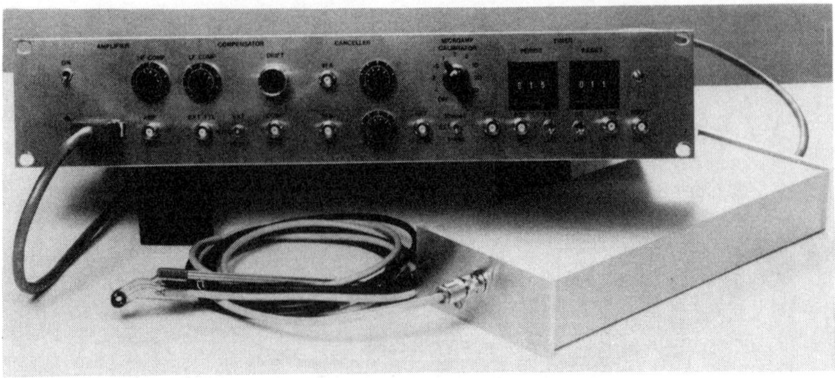

Figure 1. A photograph of the complete Neuromagnetic Current Probe System.

The present version contains several significant improvements over the first amplifier that we reported. While the prototype amplifier was single-ended, the present version has a true differential input, providing increased input common-mode rejection and rejection of power-supply fluctuations. Both the sensitivity and encapsulation of the toroids have been improved. The amplifier has enhanced drift cancellation, frequency response and sensitivity as compared to the first version. Furthermore, the improved control electronics are now integrated into a single unit that provides complete control over the probe and the measurement timing.

SYSTEM OVERVIEW

The system, shown in Fig. 1, has three major components: a toroidal pickup coil that is placed around the tissue, a low-noise amplifier that detects the currents induced in the toroid by the cellular magnetic field, and control electronics for frequency compensation, calibration, and noise rejection. The principle of operation of the system is the same as that of a clip-on ammeter for measuring AC currents. The use of 10 parallel input stages in the amplifier minimizes the voltage noise of the system.

The control electronics make the amplifier system fully self-contained, with the only needed support instruments being a stimulator, an oscilloscope, and possibly a signal averager. The control unit provides power for the amplifier, and allows adjustment of the high-frequency feedback to compensate for variations in the high-frequency response of each toroid. It also provides low-frequency compensation and drift removal. Because the toroids attenuate signals at frequencies below the R/L frequency of the toroid, which is typically 100 Hz, it is necessary to integrate the signal low frequencies. To overcome drifts, a triggered track-and-hold circuit is coupled to a triggered integrator. Another control function is the cancellation of stimulus artifacts. The monitor pulse from the stimulus generator is input to a summing sign-changer that is connected to a single-turn winding on the toroid. The sign-changer is adjusted until the stimulus artifact is minimized on the output signal. The major advantage of this approach is that the artifacts are cancelled prior to any amplification, thereby minimizing the risk of overloading the amplifier or degrading the common-mode rejection of the differential amplifier. The single-turn winding also simplifies calibration of the instrument, since the control electronics include a microamp calibrator which superimposes a calibrated current pulse on the single-turn wire through the toroid. The final control function is the timing of the stimulation, recording and compensation processes. Because of its high sensitivity, the instrument will detect strong 60 Hz magnetic fields from power lines and equipment, it is useful to synchronize the data acquisition process to the power-line in a manner that all odd harmonics of the power-line interference cancel out upon averaging.

OPERATIONAL DESCRIPTION

The detailed operation of the amplifier can be understood by referring to the simplified schematics shown in Fig. 2. The low-noise multi-stage amplifier A_{11} senses the total voltage appearing across the secondary winding of the input transformer T_{11}. This is the sum of an induced voltage due to the signal input current flowing in the primary winding, and a voltage developed across the secondary due to the negative feedback current flowing through R_{11} and R_{14}. The amplifier A_{11} has a low frequency gain of 120 dB and a high common-mode rejection ratio. The gain falls off at a rate of 20 dB per decade above 100 Hz.

Differential amplifier A_{12} in the feedback section of the control unit senses the output of A_{11} while rejecting ground-loop voltages and other common-mode interfering signals. Buffer amplifier A_{13} picks off a fraction of the output voltage of A_{12} determined by the setting of the "Loop Gain" control, potentiometer P_{11}. The output of A_{13} feeds the feedback compensation circuit which generates the push-pull negative feedback signals "Feedback A" and "Feedback B" with appropriate magnitude and phase. The feedback voltages are returned to the preamplifier section where they are applied to the resistor networks R_{11}-R_{13} and R_{14}-R_{16} which generate the feedback current flowing through R_{11} and R_{14}. Potentiometer P_{11} allows the loop gain to be adjusted to the correct value for a range of probe transformer secondary inductances.

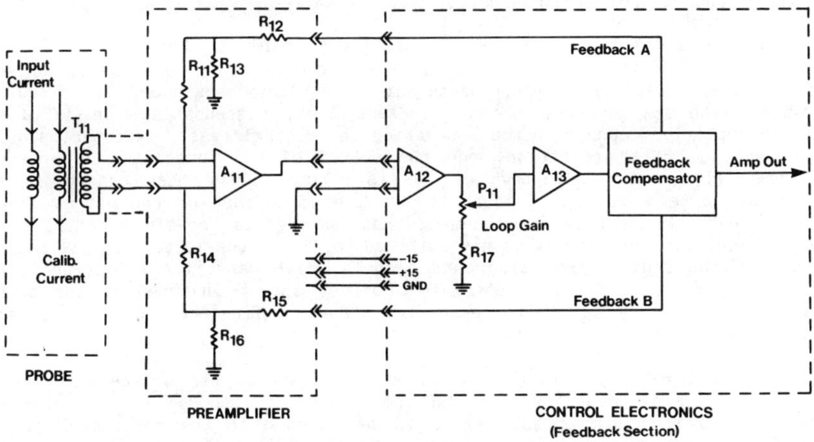

Figure 2. A functional schematic for the preamplifier and feedback circuitry for the current probe system.

In operation, shunt negative current feedback generates a very low impedance at the preamplifier input terminals. The amplifier output voltage "Amp Out" is equal to the probe transformer output current multiplied by the effective feedback resistance which is equal to $2 \times 10^6 \Omega$. Since the secondary of the probe transformer is effectively short-circuited, its output current is almost exactly equal to the input current divided by the secondary to primary turns ratio n_s. The overall transimpedance from the input current threading the probe transformer to the "Amplifier output" terminal is therefore close to the value $2 \times 10^6/n_s\ \Omega$. The gain of the system consisting of the probe, the preamplifier, and the feedback stage falls off at 20 dB per decade below a corner frequency almost exactly given by the reciprocal of the L/R time constant of the probe transformer secondary winding.

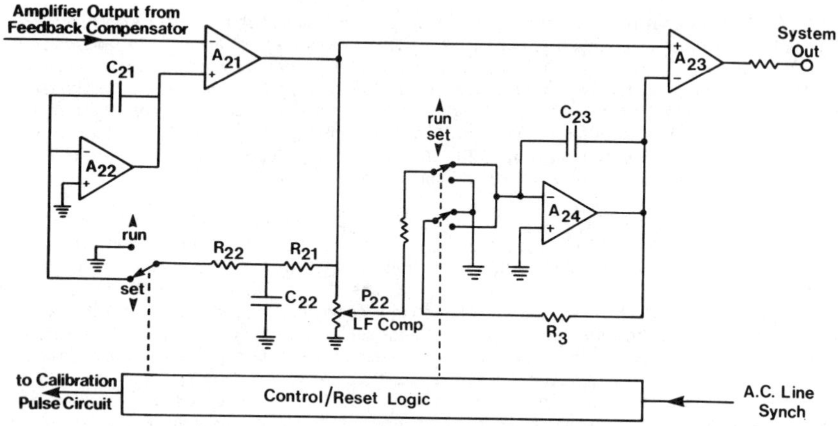

Figure 3. A functional schematic for the drift-removal and low-frequency compensation sections of the control electronics.

The circuit arrangment shown in Fig. 3 serves to minimize the effects of pre-amplifier offset drift, and to compensate for the loss in sensitivity at very low frequencies caused by the probe transformer secondary resistance. The circuitry associated with A_{21} and A_{22} is an auto-zero loop, compensating for dc offset in the preamplifier. During normal operation the control/reset logic periodically establishes the "Set" condition, and the output of the integrator A_{22} changes until the average output voltage of A_{21} is zero. Under these conditions the equilibrium voltage on C_{21} is equal to the average offset of the output signal from the feedback compensator, assuming that no signals are present during the set condition. If the circuit is now switched to "Run", the voltage on C_{21} remains constant at the equilibrium value, and signals which cause the Amplifier Output to change from the previously determined offset appear inverted at the output of A_{21}. The effect of C_{22} is to reduce the effect of line frequency interference on the operation of the auto-zero loop.

The circuitry involving A_{23} and A_{24} introduces a 20 dB per decade increase in gain with decreasing frequency. The "LF comp" potentiometer P_{22} allows the corner frequency at which this effect begins to be matched to the reciprocal of the L/R time constant of the probe, providing uniform overall sensitivity down to even lower frequencies. During the "set" period any charge which has accumulated on the integrator capacitor C_{23} is removed, zeroing the system output.

PERFORMANCE

The noise performance of the amplifier system is demonstrated in Fig. 4 by the square-root of the spectral density of the effective input noise current, referred to the primary of the toroidal input transformer. The minimum effective input current noise for the device is below 150 pA/Hz$^{1/2}$ from 2 kHz to 30 kHz. The increase in noise above 30 kHz results from the roll-off of the gain in the input stages and appearance of noise in secondary stages. From 10 Hz to approximately 1 kHz the current noise decreases linearly with frequency, consistent with white amplifier voltage noise and the fact that the toroid acts as a high-pass filter. The frequency at which the low-frequency noise levels off is determined by the impedance of the toroid and by the relative contributions of the current and voltage noise of the amplifier. At frequencies well below 10 Hz, the current noise density falls as 1/f$^{3/2}$ as the result of 1/f noise in the amplifier.

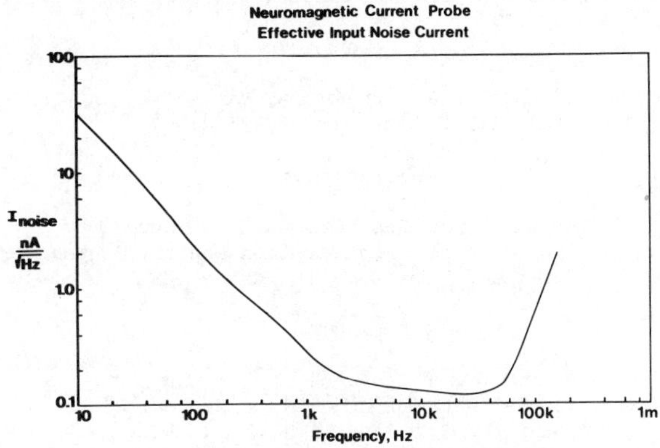

Figure 4. The effective input current noise of the system.

CONCLUSIONS

As demonstrated by companion abstracts, this system will be useful both in basic research on nerves and cardiac tissue and in surgical diagnosis of nerve damage.

ACKNOWLEDGEMENTS

This research has been supported by ONR Contract N00014-82-K-0107, NIH Grant 1-R01 NS 19794-01, the Veteran's Administration, and the Vanderbilt University Research Council.

REFERENCES

Barach, J.P., J.A. Freeman and J.P. Wikswo, Jr., (1980). Experiments on the magnetic field of nerve action potentials. J. Applied Physics, 53: 4532-4538.

Wikswo, J. P. (1982). Improved instrumentation for measuring the magnetic field of cellular action currents. Rev. Sci. Instrum., 53, 1846-1850.

Wikswo, J.P. Jr., J.P. Barach and J.A. Freeman, (1980). The magnetic field of a nerve impulse: First measurements. Science, 208: 53-55.

Wikswo, J.P., Jr., P.C. Samson and R.P. Giffard (1983). A Low-Noise Low Input Impedance Amplifier for Magnetic Measurements of Nerve Action Currents," IEEE Trans. Biomed. Eng. BME-30: 215-221.

MAGNETIC ASSESSMENT OF REGENERATION ACROSS A NERVE GRAFT

J.P. Wikswo, Jr.

Department of Physics and Astronomy, Vanderbilt University
Nashville, Tennessee, 37235 U.S.A.

G.S. Abraham

Rehabilitation Research and Development
Palo Alto Veterans Administration Medical Center
Palo Alto, California 94304, U.S.A.

V.R. Hentz

Rehabilitation Research and Development
Palo Alto Veterans Administration Medical Center
Palo Alto, California 94304, U.S.A.
Department of Surgery, Stanford University Medical Center
Stanford, California 94305, U.S.A.

ABSTRACT

The neuromagnetic current probe has been used to study regeneration of damaged peripheral nerves in primates. Preliminary data suggest that magnetic measurements may be superior to existing electrical measurements used for intraoperative assessment of peripheral nerve function.

KEYWORDS

Cellular magnetic field, nerve regeneration, neuroma in continuity

INTRODUCTION

In humans, the formation of connective tissue scar following injury or trauma to peripheral nerves usually prevents optimal functional regeneration of the nerve, and thus surgical intervention is often required to restore function to the damaged nerves. The results of surgery are poor in man because of the formation of scar. One particularly challenging problem for the surgeon is neuroma in continuity, in which the nerve is not completely or obviously severed, but is instead in continuity and with a medium-to-large and firm-to-very hard neuroma, indicating scar tissue proliferation. The difficult surgical decision is whether to dissect the neuroma apart, removing scar tissue (termed neurolysis), or to resect the neuroma and repair the nerve. If the surgeon chooses the former and the axons are actually discontinuous although the connective tissue is in-continuity, the patient is forced to wait needless months before the lack of successful regeneration indicates the actual extent of the damage to the nerve. Similarly, if many axons are actually in-continuity and the surgeon resects the site of injury, recovery may be delayed and there will be substantial risk of having poorer quality recovery because of mis-alignment of the repaired fascicular bundles. Another situation is that of the patient whose nerves have been repaired but recovery is inadequate. It is necessary to choose between neurolysis and a second resection with re-repair.

To provide a quantitative determination of the continuity of the axons at an injury site, Kline and DeJonge (1968) introduced the concept of intraoperative electrophysiological assessment performed by stimulating and recording the compound action potential (CAP) directly from whole nerves. Terzis, Dykes and Hakstain (1976) refined the technique to stimulate and record from individual nerve fasicles. Several parameters of the CAP have been proposed as a measure of the level of injury, including the shape of the waveform, the peak amplitude, the conduction velocity, and the area under the CAP. These measurements are complicated by the fact that the CAP represents the summation of many simultaneously excited single nerve fibers, so that it has proven difficult to develop quantitative criteria for assessment of nerve condition. These measurements are also limited by technical problems. For example, reliable and repeatable measurements of amplitude, conduction velocity, and shape of the CAP require very careful electrode placement. This in turn requires significant dissection of the nerve. In order for the voltages recorded extracellularly to be as large as possible, the nerve or nerve fascicles must be lifted onto electrodes that are suspended in air, risking additional nerve trauma, and the portion of the nerve between the electrodes must be kept dry during stimulation and recording, yet not allowed to desiccate to the point of cell death. Small changes in electrode position or the conductivity of the layer of moisture on the nerve make it difficult to obtain reproducible data. Electrical noise from operating room equipment is frequently a problem. The required close proximity of the stimulating and recording electrodes results in a stimulus artifact that often interferes with the recording.

Our experience with the magnetic recording of nerve action potentials (Wikswo and colleagues, 1984) indicates that because the pickup coil is fully insulated from the surrounding saline environment, the magnetic technique is much less sensitive to ground loops and external noise than are extracellular electrodes. The fact that the magnetic measurements are made with the nerve immersed in saline reduces the risk of trauma or dessication of the nerve and increases the reproducibility of the data. For these reasons, we have begun a series of preliminary measurements to determine whether magnetic recordings can overcome some of the limitations of electrical techniques.

METHODS

An animal model of human nerve regeneration and scar formation is crucial for the development of new techniques for intraoperative measurement of nerve function, and for the assessment of the efficacy of various surgical techniques designed to minimize the formation of scar and optimize restoration of nerve function. The ratio of axonal to connective tissue determines how closely the animal model relates to the human. Rats have essentially no connective tissue in their nerves and can regenerate damaged nerves without surgical intervention because little or no scar tissue is formed. Only higher animals such as primates have significant connective tissue in their nerve bundles and thereby produce sufficient scar tissue following nerve damage to serve as a model for surgical repair in humans.

The model adopted for this study uses the macaca fascicularis primate. The median nerve was divided just proximal to the wrist, a 15 mm segment was excised, and the nerve was repaired under tension with 8-0 nylon. The ulnar nerve was similarly treated except the gap was reconstructed with sural nerve grafts 3-4 in number, with the nerve ends approximated with 10-0 nylon suture. The limb was immobilized for three weeks in a cast.

Six months were allowed for the nerve axon to regenerate across the repair site and into the hand. After this time, the nerves were carefully surgically exposed using an operating microscope, both immediately proximal and distal to the repair site at the wrist, and also in the upper arm. At the wrist crease, the epinurium of the median nerve was removed over a 1 cm length. The nerve was separated into three groups, two single fascicles about 1 mm in diameter and a third group, 2.5 to 3 mm in diameter containing three to four fascicles. Just proximal to the elbow flexion crease, the median nerve was exposed over a 2-3 cm length, and at its most proximal exposure, it was sharply transected. The median nerve was threaded through the toroid and a fluid well was created by suturing tissues about the toroid. This well was filled with warm Ringer's lactate so that the nerve remained immersed throughout the study. One to two cm proximal to the toroid, the nerve was placed on platinum hook electrodes, dried, and suspended in air.

RESULTS

Figure 1 shows two recordings of both the magnetic action current (MAC) and the compound action potential (CAP) recorded at the elbow after stimulation with a 0.30 mA, 0.05 ms stimulus pulse applied at the wrist proximal to the repair. Note that the magnetic signal resembles the first derivative of a transmembrane action potential while the electric signal resembles the second derivative, consistent with theoretical and experimental studies made on single nerve axons (Barach, Roth and Wikswo, 1984; Woosley, Roth and Wikswo, 1984; Roth and Wikswo 1984).

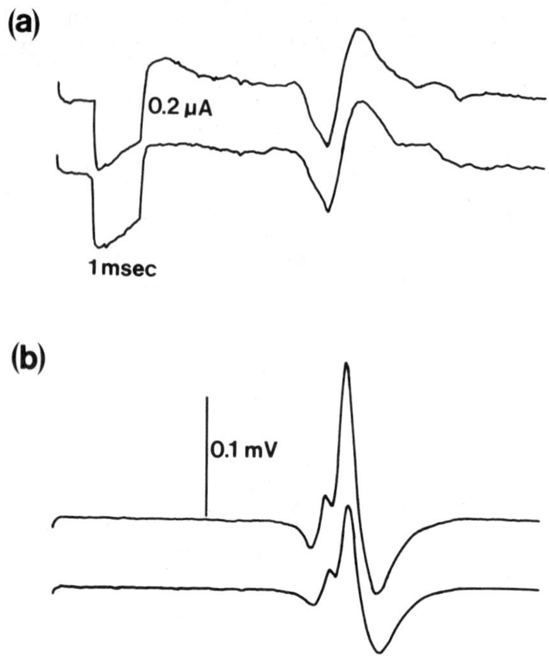

Figure 1. (a) The magnetic action current (MAC) and (b) the compound action potential (CAP) recorded simultaneously at two sequential times for the same animal. In each panel, the lower trace was recorded several minutes after the upper trace. The change in the CAP indicates the large variation in CAP recording amplitude associated with making electric recordings on nerves suspended in air. (Average of 256 recordings)

In each panel of Fig. 1, the upper and lower recordings were made under identical conditions, except that the lower recording was made several minutes after the upper one. The two MAC signals are virtually identical, while the second CAP is approximately 50% the size of the first CAP. This decrease in electrical signal size is apparently due to a decrease in the resistance of the fluid layer outside the nerve, dramatic evidence of the difficulty in obtaining reproducible results with electrical recordings made in air.

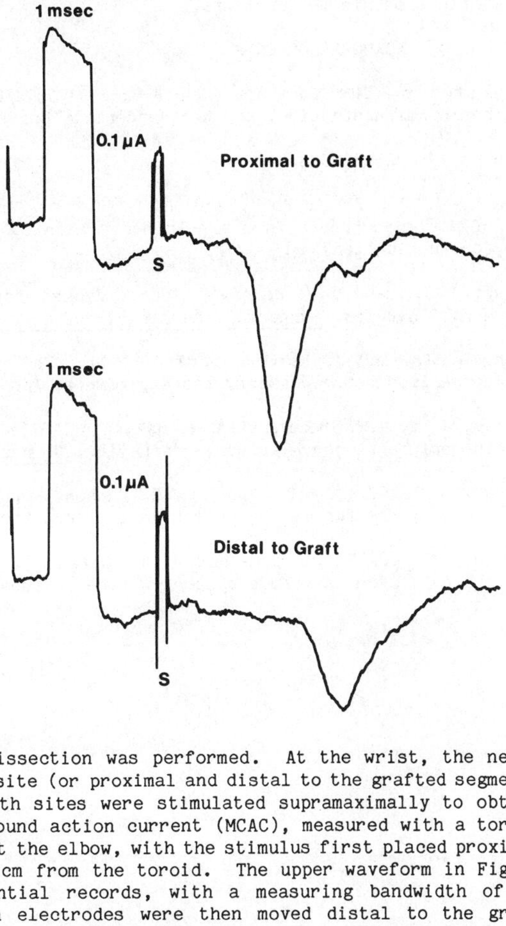

Figure 2. A magnetic recording of the action currents in a grafted right median nerve in a macaca fascicularis primate. The recording toroid was left fixed, and the stimulation pulse was applied to each side of the graft. (Average of 256 recordings)

In another animal, a similar dissection was performed. At the wrist, the nerve just proximal to the the repair site (or proximal and distal to the grafted segment) was lifted onto electrodes. Both sites were stimulated supramaximally to obtain recordings of the magnetic compound action current (MCAC), measured with a toroid placed around the median nerve at the elbow, with the stimulus first placed proximal to the graft, approximately 10 cm from the toroid. The upper waveform in Fig. 2 shows an average of 256 sequential records, with a measuring bandwidth of 10 Hz to 10 kHz. The stimulation electrodes were then moved distal to the graft site, approximately 13 cm from the toroid. The difference in the amplitude of the responses provides a measure of the proportion of the axons in the bundle that have successfully bridged both anastomoses of the graft. The time between the stimulus S and the peak response gives a measure of the nerve conduction velocity. The proximal recording yields a 45 m/s conduction velocity, while the lower gives 38 m/sec, indicating an additional delay that occurs because of the lower conduction velocity in the regenerated and scarred graft region.

CONCLUSIONS

Our measurements demonstrate the feasibility of magnetic assessment of nerve regeneration in higher mammals. Magnetic measurements offer the advantages of being made with the nerve immersed in saline, whereas extracellular electric measurements are most readily made with the nerve suspended in air, and it appears that magnetic measurements utilizing our present technique are easier to perform, more accurate, less noisy, and less susceptible to artifact than are presently-available electric measurements.

ACKNOWLEDGEMENTS

Supported by ONR Contract N00014-82-K-0107, NIH Grant 1-R01 NS 19794-01, the Veteran's Administration, and the Vanderbilt University Research Council.

REFERENCES

Barach, J. P., B. J. Roth, and J. P. Wikswo, Jr. (1984). Magnetic measurements of action currents in a single nerve axon: a core-conductor model. (submitted for publication)

Kline, D.C., and B.R. DeJonge (1968). Evoked potentials to evaluate peripheral nerve injuries. Surgery, Gynecology, and Obstetrics, 127, 1239-1248.

Roth, B. J., and J. P. Wikswo, Jr. (1984). The magnetic field of a single axon: comparison between theory and experiment. (in preparation).

Terzis, J.K., R.W. Dykes, and R.W. Hakstain (1976). Electrophysiological recordings in peripheral nerve surgery - a review. J. Hand Surgery, 1, 52-65.

Wikswo, J.P., Jr., W.P. Henry, P.C. Samson, and R.P. Giffard (1984). A current probe system for measuring intracellular action currents. These proceedings.

Woosley, J. K., B. J. Roth and J. P. Wikswo, Jr. (1984). The magnetic field of a single axon: a volume conductor model. (submitted for publication)

NEW DEVELOPMENTS IN MAGNETOCARDIOGRAPHY

P. Siltanen

Cardiovascular Laboratory,
Helsinki University Central Hospital,
Helsinki, Finland.

ABSTRACT

The development in magnetocardiography is briefly reviewed with emphasis on the more recent advances. The physiological and anatomical fundamentals of magnetocardiography are surveyed and its objectives and possibilities are appraised with special reference to future development.

KEYWORDS

Magnetocardiography; electrocardiography; Cardiac activation; His-Purkinje signal; late potentials.

INTRODUCTION

The era of biomagnetic measurements started in the early sixties when Baule and MgFee (1963), using a coil magnetometer, recorded the weak magnetic field generated by the beating human heart. Magnetocardiogram (MCG) was a natural first step in biomagnetic recording because myocardial activity generates the highest bioelectric potentials in the body, if the skeletal muscles are excluded. Myocardial depolarization is associated with surface potentials of several mV. The corresponding magnetic field generated by the myocardial activation currents is, however, extremely weak with the peak field strength generally remaining under 100 pT.

Only the introduction of the SQUID magnetometer by Cohen and others (1970) rendered it possible to accomplish a systematic investigation of the cardiac magnetic field. The most notable steps in the development of magnetocardiography are listed in Table 1. A more detailed presentation on this topic including a complete list of references will appear in the forthcoming comprehensive review on magnetocardiography (Siltanen 1985).

My intention here is to survey some physiological fundamentals inherent to magnetocardiography and the objectives and possibilities afforded by this technique. Further, I aim to briefly review the recent developments in this field and to outline the present state of the art in magnetocardiography. Finally, I shall attempt to predict the future of this method.

TABLE I Major Steps of Progress in Magnetocardiography

Year	Progress	Authors
1963	First MCG; signal averaging; coil magnetometer.	Baule and McFee.
1967	MCG in shielded room; signal averaging; coil magnetometer.	Safanov and others; Cohen.
1970	Real time MCG; shielded room; SQUID magnetometer.	Cohen and others.
1970	First MCG vector presentation.	Baule and McFee.
1971	MCG in unshielded environment; 1st order SQUID gradiometer.	Zimmerman and Frederick.
1973	First computer simulation of MCG.	Horaček.
1974	MCG in laboratory or in hospital; 2nd order SQUID gradiometer.	Opfer and others.
1974	First survey on MCG in various heart diseases.	Saarinen and others.
1974	Normal material and standard grid for MCG.	Siltanen and others.
1975	MCG of direct currents.	Cohen and Kaufman.
1978	First MCG record on His-Purkinje signal.	Farrell and others.
1981	First computer simulation of His-Purkinje -MCG.	Tripp and Farrell.
1983	First MCG record of late potentials	Erné and others.

ANATOMICAL AND PHYSIOLOGICAL CONSIDERATIONS

The heart is a muscular organ weighing some 1/3 kg, and is made up of four cavities, two thin-walled atria and two more muscular ventricles. The left ventricle, which maintains the systemic circulation, has about 10 mm thick walls which comprise the major portion of the ventricular muscle mass of the heart. The architecture of the musculature of the left ventricle is worthy of note from the electrophysiological viewpoint. It consists of a helical outer layer of long muscle fibres, which at the apex of the ventricle takes the form of a vortex, and after having turned into a reversed threaded helix it continues upwards thereby constituting an inner muscle layer. This double spiral anatomy ensures the most complete and effective emptying of the ventricle during its contraction. The right ventricular walls are about 4-5 mm thick, but they contribute less than 10 per cent to the ventricular bioelectric forces; therefore the right ventricle is usually ignored in most considerations. The atrial walls are only 2-3 mm in thickness and they envelope the cavities of the right and left atrium like a distensible shell. The atrial function consists of three components: pipe function, reservoir function, and booster-pump function. Because of the last mentioned function, the atria and the ventricles must contract in sequence in order to ensure proper filling of ventricles. The contraction sequence of the atria and ventricles is governed by differentiated muscle fibres which are specialized for controlling the impulse conduction from the atria to the ventricular musculature.

The impulse formation takes place in the sinus node which is situated in the upper part of right atrium. This tissue fires automatically and represents the normal pacemaker of the heart. When the impulse spreads itself from the sinus node the right atrium starts to contract. Simultaneously the activation propagates along Bachmann's bundle to the left atrium, and via the internodal tracts to the atrioventricular node. In this node the impulse is delayed about 150-200 ms in order to allow the ventricles to be filled up by the atrial booster kick. From the atrioventricular node the conductive tissue continues down towards the ventricles, primarily as a single stem about 10-15 mm in length, called the His bundle. It divides into right and left bundle branches. The right branch leads to the right ventricle. The left one takes the shape of a fan and interweaving with the Purkinje fibres it spreads itself to the inner surface of the left ventricle. The major portion of this fan-like mesh is constituted of Purkinje fibres which connect with the endings of the left bundle branch and, on the other hand, penetrate the inner third of the left ventricular wall. The Purkinje fibres conduct the action potential quickly, 4000 mm/s, as compared with the conduction velocity of atrial muscle fibres or the His bundle with its branches (800-1000 mm/s), ventricular muscle fibres (400 mm/s) and atrioventricular node (200 mm/s). Thus, the impulse is propagated very rapidly by the endocardial Purkinje network into the subendocardial layer of the middle and apical portions of left ventricle. At this stage the activation wavefront sweeps almost isochronously from the endocardium to the epicardium; the basal portions of the left ventricle are poor of Purkinje tissue and are depolarized with slight delay from below as the last depolarizing part of the ventricle. Thus, the ventricular depolarization comprises of a rapid tangential, a slower radial and a terminal tangential phase.

ELECTROCARDIOLOGICAL CONSIDERATIONS

In this context, it is not necessary to review profoundly the current knowledge of the electrophysiological events of depolarization and repolarization on the cellular level. However, some essential points should be mentioned. Resting transmembrane potential, created by metabolic activity of the myocardial cell membrane, is the generator of ionic currents across the membrane during myocardial activation. Depolarization of myocardial cells is associated with a rapid sodium flux and a slow calcium flux across the cell membrane along the potential gradient, and during repolarization a metabolically maintained reverse cation flux restores the transmembrane potential. Ion fluxes which are associated with depolarization and repolarization of myocardial cells constitute local ionic impressed currents, current dipoles of which can be considered as elementary biological sources of the magnetic field. In a conducting medium these dipoles, according to Ohm's law, create opposite volume currents, which in a inhomogenous conductor like the human torso, can give rise to a magnetic field which interferes with the field generated by impressed currents. However, it has been shown by Grynszpan and Geselowitz (1973) and by Cohen and Hosaka (1979), that the contribution of volume currents is insignificant in generating component B_n of the magnetic field, which is normal to the surface. When using a frontal measurement grid, the z-component comes near to component B_n, which contains all the measurable information on the field, and in which the influences of the torso's inhomogeneities are significantly smaller than in the x- and y-components.

Activation of the ventricular muscle mass has traditionally been represented as a propagation of a uniform double-layer with dipoles oriented normal to the layer and are of equal strength, and can therefore be represented approximately as a single current dipole. This model satisfactorily explains the potential distribution detectable by ECG. However, the currend dipole model does not explain as well the distribution of the magnetic field. Recent observations in our laboratory have shown, that secondary currents neither can explain this discrepancy between field measurements and experimental or computer model calculations (Varpula and others, 1984). It is possible that the distribution of the primary cardiac currents contains circular currents which influences the distribution of the z component of the magnetic field.

The myocardial fibres are interconnected although they are on the connecting surfaces, separated by "intercalated discs" which allow propagation of the impulse from one fibre to another. However, these connections offer significantly greater resistance for currents than does cytoplasm along the fibres. This situation results in a significant anisotropism of the myocardium. In fact, the velocity of the excitation wavefront moving across the fibres is only half of the velocity along the main fibre and the wavefront proceeding along the fibres can develop even 20 times more current dipole strength than a wavefront moving across fibres (Colli-Franzone and others, 1982). Considering the helical architechture of the left ventricular muscle, the anisotropism could easily result in such a behaviour of the magnetic field that could not be expected on the grounds of the uniform double-layer model of myocardial depolarization.

There is much experimental evidence suggesting that the subendocardial layer of the middle and apical part of the left ventricle is electrocardiographically silent. In spite of that it undergoes almost instantaneous activation at the very onset of ventricular contraction, it does not contribute the built-up of the QRS complex (Sodi-Pallares and others, 1961). Purkinje tissue is known to extend several millimeters beyond the endocardium into the subendocardial muscle tissue in the middle and apical part of the left ventricle. It has been postulated that the endings of the Purkinje fibres create polarized islands in the subendocardium which cannot be traced by ECG before they "open" onto the endocardial surface. Neither is the magnetometer capable of discovering these closed islands of depolarization because there is no resultant current dipole.

From a clinical viewpoint, determination of the primary sources, i.e.,the solution of the inverse problem would be of uppermost importance. Unfortunately, the inverse problem has no unique solution neither in electrocardiography nor in magnetocardiography. Therefore, the concept of equivalent generators has been introduced which represents a simplified equivalent of the true sources and can, to some extent, depict the cardiac activation process. The equivalent generator can be defined in magnetic or electric terms, either as a fixed dipole or as multipole expansion (Wikswo and others, 1979; Karp and others, 1980; Katila and Karp, 1981). The multipole expansion has the advantage in magnetocardiography, that once multipole strengths have been determined, for which no knowledge of torso or source geometry is required, the magnetic field can be calculated everywhere since the multipole expansion contains all measurable information of the cardiac magnetic field. The connections of its parameters to real cardiac events is, however, difficult to establish.

The most simple solution of the inverse problem is the determination of a single equivalent current dipole on the basis of magnetic isofield distribution (Williamson and Kaufman, 1981; Tripp, 1983). In a case of current dipole beneath a limiting surface the measured pattern of the field component normal to the surface determines the perpendicular distance of the dipole from the surface. This relationship is a simple one, since the depth $D = \frac{\Delta}{\sqrt{2}}$ where Δ is a distance between the two polarity extremes of the field distribution in a case of plain boundary. This method offers a rough means to localize bioelectric sources, for example, in the heart or brain provided that the depth is small enough that the convexness of the boundary surface can be ignored and provided that the source is not more complex one. This method has the drawback, in that in examining abnormal fields, the previous normal field distribution should be available in order to obtain, by subtraction, the field distribution inherent to the abnormality in question. However, this is usually not accessible. A comparable situation is met both in clinical electrocardiography and in clinical magnetocardiography, but it is solved by using averaged measures of tracings obtained from healthy population as references. Our laboratory is presently elaborating this type of reference material for isofield technique.

The topic of source localization will be dealth with more profoundly in a particular panel in this conference.

METHODS IN MAGNETOCARDIOGRAPHY

In instrumentation, no important new development has occurred since the introduction of the second order SQUID gradiometer. The topic of instrumentation will be dealt with extensively in a panel during this conference and can be therefore be passed over in this context. Clinical cardiologists are still looking ahead for devices which are simpler to use as regards cryotechnology and, are less sensitive to ambient noise. On the other hand, high-resolution SQUID magnetometry has recently appeared as an important branch of magnetocardiography and because beat-to-beat analysis is essential, special, more sensitive devices are desired which are also usable in the hospital environment. This discrepancy between the requirements due to small signals and the need for easy execution of recordings and clinical usability, remains to be solved by engineers and physicists. Obviously in the future a multidetector system would be necessary even in cardiology in order to obtain simultaneous records of the same signals on several precordial positions.

At the present, the measurement grid introduced by our group (Siltanen and others, 1974; Saarinen and others, 1978; Karp, 1981) has been accepted by most workers. For clinical purposes, seemingly only six of the 36 recording positions of the standard grid are needed for recording conventional MCG's (Siltanen, 1985) as most abnormalities can be distinguished by using positions B_2, B_5, C_3, C_4, D_4 and E_5 only.

NORMAL MAGNETOCARDIOGRAM

Of the 250 MCG time curve records of normal subjects which have appeared in the literature, less than one-fifth of the measurements have been carried out according to the standards discussed and adopted in the Workshops of Magnetocardiography 1976-1982 (Siltanen, 1985). Average maps of 31 healthy men and 10 healthy women recorded in supine position were published by our group in 1978 (Saarinen and others). In this report statistical cumulative distributions of various MCG amplitudes were also included. It offers, therefore, a useful reference material for evaluation normality of MCG tracings. At present this material has not been supplemented with any other comparable material found in the literature.

The normal MCG exhibits the same deflections, time relationships, and almost the same relative amplitudes as the corresponding ECG. Only repolarization events seem often to be sensed differently by MCG and ECG. For example, using a unipolar lead system and vector presentation, it has been shown that in ECG the R-T angle is on average about three times that seen in MCG (Barry and others, 1977). At times, normal subjects present biphasic T-wave of + - -type, in which the negative phase is isochronous with the U-wave in the ECG and which can be normalized by hyperventilation (Barbanera and others, 1981). The U-wave can be seen in MCG more often than in ECG and it may be confluent with the T-wave.

Isofield maps of normal subjects have appeared only sporadically and no systematical approach has been as yet reported. However, the growing interest in isofield mapping technique requires normal reference materials.

Recently, the literature on normal magnetocardiograms has been supplemented with a report of vectorcardiograms of 19 healthy subjects (Nousiainen and others, 1984 a). The reported observations are in good accordance with those of scalar MCG mentioned above. In this material there is some evidence on higher relative sensitivity of MCG during the final phase of QRS when the activation proceeds more tangentially. Data supporting this view will be presented by the same group also in this conference (Nousiainen and others, 1984 b).

ABNORMAL MAGNETOCARDIOGRAM

Mapping using standard grid

Clinical MCG materials are very few, the total number of published cases being some 200. An analysis of this pooled material will be published elsewhere (Siltanen 1985), but I shall comment on a few pertinent points in this context.

Comparison between premature ventricular beats as well as bundle branch complexes in ECG and MCG has shown that the relative amplitude of the terminal QRS is greater in MCG than in ECG. This observation augments the evidence for greater sensitivity of MCG to aberrant tangentially proceeding conduction, because in the beats mentioned above, the wavefront is propagated after the initial radial phase mainly tangentially (Barry and others, 1977). In general, time records of MCG and their morphological analysis in conduction disturbances and abnormalities in heart rhythm do not contribute much to that information obtained from ECG.

The relative amplitude of the P wave in healthy subjects is smaller in MCG than in ECG. On the contrary, in cases with atrial overload and enlargement, the relative increment of the P amplitude over the average normal value is markedly greater in MCG than in ECG. This is probably due to the rapid attenuation of the magnetic field apart from the origin, approximately as r^{-3}, when r is the distance between the source and the observation point. Therefore,in right atrial enlargement in heart diseases, the atrial wall draws nearer to the anterior thoracic wall resulting in increased P wave amplitudes. In left atrial enlargement which results usually very early from diseases loading or damaging the left ventricle, the expansion occurs backwards and therefore the P wave recorded on the subject's back is usually enlarged. Due to the anatomy of atrial walls described above, the activation wavefront propagates almost exclusively tangentially. In both left and right atrial hypertrophy, the sensitivity of MCG to this type of wave front propagation often results in a marked increase of P wave amplitude. In atrial overload, the P-R segment is often slightly elevated which probably represents the effect of atrial repolarization currents.

In cases of ventricular hypertrophy, high R deflections and ST strain patterns are found in the upper half of the standard grid in cases of right ventricular hypertrophy due to systolic overload and in the left lower quadrant of the grid in cases of left ventricular hypertrophy. This can be expected on grounds of orientation of the main ventricular vector in ECG to the right and anteriorily in the former cases and to the left and posteriorly in the latter cases.

In hypertrophic idiopathic cardiomyopathy, discrepancy between ECG and MCG is sometimes striking and involves both depolarization and repolarization. In some cases, MCG displays relatively small notched QRS complexes in spite of high R amplitudes in ECG, and vice versa. In certain cases repolarization abnormalities in MCG are extraordinarily great despite minimal ST- and T-abnormalities in ECG and vice versa. In hypertrophic idiopathic cardiomyopathy, the normal architecture of myocardial fibres is largely deranged because of local abundant hypertrophy and fibrosis, and consequently, propagation of the activation wavefront is tortuous. Discrepances between ECG and MCG therefore probably reflect the different sensitivities of these methods in various cases of aberrant activation.

Ischemic heart disease is today the most important clinical challenge in cardiology. The contribution of the conventional MCG in diagnosis of acute myocardial infarction has not been evaluated. In angina pectoris and chronic ischemic heart disease after myocardial infarction, the contribution of MCG to the diagnostics has remained small. Typical patterns of MCG in myocardial infarction has been ascertained both empirically and by using the computer simulation technique. However, in clinical diagnosis of previous myocardial infarction MCG does not seem to offer any advantage in accuracy or sensitivity as compared to ECG. The infarction dipole model is a well-working tool in localizing the ischemic myocardial lesions as was recently shown by our group (Siltanen and others, 1983), and is shown by us also in this conference using computer simulation (Saarinen and others, 1984). The infarction dipole can be demonstrated and even localized by using the isofield mapping technique (Katila and others, 1983; Gonnelli and others, 1984). However, in this approach, the problem is lack of reference. Differentiating between ischemic and non-ischemic repolarization changes in ECG is sometimes difficult or even impossible. In those cases direct current MCG would be of great importance. Using a specially constructed DC SQUID sensor in a shielded room Cohen and others (Cohen, 1983) have succeeded to differentiate between primary and secondary ST-segment changes in man. This is not possible with the conventional ECG technique. However, the MCG technique is complicated but has many obstacles some of which may be unsurmountable.

To conclude, magnetocardiographic mapping using the standard (or reduced) grid yields in most heart diseases typical tracings which can be used for diagnostic purposes. Some of the abnormal patterns even overcome those of the ECG in diagnostic sensitivity. In hospital practice, however, the drawback of magnetocardiography is the complicated technology in comparison with the relatively small advantages on certain areas of diagnostics. On the other hand, the contactless means of recording makes magnetocardiography an ideal means of mass screening for heart diseases. There its diagnostic power could be utilized if only the technical problems can be solved.

High-resolution techniques

Sudden coronary death is an important manifestation of coronary heart disease. Its annual incidence, for example, in Finland is about 2 % of men at the age of 50-60 years. Therefore, everywhere, great effort has been put on researching the possibilities to reduce this high frequency of sudden death. One of the avenues in this research present the attempts to uncover risk factors of sudden death by electrophysiological investigation. There is a bulk of conclusive evidence that in majority of sudden deaths, the primary cause is cardiac and the primary disturbance in the heart is the so called "re-entry phenomenon". In a small proportion of cases, the underlying disturbance is an increased or depressed automaticity of myocardium or their combination with re-entry.

Re-entry is a phenomenon in which the activation wavefront meets a damaged myocardial island with an abnormal property of reduced conduction velocity and unidirectional block. These prevent the activation from entering the island from the normal direction but permits the impulse to enter it retrogradely after having moved around the islet. The consequent late depolarization of the islet further spreads itself retrogradely to the normal tissue provided that the latter has already passed the refractory phase. In certain circumstances this process can initiate a short-circuit resulting in ventricular tachycardia or ventricular fibrillation and death (Wellens and Dürrer, 1976).

By using high-resolution ECG, Fontaine and others (1976, 1978) as well as El-Sherif and others (1977), using epicardial electrodes, have demonstrated these late depolarization potentials which can be seen in their tracings as small deflections superimposing on the ST-segment or T-wave. Berbari and others (1978) demonstrated these late potentials by using high-resolution surface-ECG. More recently several groups in Europe and the U.S.A. have investigated the late potential as a predictor of malignant ventricular arrhythmias (Simson, 1981; Breithardt and others, 1982; Hombach and others, 1982). In these studies, a definite association has been revealed between appearance of late potentials and malignant ventricular arrhythmias. Surgical excision of the arrhythmogenic area in selected cases has resulted in disappearance of both clinical symptoms and late potentials. Unfortunately, late potentials can sometimes be also found in people presumably healthy. For the present, it is not possible to distinguish uneqiuvocally benign late potentials from those predicting arrhythmias, although their duration may be in positive correlation with risk of sudden death (Breithardt and others, 1981). Late potentials were first recorded by means of MCG by Erné and others (1983 a). Due to the potential possibility to localize the source, MCG may offer a more useful means than ECG for investigation these cases. Our group is carrying out the screening of patients with late potentials in hospital by high-resolution ECG in order to examine all positive cases by means of high-resolution MCG in a shielded room. At present, no observation can be reported but we consider this type of approach promising. Isofield mapping technique has been proposed for discriminating patients with late potentials (Erné and others, 1983 b). Patients with late potentials in ECG show rapid changes in the isofield pattern associated with differences between maximum and minimum up to 5 pT.

Examination of depressed automaticity and disturbances of conduction in the specialized conductive tissue of the heart requires the high-resolution technique as well. Particularly, recording of the His-Purkinje signals (HPS) has been a matter of growing interest although for the present its clinical utilization has been rather limited. Scherlag and others (1969) succeeded to demonstrate HPS in man by using an intracardiac catheter technique. In their records a spike appeared about 40 ms prior to the onset of the QRS complex which was called "H" because it was considered to represent the intrinsic deflection of the His bundle. Some years later, Berbari and others (1973) demonstrated that HPS was possible to record by surface-ECG using high amplification, proper filtering and an averaging technique. Later on, the same group (Berbari and others, 1979) showed, that by using a bandwidth of 0.1-300 Hz the P-R segment takes a pattern of "ramp" which commences about 40 ms before initiation of QRS. According to them this ramp represents a continuing process of His-Purkinje system activation. These investigators considered that the spikes in the HPS records actually represent anatomically determined bends of the impulse passage in the His bundle and its branches (Berbari and others, 1983). Their observations suggest that changes in the ramp morphology may afford an estimate of the extent and location of injuries of the His bundle and the left bundle branch with its arborescences.

The first magnetic record of HPS was reported by Farrell and others (1978). These investigators recognized two notches in the P-R segment at 40 and 20 ms prior to QRS. Later on (Farrell and others, 1980), they also demonstrated the ramp pattern in man by means of high-resolution MCG. In the same year, Fenici and others (1980) as well found the ramp and spike pattern of the PR segment in man by using high-resolution MCG, 2nd order SQUID gradiometer without shielding and a bandwidth of 0.1-130 Hz. In their recordings, ramp commenced at 40-45 ms prior to QRS and showed at its outset a small but distinct notch of 200-400 fT which was isochronous with the H-spike of HPS-ECG recorded by intracardiac catheter. Both groups considered the ramp to be due to continuous activation of the His-Purkinje system. This hypothesis is strongly supported by later computer studies (Tripp and Farrell, 1981; Leoni and Romani, 1982), in which modelling and simulation of the His-Purkinje system revealed a ramp pattern very similar to those found empirically. Both experimental and computer simulation data display a ramp pattern which is upsloping above a diagonal line running along the cardial apex-base axis, and downsloping below that line. However, this symmetrical slope pattern has not been found in all studies (Erné and others, 1983 b; Lorenzana and others, 1984), Obviously the structure of the PR segment is complicated and variable and the sources of its signals are difficult to determine. On the other hand, Leifer and others (1981, 1983) have presented data suggesting that atrial repolarization currents in animal experiments explain the ramp pattern entirely. In fact, Riff and Carleton (1971) had already showed over ten years earlier that atrial repolarization potentials lag until 250-380 ms past the outset of the P wave in the normal ECG. Presently, it is uncertain as to what are the contributions of HPS and atrial repolarization currents to the ramps and spikes of the PR segment. Data obtained by using a special subtraction technique and presented by Erné and others (1984) in this conference support the view that both signals are involved. If differentiation of HPS and atrial repolarization become possible, high-resolution MCG will be a favourable means for HPS-studies because of the good spatial resolution property of MCG.

Examination of HPS and late potentials requires high quality recording on beat to beat basis because the averaging technique is not suitable for the sake of moving of the signal in relation to the triggering QRS complex and because the appearance of the signals is often random.

FUTURE OUTLOOK

The place of MCG as a clinical tool has not yet been clarified. Probably it will never replace ECG which has established its place as a technically developed simple method with long traditions and an enormous body of clinical empirical data. On the other hand, there are several circumstances in which the MCG displays better resolution and even better diagnostic power. Particularly its property to afford estimates of source locations may offer MCG special tasks which are beyond the possibilities of ECG. For example, late potentials, after-potentials and pre-excitation syndrome, occasionally necessitate surgical intervention and, therefore, an exact preoperative localization of the abnormal signal is required.

In clinical cardiology there are today several methods in use for localizing structures and even metabolic processes of the myocardium. Anatomic details and mechanic cardiac function can be depicted invasively by cardiac catheterization and angiocardiography, and noninvasively by radiotomography, nuclear angiocardiography and echocardiography. Metabolic myocardial details as ischemia can be examined and localized by using Thallium-201 and Technetium-99m radioisotopes and gamma-camera tomography. By means of a positron emission camera and appropriate radioisotopes one can accurately localize complicated metabolic processes in selected myocardial segments. Nuclear magnetic resonance imaging is an additional example of a three-dimensional localizing means for metabolic myocardial events. In electrocardiology, however, there is no comparable method for localization of the sources. It would be highly desirable to be able to correlate anatomic, metabolic, pharmacologic and electromagnetic phenomena in selected myocardial segments which are involved by disease and/or some kind of intervention. Time will show to which extent by combination of magnetocardiographic and electrocardiographic data this goal can be achieved.

The non-invasive no contact means of recording can be considered advantageous in some special purposes as in mass screening for heart diseases. On the other hand, the necessity of cryotechnology and the problems due to the ambient magnetic field certainly will limit the usefulness of the current magnetocardiographic devices in clinical practise.

REFERENCES

Barbanera, S., P. Carelli, R. Leoni, G.-L. Romani, F. Bordoni, I. Modena, R. Fenici, and P. Zeppilli (1981). Magnetocardiographic study of some human cardiac electrophysiological phenomena: preliminary observations. In S. Erné, H.D. Dahlbohm and H. Lübbig (Ed.), Biomagnetism. DeGruyter, Berlin. pp. 283-290.

Barry, W.H., W. M. Fairbank, D. C. Harrison, K. L. Lehrman, J. A. V. Malmivuo, and J. P. J. Wikswo (1977). Measurement of the human magnetic heart vector. Science, 198, 1159-1162.

Baule, G. M., and R. McFee (1963). Detection of the magnetic field of the heart. Am. Heart J., 66, 95-96.

Baule, G. M., and R. McFee (1970). The magnetic heart vector. Am. Heart J., 79, 223-236.

Berbari, E. J., S. M. Collins, Y. Salu, and R. Aerzbaecher (1983). Orthogonal surface lead recordings of His-Purkinje activity: comparison of actual and simulated waveforms. IEEE Trans. bio-med. Engng., BME-30, 160-167.

Berbari, E. J., R. Lazzara, P. Samet, and B. J. Scherlag (1973). Noninvasive technique for detection of electrical activity during the PR segment. Circulation, 48, 1005-1013.

Berbari, E. J., R. Lazzara, and B. J. Scherlag (1979). The effects of filtering the His-Purkinje system electrocardiogram. IEEE Trans. bio-med. Engng., BME-26, 82-84.

Berbari, E. J., B. J. Scherlag, R. R. Hope, and R. Lazzara (1978). Recording from the body surface of arrhythmogenic ventricular activity during the S-T segment. Am. J. Cardiol., 41, 697-702.

Breithardt, G., M. Borggrefe, J. Schwarzmaier, U. Korbenn, H. L. Yeh, and L. Seipel (1982). Klinische Bedeutung nicht-invasiv registrierter ventrikulärer Spätpontentiale. D. Med. Wschr., 107, 643-648.

Breithardt, G., J. Szchwarzmaier, R.-R. Abendroth, M. Borggrefe, and L. Seipel (1981). Prospective study on the incidence of late potentials in postmyocardial infarction patients. Circulation, Suppl. IV, abstr. 1261.

Cohen, D. (1967). Magnetic fields around the torso: production by electrical activity of the human heart. Science, 156, 652-654.

Cohen, D. (1983). Steady fields of the heart. In S. J. Williamson, G.-L. Romani, L. Kaufman, and I. Modena (Ed.), Biomagnetism. An Interdisciplinary Approach. Plenum Press, New York. pp. 265-274.

Cohen, D., E. A. Edelsack, and J. E. Zimmerman (1970). Magnetocardiograms taken inside a shielded room with superconducting point-contact magnetometer. Appl. Phys. Lett., 16, 278-280.

Cohen, D., and H. Hosaka (1976). Magnetic field produced by a current dipole. J. Electrocardiol., 9, 409-417.
Cohen, D., and L. A. Kaufman (1975). Magnetic determination of the relationship between the S-T-segment shift and the injury current produced by coronary artery occlusion. Circ. Res., 36, 414-424.
Colli-Franzone, P., L. Guerri, C. Viganotti, E. Macchi, S. Baruffi, S. Spaggiari, and B. Taccardi (1982). Potential fields generated by oblique dipole layers modeling excitation wavefronts in the anisotropic myocardium. Circ. Res., 51, 330-346.
El-Sherif, N., B. J. Scherlag, R. Lazzara, and R. R. Hope (1977). Reentrant ventricular arrhythmias in the late myocardial period. 1. Conduction characteristics in the infarction zone. Circulation, 55, 686-696.
Erné, S. N., R. R. Fenici, H.-D. Hahlbohm, W, Jaszczuk, H. P. Lehmann, and M. Masselli (1983a). High-resolution magnetocardiographic recordings of the S-T segment in patients with electrical late potentials. Nuovo Cimento, 2D. 291-293.
Erné, S. N., R. R. Fenici, H.-D. Hahlbohm, W. Jaszczuk, H. P. Lehmann, and M. Masselli (1983b). High-resolution isofield mapping in magnetocardiography. Nuovo Cimento, 2D, 291-300.
Erné, S. N., R. R. Fenici, H.-D. Hahlbohm, J. Korsukewitz, H. P. Lehmann, and Y. Uchikawa (1984). Magnetocardiographic study of the PR segment of normals. V World Conference of Biomagnetism, Vancouver.
Erné, S. N., R. R. Fenici, H.-D. Hahlbohm, M. Masselli, H. P. Lehmann, and Z. Trontelj (1983c). High-resolution recordings of the PR segment in magnetocardiography. Nuovo Cimento, 2D, 248-254.
Farrell, D. E., J. H. Tripp, and R. Norgren (1978). Non-invasive information on the cardiac cycle: an assessment of the clinical potential of the electric and magnetic methods. Proc. SPIE, 167, 173-177.
Farrell, D.E., J. H. Tripp, R. Norgren (1980). Magnetic study of the His-Purkinje conduction system in man. IEEE Trans. bio-med. Engng., BME-27, 345-349.
Fenici, R. R., G.-L. Romani, S. Barbanera, P. Zeppilli, P. Carelli, and I. Modena (1980). High resolution magnetocardiography: noninvasive investigation of His-Purkinje system activity in man. G. Ital. Cardiol., 10, 1366-1372.
Fontaine, G., R. Frank, F. Gallais Hamonno, I. Allali, H. Phan-Thuc, and Y. Grosgogeat (1978). Electrocardiographie des potentials tardifs du syndrome de post-excitation. Arch. Mal. Coeur, 71, 854-864.
Fontaine, G., G. Guirandon, R. Frank, C. Coute, and C. Dragodanne (1976). In H. J. J. Wellens, K. I. Lie, and M. J. Janse (Ed.), The Conduction System of the Heart. Structure, Function and Clinical Implications. Leiden.
Gonnelli, R., P. Galeoni, M. Sicuro, and A. Tartaglia (1984). Magnetocardiographic isofield mapping in the characterization of the infarcted area. V World Conference of Biomagnetism, Vancouver.
Grynszpan, F., and D. B. Geselowitz (1973). Model studies of the magnetocardiogram. Biophys. J., 13, 911-925.
Hombach, V., V. Braun, H.-W. Höpp, D. Gil-Sanchez,H. Scholl, D. W. Behrenbeck, M. Tauchert, and H. H. Hilger (1982). The applicability of the signal averaging technique in clinical cardiology. Clin. Cardiol., 5, 107-124.
Horaček, B. M. (1973). Digital model for studies in magnetocardiography. IEEE Trans. Mag., MAG-9, 440-444.
Karp, P., T. E. Katila, M. Saarinen, P. Siltanen, and T. Varpula (1980). The normal human magnetocardiogram II: A multipole analysis. Circ. Res., 47, 117-130.
Karp, P. (1981). Cardiomagnetism. In S. Erné, H.D. Hahlbohm, and H. Lübbig (Ed.), Biomagnetism. DeGruyter, Berlin. pp. 219-258.
Katila, T., and P. Karp (1983). Magnetocardiography: morphology and multipole presentations. In S. J. Williamson, G.-L. Romani, L. Kaufman, and I. Modena (Ed.), Biomagnetism. An interdisciplinary Approach. Plenum Press, New York.

pp. 237-263.

Katila, T., T. Poutanen, M. Seppänen, T. Varpula and P. Siltanen (1983). On the ECG and MCG changes in myocardial infarction. X International Congress on Electrocardiology, Bratislava.

Leifer, M., N. Capos, J. Griffin, and J. Wikswo (1983). Atrial activity during the PR-segment of the MCG. Nuovo Cimento, 2D, 266-279.

Leifer, M., J. Griffin, E. Iufer, J. Wikswo, W. Fairbank, and D. Harrison (1981). An intergrated system for magnetic assessment of cardiac function. In S. Erné, H. D. Hahlbohm, and H. Lübbig (Ed.), Biomagnetism. DeGruyter, Berlin, pp. 123-137.

Leoni, R., and G.-L. Romani (1982). Computer simulation of magnetic field patterns from the human cardiac conduction system. Nuovo Cimento, 1D, 737-750.

Lorenzana, H. E., P. B. Pipes, M. P. Zaitlin, and D. James (1984). Structures in the P-R interval of the human magnetocardiogram. V World Conference of Biomagnetism, Vancouver.

Nousiainen, J. J. O., J. O. Lekkala, and J. A. V. Malmivuo (1984a). Comparative study of the normal vector magnetocardiogram and vector electrocardiogram. Tampere University of Technology, Electronics Laboratory; Research Report No. 3 / 1984, Tampere.

Nousiainen, J., J. Lekkala, and J. Malmivuo (1984b).Statistical analysis of the relationship between the VMCG and VECG. V World Conference of Biomagnetism, Vancouver.

Opfer, J. E., Y. K. Yeo, J. M. Pierce, and L. H. Rorden (1974). A superconducting second-derivate gradiometer. IEEE Trans. Mag., MAG-10, 536-539.

Riff, D. P., and R. A. Carleton (1971). Effects of exercise on the atrial recovery wave. Am. Heart J., 82, 759-763.

Saarinen, M., T. Katila, J. Nenonen, M. Seppänen, and P. Siltanen (1984). Simulated and measured magnetocardiograms in localized myocardial abnormalitis. V World Conference of Biomagnetism, Vancouver.

Saarinen, M., P. J. Karp, T. E. Katila, and P. Siltanen (1974). The magnetocardiogram in cardiac disorders. Cardiovasc. Res., 8, 820-834.

Saarinen, M., P. Siltanen, P. J. Karp, and T. Katila (1978). The normal magnetocardiogram: I Morphology. Ann. Clin. Res., 10, Suppl. 21, 1-43.

Safonov, Y. D., V. M. Pravotorov, V. M. Lube, and L. J. Yakimenkov (1967). Method of recording the magnetic field of the heart (magnetocardiography). Bull. Exp. Biol. Med., 64, 1022-1024.

Scherlag, B. J., S. Law, R. H. Helfant, W. D. Berkovitz, E. Stern, and A. N. Damato (1969). Catheter technique for recording His bundle activity in man. Circulation, 39, 13-18

Siltanen, P. (1985). Magnetocardiography. In P. W. McFarlane, and T. D. W. Lawrie (Ed.), Electrocardiography. Baillière Tindall, London, Chap. 37.

Siltanen, P., T. Poutanen, T. Katila, M. Seppänen, and T. Varpula (1983). Comparison between measured and simulated magnetocardiograms in a case of antero-lateral myocardial infarction. Nuovo Cimento, 2D, 301-310.

Siltanen, P., M. Saarinen, T. Katila, and J. Ahopelto (1974). Magnetocardiogram in healthy men and women. Proc. VII World Cong. Cardiol., Buenos Aires.

Simson, M.B. (1981). Identification of patients with ventricular tachycardia after myocardial infarction from signals in the terminal QRS complex. Circulation, 64, 235-242.

Sodi-Pallares, D., G. A. Medrano, A. DeMicheli, M. R. Testelli, and A. Bisteni (1961). Unipolar QS morphology and Purkinje potential of the free left ventricular wall: the concept of electrical endocardium. Circulation, 23, 836-845.

Tripp, J. H. (1983). Physical concepts and mathematical models. In S. J. Williamson, G.-L. Romani, L. Kaufman, and I. Modena (Ed.), Biomagnetism. An Interdisciplinary Approach. Plenum Press, New York. pp. 101-139.

Tripp, J. H., and D. E. Farrell (1981). Theory of the PR segment of the human magnetocardiogram. In S. N. Erné, H. D. Hahlbohm, and H. Lübbig (Ed.),

Biomagnetism. DeGruyter, Berlin. pp. 259-271.
Varpula. T., T. Katila, T. Poutanen, and M. Seppänen (1984). In vivo study of the secondary currents of the MCG. V World Congress of Biomagnetism, Vancouver.
Wellens, H. J. J., and D. R. Dürrer (1976). Observations on mechanisms of ventricular tachycardia in man. Circulation, 54, 237-247.
Wikswo, J. P., J. A. V. Malmivuo, W. H. Barry, M. C. Leifer, and W. M. Fairbank (1979). The theory and application of magnetocardiography. Adv. Cardiovasc. Phys., 2, 1-67.
Williamson, S., and L. Kaufman (1981). Magnetic fields of the cerebral cortex. In S. N. Erné, H. D. Hahlbohm, and H. Lübbig (Ed.), Biomagnetism. DeGruyter, Berlin. pp. 353-402.
Zimmerman, J. E., and N. V. Frederick (1971). Miniature ultrasensitive supraconducting magnetic gradiometer and its use in cardiography and other applications. Appl. Phys. Lett., 19, 16-19.

MAGNETOCARDIOGRAPHY:
CHALLENGING CLINICAL PROBLEMS AND PROMISING ANALYTIC TECHNIQUES

[1]S.N. Erne, R.R. Fenici[2], T.E. Katila[3], P. Siltanen[4], and J.P. Wikswo, Jr.[5]

[1]Physikalisch-Technische Bundesanstalt, Institut Berlin, Berlin, F.R.G.
[2]Dept. of Clin. Physiol., Catholic University of the Sacred Heart, Rome, Italy
[3]Dept. of Technical Physics, Helsinki Univ. of Technology, Helsinki, Finland
[4]Cardiovascular Laboratory, Helsinki Univ. Central Hospital, Helsinki, Finland
[5]Dept. of Physics and Astronomy, Vanderbilt Univ., Nashville, Tennessee, U.S.A.

ABSTRACT

This paper presents the views of the panel moderator and the four panelists regarding a critical assessment of the MCG research of the past and the possibilities for future research applying the MCG to challenging problems in basic and clinical cardiac electrophysiology.

KEYWORDS

Magnetocardiogram, electrocardiogram, cardiac electrophysiology

INTRODUCTION BY THE MODERATOR - J.P. WIKSWO, JR.

Since its introduction in the 1960's, the field of magnetocardiography has not been as exciting or productive as many investigators might have expected. A testimony to this fact is that the only MCG study regularly quoted in the conventional cardiac electrophysiology literature is the detection of the diastolic injury current by Cohen and Kaufman (1975). There are several possible explanations for the lackluster performance of MCG research. Much of the work so far has had to be directed towards instrument development. Complex problems in magnetometer design, noise attenuation or cancellation, and data acquisition have had to be solved. Mathematical models already developed for the ECG had to be extended to the MCG, and many approximations and mathematical techniques already justified for the ECG had to be reexamined with respect to the MCG. Possibly a fundamental reason for the lack of exciting results from the limited number of MCG studies conducted to date is that the majority of these studies have been made on either normal subjects or patients whose ECGs are well understood. Since the clinical understanding of the ECG is a well-developed art, these studies have produced no surprising or exciting results. The instrumentation required to record the simplest MCG is more than an order of magnitude more costly, complex and difficult to use than an ECG instrument producing the same quality of signals. From the perspective of a clinical diagnostician, these factors are compelling reasons not to invest the man-years of effort required to develop and validate a set of standardized MCG waveforms that would merely compliment, at the best, the long established set of ECG waveforms that provide the basis of clinical cardiology.

In all fairness to the skilled researchers who have examined the MCG over the past two decades, such a critical review of magnetocardiography is best followed by a detailed discussion of how our past experience with the MCG can guide physicists and physicians towards new, challenging, and productive areas of MCG research. Whether or not the MCG and ECG contain different information, the two techniques clearly have somewhat different capabilities and either technique might be applied to solve problems that are the focus of present interest. One way to ensure that few surprises will be found is to conduct research in barren areas of cardiology where few exciting results remain to be discovered. Alternatively, possibly the best way to ensure new and exciting MCG results is to concentrate one's efforts on those challenging problems where either cardiac electrophysiology or the interpretation of the ECG is poorly understood and is still the source of major controversy. This, however, will undoubtedly require recording the MCG from critically ill patients, recordings which must be performed within the hospital. Only one or two research groups in the world are so equipped.

For our panel discussion and this written summary, I have challenged the cardiologists on our panel to attempt to identify those areas of clinical and basic cardiac electrophysiology that are presently uncharted or embroiled in controversy and might also be well suited for study with the MCG. I have challenged the physicists to discuss how new measurement and analysis techniques might best be applied to address such problems.

REPLY BY R.R. FENICI

I agree that magnetocardiography should be used to study both basic cardiac electrophysiology and clinical situations where conventional electrocardiography has failed to give information of diagnostic interest. On the other hand, my firm opinion is that in order to interpret pathologic data, a reference goldstandard of normality has to be provided. Following this guideline, I've worked in parallel on two lines of research during the last two years: 1) to determine whether the best level of resolution achievable in our unshielded hospital laboratory is good enough to allow the use of magnetocardiography for the non-invasive study of phenomena of electrophysiological interest under clinical conditions; and 2) to set up a biomagnetism laboratory adequate for the study of critically ill patients. In theory these targets seem very simple but in reality it has been very difficult, at least in our institution, to get the necessary permissions and appropriate instrumentation to be able to start the job!

With respect to the first line of research, although the specific software has to be refined and fully adapted to our new computer, I'm rather satisfied by the first results, summarized in our accompanying paper (Fenici and colleagues, 1984). I think we have demonstrated that if the overall level of electromagnetic noise does not exceed 50 $fT/\sqrt{Hz}$, magnetocardiography can be used adequately to study non-invasively in humans the electrical activity even of small cardiac structures like the His-Purkinje conduction system (HPS). We decided to begin studying the HPS because we could assume, at least under physiological conditions, that this structure could be considered the simplest *in vivo* model of a microscopic cardiac source generating a magnetic field with only minor differences between individuals. In order to validate the results of the hospital measurements we did the same kind of study on normal subjects in the Berlin shielded room, in collaboration with Erne and his coworkers (Erne and colleagues, 1984a). The quality of the Berlin measurements is obviously better, but the results are substantially identical. Now we are moving to study specific pathologies. This can be done only in the hospital and unfortunately we are too late to have data for this conference!

With respect to the problems that represent challenges for the clinical cardiac electrophysiologist and could be addressed by the biomagnetic method, I think that it would be useful to attempt to apply magnetocardiography to a non-invasive study of the electrogenic mechanisms of cardiac arrhythmias (Hoffman and Rosen, 1981), with particular regard to the life-threatening ventricular arrhythmias due to coronary artery disease or to other cardiac abnormalities. At present, although the electrogenic mechanisms of cardiac arrhythmias have been deeply investigated both in vitro and in the intact animal, neither surface electrocardiography nor invasive study are effective in diagnosing in individual patients whether a ventricular arrhythmia is due to altered automaticity, re-entry phenomena, or both mechanisms acting simultaneously. This is even more difficult after taking into account that recently triggered activity (a special kind of abnormal pacemaker activity which is not spontaneous, but initiated by previous spontaneous or stimulated beats) has been demonstrated in myocardium removed from patients with atrial and ventricular arrhythmias (Mary-Rabine and colleagues, 1980; Gilmour and colleagues, 1983) and, therefore, is considered one more mechanism possibly active in clinical arrhythmias. Whether or not magnetocardiography will improve our diagnostic power in this field is unknown. However, four observations suggest that in principle it could be possible.

First, from electrophysiology we know that ventricular arrhythmias due to a micro re-entry mechanism are caused by local conduction disturbances which induce delay of depolarization of small myocardial areas close to each other. The electrocardiographic expression of this phenomenon is the presence of delayed depolarizations (also called "late potentials") or even a "continuous electrical activity" after the end of the QRS and along the S-T segment. These pathological signals, generated within the damaged area where the arrhythmia most likely originated, are usually of subendocardial origin. Their detection from the body surface is possible with ECG techniques. However, comparative investigations with invasive techniques have demonstrated that surface ECG recordings cannot give any reliable information about the localization of the related myocardial lesion. Moreover, it is not uncommon to observe false negative surface recordings in patients with endocardially documented late potentials. This is easily explainable by the possibility that the geometry of the recording leads is inadequate in individual cases for detecting the spatial orientation of the late potential vectors. In fact, the MCG has already proven useful even in discriminating false positive electrical recordings (Erne, 1984b). In these cases the MCG could be more sensitive. Moreover, MCG mapping could provide a three-dimensional localization of the pathologic source. This is a typical case for which a solution by a simple dipole model would probably be inadequate. Perhaps a current multipole solution would be better.

Second, a re-entry phenomenon could also be determined, in theory, by the presence of a significant non-uniformity of repolarization between different myocardial areas. A pathological dispersion of ventricular repolarization, therefore, is considered suggestive of the presence of ventricular vulnerability to lethal arrhythmias. On the basis of interesting results of experimental animal studies, computer analysis of surface electrical recordings in the form of isoarea maps has been proposed for several years as a method for non-invasive detection of ventricular vulnerability to serious arrhythmias. The method has not yet been applied extensively to clinical use despite the fact that it is still difficult in many cases to distinguish between repolarization abnormalities which, although apparently similar, implicate a completely different prognosis. The MCG could be an independent method for studying this problem.

Third, triggered activity is a peculiar sort of non-automatic impulse generation caused by after-hyperpolarization, which can occur either before the repolarization of a previous normal action potential is completed (early after-depolarization) or after repolarization is completed (delayed after-depolarization) (Hoffman and Rosen, 1981). On the basis of experimental and clinical observation it has been suggested that both kinds of after-depolarizations can, upon reaching threshold, induce sustained ventricular rhythms and can be responsible, as an example, for the ventricular arrythmias observed during re-perfusion following a prolonged transitory myocardial ischemia as with coronary artery spasms. The same can occur in other situations that induce an intracellular calcium overload. At present, there are neither invasive nor surface methods to detect such arrythmiagenic mechanisms in the clinic, although after-depolarizations are signals that should be detectable during diastole. As MCG resolution has been proven to be good enough to detect fields as weak as 200 fT even in unshielded hospital laboratories (Fenici and colleagues, 1984), this method could be used to scan the diastolic phase in order to find subthreshold oscillatory activity.

Fourth, in high risk patients with ventricular arrythmias refractory to medical treatment, surgical ablation of the arrythmiagenic myocardium can be attempted. For effective surgery, a precise localization of the arrythmiagenic focus is necessary. At present preoperative endocardial catheter mapping is used to guide the surgeon and reduce the time of localization procedures during surgery. The capability of the MCG for non-invasive localization of intracardiac sources could perhaps be useful in approaching this problem.

There are undoubtably many additional diagnostic problems in which the MCG could be used to overcome the limitations of electrical measurements that have proven unsatisfactory for clinical use.

REPLY BY P. SILTANEN

I agree entirely with the moderator's statement regarding the comparison of the MCG with the ECG. Apparently we try to read the MCG as if it were written in the same language as the ECG. Undoubtedly this is due to the long-established clinical ECG tradition. Confounding the problem is the fact that the MCG waveform looks like that of the ECG, with the same deflections and time relationships. Our equalizing these two methods is justified to some extent, because both of them depict the same bioelectric phenomena, but in interpreting them one has to keep in mind that the ECG is a record of voltages while the MCG is a record of currents.

Personally, at the beginning our our studies I had expected many more quantitative and qualitative differences between ECG and MCG waveforms than I have observed from patients with cardiac diseases. In spite of this disappointment, I still believe that important "new information" is concealed in the traditional PQRST waveform presentation of the cardiac magnetic field, hidden only because clinical experience with the MCG is so far very limited and a perspective about specific patterns of the MCG in clinical cases has not yet developed. For example, the possible clinical usefulness of the well-known sensitivity of the MCG for tangential propagation of activation cannot be evaluated on the basis of only some 200 abnormal MCG's published in the literature. According to my view, the usefulness of the traditional morphological and quantitative analysis of MCG maps will be resolved only after thousands of abnormal tracings have been recorded and analyzed. This was also the case in electrocardiography. Therefore, we are not yet prepared to judge this type of approach in magnetocardiography. The same applies also to magnetic vectorcardiography. Presently, MCG isofield maps seem to offer a new dimension in electrocardiology that cannot be reached by the ECG, and which render it possible to determine roughly the location of the sources.

I also agree with the importance of MCG investigation of cases where the interpretation of ECG is controversial or in which the bioelectric process is not understood. I can list at least ten ECG problems in this category.

1. Atrial Depolarization: There is still much controversy on the route and sequence of atrial depolarization, especially in P wave abnormalities due to different atrial overloadings. The role of three preferential pathways and the Bachmann's bundle is still more or less unclear, and their selective dysfunction might explain the various ECG patterns of atrial hypertrophy and/or dilatation. Due to rather tangential propagation of the excitation wave in the shell-like atrial musculature, the MCG might be more apt for its external recording and analysis than the ECG.

Chaotic atrial activity, called "atrial fibrillation," can easily be diagnosed by the ECG, but the propagation of excitation in the fibrillating atria cannot be traced with the ECG. In practice this is not important, but theoretically the problem still persists as to whether this arrhythmia (and some other atrial arrhythmias too) is unifocal in its origin or is due to circus movement. For the same reason as mentioned in the preceding paragraph, the MCG could contribute to solving this problem.

2. P-R Segment Phenomena: Whether the "ramp pattern" of the P-R segment in high-resolution records is due to atrial repolarization, to His-Purkinje signals, or to their superposition is still a matter of controversy. Several excellent and controversial papers on this topic are being presented in these proceedings. There is evidence, however, that at least some proportion of the magnetic field during the P-R segment has its origin in the activation of the His-Purkinje system, and that anatomical and functional abnormalities in this system may be to some extent traced by the MCG in a totally non-invasive manner. This type of approach yields information about the subendocardium, which is subtle to examine electrically. The surface ECG demonstrates only indirectly some gross abnormalities of HPS. If MCG analysis of the HPS can be introduced into clinical practice in the future, it will represent a new and important tool for investigation of conduction disturbances.

3. Subendocardial Activation: The ECG does not adequately trace the subendocardial activation of the left ventricular wall. On the other hand, the left ventricular subendocardium is the most commonly jeopardized part of the myocardium in various pathological conditions. Theoretically, it is probable that the MCG does not contribute to the information on subendocardial activation, but it is possible that to some extent it could be different from the ECG as regards the QRS build-up at the very onset of ventricular activation. Comparative ECG and MCG investigations should be carried out in cases with subendocardial lesions in order to evaluate the importance of the MCG in this special situation.

4. Pre-excitation of Ventricles (Wolf-Parkinson-White): Ventricular pre-excitation is due to precocious propagation of excitation from the atria to the ventricles via an abnormal atrio-ventricular muscle connection. It can be diagnosed easily with the aid of the surface ECG, but the MCG may help in more accurate localization of the anomalous bundle in case of surgery, which is needed in some instances.

5. Non-uniform Double-layer Activation: This can be encountered in cases where part of the conductive tissue (His-Purkinje) or of the ventricular muscle wall is replaced by patchy or segmental fibrous tissue. Theoretically, it is probable that the MCG in these cases could demonstrate information weighted differently than in the ECG. The possible usefulness of the MCG in demonstrating intra-

ventricular conduction abnormalities and the ratio of fibrous tissue to muscle is still unsettled.

6. Ventricular Re-entry Potentials ("Late Potentials"): The ventricular re-entry phenomenon is an important cause of malignant ventricular arrhythmias, which often lead to sudden death. Sudden cardiac death is an important and not uncommon cause of death, and it represents one of the major challenges in cardiology. Re-entry signals may appear both in the ECG and in the MCG as small deflections after the QRS complex, usually discernible in high resolution recordings as notches and bumps in the ST-segment or T-wave. The MCG can be used for detection of the tissue area responsible for generation of the late potentials in order to use surgery in cases resistant to drug therapy. Excision of the re-entry loop may sometimes be mandatory.

7. Diastolic Depolarization ("After Potentials"): Slow depolarization of a damaged ventricular muscle segment may sometimes occur early in diastole, triggered by normal ventricular activation. In some conditions this may lead to malignant ventricular arrhythmias. One is known by name of "torsades de pointes," ventricular tachycardia which usually is found as a side-effect of some antiarrythmic drugs. This type of arrythmia may also result in ischemia induced by coronary artery spasm. After potentials can be seen in the ECG following the T-wave as a pronounced U-wave-like deflection. As a local current it could be better traced and localized by using the MCG. Demonstrating its anatomical origin is important in recurrent cases, because surgical treatment may be necessary in cases resistant to other therapy.

8. ST-segment Shifts: There are primary ST-shifts and ST-shifts secondary to TP-segment shifts. Distinguishing between these diastolic and systolic currents is not possible with the surface ECG, but under certain conditions it is possible by using the MCG. The development of the MCG techniques that are required for DC recordings but are yet simple enough for clinical use would be of great importance for differentiating between ischemic and non-specific ST-segment shifts.

9. T-wave Abnormalities: The greatest differences between the ECG and MCG complexes exist during the repolarization phase of the cardiac cycle. The T-wave belongs to the terminal part of repolarization and is often differently displayed by the MCG and ECG. The T-wave abnormalities in the ECG are often puzzling and may be influenced by different physiological or pharmacological interventions. A small proportion of these abnormalities are malignant and are associated with an increased risk of sudden death. It can be anticipated that the MCG could contribute in distinguishing between benign and malignant T-wave abnormalities.

10. U-wave: The U-wave is generally more conspicuous in MCG waveforms than in ECG ones, and rather large U-waves can be found in some cases of cardiac abnormalities. The clinical significance of the U-wave is obscure and its genesis is more or less unknown. Therefore, analysis of the U-wave in different cardiac conditions would be a challenging problem to be addressed with the MCG.

A distinguishing property of the MCG is the ability to localize sources with a reasonable accuracy, which is not possible with a surface ECG. Development of MCG recording devices should be focused on utilizing this special property, because in cardiology a general trend at the present is towards three-dimensional imaging of cardiac structures and events. So far bioelectric phenomena have not been presented in this fashion. Vectorcardiography is the only effort in this direction, but it offers very general and undetailed information which is not suitable for demonstration of local sources. Multiple detectors would be necessary for satisfactory analysis of transient, temporary fields of short duration. Therefore, development of a system of twenty to thirty detectors for simultaneous recording would be mandatory.

REPLY BY S.N. ERNE

In the last year we have concentrated our activity towards examining the cardiological problems associated with predicting life-threatening arrythmias which are the major cause of sudden cardiac death. This is an area of clinical cardiology where electrical measurements are poor and there exists a need for new methods. There are two reasons why the MCG might contribute to the solution of these problems. First, epicardial and endocardial measurements suggest that fractionated ventricular activity in the ST segment will be a marker for the high risk of sudden death. Secondly, electrical measurements of such late potentials have not produced encouraging results because averaging techniques increase the number of false negatives due to the intrinsic instability of the major components of the late potentials (Erne, 1982), because beat-to-beat recording techniques result in an increase in the number of false positives due to the intrinsic patient noise, and because specialized algorithms such as the Simson method (Erne and colleagues, 1984c) do not offer sufficient improvement.

Because there is a lack of models for arrythmia sources, we have developed an empirical MCG method (Erne, 1983) that promises to discriminate high-risk patients on the basis of the time evaluation and structure of the isofield contour maps in the ST-segment. Our first tests on 19 patients lead us to be optimistic.

In the special case when tachycardia is a consequence of the WPW syndrome, the source of the disorder is known and can be modeled. It is then possible to apply the procedure developed for MCG studies of the normal cardiac conduction system to the problem in WPW of isolating the signals produced by the accessory pathway and finding the location of the pathway. This could be a good application of the MCG because this localization is needed by the surgeon and it is impossible to obtain it by body surface electrical measurements. The technique has been applied to a few cases of WPW, and recent intracardiac electrical measurements in Berlin have provided the first confirmation of the magnetic determination.

There are in my opinion promising applications of the MCG on the basis of actual instrumentation, data evaluation, and modeling. The most important restriction of the MCG in pathological studies is the impossibility of making maps of transient phenomena. For this reason, it will be of fundamental importance to develop multisensor systems.

REPLY BY T.E. KATILA

It is certainly correct to say that the ECG is a well understood art and, as compared with the ECG, the MCG has not produced great surprises. When you consider how much more difficult the MCG is to measure than the ECG, it is fully understandable that it did not come into standard clinical use in spite of the use of SQUIDs. The MCG would be superior for mass screening only with a big multisensor system. Ideally this system should be usable in every clinic without magnetic shielding. It is worthwhile considering the possibility of building such a system in the future, or even of building a similar one that required using magnetic shielding.

Until now the ECG (and also the MCG) have had an almost purely empirical basis. The reading of the 12-lead ECG is based on the long experience of the clinicians, and even the computer programs serving the same purpose are primarily "data banks." In comparison to the ECG, the MCG has been studied comparatively little. Most people working on the MCG seem to believe that the sensitivities of both methods are rather similar. Extensive routine studies will be needed before we will have as much experience with the MCG as is available for the ECG. To my knowledge the largest number of normal MCGs has been acquired by our group in Helsinki, and even so, that is only for about 50 persons. Only very small groups of pathological cases have been studied.

Clearly another approach to the development of MCG research is not to compete with the ECG in general, but to seek cases solvable especially by use of the magnetic method. Magnetic measurements have clearly been successful in one type of problem: the location of tangential current sources in the brain. This is important because, especially for humans, there are not many methods for obtaining information on the location of the activity in the brain at the millimeter level. Due to the structure of the cortex, the very simple current dipole approximation has somehow worked. The same method is now used for His-Bundle studies and it is clearly useful also for late potentials. In some of these cases the data acquisition can be simplified by use of an external trigger. The signals mentioned can often be well separated from other signals due to the cardiac electric activity. This makes the localization problem easier (e.g. the late potentials may be in an otherwise silent section between QRS and T), but unfortunately such signals are small in magnitude and their study requires the best sensitivities available. With these qualifications, the MCG clearly gives highly desired information on localization of cardiac current sources of clinical interest.

We also have tried something similar in studies of small localized myocardial infarctions. The difference in the signals between the pre- and post-infarction cases should contain information not only on the size but also on the localization of the infarction area. To a certain extent this information can be obtained from the ECG, but according to our studies the MCG gives valuable additional knowledge (Saarinen and coworkers, 1984). One problem is that the difference signal is not readily available since few people have had their magnetocardiograms measured prior to an infarction. Thus we have to use the normal MCG for subtraction. This causes considerable uncertainty in the difference maps. Our work also demonstrates the value of a reliable computer model for simulating the MCG and the ECG changes (Geselowitz, 1980). The computer calculations support the view that the MCG is efficient in localizing the sources.

In vectorcardiography, the current dipole source is deduced from ECG measurements and is used to explain the temporal and spatial dependences of the ECG. We have studied the use of the current multipole expansion, termed the CME (Katila and Karp, 1983), to extract from MCG measurements both the same current dipole term and the higher order terms. Especially during the QRS complex there seems to be a discrepancy between the simple dipole model and the MCG measurement results with respect to the localization of the source. Even the best computer models presently available give a near-zero signal at a location directly above the heart, whereas the experimental signal recorded at that location is nearly at its maximum. This "shift of the zero-crossing line" cannot, according to our results, be explained solely with the addition of a surface contribution (Varpula and coworkers, 1984). This may indicate the capability of the MCG to measure vortex sources essentially not seen in the ECG. However, the dipole term of the CME, as obtained from magnetic measurements, resembles closely the corresponding components of the cardiac electric vector obtained from electric measurements. Such tests should provide information on the interdependence of the sources of the ECG and MCG.

REFERENCES

Cohen, D., and A.L. Kaufman, (1975). A magnetic determination of the relationship between the S-T segment and the injury produced by coronary occlusion. Circ. Res., 36, 414-424.

Erne, S.N., R.R. Fenici, H.-D. Hahlbohm, H.P. Lehman, E.R.v. Leitner, and M. Oeff, (1982). Beat to beat surface recording of electrical late potentials in man., Proc. of the Ninth Int. Congress on Electrocardiology, Japan Heart J., 23, suppl. 1, 703-705.

Erne, S.N., R.R. Fenici, H.-D. Hahlbohm, W. Jaszczuk, H.P. Lehman, M. Masselli (1983). High resolution isofield mapping in magnetocardiography. Il Nuovo Cimento 2D, 291-300.

Erne, S.N., R.R. Fenici, H.-D. Hahlbohm, J. Korsukewitz, H.P. Lehmann, and Y. Uchikawa, (1984a). Magnetocardiographic study of the PR segment of normals, these proceedings.

Erne, S.N., H.-D. Hahlbohm, J. Korsukewitz and H.P. Lehman, (1984b) Discriminating late potentials using biomagnetic techniques, these proceedings.

Erne, S.N., H.-D. Hahlbohm, J. Korsukewitz, H.P. Lehman, (1984c). The evaluation of normal subjects for late potentials at the end of the QRS complex. Proc. of the Eleventh Int. Congress on Electrocardiology, in press.

Fenici, R.R., M. Masselli, S.N. Erne, and H.-D. Hahlbohm, (1984). Magnetocardiographic mapping of the P-R Interval phenomena in an unshielded hospital laboratory, these proceedings.

Geselowitz, D.B., (1980). Computer simulation of the human magnetocardiogram. IEEE Trans. Magn. Mag-16, 812-817.

Gilmour, R.F., Jr., J.J. Heger, E.N. Prystowsky and D.P. Zipes, (1983). Cellular electrophysiologic abnormalities of diseased human ventricular myocardium, Am. J. Cardiol., 51, 137-144.

Hoffman, B.F. and M.R. Rosen, (1981). Cellular mechanisms for cardiac arrythmias, Circ. Res., 49, 1-15.

Katila, T. and P. Karp, (1983). Magnetocardiography: Morphology and multipole presentations. In S.J. Williamson, G.-L. Romani, L. Kaufman and I. Modena (Eds.) Biomagnetism. Plenum Press, New York, pp 237-263.

Mary-Rabine, L., A.J. Hordof, P. Danilo, Jr., J.R. Malm, and M.R. Rosen, (1980). Mechanisms for impulse initiation in isolated human atrial fibers, Circ. Res. 47, 267-277.

Saarinen, M., T. Katila, J. Nenonen, M. Seppänen, and P. Siltanen, (1984). Simulated and measured magnetocardiograms in localized myocardial abnormalities, these proceedings.

Varpula, T., T. Katila, T. Poutanen, and M. Seppänen, (1984). In vivo study of the effect of the secondary currents on the MCG, these proceedings.

SIGNAL ANALYSIS OF MAGNETOCARDIOGRAMS AND ELECTROCARDIOGRAMS TO TEST THEIR INDEPENDENCE

C. E. MacAulay*, G. Stroink*, and B. M. Horacek**

*Department of Physics,
**Physiology and Biophysics, and Medicine
Dalhousie University Halifax, Nova Scotia, Canada B3H 3J5.

ABSTRACT

To study the intrinsic electrocardiogram (ECG) and magnetocardiogram (MCG) relationship, ECG and MCG recordings of five normal subjects were studied. For each subject 120 simultaneous ECG recordings and 56 sequential MCG recordings were obtained. For both the ECG and the MCG families of signals of each subject, the data were organized into sets of temporal signals and into sets of spatial maps. For each of the 20 sets of data the principal components of the signals were represented in an orthogonal signal space by sets of eigenfunctions. The eigenfunctions could then be used to fit individual ECG/MCG signals or maps. The results revealed: 1) the integrity of both the ECG and MCG data; 2) a relatively poor ability of ECG temporal eigenfuntions to fit MCG data; 3) a relatively good ability of MCG temporal eigenfunctions to fit ECG data; 4) the relative uniformity among the ECG spatial data and eigenfunctions of the subjects; 5) the relatively large differences between the MCG spatial data and eigenfuctions of the subjects.

KEYWORDS

Magnetocardiogram, electrocardiogram, eigenfunction analysis, data comparison, isofield maps, isopotential maps.

INTRODUCTION

Since the ECG and the MCG are both the result of the electrical activity of the heart, their independence/dependence is a question which arises naturally. The current distribution in the conductive body, generated by the cardiac electrical activity, can be represented as the sum of an impressed current density (primary sources) and an induced current density (secondary sources). Both current densities are involved in the genesis of the ECG and the MCG. The impressed current distribution can be expressed (using the Helmholtz equation) as the sum of a vortex distribution and a flow distribution. The independence of the MCG and the ECG would be assured if the impressed vortex and the impressed flow distributions (Wikswo, 1983) were independent. In this work we present experimental evidence concerning the intrinsic dependence/independence of the ECG and the MCG based on the analysis of ECG and MCG signals in five subjects.

METHODS

For each subject, we: 1) Recorded 120 simultaneous body surface ECGs of 15 seconds duration (Horacek and co-workers, 1977); 2) Collected MCG recordings, of 30 seconds in duration, sequentially at 56 locations (eight rows of seven, one inch grid spacing) in the precordial region. The MCG measurements were performed in an eddy current shielded room (Stroink and co-workers, 1981,1983), using a CTF asymmetric second order gradiometer with an ambient background noise of 26fT/√Hz. The gradiometer system was balanced using a four square coil system that produces a uniform field (Meritt and co-workers, 1983). Also used, for the first subject only, was an inhouse symmetric second order gradiometer having a noise level of 60fT/√Hz. The MCG sampling frequency was 500 Hz and the bandwidth was 0.15 Hz to 125 Hz. No other filters were used on the MCG data. 3) Each recording was averaged off-line to produce a waveform with a superior signal to noise ratio and then subdivided into two parts, the QRS waveform and the STT waveform, to facilitate the calculations which followed.

The ECG and MCG recordings contain a substantial amount of redundant information. The redundancy reduction of the data in body surface potential maps (BSPM) has received attention in recent years (Lux and co-workers, 1981; Evans and co-workers, 1981). Using the temporal compression method detailed by Evans and co-workers (1981), the four sets of data for each subject were reduced to 4 sets of 15 optimal basis functions/eigenfunctions (eigfns). A spatial compression method detailed by Lux and co-workers (1981) was used to perform a similar spatial reduction on the four spatial data sets (2 QRS and 2 STT) of each subject. The number and location of the ECG leads making up the the spatial data sets were constrained so that they covered the same section of the torso as the 56 MCG locations.

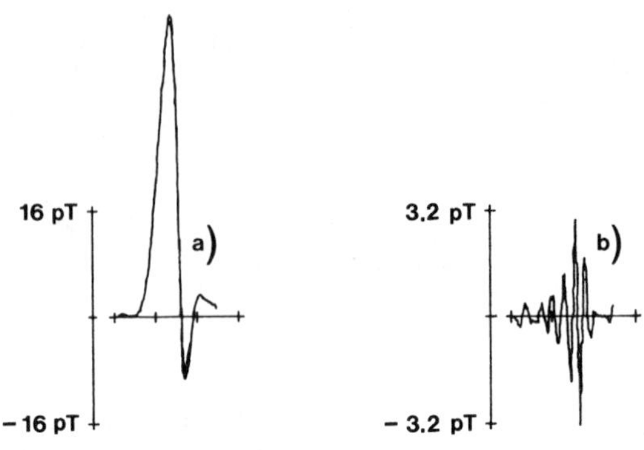

Fig. 1 A plot of a) a MCG QRS signal and a similar reconstructed signal, and b) the difference between the two signals in part a).

The eigs generated by these reduction processes represent the set of waveforms (temporal reduction) or the set of body surface distributions (spatial reduction) which can best describe the information of the original data sets. These eigfns were used to reconstruct the various waveforms or distributions of individual signals of the same sets or of other sets. In Fig. 1 part a) are displayed an averaged MCG signal and a similar signal reconstructed from a set of ECG eigfns and in part b) is displayed the difference between the two signals of part a). A reconstructed waveform or distribution is a weighted sum of the appropriate 15 basis functions. Different reconstructed signals can be obtained by using different coefficients in the weighted sum.

Some definitions are necessary before proceeding to the data reduction results. The normalized error total (NET) for a set of signals fit by a set of eigfns is the sum of the squared (value by value) differences between the individual regenerated signals and the recorded signals, normalized by the sum of the recorded signals squared. The agreement obtained when a signal set is fitted by the basis functions generated from the same signal set is called the self-fit. The agreement obtained when the family (MCG or ECG) of signals of a subject is fitted by the eigfns generated from the other family of the same subject is called the cross-fit (e.g. the cross-fit for a MCG QRS signal set of a subject is the agreement obtained when the MCG QRS signals are fitted by the eigfns generated from the ECG QRS signals of the same subject). The agreement obtained when a set of signals of one subject is fitted using the eigfns generated from a similar set of another subject is called the type-fit (e.g. a QRS MCG type-fit is the ageement obtained when a QRS MCG signal set of one subject is fitted using the eigfns generated from the QRS MCG signal set of another subject).

TEMPORAL RESULTS AND DISCUSSION

We found that at least 13 to 15 eigfns per set are necessary to reduce the NET of the self-fits of the MCG sets to the same level as the signal noise, and hence describe all the information found in the signal sets. For the self-fits of the ECG sets it appears that 13 eigfns per set are more than enough to reduce the NET to the level of the signal noise. This is probably due to the high background noise of the ECG recordings (10 bit A/D conversion and 15 sec averaging) relative to the MCG recording`s noise level (12 bit A/D conversion and 30 sec averaging). To keep the ECG and MCG situations as equivalent as possible, 15 eigfns were always used. Using the extra (last 3 or 4) eigfns of the ECG sets should not invalidate the MCG-ECG comparison since the other ECG eigfns contain all the information present in the ECG signal sets.

The average temporal NET for the self-fit of the five subjects was 2×10^{-5} for the MCG QRS sets, 1×10^{-5} for the ECG QRS sets, 30×10^{-5} for the MCG STT sets, and 27×10^{-5} for the ECG STT sets. The average cross-fit NET for the five subjects was: 19×10^{-4} for the MCG QRS signal sets, 8×10^{-4} for the MCG STT signal sets, 45×10^{-4} for the MCG STT signal sets and 15×10^{-4} for the ECG STT signal sets. The cross-fit NETs are significantly larger than the self-fit NETs, indicating some ECG and MCG signal independence. Also, from the cross-fit results, we see that on the average the MCG basis functions are better able to regenerate the ECG signal sets (resulting in a lower average cross-fit NET) than the ECG basis functions are able to regenerate the MCG signal sets. At the individual subject level, this was also true for all the temporal MCG-EGC signal set pairs except one. (See Table 1 for the individual subject temporal cross-fit results.) The subject to whom this did not apply had a partial right bundle branch block (RBBB). The unique cross-fit results of this subject may be due to his unique heart physiology or to the natural variability of the results themselves. To determine this point a study involving more subjects is clearly necessary.

SPATIAL RESULTS AND DISCUSSION

The spatial self-fit results were similar to the temporal results with 15 basis functions being more than enough to reduce the NET to less than the level of the signal noise. Since there is an inherent 90° rotation between the magnetic and the electric fields of current dipoles the form of, and the interpertation of, the spatial cross-fit analysis is not simple. No cross-fit analysis was done.

TABLE 1 Results of the Cross-Fit for the QRS and the STT Temporal Signal Sets

Subject	Signal Family	NET for QRS Cross-Fit ($X 10^{-4}$)	NET for STT Cross-Fit ($X 10^{-4}$)
	MCG*	17.0 + .3	57 + 1
1	MCG	14.7 + .1	31 + 4
	ECG	*14.3 + .4/5.0 + .4	*9.8 + 1 / 4 + 1
2	MCG	19.0 + .4	80 + 2
	ECG	6.4 + 1	26 + 8
3	MCG	31.7 + .2	31.9 + .8
	ECG	6.9 + 1	8.6 + 1.5
4	MCG	11.3 + .2	62.3 + .9
	ECG	8.4 + .5	17 + 8
5	MCG	17.8 + .2	10 + 3
	ECG	7.5 + .4	24 + 8

*MCG data recorded using symmetric gradiometer system

The type-fit analysis of the spatial data sets does provide some useful information. Using the two set of eigfns generated from the ECG and MCG QRS spatial data of subject 2, the average NET for the possible QRS MCG type-fits is 1.28×10^{-1} and the average NET for the possible QRS ECG type-fits is 2.7×10^{-3}. Using the two sets of eigfns generated from the STT spatial data of suject 3, the average NET for the STT MCG type-fits is 9.3×10^{-2} and the average NET for the STT ECG type-fits is 5.6×10^{-3}. These results are quite remarkable. They show that there is very little variation in the ECG BSPM information content among the five subjects. They also show that there is 47 times more subject specific information (as defined by the NET values) in the QRS MCG spatial data than in the QRS ECG spatial data and 16 times more subject specific information in the STT MCG spatial data than in the STT ECG spatial data. The large difference between the QRS situation and the STT situation is probably due to the QRS being generated by the depolarization process (which is a very time and location specific process), while the repolarization process (the STT generating process) is much more uniform over the ventricles. The subject unique MCG source distributions, prominent in the QRS data, are probably buried under the large common (between subjects) source distributions in the STT data due to the uniform repolarization of the ventricle.

Figure 2 presents the ECG and MCG distributions in the precordial region during one instant of the QRS period. In the QRS maps the 90° twist between the ECG and the MCG distributions is easily observable. Each figure corresponds to an expanded precordial region (our 8 x 7 one inch grid) on the front of the torso.

The contour lines connect isofield or isopotential points. the solid lines indicate areas of positive potentials (ECG) or flux directed out of the torso (MCG). The contours progress logarithmically, with each decade numerically identified as are a few others. Extrema are numerically identified (in mV for ECG and in pT for MCG). Figure 3 shows the first STT MCG and the first STT ECG spatial eigfns. Again the 90° twist is immediately observable between the first ECG and MCG eigfns.

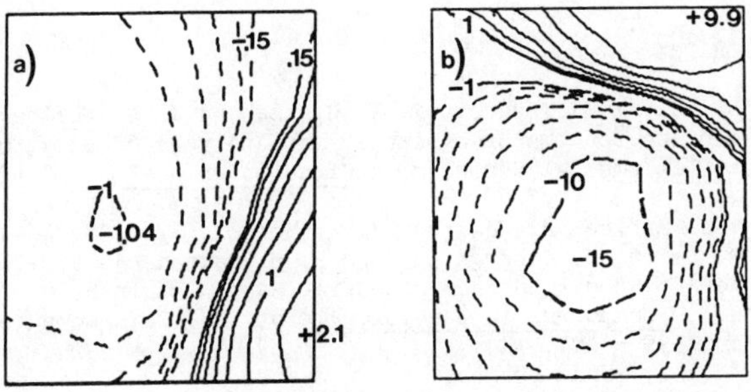

Fig. 2 Plots of a) an ECG isopotential map 9 ms before the VCG X-component maximum, and b) an MCG isofield map 9 ms before the VCG X-component maximum. The values displayed in a) are mV and in b) are pT.

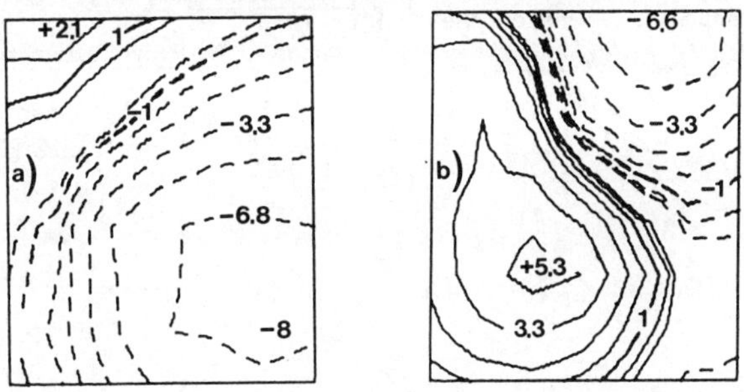

Fig. 3 Plots of normalized maps of a) the first ECG STT eigenfunction and b) the first MCG STT eigenfunction.

CONCLUSIONS

We find in the temporal results an ECG independent MCG component and a MCG independent ECG component. Our results also show the relatively superior ability of the MCG eigenfunctions to fit the ECG data and the inferior ability of the ECG eigenfunctions to fit the MCG data. Therefore, we conclude that the temporal MCG signals of the five subjects contain more information than the temporal ECG signals of these subjects.

From the spatial results we found that the MCG maps contain a substantial amount of subject specific information, while the information content of the BSPMs is much more uniform from subject to subject.

ACKNOWLEDGEMENTS

This research has been supported by the National Science and Engineering Research Council of Canada.

BIBLIOGRAPHY

Evans, A. K.m R. L. Lux, M. J. Bugess, R. F. Wyatt, and J. A. Abildskov (1981). Redundancy Reduction for Improved Display and Analysis of Body Surface Potential Maps, II Temporal Compression.Circulation Res., Vol. 49, No. 1, 197-203.

Horacek, B .M., W. J. Eifler, H. Gewirtz, R. K. Helppi, P. B. MacAulay, J. P. Sherwood, E. R. Smith, J. Tiberghien, and P. M. Rautaharju (1977). An Automated System for Body- Surface Potential Mapping. In H. G. Ostrow and K. L. Ripley (Eds.), Computers in Cardiology, Vol. IV. IEEE Computer Society, California. pp. 399 - 407.

Lux, R. L., A. K. Evans, M. J. Burgess, R. F. Wyatt, and J. A. Abildskov (1981). Redundancy Reduction for Improved Display and Analysis of Body Surface Potential Maps, I. Spacial Compression. Circulation Res., Vol. 39, No. 1, 186-196.

Meritt, R., C. Purcell and G. Stroink (1983). Uniform magnetic field produced by three, four and five square coils. Rev. Sci. Instrum. Vol. 54, No. 7, 879-882

Stroink, G., B. Blackford, B. Brown, and B. M. Horacek (1981). Aluminum shielded room for biomagnetic measurements. Rev. Sci. Instrum., 52, 463-468.

Stroink, G., C. Purcell, F. Brauer and B. Blackford (1983) An Eddy-Current-Shielded Room with a Partially Closed Entrance. Il Nuovo Cimento, Vol.2D. N.2, 195-202.

Wikswo, J. P. (1983). Theoretical Aspects of the ECG-MCG Relationship. In S.J. Williamson, G.L. Romani, L. Kaufman, and I. Modena (Eds.), Biomagnetism, An Interdiscipolinary Approach, Plenum Press, New York. pp. 311-326.

MAGNETIC MEASUREMENT OF PROPAGATING ACTION POTENTIALS IN ISOLATED, ONE-DIMENSIONAL CARDIAC TISSUE PREPARATIONS

J. P. Wikswo and B. J. Roth

Department of Physics and Astronomy
Vanderbilt University
Nashville, Tennessee 37235, U.S.A.

ABSTRACT

The propagating action currents in cardiac tissue are directly measured by detecting their magnetic field. The effects of extracellular return currents are examined using the core-conductor and volume-conductor models.

KEYWORDS

Cellular magnetic field; cardiac action potential, rabbit papillary muscle

INTRODUCTION

The magnetic field produced by an action potential propagating in a one-dimensional strand of cardiac muscle can be measured with the same instrumentation used to record the magnetic field of a nerve axon. Such magnetic measurements are useful because they provide a direct measurement of action currents, which cannot be measured electrically without knowledge of the impedance of the tissue. The first magnetic measurements of cardiac tissue were made on canine Purkinje Fibers (Wikswo and colleagues, 1983a). Recently we have made preliminary measurements using improved instrumentation, in anticipation of a new series of experiments on rabbit cardiac papillary muscle. We present here a description of our latest work and discuss how these results are similar to and differ from our first measurements on Purkinje fibers and our work on nerve axons.

METHODS

The magnetic signal was measured from an isolated rabbit cardiac papillary muscle. A papillary muscle approximately 5 mm long and 1 mm in diameter was dissected from the heart and bathed in Ringer's solution. An action potential was generated by applying a 500 μs voltage pulse to the tissue. The strand was threaded through a wire-wound ferrite-core toroid (1.1 mm inner radius, 1.9 mm outer radius, 1.2 mm width) which sensed the magnetic field. The output of the toroid was detected by our room-temperature low-noise amplifier (Wikswo and colleagues, 1983b, 1984). This amplifier eliminates the need for a SQUID magnetometer and liquid helium, which were used in the original measurements of cardiac magnetic fields (Wikswo and collegues, 1983a). Since the toroid senses the time derivative of the magnetic field, its output must be corrected using frequency compensation (Wikswo, 1982). In the original work on Purkinje fibers this compensation was performed after the experiment on a digital computer. In this experiment the compensation was carried out with analog electronics in real time.

The simultaneous measurement of the transmembrane potential with a glass microelectrode represents another advantage of our recent experiment over our first work because it allows us to make comparisons between model predictions and our data. The electric and magnetic data are shown in Figs. 1a and b. The electric signal had a depolarization phase lasting approximately 5 ms, while the repolarization phase lasted much longer, about 350 ms. The conduction velocity was measured to be 0.3 m/s. The observed magnetic signal is primarily due to the depolarization phase, although there is a marked DC offset in the baseline of the magnetic signal which may represent a signal from the repolarization phase of the action potential. Present in the raw data is a significant amount of powerline noise which we are able to eliminate using a digital lock-in filter (Wikswo and colleagues, 1983a). The filter calculates the Fourier component at the noise frequency over a portion of the signal where there is noise but no action potential, and then subtracts this component from the entire signal. In Fig. 1c we see that the 60 Hz power line noise can essentially be eliminated from our data.

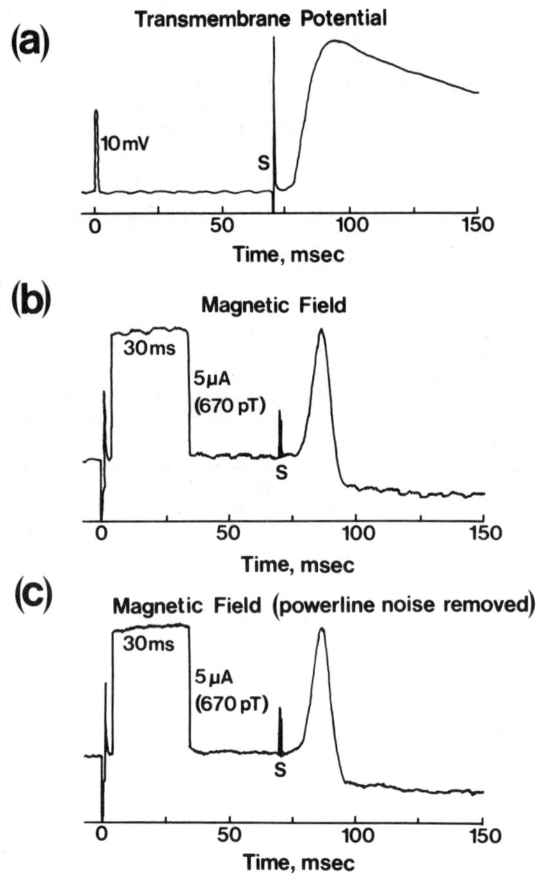

Figure 1. a) The measured transmembrane potential and b) the measured magnetic field of a rabbit cardiac papillary muscle (64 averages). c) The magnetic signal after removal of powerline noise.

THEORETICAL PREDICTIONS

The core-conductor model (Barach, Roth and Wikswo, 1984) assumes that the magnetic field is proportional to the current inside the fiber, and therefore predicts that the magnetic signal should be proportional to the first derivative of the transmembrane potential. This result immediately explains the qualitative features of the magnetic field. The signal is largest when the transmembrane potential changes most quickly, in the depolarization phase. The ratio of the sizes of the magnetic signal between the repolarization and depolarization phases should be approximately equal to the inverse ratio of their durations: 5 ms to 350 ms, or .015. Based on the core-conductor model, we therefore expect that the repolarization phase should have a magnetic signal only one to two percent as large as the depolarization signal.

The core-conductor model ignores the return currents, currents flowing outside the cardiac tissue but through the toroid. These return currents can be estimated using the volume conductor model for a nerve axon (Woosley, Roth and Wikswo, 1984), which indicates that their contribution to the magnetic field will be negligible if ρ, the inner radius of the toroid, is small compared to d, the spatial distance over which the action potential changes significantly. In cardiac muscle, there are two characteristic distances: the length of the depolarization phase, which extends over one to two millimeters, and the length of the repolarization phase, which is about 100 mm long (assuming a long enough strand of muscle). Therefore, the condition $\rho \ll d$ is not fulfilled for depolarization, so that consideration of the return current is essential in quantitative studies of the depolarization phase of the magnetic signal; the actual signal will be wider and have a smaller amplitude than predicted by the core conductor model. However, the length of the repolarization phase is long compared to the size of the toroid, so the second phase of the magnetic signal will have relatively little contribution from the return current. This will tend to increase the size of the magnetic field due to repolarization relative to that produced by depolarization. Figure 2 shows the predictions of the core conductor and the volume conductor models. The calculation was based on a transmembrane potential measured during the experiment, also shown in Fig. 2. The difference between the two signals represents the contribution of the return currents to the magnetic field.

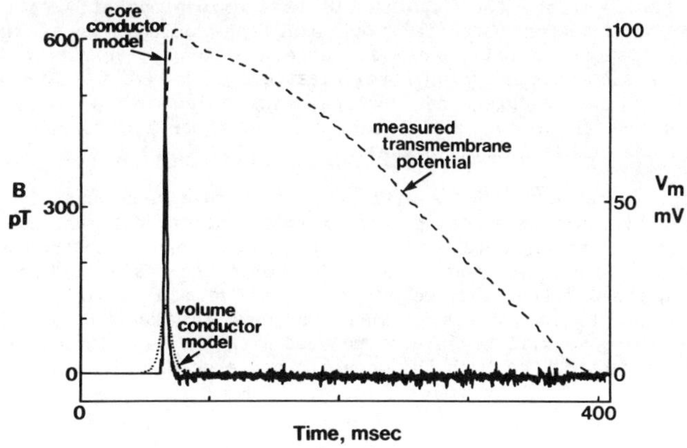

Figure 2. The measured transmembrane potential (dashed) and the corresponding magnetic field calculated using the core conductor model (solid) and the volume conductor model (dotted).

DISCUSSION

The present study differs from our previous one in several ways. Most importantly, we simultaneously recorded the magnetic signal and the transmembrane action potential, thereby providing us with the data necessary for model calculations. Secondly, the instrumentation is much improved, providing us with better low-frequency response and lower noise. Finally, we examined papillary muscle, typical of ventricular myocardium, whereas previously we examined Purkinje fibers from the higher conduction velocity and less contractile tissue of the cardiac conduction system.

Based on these preliminary experiments, we recognize that although magnetic measurements have certain limitations, these limitations are sufficiently different from those of electrical measurements using microelectrodes so that the two techniques are quite complementary. In particular, we find that the magnetic recordings are substantially easier to make and much more stable than electrode measurements, which require penetrating the cell membrane with a glass microelectrode to measure the transmembrane potential.

The studies also provide us with input for our models. Some aspects of cardiac muscle can not be explained by using models derived for a nerve axon. For instance, the cardiac cells are not isolated from each other but are coupled electrically to form a functional syncytium. The interstitial return currents flowing in the extracellular space may change the magnetic signal. Also, unlike nerve axons which can be accurately modeled as an infinitely long cylinder, cardiac cells have a length only a factor of ten larger than their diameter. The ends of the cells are connected by gap junctions at the intercalated disks. We are presently extending our volume conductor model to account for attributes which are specific to cardiac muscle.

The differences between cardiac muscle and nerve tissue are also reflected in the requirements for the instrumentation. While cardiac papillary muscles from the rabbit are five or ten times longer than the spatial length of the depolarization phase of the cardiac action potential, they are much shorter than that of the repolarization phase. Thus, we must account for the effect of the short length of tissue. Also, the bath should have dimensions at least as large as the length of the action potential, but since the dimensions of the bath used in these experiments was not nearly as large as the repolarization length, bath effects may has contributed during repolarization. Finally, the frequency compensation of our signal exaggerates low frequency noise in our data. Therefore we cannot measure the slow repolarization of the cardiac tissue with as much confidence as the depolarization signal.

In conclusion, we believe that we will be able to measure accurately the magnetic field produced by one-dimensional cardiac muscle. As our data and models improve, we hope to begin using magnetic techniques to increase our understanding of cardiac muscle behavior. Our primary goal is to determine the resistance and capacitance of the intercalated disks, but we also hope to gain a better understanding of the distribution of interstitial return currents in the cardiac tissue. Upon accomplishing this, we will be able to proceed with the examination of the effects of various interventions on both the magnetic and electric signals.

ACKNOWLEDGEMENTS

This research has been supported by ONR Contract N00014-82-K-0107 and the Vanderbilt University Research Council.

REFERENCES

Barach, J. P., B. J. Roth and J. P. Wikswo, Jr. (1984). Magnetic measurements of action currents in a single nerve axon: a core-conductor model. (submitted for publication).

Wikswo, J. P. (1982). Improved instrumentation for measuring the magnetic field of cellular action currents. Rev. Sci. Instrum., 53, 1846-1850.

Wikswo, J. P., J. P. Barach, S. C. Gundersen, M. J. McLean, and J. A. Freeman (1983a). First magnetic measurements of action currents in isolated cardiac Purkinje fibers. Il Nuovo Cimento, Vol. 20, 368-378.

Wikswo, J. P., W. P. Henry, P. C. Samson and R. P. Giffard (1984). A current probe system for measuring cellular action currents. (These proceedings).

Wikswo, J. P., P. C. Samson and R. P. Giffard (1983b). A low-noise low input impedance amplifier for magnetic measurements of nerve action currents. IEEE Trans. Biomed. Eng. BME-30, 215.

Woosley, J. K., B. J. Roth and J. P. Wikswo (1984). The magnetic field of a single axon: a volume conductor model. (submitted for publication).

MODELLING OF THE HIS-PURKINJE HEART CONDUCTION SYSTEM

S.N. Erné, H.P. Lehmann, M. Masselli and Y. Uchikawa

Physikalisch-Technische Bundesanstalt, Institut Berlin,
Abbestrasse 2 - 12, D-1000 Berlin 10, Germany

ABSTRACT

A three-dimensional model of the His-Purkinje conduction system based on a realistic heart model has been developed. The intensity of the magnetic field perpendicular to the chest is calculated as measured with a magnetometer coil. By slight modifications of the model, i.e. slight anatomical changes, the calculated magnetic fields show strong variations of morphological fine structures. However, the fundamental structure of the model is always traced correctly by calculating the propagation pathway on the basis of equivalent current dipoles.

KEYWORDS

His-Purkinje modelling, biomagnetic field, equivalent current dipole.

INTRODUCTION

With high resolution magnetocardiographic recording technique the morphological structure of the PR segment has been studied, in particular to detect His-Purkinje conduction system (HPS) activity (Farrell and co-workers, 1980; Fenici and co-workers, 1980, 1983; Erné and co-workers, 1983a, 1983b). In all cases a ramp pattern has been found, superimposed in some cases by one or more definite deflections. It has been confirmed by our investigations of the PR segment of 35 normals (to be published), that as a general feature, the morphology of the PR segment varies strongly from subject to subject.

Fenici and co-workers (1983) and Erné and co-workers (this volume) used time recordings, measured at different positions, to construct isofield contour mappings. In combination with a suitable subtraction technique a dipolar activation pattern was observed which was attributed to HPS depolarization and served to calculate the location of equivalent current dipoles (ECDs).

For a better understanding of these observations HPS models have been constructed by several authors. In a first model, substituting the HPS conduction system by three parallel straight segments, the ramp structures could be simulated (Farrell and co-workers, 1980). In a more detailed two-dimensional model by Leoni and Romani (1982) the ramp structure as well as a small spike-like signal could be reproduced.

In order to find out causes for the always observable ramp structure combined with a strong variability of morphology with regard to fine structure, a three-dimensional model of the HPS conduction pathway has been constructed and the effect of small modifications of the model has been studied. Furthermore the simulated model field patterns have been used to calculate the locations of ECDs representing HPS activity and to compare them with the geometry of the model.

METHODS

A three-dimensional model of the HPS conduction system, which serves as a pathway for the excitation of HPS has been developed. The model consists of 19 segments or sections representing in a simplified manner the three-dimensional anatomical structure of a real HPS (which actually varies strongly from subject to subject). It has been obtained by adapting a wire model to a realistic heart model. A projection of the final configuration onto the x, y-plane, tangential to the anterior chest wall is shown in Fig. 1.

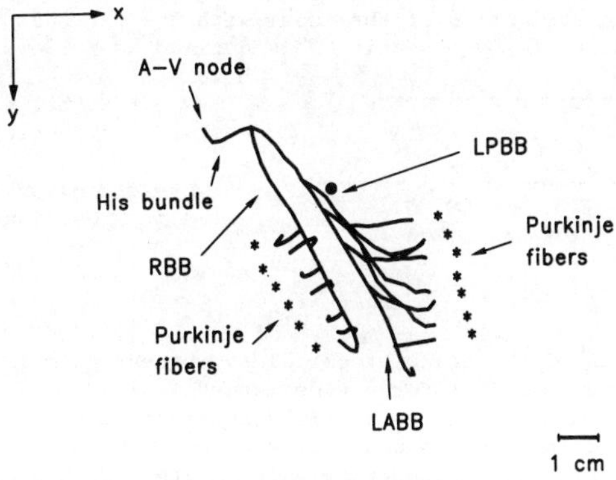

Fig. 1 Projections of the three-dimensional model of the His-Purkinje conduction system onto a plane tangential to the anterior chest wall. LABB = left anterior bundle branch, LPBB = left posterior bundle branch, RBB = right bundle branch. For the dot and stars see text.

Corresponding to the HPS in a normal human heart the conduction pathway of the model consists of segments representing the His bundle, the right bundle branch (RBB), the left anterior bundle branch (LABB), the left posterior bundle branch (LPBB) and the Purkinje fibres.

The activation starts at the AV node and ends at the dendritic Purkinje network. During spreading of the activation several current dipoles corresponding to the different segments are active at the same time instant. To simulate the dendritic character of a network of Purkinje fibres each dipole corresponding to segments indicated by stars in Fig. 1 has been substituted by an ensemble of dipoles. The location of these dipoles is obtained by adding small random displacements to the coordinates of the original dipole.

Intrinsic parameters are the propagation velocity and the intensity of the current dipoles along each of the segments. These parameters have been chosen on the basis of physiological data (measured velocities, cross section of the conduction fibres, see for example Massing and James, 1976). Scaling the overall size of the model warrants full activation of HPS within 45 ms. Once chosen, these parameters as well as the structure of the model have been fixed.

The remaining parameters are the heart location and its orientation with respect to the "measuring plane", i.e. the x, y-plane where the magnetic field component is calculated.

During the activation of each current dipole the projection of the dipole onto the x, y-plane is calculated in steps of 1 ms. The intensity of the z-component or the magnetic field then is calculated taking into account the finite diameter of the magnetometer coil (Williamson and Kaufman, 1981) for the 36 positions of the standard grid (Saarinen and co-workers, 1978), considering the dipole embedded in a conductive semi-infinite medium.

In all cases the model data have been used to construct isofield contour maps and to calculate the location, direction and intensity of ECDs. This is done by comparing the field distribution of the model with the field distribution of one dipole and minimizing σ by best fit. σ is defined as follows:

$$\sigma = \frac{\sqrt{\frac{1}{36}\sum_{j=1}^{36}(F_{Tj} - F_{Dj})^2}}{F_{T_{max}} - F_{T_{min}}}$$

F_{Tj} = actual total field at position j,

F_{Dj} = dipolar field at position j

RESULTS

The time dependence of the calculated field component normal to the x, y-plane is displayed in Fig. 2 (model A) for a time interval of 40 ms as measured with a magnetometer (coil diameter 3 cm) at the 36 standard grid positions. The location of the main bundle branching point was selected to be 10 cm below the point marked by a cross in the Fig. 1. The heart axis is rotated by 30 degrees around the z-axis. A ramp pattern and additional deflections (bumps) are visible.

In a second example Fig. 2 (model B) the geometry has been left unchanged, but in the LPBB, indicated in Fig. 1 by a dot, the intensity of the current dipoles has been reduced by a factor of 3 to simulate a different anatomical situation. In Fig. 2 (model B) the fields calculated under the same conditions are displayed. Direct comparison shows, that the morphology of the calculated time recording is changed. In contrast, the overall ramp structure is almost unaffected. In general the fine structure of the time recording will strongly depend on the anatomical details of the HPS system. The bumps are mainly determined by bendings of the propagation pathways as well as by the branching points.

It is interesting to note that the activation of the dendritic parts of the model give a terminal deflection of the PR waveform (see Fig. 2) corresponding with experimentally recorded onsets of ventricular activation.

The calculated magnetic field intensities at a specified time have been used to construct isofield contour maps. An example for four successive instants is shown in Fig. 3, the ECDs calculated on the basis of these field distributions are indicated by solid arrows.

In Fig. 4 projections of the basic HPS-model are combined with projections of the pathways of the ECDs. In Fig. 4A the projection plane is the x, y-plane, while in Fig. 4B a different view is presented in order to demonstrate the correct three-

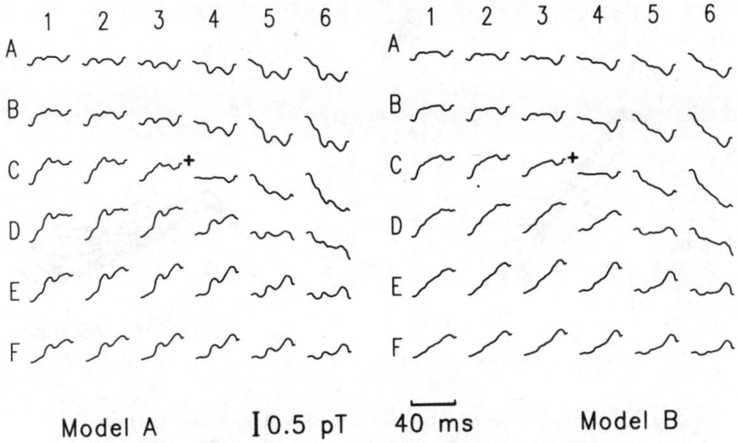

Fig. 2 Time dependence of the calculated magnetic field (z-component) for two model configurations as detected by a magnetometer with a coil of 3 cm diameter. A 1 to F 6 indicate grid positions (Saarinen and co-workers, 1978). The AV-node is located below the crosses. Model B is a slight modification of the basic model A, see text.

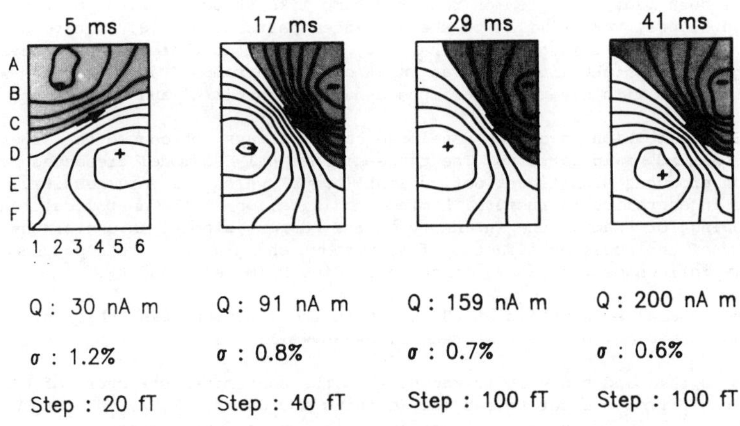

Fig. 3 Isofield contour maps of model A for four successive time instants. Location and direction of the equivalent current dipoles are indicated by solid arrows. Q is the intensity of the ECDs and the deviations between model field and field of the ECD.

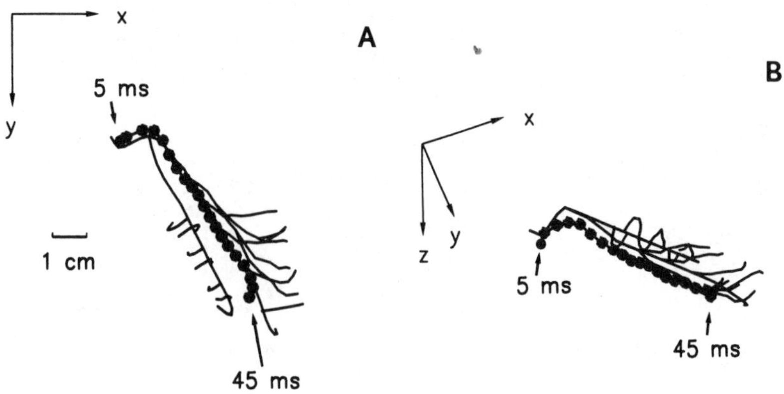

Fig. 4 Projection of the basic model A and the pathway of equivalent current dipoles (time step 3 ms). In panel B a different viewing direction has been chosen to show the depths of the calculated ECD locations with respect to the model.

dimensional location of the ECDs with respect to the model.

DISCUSSION

Whenever high resolution magnetocardiographic measurements of the PR interval in man have been made, as a common result a ramp like structure with positive and negative slopes, very similar to the simulated patterns of Fig. 2, have been reported. Apart from this general behaviour, different morphologies have been found ranging from straight lines with an onset of smooth ramps (Farrell and co-workers, 1980) to ramps with several superimposed deflections (Erné and co-werkers, 1983 a).

The HPS models which have been developed so far always reproduced the experimentally observed ramp patterns. The three-dimensional HPS model presented here in addition produces significant deflections (bumps) which are very sensitive to changes in anatomical structure. A minor modification of the anatomical features by reducing for example the intensity of the current dipoles in one of the 19 segments of the model by a factor of 3, without changing any other parameter, strongly influences the bumps, as shown in Fig. 2 (model A and B).

When the general orientation of the heart in the chest changes, this influences the detailed morphology of the time recordings as well.

Because of this and since it is known, that the anatomical structure of HPS varies considerable from subject to subject (Massing and James, 1976), the morphological fine structure of the PR segment has to be interpreted with great care.

Despite to the fact that in general many dipoles are activated simultaneously, the total resulting field pattern can be simulated by a single ECD with a σ of less than 1 %. The results of the three-dimensional localization algorithm for the ECD during the activation time show a structure which resembles the geometry of the original model in a simplified manner. In the case when more than one dipole is activated, the ECD is located in a weighted position; this is important when the dentritic part of the HPS is activated. In the framework of the model, the moving ECD traces the spatial propagation of the excitation in the HPS conduction system.

ACKNOWLEDGEMENT

The authors would like to express their gratitude to Prof. R.R. Fenici, Prof.B. Taccardi and Z. Trontelj for stimulating discussions.

REFERENCES

Erné, S.N, Fenici, R.R., Hahlbohm, H.-D., Masselli, M., Lehmann, H.P. and Z. Trontelj (1983a), Il Nuovo Cimento, 2 D, 248 - 254.

Erné, S.N., Fenici, R.R., Hahlbohm, H.-D., Lehmann, H.P. and Z. Trontelj (1983b), Il Nuovo Cimento, 2 D, 1110 - 1118.

Erné, S.N., Fenici, R.R., Hahlbohm, H.-D., Korsukewitz, J., Lehmann, H.P. and Y. Uchikawa, this volume.

Farrell, D.E., Tripp, J.H. and R. Nordgren (1980), IEEE Trans. Biomed. Eng. BME-27, 345 - 350.

Fenici, R.R., Romani, G.L., Barbanera, S., Zeppilli, P., Carelli, P. and I. Modena (1980), G. Ital. Cardiol. 10, 1366.

Fenici, R.R., Romani, G.L. and R. Leoni (1983), Il Nuovo Cimento, 2 D, 280 - 290.

Leoni, R. and G.L. Romani (1982), Il Nuovo Cimento, 1 D, 737, 262.

Massing, G.K. and T.N. James (1976), Circulation 53, 609.

Saarinen, M., Siltanen, P., Karp, P.J. and T.E. Katila (1978), Ann. Clin. Res. 10 Suppl.21, 1.

Williamson, S.J. and L. Kaufman (1981). In S.N. Erné, H.-D. Hahlbohm and H. Lübbig (Eds.), Biomagnetism, de Gruyter, Berlin-New York, 353-402.

MAGNETOCARDIOGRAPHIC STUDY OF THE PR SEGMENT OF NORMALS

S.N. Erné, R.R. Fenici*, H.-D. Hahlbohm, J. Korsukewitz**
H.P. Lehmann and Y. Uchikawa

Physikalisch-Technische Bundesanstalt, Institut Berlin,
Abbestrasse 2 - 12, D-1000 Berlin 10, Germany
* Cardiov. Lab. Biomagnetic Unit, Med. Pathology, Inst.,
Catholic University, Lg. A. Gemelli 8, 00168 Rome, Italy
**Klinikum Charlottenburg, Freie Universität Berlin,
Spandauer Damm 130, D-1000 Berlin 19, Germany

ABSTRACT

To get a better insight in the details of the PR segment of healthy man, high resolution magnetocardiographic recordings of the PR segment have been performed on 32 healthy subjects. For most of them it has been possible to trace the pathway of the HPS conduction system and to define quantitative MCG parameters on the basis of the equivalent current dipole.

KEYWORDS

Biomagnetism, magnetocardiography, His-Purkinje system, high resolution MCG, equivalent current dipole.

INTRODUCTION

The PR segment has been studied with high resolution MCG recording technique since 1980. As a general feature a ramp pattern with a positive or a negative slope depending on the location of the magnetometer has been found. The morphological fine structure which has been reported varied between smooth ramps (Farell and co-workers, 1980) and ramps with superimposed deflections (Barbanera and co-workers, 1981; Erné and co-workers, 1983). On the basis of the morphology of time recordings it was not possible to distinguish between the contribution of the atrial repolarization and the HPS depolarization to the genesis of the PR waveform. In this paper the PR segment is analysed on the basis of the magnetic mapping, i.e. on the measurement and evaluation of the time evolution of the magnetic field distribution over the chest.

METHODS

After informed consent, 32 volunteers (15 male,17 female, aged between 22 and 37 years) have been studied. All subjects have been found normal according to clinical history, physical examination, stress ECG and 24 h Holter monitoring.

The measurements have been performed in the Berlin Magnetically Shielded Room using a magnetometer (3 cm coil diameter) with an equivalent input white noise less than 15 $fT/\sqrt{Hz}$ under measuring conditions. The field component normal to a plane tangential to the chest has been recorded in 36 positions according to the standard grid (Saarinen and co-workers, 1978). The passband for the signals was 0.016 Hz to 250 Hz with a RC high pass filter of first order and a Bessel low pass filter of 8th order. The signals have been digitized with 12 bit resolution and 1000 Hz sampling rate.

To compensate the disadvantage of using a single channel magnetometer and measuring subsequentially, the signals have to be preprocessed to select characteristic beats at any position. This selection procedure is performed on the simultaneously recorded limb ECG: only such beats are accepted for which the QRS matches the template inside of 1σ (variance of the optimal cross-correlation) and the beat rate matches the averaged beat rate inside of 1σ (variance of beat rate).

After a standard editing procedure which eliminates artefacts, the selected beats are averaged at each position. Finally the signals of all positions are realigned by cross-correlation technique to obtain the synchroneous condition. The final data evaluation allows to study the data according to the following procedure:

a) Examination of high resolution MCG time recordings at the different positions for morphological analysis.
b) Examination of isofield contour maps at different time instants during P wave and PR intervall until the onset of ventricular activation.
c) Interpretation of the field distribution in terms of equivalent current dipoles (ECD), whenever this is possible. The field distribution can be described in many cases by an ECD. The parameters of the ECD are calculated according to the best fit of the measured data, for further information see Erné and co-workers (this volume).
d) Determination of the three-dimensional propagation pathway of the HPS system after separation of the contribution due the atrial repolarization.

As criterion for the best fit can be used either

$$\sigma = \frac{\sqrt{\frac{1}{N}\sum (F_{mj} - F_{mc})^2}}{F_{m_{max}} - F_{m_{min}}}$$

F_{mj} = measured field at position j

F_{cj} = calculated field at position j

or

$$\chi^2 = \frac{\sum (F_{mj} - F_{cj})^2}{E_j^2} .$$

E_j = measurement error at position j

χ^2 is used, when the measuring conditions during sequential recording are unstable, resulting in strong differences of the number of selected beats and final noise.

RESULTS

In all cases ramp patterns have been found in the time recordings and in general still additional bumps; Fig. 1 shows a map of time recordings of the PR segment for a typical subject. The detailed structure (ramp onset, number of bumps, timing of bumps, etc.) varies strongly from subject to subject, however it seems to be a characteristic of each subject. Studying the isofield contour maps in the PR segment, the following can be observed:

a) During the P wave a dipolar pattern, which can be reasonably interpreted as the result of the atrial depolarization.
b) In contrast, the opposite polarity in the field distribution after the end of the P wave suggests that the field is mainly due to the atrial repolarization.

Fig.2 shows this effect for the same subject as in Fig. 1.

In 17 of 32 cases to the atrial depolarization pattern an ECD was as assigned with

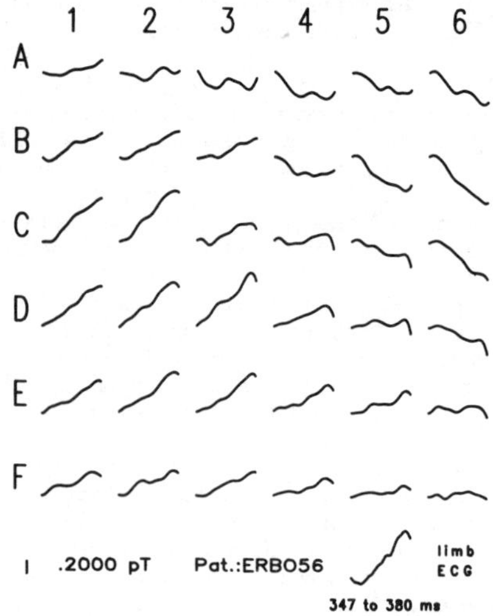

Fig. 1 Time recordings of the magnetic signals in the PR segment in all measuring positions of the standard grid.

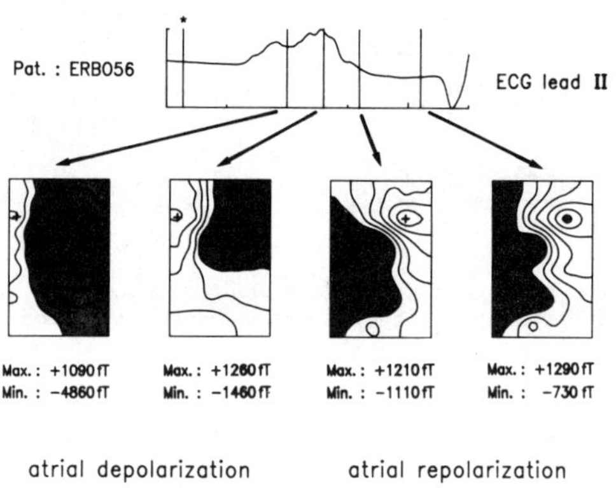

Fig. 2 Isofield contour maps of the field distribution at different instants during the P wave and the PR interval. The star denotes the instant in the pre-P region where the reference distribution has been defined.

sufficient confidence. The calculated ECDs average intensity is 950 + 500 nA m. On the other hand, only in 14 of 32 cases it was possible to calculate the ECD for the atrial repolarization giving an average dipole intensity of 360 $\pm$ 180 nA m.

On the basis of the time stability of the atrial repolarization pattern a subraction technique has been used to evidence the contribution of the HPS depolarization. In Fig. 3 are shown the field distributions obtained by subtracting the background correlated to the atrial repolarization. The clear dipolar pattern allows to calculate easily the parameters of the ECDs, indicated by arrows in the same figure.

In 22 of 32 cases this procedure has been used successfully (for subtraction, atrial repolarization may have a non-dipolar structure). The ECD of the initial HPS depolarization has been found to be 50 $\pm$ 40 nA m, the final depolarization just before

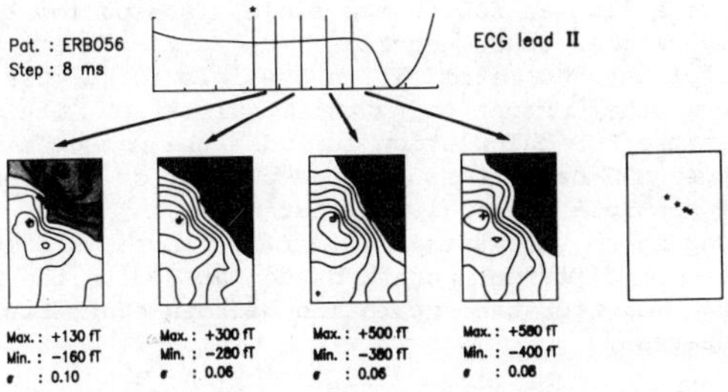

Fig. 3 Field distribution at four different instants in the PR interval after subtraction of the atrial repolarization (calculated at the instant denoted by a star in the limb ECG). The arrows represent the ECDs. The right panel summerizes the motion of the ECD.

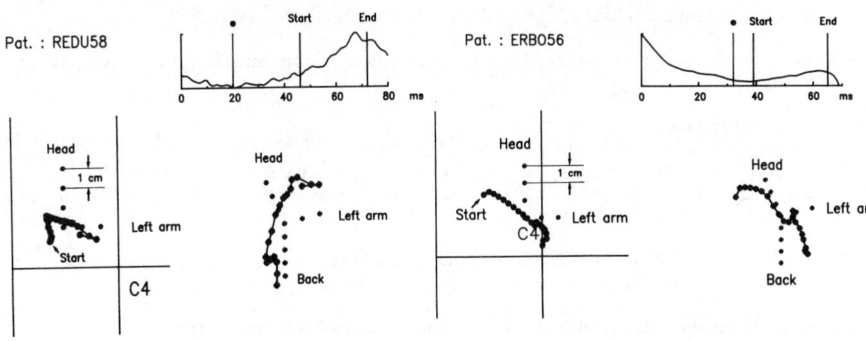

Fig. 4 Two examples of calculated traces of HPS activation pathways. The upper panels shows the limb ECG with the time interval, during which the propagation has been calculated. The dots indicate the reference for atrial repolarization subtraction. The lower panels show two different views of the 3D-traces. C4 denotes the corresponding position in the standard grid.

onset of ventricular activation 190 ± 110 nA m.

In 22 of 32 cases traces of the propagation pathway of the HPS excitation have been obtained by applying the 3D-localization algorithm for the ECD. In Fig. 4 two examples are given.

DISCUSSION

As suggested by model studies (Erné and co-workers, this volume) the fine structure of the high resolution MCG recordings in the PR segment probably take into account the anatomical details of the heart conduction system. Until now a general classification or interpretation of the fine structure details within the PR segment is not possible.

The possibility to assign an ECD to the field distribution related to the atrial activity has some notable consequences:

a) On the basis of the ECD method it is possible to calculate new quantitative MCG parameters: the ECD intensity and its location within the heart. Following the fact that the ECD calculation includes the measurement grid and the sensor geometry, these MCG parameters are independent on the measuring procedure and therefore of possible clinical application.
b) The ECD belonging to the atrial repolarization shows only minor changes in time and is located at different positions compared with the HPS depolarization. This gives the basis for the separation of both contributions to the total field in the PR interval.

According to the model studies (Erné and co-workers, this volume) it seems possible to interprete the ECD 3D-pathways as traces of the HPS activation. However in some cases the ECD pathways show a complicated structure which cannot be explained by the anatomical structures. Two sources of error have been identified: First the stability of the localization algorithm requires a minimal signal to noise ratio of 3. Second, at the level of precision required it became noticeable that the measurements have been performed sequentially and therefore the field distribution is not strictly related to one event.

ACKNOWLEDGEMENT

The authors would like to express their graditude to Z. Trontelj for stimulating discussions.

REFERENCES

Barbanera, S., Carelli, R., Leoni, R., Romani, G.L., Bardoni, F., Modena, I., Fenici, R.R. and P. Zeppilli (1981). In S.N. Erné, H.-D. Hahlbohm and H. Lübbig (Eds.), Biomagnetism, de Gruyter, Berlin-New York, 283 - 290.

Erné, S.N., Fenici, R.R., Hahlbohm, H.-D., Masselli, M., Lehmann, H.P. and Z. Trontelj (1983), Il Nuovo Cimento **2D**, 248 - 254.

Erné, S.N., Lehmann, H.P., Masselli, M. and Y. Uchikawa, this volume.

Farrell, D.E., Tripp, J.H. and R. Norgren (1980), IEEE Trans. Biomed. Eng. BME-27, 345 - 350.

Saarinen, M., Siltanen, P., Karp, P.J. and T.E. Katila (1978), Ann. Clin. Res. 10 Suppl.21, 1.

MAGNETOCARDIOGRAPHIC MAPPING OF THE P-R INTERVAL PHENOMENA IN AN UNSHIELDED HOSPITAL LABORATORY

R.R. Fenici[^], M. Masselli[^], S.N. Erné[°], H.D. Hahlbohm[°]

[^]Cardiovascular Biomagnetism Unit, Clinical Physiology, Medical Pathology Inst, Catholic University, Rome, Italy.
[°]Physikalisch-Technische Bundesanstalt, Institut Berlin, West Germany.

ABSTRACT

In this paper the magnetic field distribution during the P-R interval of normal subjects is described. The recordings have been performed in an unshielded hospital laboratory. Signal processing has been carried out with the aim to differentiate between fields generated by atrial depolarization, atrial repolarization and His-Purkinje system activation.

KEYWORDS

Atrial repolarization, His-Purkinje system, Magnetocardiographic mapping, Equivalent current dipole.

INTRODUCTION

Since 1973 surface electrical recordings of the activity of the His-Purkinje system (HPS) have been attempted with different lead configurations and the signal averaging technique (Hombach and Hilger, 1981). More recently surface electrical mapping has been used as well to study in more detail the distribution of the electrical field generated during the P-R interval (Spach and others, 1979). Horan and others (1982) have demonstrated that, after the subtraction of the field generated by atrial repolarization, a residual field can be evidenced which is consistent with the activation of the bundle of His. In that study a tentative three-dimensional localization of the Equivalent Current Dipole (ECD) of the HPS was partially prevented by the so-called "back tracking phenomenon". This drawback, probably linked to the intrinsic limitations of the electrocardiographic method, could theoretically be overcome by recording the magnetic fields generated during the P-R interval (Wikswo, 1983). High resolution magnetocardiography has been recently used to attempt the calculation and localization of ECDs generated by the human conduction system (Fenici and others, 1983). The rationale for the application to cardiac electrophysiology of the biomagnetic source localization method, originally developed for brain studies (Williamson and Kaufman, 1981), had been provided by the experimental observation that cardiac fields are usually of dipolar configuration, and by the mathematical model of Leoni and Romani (1982); nevertheless the interpreta-

tion of our preliminary measurements (Fenici and others, 1983) had been matter of discussion because other investigations, where the high resolution magnetic P-R waveforms had been examined only in function of the time, had been partially in disagreement (Leifer and others, 1983; Patrick and others, 1983). In particular it was unclear whether the typical magnetic P-R waveform was mainly generated by the atrial repolarization or HPS activation.
The main purpose of the present study has been to assess if the level of resolution achievable in our unshielded hospital laboratory was sufficient to detect the magnetic field distribution during the P-R interval in normal subjects under clinical conditions. The respective contribution of atrial and HPS activities to the genesis of the P-R field distribution have been studied as well to determine whether or not the magnetic field of the HPS could be reproducibly unmasked with a proper subtraction of the atrial repolarization. In order to avoid misinterpretation, the results of the hospital measurements have been compared with those obtained with a similar procedure in the Berlin Magnetically Shielded Room (BMSR) (Erné and others, in this book).

MATERIAL AND METHOD

The recordings were carried out with the prototype of the BIOMAG I (Elettronica S.p.A., Rome, Italy), an instrumentation based on a r.f. SQUID coupled to a second order gradiometer (coil diameter = 2.9 cm; baseline = 5 cm) with the special feature to be balanciable also against spatially uniform magnetic field gradients. The typical level of noise of the system, at the output of the Comb filter used to reduce power interference, was about 50 fT/Hz$^{\frac{1}{2}}$, with fluctuations related to the level of activity of other hospital instrumentations. MCG and ECG have been simultaneously recorded in a bandwidth of 0.016-250 Hz.
The component of the magnetic field perpendicular to the anterior chest wall was mapped according to the standard 36 points grid (Karp, 1981). 15 normals were studied. The signals were stored on analog tape and digitized off line at a sampling frequency of 1 KHz and 12 bits resolution. Digital processing to obtain MCG isofield contour maps and to calculate the parameters of the ECDs was carried out on a 1000F HP computer with the same software used at the Physikalisch-Technische Bundesanstalt, described in detail elsewhere in this book (Erné and others). The ECD is defined as the current dipole for which the calculated magnetic field distribution fits under least square conditions the experimental one. Sequential isofield contour maps were automatically drawn by the plotter with a time interval between maps preselectable by the operator.

RESULTS

In fig. 1 an example of sequential magnetic mapping during the P-R interval of a normal subject studied with an RMS noise level of 50 fT/Hz$^{\frac{1}{2}}$ is shown. The mapping sequence starts with the onset of the P wave. The time interval between maps is 20 msec. In coincidence with the peak of the electrical P wave the field distribution, of dipolar configuration, is characterized by a negative area (i.e. a field leaving the chest) on the upper left and a positive area (which indicates a field entering the chest) on the lower right of the anterior chest wall. The ECD of this field distribution, due to atrial depolarization, is localized approximately over the anatomical position of the right atrium; its intensity is 1300 nA · m. In coincidence with the descending limb of the P wave, the polarity of the magnetic field distribution reverts, remaining unchanged during the rest of the P-R interval. The latter field, which ECD localization on the grid is the same of atrial depolarization's one (C_2-C_3), is due to atrial repolarization. Its intensity is 400 nA . m.

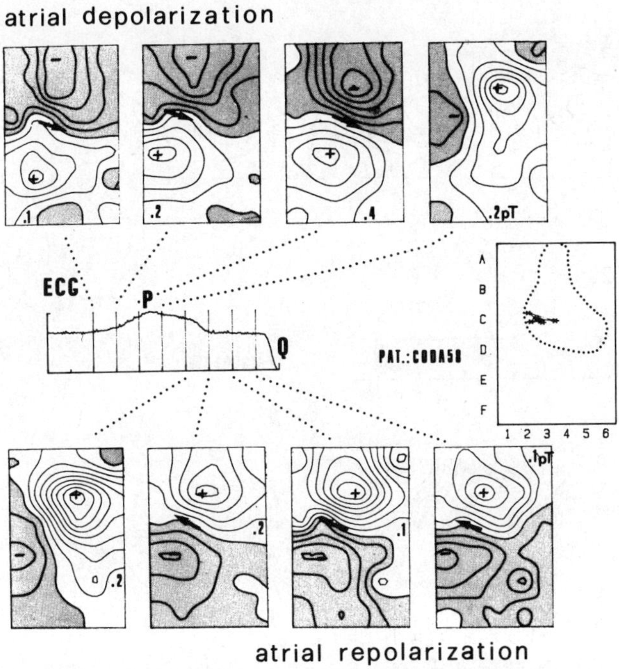

Fig.1: MCG mapping during the P-R segment. Time interval between maps=20 msec. Contour line values (in pT) are indicated by the numbers in each map. Negative areas (field leaving the chest) are shaded. Positive areas (field entering the chest) are clear. Bidimensional ECDs localization is marked by arrows in each map and by asterisks in the heart siluette inferred by P-A X-ray.

In fig.2 the magnetic field distribution of the last 40 msec of the P-R interval of the same subject is shown, with a time interval between maps of 10 msec. In fig. 2A the first two maps are consistent with atrial repolarization. The last map is due to ventricular septal activation. The third map is of transitional configuration and its ECD localization is out of the anatomical location of the heart. In fig. 2B the same sequential mapping is shown as in fig.2A, but after the subtraction of the atrial repolarization field, computed at the instant marked by a triangle on the limb ECG. The residual field distribution appreciable in the first three maps is of rough dipolar configuration and opposite polarity with respect to atrial repolarization's one. This field, which intensity increases in function of the time from 120 to 350 nA·m, merges with the stronger field due to the ventricular septal activation (fourth map). The time-dependent movement of the ECDs of these four maps gives a rough two-dimensional image of the propagation pathway of the electrical impulse along the longitudinal axis of the heart and is consistent therefore with the activation of the HPS.

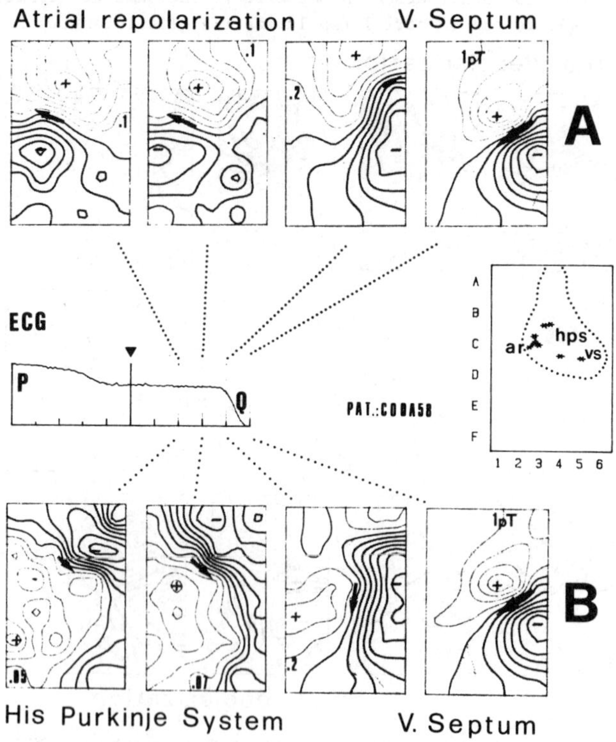

Fig.2: Same subject as in fig.1. MCG mapping of the last 40 msec of the P-R segment. Time interval between maps=10 msec
A- First two maps: atrial repolarization (ar); last map: ventricular septal activation (vs).
B- Subtraction of the atrial repolarization computed at the instant marked by ▼. First three maps: activation of the His-Purkinje system (hps); last map: ventricular septal activation (vs).
Bidimensional localizations of ar, hps, and vs ECDs are marked by asterisks on the real size heart siluette.

DISCUSSION

Effective electromagnetic shielding and sofisticated digital signal processing have improved the possibility of recording reproducible low amplitude electrocardiographic signals from the body surface (Flowers and others,1981; Berbari and others, 1983; Erné and others, 1983). However, even working under ideal recording conditions, neither surface bipolar ECG recordings nor surface electrical mapping (Spach and others, 1979) have been effective so far to separate signals related to the different electrophysiological phenomena overlapping during the P-R interval (e.g. atrial repolarization and HPS activation). Moreover the tentative localization of HPS equivalent sources, on the basis of surface electrical measurements, had been only partially successful probably because of the intrinsic limitations of this method (Horan and others, 1982). On the other hand, as expected from theory (Wikswo 1983), our preliminary experimental high resolution MCG recordings in humans had demonstrated that a magnetic field distribution of dipolar configuration could be recorded during the P-R interval and that cardiac sources consistent with the anatomical localization of the HPS could be approximately localized (Fenici and others, 1983).

The results of the present investigation, carried out parallel in the unshielded hospital laboratory and in the BMSR in order to verify the reproducibility of this kind of measurements, have provided the following new informations: 1)under physiological conditions the fields generated by atrial depolarization, atrial repolarization, HPS and ventricular septal activations are usually of dipolar configuration. 2)Atrial repolarization starts in coincidence with the descending limb of the electrical P wave, partially overlaps terminal atrial depolarization and is the only field directly detectable during the P-R interval. This implies that, in order to unmask the field generated by the HPS, atrial repolarization contribution has to be properly subtracted. 3)Provided that the hospital measurements were performed with an overall RMS noise level not exceeding 50 $fT/Hz^{\frac{1}{2}}$, the similarity with the BMSR recordings demonstrated that the resolution achievable in the hospital is good enough to detect HPS fields and to attempt a reasonable reconstruction of the HPS activation pathway. Finally, although the localization procedure is not yet refined (fig.2A, third map), MCG mapping seems to provide a good spatial discrimination between ECDs related to atrial, HPS and ventricular septal activities.

REFERENCES

Berbari, E.J., S.M. Collins, Y. Salu and R. Arzbaecher (1983). IEEE Trans. biomed. Engng. BME, 30, 160-167.

Erné, S.N., R.R. Fenici, H.D. Hahlbohm, H.P. Lehmann and Z. Trontelj (1983). J. of Electrocardiology, 16, 355-362.

Erné, S.N., R.R. Fenici, H.D. Hahlbohm, J. Korsukewitz, H.P. Lehman and Y. Uchigawa (1984). In this book.

Fenici, R.R., G.L. Romani, R. Leoni (1983). Il Nuovo Cimento D, 2, 280-290.

Flowers, N.C., V. Shvatsman, B.M. Kennelly, G.S. Sohi and L.G. Horan (1981). Circulation, 63, 968.

Hombach, V. and H.H. Hilger (1981). Signal averaging technique in clinical cardiology, F.K. Schattauer Verlag, Stuttgart-New York.

Horan, L.G., N.C. Flowers and G.S. Sohi (1982). Circul. Res., 50, 47-54.

Karp, P.J. (1981). In S.N. Erné, H.D. Hahlbohm and H. Lubbig (Ed.), Biomagnetism, Walter de Gruyter, Berlin-New York, pp. 219-258.

Leifer, M., N. Capos, J. Griffin and J. Wikswo (1983). Il Nuovo Cimento D, 2, 266-279.

Leoni, R. and G.L. Romani (1982). Il Nuovo Cimento D, 1, 737-750.

Patrick, J.L., D.W. Hess, J.H. Tripp and D.E. Farrell (1983). Il Nuovo Cimento D, 255-265.

Spack, M.S., R.C. Barr, R.B. Warren, D.W. Benson, A. Walston and S.B. Edwads (1979) Circulation, 59, 805-821

Wikswo, J.P. (1983). In S.J. Williamson, G. L. Romani, L. Kaufman and I. Modena (Ed) Biomagnetism: An interdisciplinary approach, Plenum Press, New York-London, 311-326.

Williamson, S.J. and L. Kaufman (1981). J. Magn. Mag Mat., 22, 2, 129-202.

A Study of the P-R Segment in the Human Magnetocardiogram

H. E. Lorenzana*, P. B. Pipes*, M. P. Zaitlin* and D. James**

* Physics Dept., Dartmouth College, Hanover, NH, USA 03755
**Cardiology Dept., Mary Hitchcock Hospital, Dartmouth College, Hanover, NH USA 03755

ABSTRACT

Subjects with prolonged P-R segments have been studied. The human atrial repolarization wave, Ta, has been characterized magnetically for the first time. Evidence is presented demonstrating that ramps attributed to activity of the His-Purkinje system (HPS) (Farrell, 1980) may be explained by the Ta. Lastly features smaller than the ramps have been seen as signals possibly produced by the HPS.

KEYWORDS

Magnetocardiogram; P-R segment; atrial repolarization; His bundle; His-Purkinje system.

INTRODUCTION

Recently, Farrell (1980) pioneered the use of the SQUID based magnetometer to measure signals in the P-R segment of the cardiac cycle. He observed "ramps" about .6 pT that lasted about 35 ms. Furthermore, he showed that these ramps demonstrated a symmetry about a line running from the upper right to the lower left of the chest. All the ramps had a negative slope on the right side and a positive slope on the left side of the line, indicating that the magnetic fields were directed into the body and out of the body respectively. A model based on an equivalent moving current dipole predicted ramps similar to those measured (Tripp, 1981; Leoni, 1982). These signals have been verified by several researchers (Fenici, 1980; Erne, 1983). Both Erne and Fenici have also found bumps of about .3 pT in amplitude superimposed on the ramps. The bumps seem to coincide with the HPS signals as measured by high resolution chest electrocardiograms (EKG's) and with His bundle electrograms.

In contrast, Leifer (1983) has published results suggesting that the ramps are explained in dogs by the Ta wave. Leifer pharmacologically induced 2nd degree blocks in dogs and thus could measure the Ta without interference from the QRS complex. Comparing normal beats with blocked beats, he found that the ramps in dogs were explained by the Ta.

The purpose of our study has been to characterize the signals in the P-R segment of the human magnetocardiogram (MCG). The goal was to identify and characterize signals associated with the atrial repolarization and the HPS activity.

METHODS

The magnetic measurements were made using a commercial 2nd order gradiometer, SHE biomagnetic probe model BMP-300. The superconducting coils are 2.36 cm in diameter, and the baseline is 3.18 cm. The signal was filtered by both the SHE analog 60 Hz notch and a 150 Hz low pass filter. A high pass filter was deliberately not used to minimize possible distortion of the data. A microcomputer then digitized the signal at the rate of 1 kHz and with a resolution of 10 femtotesla (fT). The data were later further filtered digitally to remove 120, 180, and 300 Hz and were then averaged using QRS triggering. It is estimated that the QRS triggering for most locations introduced a jitter of no more than 5 ms, but some locations, because of the nature of the QRS peaks may have a jitter of up to 8 ms. The RMS noise level of the output signal was, after averaging, about 26 fT. The measurements were taken on a grid suggested by Saarinen (1978). For most subjects, four minutes of data were taken at each location on the grid whenever the signal was strong enough to be measured.

We examined in detail three normal subjects and three patients with first degree blocks, GP, GM, and RD. GP is a 73 year old man with a 1st degree heart block and a right bundle branch block but otherwise normal. GP was our first abnormal patient, and we did not take enough data to allow us to average sufficiently to see small details in his MCG's. The second abnormal subject, GM, is normal except for the 1st degree block. His data were recorded during a weekday, increasing the noise level slightly. Lastly, RD underwent valvular surgery recently. Because some stainless steel sutures in the sternum caused localized magnetic interference, we could not get a full map of his MCG's. We did use RD's and GP's data, whenever possible, to confirm those results that we obtained from GM. Both GM and RD were males in their early 70's. The location of the blocks was not established in any of the patients.

RESULTS

Full grid maps of normal subjects were taken. We observed significant ramps in the P-R segment of all the normal subjects.We also observed significant long-lasting waveforms in the P-R segment of all the patients with the 1st degree blocks. These waveforms are seen immediately following the end of the P wave and much earlier than any known activity of the HPS system. In GP, the activity lasted for the duration of his long P-R segment, about 220 ms, until the onset of the QRS complex. Similar long lasting waves were seen in GM and RD, who had P-R segments of about 140 ms and 200 ms respectively (Fig. 1). Also features above the noise level are seen about 50 ms before the onset of the QRS in all the patients, including GP.

The features occurring just before the QRS onset could be identified visually in GP's records, but insufficient data were available to accurately analyze them. In the data of GM, the noise level was such that detailed analysis could be done on only five readings. All these readings were located in the two top rows of the map. The average duration of the features was 50 ms with a standard deviation (SD) of 8 ms and extrema 43 and 65 ms. The average amplitude was 210 fT with a SD of 70 fT and extrema of 125 and 330 fT. RD's data was less noisy because the readings were taken during a quiet weekend. The features were definitely seen in 6 readings of the MCG map. The average duration was 51 ms with a SD of 3 ms and extrema of 48 and 58 ms. The average amplitude was 150 fT with a SD of 45 fT and extrema of 90

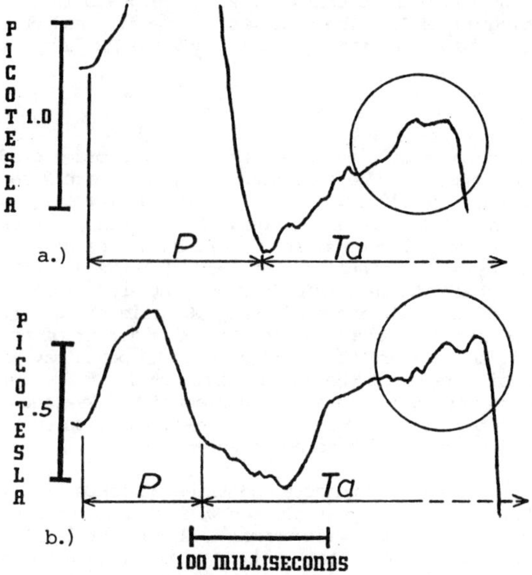

Fig. 1 P-R intervals of two subjects with 1st degree blocks. Circled are the possible HPS signals. The Ta is the long-lasting waveform.
a.) GM -- location B4, 181 averaged
b.) RD -- location C5, 251 averaged

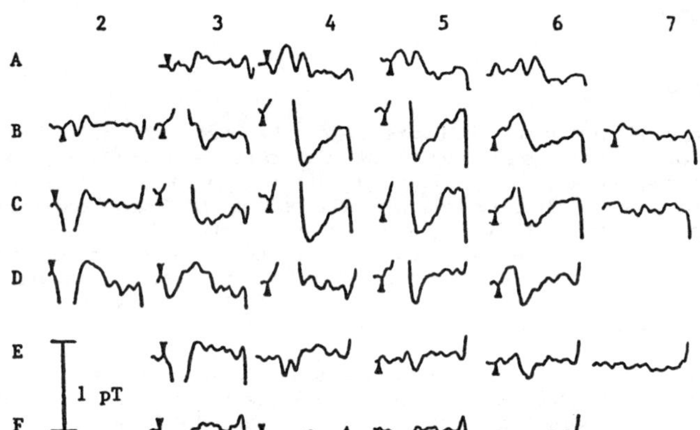

Fig. 2 Full P-R interval map of abnormal subject GM. Arrows indicate onset of the P-wave. Each record is 300 milliseconds long and is an average of about 180 heartbeats.

and 230 fT. RD's data showed features in the top four rows only, and they were for the most part plateaus as in part a of fig. 1. Some showed an almost ramp-like structure as in part b of fig. 1.

Most of the data from abnormal subjects did not show any features before the onset of the QRS complex, unlike the MCG's of the normal subjects, which showed ramps in almost all the places of the map. On the other hand, the long-lasting waveform was clearly visible after the end of the P-wave in almost all the measurements of the abnormal subjects. In some cases, this waveform was long-lasting and gradual as in part a, fig. 1., but in other cases, however, it showed some structure as in part b, fig. 1. The magnitude of the waveform varied from not detectable to about 1 pT, and a full MCG map showing its spatial variation in the patient, GM, is shown in fig. 2.

DISCUSSION

We confirmed the ramps and the symmetry that has been described by Farrell and co-workers. As pointed out by Erne, however, the symmetry is not observed in all subjects. We confirm Erne's results since we had two normal subjects who did not demonstrate the symmetry . The lack of symmetry is evidence that the proposed model by Tripp (1981) and further developed by Leoni (1982) is not valid in all cases.

We believe that the long-lasting waveform observed in the abnormal subjects is the atrial repolarization wave, Ta. The wave is different from what has been recorded with the EKG. Tranchesi [1960] looked at the Ta from patients with 2nd degree blocks. He saw a long lasting wave, 300 ms, with a relatively low amplitude and no remarkable structure. Little is known about this signal because it occurs at the same time as the much stronger QRS signal in normal individuals. Invasive studies in dogs have been done, however, that show that the atrial repolarization begins even while the depolarization occurs (Scher, 1974). Scher also reports that the atrial repolarization surface isopotential maps show a maximum on the upper right and a minimum on the lower left of the canine chest. This supports the view that the source of the signals is a flow of current directed diagonally across the chest from the upper right maximum to lower left minimum. This simplification may explain the rough symmetry that the possible Ta's of GM demonstrate in the full grid map. The gross symmetry we observe in the Ta can produce ramps with the symmetry seen by Farrell et al. This symmetry has been confirmed by the incomplete maps of RD and GP. Leifer's magnetic studies of the Ta in dogs show that the atrial repolarization is a long lasting slope as in part a, fig. 1, which accounts for the ramps in the normal subjects. There are two uncertainties with Leifer's evidence, however. Extrapolation of canine results to human may not be justified. Also, the work was performed with the chests open, perhaps altering volume currents and causing significant distortions in the observed signal. Our data indicate these are not significant concerns.

At several locations in all of the patients studied, the long-lasting waveform showed significant structures earlier than any known HPS activity (Fig. 1, part a). Because these structures were not expected, Farrell ruled out the possibility that the ramps that he had attributed to the HPS could have been caused by the Ta. Figure 3 shows that the Ta nicely accounts for the ramps and large structures seen in the P-R segment of normal individuals. Our data show that most of the changes in the Ta occur early, in the first 50 to 100 ms, and changes thereafter are small and diffuse. We cannot ascertain how long the Ta lasts since the QRS obscures the latter part in the abnormal subjects.

As mentioned before, most of the data on abnormal subjects showed no significant features that could be distinguished from the Ta just before the onset of the QRS. However, when features could be seen, they were of about .2 pT (Fig. 1). This is reasonable since the HPS has a much smaller mass than the atria, and one would expect the HPS signal to be on par with or perhaps smaller than the Ta even though the HPS signal is highly coherent. The high resolution EKG shows that the early Ta is indeed of the same size as the HPS activity (Horan, 1982), though the structures not associated with the HPS that we have observed in the Ta are absent in the EKG measurements. One would also expect, if the HPS signals are weak, to measure them only near the anatomical HPS, which may explain why the possible HPS features were only measured in the top four rows and not in the bottom, distant rows in all the abnormal subjects. These features are roughly the same amplitude as the structures and bumps observed by Fenici and Erne. It is likely that the features that we measured are the same that they have studied.

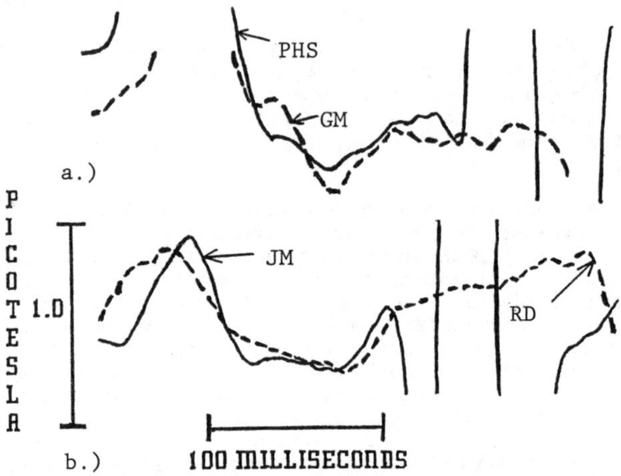

Fig. 3 Structures within the possible Ta explain the "ramps" reported by others. Solid are normal and dashed are abnormal subjects.
a.) PHS -- location B3, 206 averaged
GM -- location B4, 264 averaged
b.) JM -- location C5, 262 averaged
RD -- location C5, 251 averaged

The shape of these possible HPS features is highly dependent on jitter. The shape of most that we saw were plateaus. It is argued by Peper (1982) that the signal should be a plateau in the electrical case. As the signal travels through the A-V node, signals at the surface should be incoherent because of the multidirectional nature of the A-V activity. When the signal reaches the His-Bundle, however, the signal suddenly becomes coherent and the signal should present itself at the surface as a step. Models developed by Tripp and Fenici, however, predict a ramp. We did see features that resembled ramps, but these were few. More work needs to be done.

CONCLUSIONS

We believe that the ramps seen by Farrell and co-workers and thought to be a by-product of HPS activity can be explained by the unexpected finding that the Ta has structures that exhibit a symmetry similar to the ramps. Furthermore, we believe that we have identified both the Ta and the HPS activity separately in patients with 1st degree block. The possible HPS signals that we have seen probably are the same that Fenici and Erne have reported measuring within the ramps. We hope that the characterizing of the two signals will be useful in eliciting information about the HPS in normal subjects.

ACKNOWLEDGEMENTS

This research has been supported by a grant from the Whitaker Foundation.

REFERENCES

Erne, S. N., R. R. Fenici, H. D. Hahlbohm, M. Masselli, H. P. Lehmann, and Z. Trontelj (1983). Il Nuovo Cimento, 2 D, 248-254.
Farrell, D. E., J. H. Tripp, and R. Norgren (1980). IEEE Trans. Bio-med. Eng. BME-27, 345-350.
Fenici, R., G. L. Romani, S. Barbanera, P. Zeppili, P. Carelli, I. Modena (1980). G. Ital. Cardiol., 10, 1366-1370.
Horan, L., N. C. Flowers, and G. S. Sohi (1982). Circulation Res., 50, 47-54.
Leifer, M., N. Capos, J. Griffin, and J. Wikswo (1983). Il Nuovo Cimento, 2 D, 266-279.
Leoni, R., and G. L. Romani (1982). Il Nuovo Cimento, 1 D, 737-750.
Peper, A., R. Jonges, T. G. Losekoot and C. Grimbergen (1982). Med. & Biol. Eng. & Comput., 20, 195-201.
Scher, A. M., and M. S. Spach (1979). In K. M. Berne (Ed.), Handbook of Physiology, Williams & Wilkins, Baltimore, 357-392.
Tranchesi, J., V. Adelardi, J. M. de Oliveira (1960). Circulation, 22, 635-644.
Tripp, J. H., and D. E. Farrell (1981). J. Appl. Phys., 52, 2560-2564.

DISCRIMINATING LATE POTENTIALS USING BIOMAGNETIC TECHNIQUE

S.N. Erné, H.-D. Hahlbohm, J. Korsukewitz* and H.P. Lehmann

Physikalisch-Technische Bundesanstalt, Insitut Berlin,
Abbestrasse 2 -12, D-1000 Berlin 10, Germany
*Klinikum Charlottenburg, Freie Universität Berlin,
Spandauer Damm 130, D-1000 Berlin 19, Germany

ABSTRACT

In normals sometimes rapid deflections in the T wave of the body surface ECG have been observed. These deflections are similar in their characteristics to diastolic late potentials. Using magnetocardiographic techniques it is possible to suggest an explanation for the appearance of such signals and in this way to offer a method to discriminate such signals from recordings of late potentials with prognostic significance with respect to a high risk of sudden death.

KEYWORDS

Biomagnetism, Magnetocardiography, Late potentials.

INTRODUCTION

Disturbances in the propagation of the ventricular depolarization front due to locally jeopardized heart muscle tissue have been measured electrically on the epicardium (Durer and co-workers, 1961), on the endocardium (Horowitz and co-workers, 1980) and on the body surface (Berbari and co-workers, 1978). The fundamental problem of measurements on the body surface is the identification of measured signals as late potentials in the definition given on the basis of the epicardial and endocardial measurements. In general, all irregular rapid deflections within the ST segment until into the diastolic interval may be produced by delayed depolarization and thus may be identified as late potentials. Until now all identification criteria were based on the assumption that in normals no such deflections occur. The prognostic value of electrically measured late potentials with respect to malignant ventricular tachycardia has been studied by several authors as well as the incidence of late potentials in normal subjects which resulted in approximately more than 5% false positive findings.

In this paper we present a possible application of magnetocardiographic recordings, helping to discriminate electrical recordings of late diastolic potentials with prognostic significance from similar signals which could be recorded in normal subjects too.

METHODS

After informed consent 22 healthy subjects (8 female, 14 male, age between 22 and 34 years) were studied. All subjects were free of cardiovascular diseases and complaints according to clinical history and physical examination. No arrhythmias were found in the exercise stress test and 24 h Holter monitoring.

The electrical measurements were performed using the procedure previously applied to patients effected by chronic infarction and with history of ventricular tachycardia (Erné and co-workers, 1982). The data evaluation was concentrated on the bipolar lead recording V4 - V7, which has been demonstrated to be the most sensitive lead combination for the detection of late potentials.

The magnetic measurements were performed using the standard instrumentation in the Berlin Magnetically Shielded Room. A description of the measurement procedure as well as of the signal evaluation technique is given elsewhere (Erné and co-workers, this volume). After signal processing the magnetic data were used to calculate the field distribution and the position of the Equivalent Current Dipole (ECD) in correspondence with the apex of the T wave in the limb ECG (standard lead II). By definition the ECD is the macroscopic current dipole which generates a field distribution which matches the measured field distribution by a least square fit, considering the ECD embedded in a conductive semi-infinite medium.

RESULTS

We found electrical diastolic late potentials in 6 out of 22 normal subjects in the bipolar lead combination V4 - V7, coinciding with the apex of the T wave in the standard limb ECG lead II. As an example in Fig. 1 panel A is shown a beat sequence of one of the 6 healthy subjects with deflection in the diastolic phase and in panel B a sequence for a patient with history of sustained ventricular tachycardia. In both cases the signals coinciding with the apex of the T wave are indicated by arrows.

In normals as well as in patients this signals show a remarkable beat to beat variability, as shown in Fig. 2 for the same subject as in panel A of Fig. 1. A further characteristic common to the diastolic late potentials in normal and in high risk patients is the strong sensitivity to the electrode position. Fig. 3 shows the drastic effect of moving one electrode around the standard position V4 (again the same normal subject as in Fig. 1, panel A).

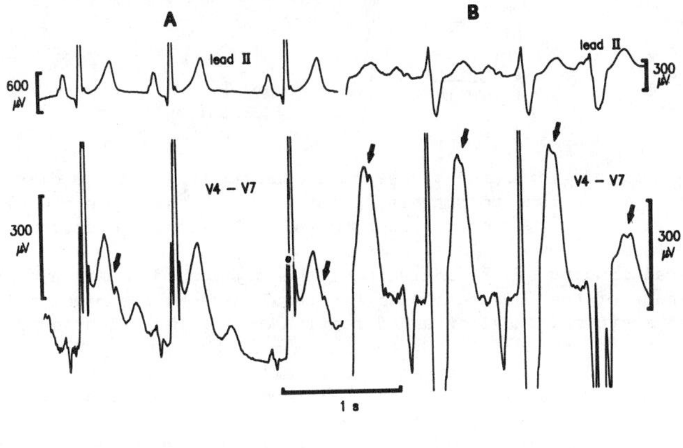

Fig. 1 Beat to beat sequences of electrical V4-V7 recordings in a healthy subject (panel A) and in a patient (panel B). Arrows indicate signals in the diastolic phase.

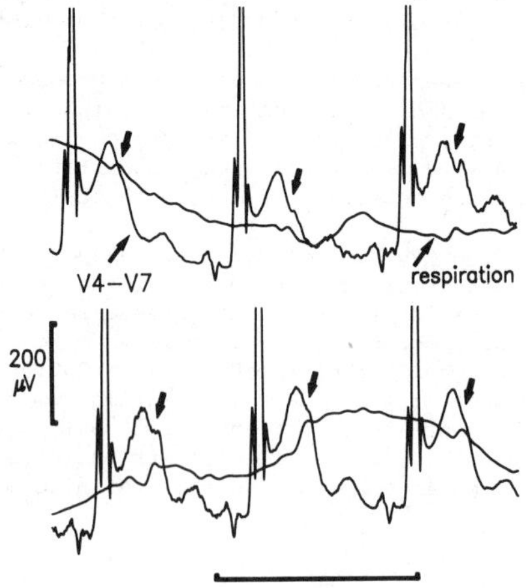

Fig. 2 Beat to beat variability of signals in the diastolic phase (same healthy subject as in Fig. 1, panel A).

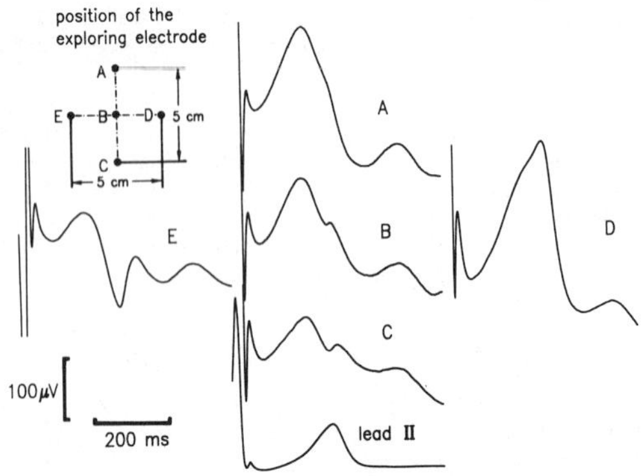

Fig. 3 Influence of the electrode position on the diastolic late potentials; position B for the exploring electrode corresponds to V4, reference is electrode V7.

The principal finding of the analysis of the magnetic data was a correlation between the appearance of the indentation within the T wave in the bipolar lead combination V4 - V7 and a typical location and direction of the equivalent current dipole calculated at the instant defined by the maximum of the T wave. Fig. 4 shows two examples of isofield contour maps for two normal subjects. Panel A refers to a subject with smooth T wave in all electrical recordings; panel B refers to a normal subject with electrical recordings V4 - V7 showing the indentation in the T wave. In this case the ECD, indicated by an arrow, is located in a high lateral position; in the general case (panel A) the BCD is located in a lower, more median position.

In the subject referred in Fig. 4 panel A it has been possible to induce high fre-

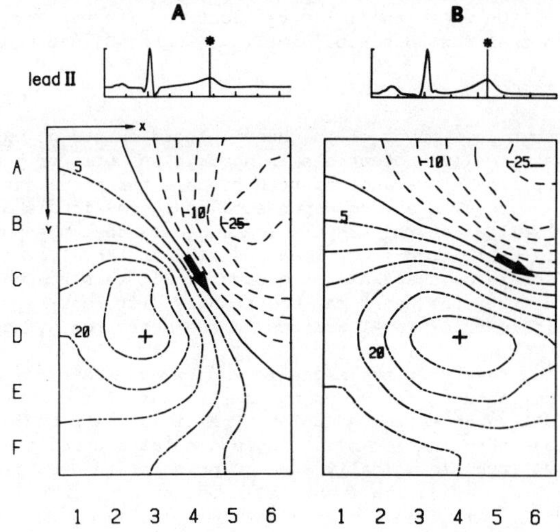

Fig. 4 Examples of isofield contour maps for healthy subjects. The maps are calculated in correspondence of the apex of T wave in the limb ECG. Panel A refers to a subject with smooth electrical recordings; panel B to the same subject as in Fig. 1 (A)., 2 and 3. The arrows indicate position and direction of the corresponding ECDs. The levels are given in pT.

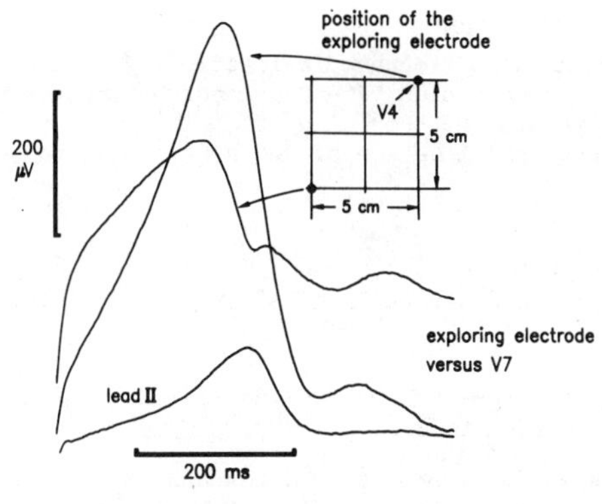

Fig. 5 Effect of positioning an electrode instead of in the standard position V4 in a lower and more right position with respect to the ECD location. In the recordings with this electrode combination an indentation appears in the diastolic phase.

quency deflections within the T wave by moving the electrode from the original location V4 to a position where the relative location with respect to the ECD is similar to the abnormal case described in Fig. 4 panel B (see Fig. 5).

DISCUSSION

On the basis of known physiology of the heart, there exists no explanation for the appearance of diastolic late potentials in normals in terms of physiological processes in the heart. The appearance of such signals therefore can only be due to the recording procedure. The strong dependence of these signals on the position of the electrodes as well as on the respiration suggests that the source could be searched in the measurement geometry. The position of the electrodes V4 and V7 is determined on the basis of the skeleton geometry. For this reason the position of the electrode V4 near to the heart may be strongly different in relation to the position of the bioelectric signal sources in the heart for different patients.

It is an interesting result of the magnetic measurements that the magnetic field distribution at the instant corresponding to the apex of the T wave in the limb recordings is in all subjects well dipolar. The field distribution can be represented in very good approximation by a single dipolar source, the equivalent current dipole (ECD). Physiologically a description of the ventricular repolarization in terms of a repolarization front is inappropriate, thus it is impossible to give a physiological interpretation of the ECD on the basis of a simple model. Nevertheless the ECD could be considered to be a representation of the total electrical activity of the heart and located in a weighted center of this activity.

On this basis the appearance of the diastolic late potentials in normals can be explained in terms of the measurement geometry: if the ventricular repolarization ECD is located higher and more left than the position V4 then in the recordings of the dipolar lead combination V4 - V7 indentations will appear in correspondence of the apex of the T wave in the limb ECG. This suggestion is confirmed by the fact that forcing the electrode from V4 to a suitable position predicted by the above defined condition it is always possible to induce such indentation in any normal (see for example Fig. 5).

In conclusion, irregular high frequency deflections during the diastolic phase in recordings of the lead combination V4 - V7 cannot be interpreted in general as indicator for late potentials. Further diagnostic informations are necessary, as for instance knowledge about the location of the ventricular repolarization ECD by biomagnetic measurements.

REFERENCES

Berbari, E.J., Scherlag, B.J., Hope, R.R. and R. Lazzara (1978). Am. J. Cardiol. 41, 697.

Durrer, D., Fromijne, F., Dam, R.T.H. van, Buller, J., Lier, A.A.W. van and F.L. Meijler (1961). Am. Heart J. 61, 303.

Erné, S.N., Fenici, R.R., Hahlbohm, H.-D., Lehmann, H.P., Leitner, E.-R. and M. Oeff (1982). Japan Heart J. 23 Supp 1, 703.

Erné, S.N., Fenici, R.R., Hahlbohm, H.-D., Korsukewitz, J., Lehmann, H.P. and Y. Uchikawa, this volume.

Horowitz, L.H., Josephson, M.E. and A.H. Herken (1980). Circulation 61, 1227.

MAGNETOCARDIOGRAPHIC ISOFIELD MAPPING IN THE CHARACTERIZATION OF THE INFARCTED AREA*

R. Gonnelli°, P. Galeone°, M. Sicuro°° and A. Tartaglia°

° Dipartimento di Fisica, Politecnico di Torino, Torino, Italia.
°° Divisione di Cardiologia, Ospedale San Giovanni Battista e della città di Torino, Torino, Italia

ABSTRACT

Some experimental MCG maps obtained using a superconductive gradiometer and a data acquisition system of original design are presented. Both normal and pathological subjects were considered. The experimental results show that the MCG isofield mapping approach can give useful informations about localization, depth and intensity of the cardiac generators. In particular, in a case of high lateral myocardial infarction it seems to be possible to identify a characteristic dipolar structure associated to the infarcted area.

KEYWORDS

Magnetocardiography; superconducting magnetometers; current dipole; myocardial infarction; infarcted area.

INTRODUCTION

Baule and Mc Fee (1970) were the first researchers that studied myocardial infarction using the magnetocardiographic technique. Since 1971 several others have been working on this field (Cohen and co-workers, 1971, 1975, 1976; Saarinen and co-workers, 1974; Denis and others, 1978; Siltanen and co-workers, 1983) both in view of the potential utility of magnetocardiographic recordings in the detection of steady injury currents, and for the higher spatial resolution of this technique compared with ECG.

Despite the number of papers appeared in the past years on the comparison between MCG map recordings of normal and pathological subjects, apparently it has not yet been possible to clarify the real importance and utility of this investigation tool in the diagnosis of myocardial infarction.

The purpose of this work is to evaluate the possibilities of MCG isofield maps in defining position, depth and size of the infarcted area. Data obtained from customary diagnostic tools (ECG, echocardiography, radionuclides, ventricular cineangiography) are used in the determination of some geometrical heart parameters and for comparison with our results. Our approach is based on two

*Work completely supported by Assessorato alla Sanità, Regione Piemonte

hypotheses:

1) In a first, simple approximation, the heart can be represented, at the QRS peak, as a current dipole in a homogeneus conducting half space (Williamson and Kaufman, 1981). The strenght of such a dipole can be correlated to the heart size.
2) The infarcted region can be represented as a current dipole whose direction is opposite to that of the excitation front in the normal heart in the same region. The strenght of this dipole can be correlated to the infarct size (Siltanen and others, 1983).

INSTRUMENTATION AND METHODS

The instrumentation used in detecting the magnetocardiographic maps of the next section is made of a superconductive SQUID based gradiometer with the usual electronic control equipments and a data acquisition system that we designed specifically for this purpose.
A layout of our experimental equipment is shown in Fig. 1.
The flux transformer we designed is a second derivative asymmetrical gradiometer. It is well known that this geometrical configuration allows the best reduction of background magnetic noise in a noisy urban laboratory, whereas the asymmetrical shape gives rise to a considerable increase of the relative sensitivity (Zimmerman, 1977). We checked that the relative sensitivity of our gradiometer changed from about 0.2 to about 0.37 on passing from the symmetrical to the asymmetrical configuration. The baseline of the gradiometer is 8 cm and the radius of the upper and middle coils is 2.42 cm, while the radius of the lower coil (pick-up coil) is 1.21 cm.
The cryogenic dewar is a common non-metallic, non-magnetic superinsulated fiberglass dewar (S.H.E. Model BMD 5) and the superconducting device is a R.F. toroidal SQUID (S.H.E. Model TSQX). The gradiometer balancing mechanism is made of three small superconducting plates placed perpendicularly to the x, y and z directions and movable from outside the dewar by means of three plexiglass rods emerging from the top flange. The electronic equipment consists of a SQUID control unit (S.H.E. Model 330X), some active filters (notch and band pass), an electrocardiograph, an analogic tape recorder and an oscilloscope.
The data acquisition system is low priced but provides quite good performances. It consists of an A/D converter at 8 bits and a microcomputer APPLE IIe with a 128 Kbytes memory and a dual floppy disks unit. The graphic output is realized by a plotter and by a graphic printer. The MCG signal and the ECG lead DII reference signal are simultaneously sampled at a frequency of 1000 Hz and for a time duration selected by the operator (from 1 sec to 32 sec for every chest position). The electronic delay between ECG and MCG signals is evaluated at the start of the acquisition process and then, at the end, the main program computes the average MCG signal over a number of beats variable from 2 to about 32 using as a trigger the QRS peak in the ECG.
The successive signal processing performs: a) the display of the average MCG for every position on the chest, b) the spatial smoothing of isochronous data on the thoracic grid using cubic spline functions, c) the plot of MCG isofield contour maps at every selected instant of cardiac activity. Three experimental MCG maps obtained using such equipment are shown in the following section. All the recording points of these maps have been chosen according to the standard anterior grid of each subject as described by Karp (1981) and the maps are scale plotted. They have been recorded at the end of a normal expiration and with the bottom part of the dewar tail in correspondence of an horizontal plane at a distance of about 13 millimeters from the xiphoid. The overall distance between pick up coil and chest surface was 30 $\pm$ 3 millimeters. The experimental equipment and the subjects were placed in a R.F. shielded room and there was a mechanical de-coupling between dewar support and patient wooden bed.

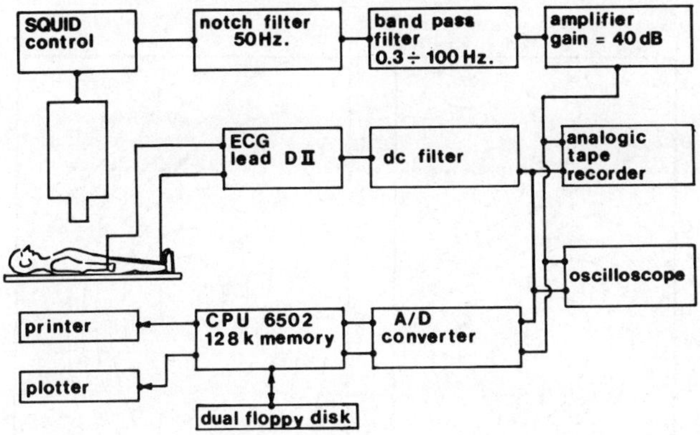

Fig. 1. Layout of our experimental equipment

RESULTS

Isofield Maps of Normal Subjects

Two maps of normal healthy subjects, at the instant of QRS peak in ECG lead DII, are shown in Figs. 2 a) and 2 b). They have been obtained from MCG signals averaged over about 16 heartbeats after a spatial smoothing with cubic spline functions and show clearly a dipolar structure. On the basis of the simplified hypothesis described in the introduction, we may calculate the approximate depth and intensity of the current dipole representative of the heart activity at the QRS peak instant. These approximate depths of the current dipoles, that are reported for both maps in the tables at the bottom of Figs. 2 a) and 2 b), were transferred on the echocardiographic bidimensional and monodimensional images taken along the parasternal long axis. In both cases they practically coincide with the barycentres of the ventricle sections in the plane of echocardiographic monodimensional scanning, in diastolic phase. The depths of these barycentres calculated on monodimensional images are about 57 and 68 mm respectively.
In the hypothesis, quite reasonable, that the current dipole intensity is correlated to the myocardium mass activated at the QRS peak instant, it would be necessary to make some comparisons with other techniques for a proper evaluation of the ventricle mass. It is quite clear that such a comparison is possible only in a average statistical way studying a large number of subjects. There are several difficulties in doing this type of analysis because there does not exist an universally accepted formula for the calculation of the ventricle mass by echocardiographic and angiographic measurements. The proposed formulae very often lead to completely incompatible results. On the other hand, the detected MCG signal is proportional to the component of the current dipole on the frontal plane that may present appreciable differences from the actual current dipole.
In the map relative to our subject P.G. the axis of the current dipole appears somewhat rotated clockwise compared with that of the other subject. This fact could be explained with increased dimensions of the right ventricle. In fact the echocardiographic 2D image shows a sensible growth of the right ventricle.

Isofield Map of a Pathologic Subject

Figure 3 a) shows the MCG isofield map of a subject with a myocardial infarction.

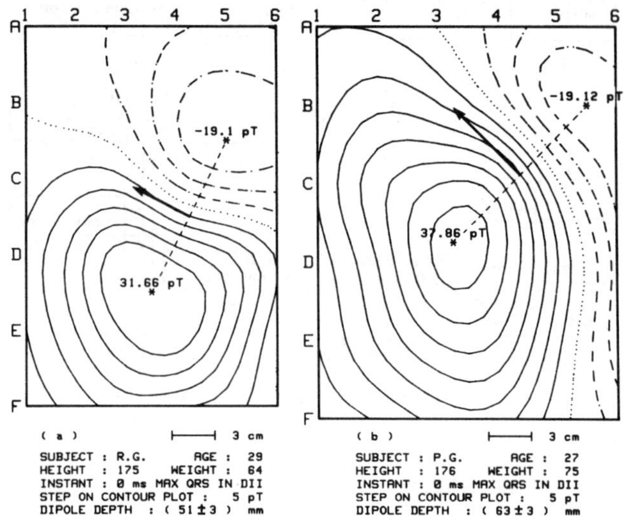

Fig. 2. MCG isofield contour maps of two normal subjects.

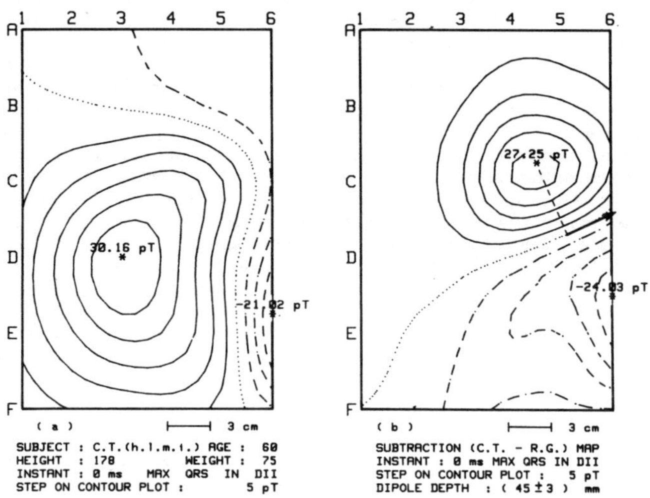

Fig. 3. a) MCG isofield contour map in a case of high lateral myocardial infarction.
b) Isofield contour map obtained by subtraction of map of Fig. 2 a) from that one of Fig. 3 a).

The data obtained by electrocardiographic, scintigraphic and angiographic routine analyses were in good agreement in defining the infaction as a small high lateral myocardial infarction. The map obtained by an averaging over about 10 heartbeats shows the isofield contour plot of the component of the magnetic field perpendicular to the frontal plane at the QRS peak instant. The isofield pattern exhibits a remarkable distortion of the dipolar structure with the minimum of the field shifted to a lateral position of the standard grid and, in the right side of the map, the electric axis (approximatively represented by the zero valued isofield line) rotated clockwise. From a qualitative point of view, the distortions of the isofield map can be explained, according to our preliminary hypothesis, by the superimposition, on the dipolar structure of the healthy heart, of a dipolar structure produced by an "infarction current dipole" representative of the necrotic region.

In order to obtain quantitative information from this MCG map it would be necessary to know or to reconstruct with appropriate criteria, the healthy map of the same subject. With the purpose of testing qualitatively the correctness of our hypothesis and to obtain some semi-quantitative results we substracted from the MCG map of the phatological subject the MCG isofield map of a normal subject taken as a reference (subject R.G., Fig. 2 a).

This particular healthy subject was chosen as a reference because he had the following characteristics quite similar to those of the pathologic one:

1) Standard grid dimensions.
2) Depth of the barycentre of the ventricle section as obtained by echocardiographic monodimensional image.
3) Isofield lines pattern in the zones of the map apparently less disturbed by the infarction dipole (lower left corner).

The subtraction map is shown in Fig. 3 b). One can see immediately that the isofield lines show, at least in the positive part of the map, quite a good dipolar structure. Unfortunately the negative part does not contain the absolute minimum, but it is possible to connect the absolute maximum with the zero-value line along the steepest descent line to obtain the approximate direction and point of application of the infarction dipole (see Fig. 3 b). Such double information permits, in this case, to conclude that the infarction occurred on the left lateral wall of the heart. Approximating the separation between absolute maximum and absolute minimum with twice the distance between maximum and zero line we can obtain a rough evaluation of the depth of the infarction dipole. For our subject this depth is about 45 $\pm$ 3 mm and therefore we may deduce the infarction took place on the high lateral wall of the left ventricle. Our conclusions are quite in agreement with the results of common diagnostic tools.

The intensity of the infarction dipole should be connected with mass and hence size of the necrotic region but in this case the results are completely unreliable. Within the limits of our hypotheses, it should be necessary to apply some fundamental corrections concerning the heart position on the standard grid and the ventricle mass of the pathologic subject. Whereas the first one seems to be simple enough, the second is rather difficult for the above mentioned reasons.

CONCLUSION

This work certainly contains a few approximations and simplifications. We ignored, in first approximation, the perturbation caused by the torso boundary and by the torso inhomogeneities. Moreover the heart activity at the QRS peak instant may be represented only in a rough way by a single current dipole. Other important sources of errors in our approach are certainly due to phisiological variation among the different subjects of the current dipole direction (direction of activation front) and of the size and position of the heart. We are studying now a simplified computer model of the heart magnetic activity taking into account the torso boundary shape and the other geometrical features of the heart, in order to correct the experimental data. At the same time, we started a program for the detection of a large number of MCG maps of normal and pathological subjects for the necessary statistical analysis. Nevertheless, the case studied seems to demonstrate that our simplified approach is able to provide useful clinical informations about the position and the depth of the infarted region.

ACKNOWLEDGEMENTS

We wish to thank very much Dr.Andreone and Dr.Boella of the Istituto Elettrotecnico Nazionale "G.Ferraris". Without their patient and very competent collaboration this work should not have been possible. We are very grateful to Prof. F. Demichelis and to Prof. P.F. Angelino for their encouragements. Many thanks also to Mrs. M. Di Tonto for typing the manuscript.

REFERENCES

Baule, G.M. and R. McFee (1970). The magnetic heart vector. Am. Heart J., 79, 223-236.

Cohen, D., Norman, J.C., Molokhia, F. and Hood, W. (1971). Magnetocardiography of direct currents: S-T segment and baseline shift during experimental myocardial infarction. Science, 172, 1329-1333.

Cohen, D. and Kaufman, L.A. (1975). Magnetic determination of the relationship between the S-T segment shift and the injury current produced by coronary artery occlusion. Circulation Res., 36, 414-424.

Cohen, D. Lepeschkin, M.D., Hosaka, H., Massell, B.F. and Myers, G. (1976). Abnormal patterns and physiological variations in magnetocardiograms. J. Electrocardiology, 9, 398-409.

Denis, B., Machecourt, J., Favier, C. and Martin-Noel, P. (1978). Interpretation du magnetocardiogramme normal et pathologique a partier de 140 cartographies thoraciques. Ann. Cardiol. Angeiol., 27, 81-86.

Karp, P.J. (1981). Cardiomagnetism. In Ernè, S.M.; Hahlbohm, H.D. and Lubbig, H. (Ed.), Biomagnetism, Walter de Gruyter, Berlin. pp. 219-258.

Saarinen, M., Karp, P.J., Katila, T.E. and Siltanen, P. (1974). The magnetocardiogram in cardiac disorders. Cardiovascular Res., 8, 820-834.

Siltanen, P., Katila, T.E., Poutanen, T., Seppanen, M. and Varpula, T. (1983). Comparison between measured and simulated magnetocardiograms in a case of anterolateral myocardial infarction. Il Nuovo Cimento, 2D, 2, 301-310.

Williamson, S.J., and Kaufman, J. (1981). Magnetic field of the cerebral cortex. In Ernè, S.N., Hahlbohm, H.D. and Lubbig, H. (Ed.), Biomagnetism, Walter de Gruyter, Berlin. pp. 353-402.

Zimmerman, J.E. (1977). SQUID instruments and shielding for low-level magnetic measurements. Jour. Appl. Phys., 48, 702-710.

SIMULATED AND MEASURED MAGNETOCARDIOGRAMS IN LOCALIZED MYOCARDIAL ABNORMALITIES

M. Saarinen*, T. Katila, J. Nenonen, M. Seppänen and P. Siltanen*

Department of Technical Physics, Helsinki University of Technology, SF-02150 Espoo, Finland
* Cardiovascular Laboratory, Helsinki University Central Hospital, SF-00290 Helsinki, Finland

ABSTRACT

In this study, simulated magnetocardiograms (MCGs) and electrocardiograms (ECGs) for different myocardial infarctions were studied using the concept of infarction vector. The simulated MCGs were also compared with measured clinical data.

KEYWORDS

Magnetocardiogram simulation, electrocardiogram simulation, infarction vector, current dipole, inverse problem

INTRODUCTION

One useful way in studying local myocardial infarctions in electrocardiography is the infarction vector. It is equivalent to the current dipole, defined as the difference of the current dipoles due to the activation waves between pre- and post-infarction. The objective of our study was to assess in which extent the infarction vector can be used to predict approximate MCG changes for a localized myocardial lesion. For a theoretical approach we have utilized the simulation program developed by Geselowitz (1980). We have also made comparisons between simulated data and data recorded in clinical cases. Such a comparison for one type of infarction has been made previously by our group (Siltanen and others, 1983).

It was anticipated that similar approximations for changes in MCG signals as in the ECG can be obtained using the infarction vector and that it gives some information on the location of the injury (inverse problem). From the clinical point of view it is important to compare the information power of the MCG and ECG regarding the location and extent of the infarction. The simulation technique offers a useful means to investigate this relationship.

MATERIAL AND METHODS

We simulated six types of myocardial infarction: septal, anteroseptal, anterior, anterolateral, posterolateral and inferoposterior. In each type a necrotic scar area was placed in the basal, middle or apical area of the left ventricle. The size of the lesion was about 10-15 % of the muscle mass of the left ventricle. Computer simulations for the conventional 12-lead ECG and anterior mapping of the ECG and the component of the MCG perpendicular to the frontal plane were performed. The ECG and MCG maps were calculated using the standard MCG grid (Saarinen and others, 1978).

In the simulation model the portion of the signal caused by the infarction vector can be found out by calculating the difference between the abnormal and normal ECGs and MCGs. In this way we produced difference maps for the 18 different infarction cases by subtracting the simulated normal curves from the abnormal ones. The amplitudes and polarities of these difference signals during the QRS complex depend on the orientation of the infarction dipole which, on the other hand, depends on the location of the scar area. The difference maps for the MCGs give information mainly about the tangential component of the dipole whereas the normal component gives an additional contribution to the ECG differences. Using these principles the approximate orientation of the infarction current dipole was estimated from the MCG and ECG difference maps.

Our clinical patients were chosen by typical changes in their 12-lead ECG caused by the infarction (septal, anteroseptal, anterior, inferoposterior and anterolateral infarction). They were from a consecutive series admitted to the hospital for cardiac evaluation - coronary angiography and possible by-pass surgery. The ECG and MCG mapping was done according to the standard MCG grid. The component of the magnetic field perpendicular to the frontal plane was measured with a first-order SQUID gradiometer (pickup coil ϕ 3.9 cm, baseline 10 cm). The other details of the experimental setup are the same as in one of our previous studies (Saarinen and others, 1978). The Q, R, S and R' waves of the measured MCGs were compared with our normal material (Saarinen and others, 1978) using the statistical distributions of the corresponding normal waves. This was done in order to find out the distribution of the changes in the magnetic QRS complexes. These distributions were compared with the calculated difference maps.

RESULTS

When comparing the simulated normal ECG and MCG maps with our normal material some distinct differences in the relative amplitudes of the different deflections can be observed. These are due to the biological variability e.g. in the torso geometry. However, these differences do not interfere with the interpretation of the maps to understand the general trends in the MCG changes due to the lesions in different locations.

In the different simulated types of myocardial infarction the following MCG changes during the QRS complex were found with the aid of the computer model of Geselowitz:

Septal lesions: enhancement of the magnetic field in the lower right part of the grid and diminuation in the left upper part. This distribution of the MCG difference field is consistent with the orientation of the infarction current dipole left-, back- and downwards. This pattern was found in the basal and middle septal lesion, but in the apical lesion the changes were small.

Anteroseptal lesions: enhancement in the central and right middle part of the grid and reduction in the left lower quadrant; infarction vector directed back- and downwards.

Anterior lesions: enhancement in the right upper quadrant of the grid and diminuation in the left lower quadrant; infarction vector directed backwards, downwards and to the right.

Anterolateral lesions: enhancement in the upper half of the grid and reduction in the lower part; infarction vector directed from left to right and anteriorly.

Posterolateral lesions: slight enhancement in the left upper quadrant of the grid and marked reduction in the lower half; infarction vector directed from left to right and anteriorly.

Inferoposterior lesions: slight enhancement in the left lower quadrant of the grid and marked reduction in the right half; infarction vector directed upwards, to the left and anteriorly.

The measured MCG patterns of the clinical cases followed satisfactorily the principles demonstrated by the simulation studies. There is a good morphological similarity in the simulated MCGs and those of the patient when taking into acount the difference of the simulated normal MCG map and that of our normal material. An example of these cases is shown in Fig. 1. In Fig. 1a are shown the simulated MCGs of an inferoposterior infarction (solid lines). The broken lines indicate the changes as compared with the calculated normal MCGs (the difference map). Fig. 1b shows the MCGs of the inferoposterior patient and a comparison of the infarction MCGs with our normal material is demonstrated. The changes in the magnitudes of the QRS complex are indicated by letters. Capital letters are used for those waves with a significant increase and small letters for those waves with a significant decrease (e.g. R for above the 75 % limit and RR for above the 90 % limit and r for below the 25 % and rr for below the 10 % limit of the cumulative distribution of the R wave). The simulation gave a decrease in the right middle part of the grid and a small increase in the left lower quadrant. The infarction vector orients transversally upwards and anteriorly to the left. When comparing the MCGs of our patient with our normals the same decrease was seen on the right side of the grid and a small area with increased amplitudes on the left side of the grid. The simulated and measured 12-lead ECGs are shown in Fig. 2.

A general feature in the simulated MCG maps was that in cases with the infarction vector directed normal or nearly normal to the frontal plane the difference amplitudes were rather small and consequently the findings were close to the normal MCG. Correspondingly the difference signals of the ECG maps show a unipolar maximum above the infarction vector. In these cases the distribution of the MCG changes depends on the orientation of the small tangential component of the infarction vector and it may be difficult to compare the simulated distributions with the measured data.

DISCUSSION

The infarction vector can be succesfully used as a first approximation to predict the MCG changes due to localized myocardial infarctions. There were some distinct differences between the recorded and simulated data which may arise from the cardiac and torso geometry in the simulation model and from the difficulties in simulating the later phase of the propagation front of depolarization. This is especially true when the heart muscle contains damaged regions. However, the general patterns were parallel in the simulated and measured MCG and ECG data.

In order to obtain clinically distinct morphological changes a scar area of about 10 % of the left ventricular mass had to be used in the simulation model. The usefulness of the MCG in diagnosis and localization of myocardial lesions is limited by the fact that in individual cases the pre-infarction MCG pattern is not known. However, the situation is the same in electrocardiography, too. In both methods the normal reference is a statistical distribution of different waves obtained from healthy subjects. Thus, the comparison of experimental MCG data can

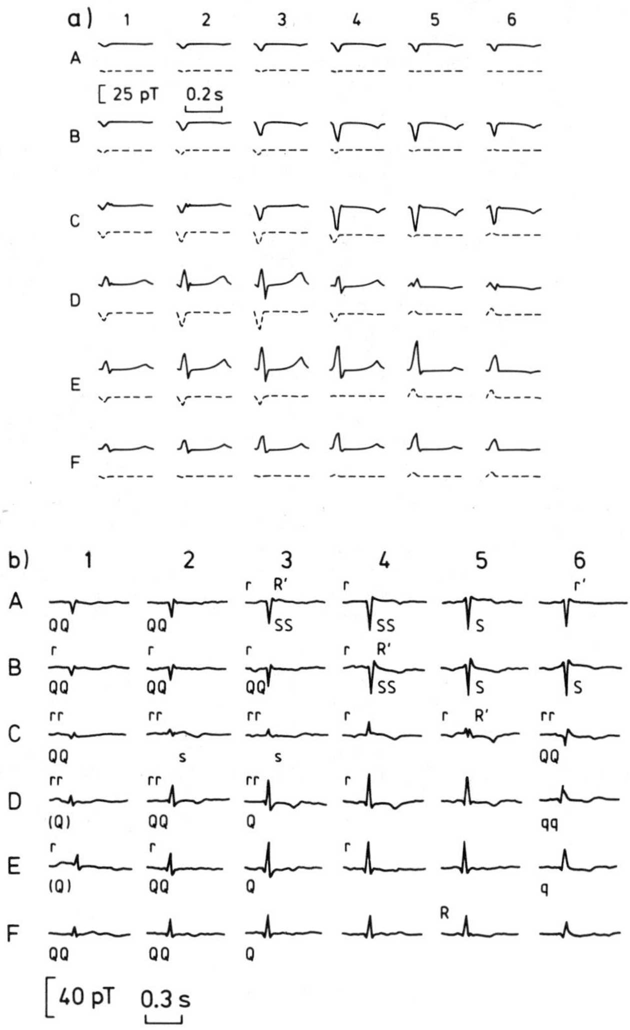

Fig. 1. a) The simulated MCG map for an inferoposterior myocardial infarction. A positive deflection corresponds to a field directed into the chest. The figure also shows the difference as compared with the normal case (broken lines).
b) The measured MCG map for a patient with an inferoposterior myocardial infarction. The band is 0.05 - 100 Hz. For more details, see text.

in most cases be made only with statistical normal material. This difficulty is not present when studying the origin of such abnormal signals which appear as distinct additional deflections without being superimposed on the other waves. Abnormal signals of this type might be caused e.g. by late ventricular activity.

The relative changes in terms of amplitude are of the same order in both ECG and MCG. However, due to the different weighting of the sources these two methods give complementary information on the location and orientation of the sources.

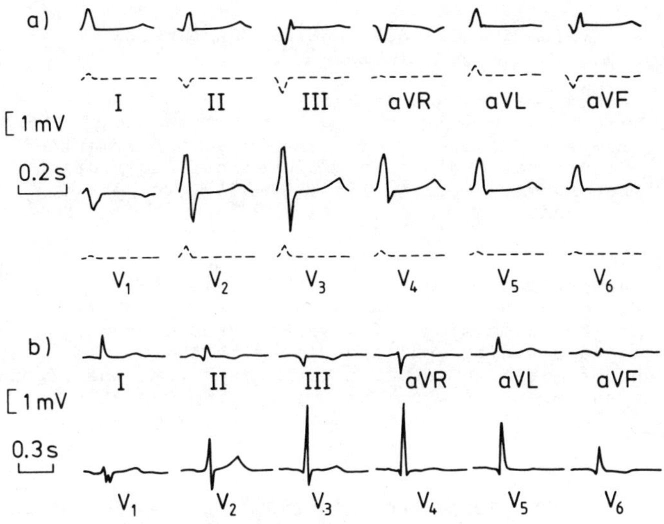

Fig. 2. a) The simulated 12-lead ECG for the infero-posterior myocardial infarction (solid lines). The broken lines indicate the difference as compared with the simulated normal case.
b) The measured 12-lead ECG for a patient with an inferoposterior myocardial infarction.

REFERENCES

Geselowitz, D. B. (1980). IEEE Trans. Magn., MAG-16, 812-817.

Saarinen, M., P. Siltanen, P. J. Karp, and T. E. Katila (1978). Ann. Clin. Res., 10, S21, 1-43.

Siltanen, P., T. Poutanen, T. Katila, M. Seppänen, and T. Varpula (1983). Nuovo Cimento D, 2, 301-310.

OPTIMIZATION OF SQUID VECTOR GRADIOMETER

Jukka Lekkala and Jaakko Malmivuo

Tampere University of Technology, Department of Electrical Engineering, Biomedical Engineering Laboratory, P.O. Box 527, SF-33101 Tampere 10, Finland.

ABSTRACT

The structure and dimensions of a vector gradiometer have been optimized for magnetocardiographic studies with the unipositional lead system. The optimization was made with respect to the signal-to-noise ratio and accuracy when determining the magnetic heart vector.

KEYWORDS

Magnetocardiography, SQUID-vector gradiometer, optimization.

INTRODUCTION

We have designed a vector gradiometer for magnetocardiographic studies with the unipositional lead system (Malmivuo, 1976). In this system the magnetic heart vector (MHV) is determined from the recording of the magnetic field vector (MFV) at a single point on the anterior side of the chest. MHV is the equivalent magnetic source of the distributed current in the heart muscle and assumed to locate in the geometric center of the heart. The purpose of this study was to find a vector gradiometer structure which would give the maximum signal-to-noise ratio together with minimum error when determining the MHV.

OPTIMIZATION PROCEDURE

The optimization of the gradiometer structure was made with respect to the signal-to-noise ratio when the space for the gradiometer in the cryostat was limited and the distance between the source and the inner bottom of the cryostat was fixed. It was also examined the effect of the dimensions and number of turns of the pick-up coil on the accuracy of the measurement when determining the MHV from the measured magnetic flux.

In the basic structure of the unipositional vector gradiometer the x-gradiometer which measures the field gradient perpendicular to the chest (dB_x/dx) is constructed coaxially. In the y- and z-directions there are coplanar gradiometers which measure the off-diagonal component of the magnetic field gradient ($dB_y/dx, dB_z/dx$) (Lekkala and Malmivuo, 1984). In all cases three different structures were optimized, one symmetric and two asymmetric gradiometers where the ratio η between the areas of the second and pick-up coil was 2 and 3, respectively. The signal source was modelled with a magnetic dipole whose dipole moment was set to 1 μAm^2.

The effective inductance of a SQUID depends on the inductance of the detection coil and the coupling coefficient k_s between the signal coil and the SQUID. The dependence of the system noise from SQUID's effective inductance L_e and the effect of that on the S/N-ratio were studied with different values of the coupling coefficient. Fig. 1 shows the difference in maximum S/N-ratios as a function of the coupling coefficient k_s in three cases: the detection coil is optimized for maximum S/N-ratio when the effect of the SQUID inductance on the noise
1) is not taken into consideration, and when the noise is proportional 2) to L_e and 3) to $L_e^{1/2}$. Here we have used the value of 20 μH for the signal coil inductance. In the calculations we have used the coupling coefficient of 0.3 and assumed that the equivalent flux noise of the system is proportional to $L_e^{1/2}$. The calculations were performed with VAX 11/780 computer.

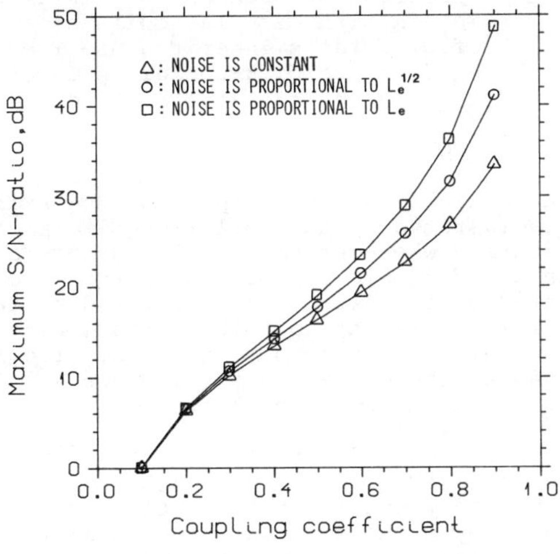

Fig. 1. Difference in maximum signal-to-noise ratio as a function of the coupling coefficient k_s in three cases. The 0 dB level is set to the case when the coupling coefficient is 0.1.

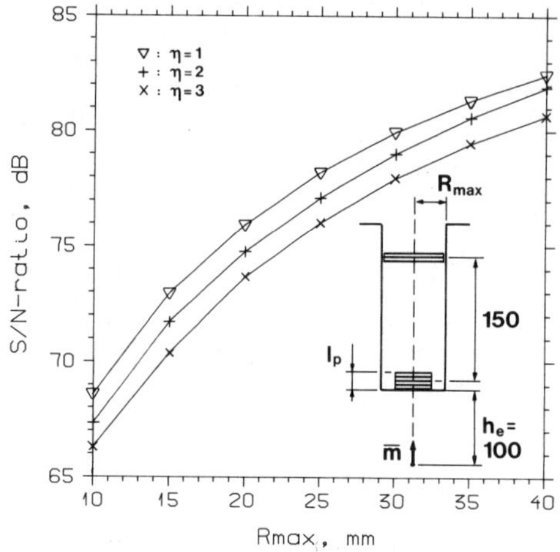

Fig. 2. Maximum available S/N-ratio of three different structures of coaxial x gradiometers with circular coils as a function of cryostat's inner radius. The distance h_e from the pick-up coil's lower end to the source is fixed to 100 mm.

SPATIAL RESOLUTION

When measuring a coaxially aligned dipole on the axis of a coil the optimum length/radiusratio of the coil (x-coil) is of the order of 1.7. Then the accuracy when determining the source dipole is highest. For the case where the dipole is pointed in the direction of the coil axis and is located on the extension of its radius the length/radius-ratio of the coil (y-coil) is about 1.5 for the best accuracy. Using these dimensions it is possible to measure the stregth of a magnetic dipole as close as at the distance of 2R, where R is the radius of the coil, with the maximum error of 3 %.

However, this is valid only for coils with high turn density. When the turn density is lower the optimum length/radius-ratio decreases. For the x-coil with only two loops the optimum ratio is 1 or the loops are separated by R where as for the y-coil the ratio is lower or around 0.9. As a conclusion, when the coil radius is greater than 10 % of the measurement distance the coil dimensions should be designed carefully for accurate results. Calculations with rectangular coils show that it is not possible to obtain the accuracy of circular coils with as short measurement distances, on the contrary, the corresponding distances are 10-20 times longer depending on the form of the coil. We have used the baseline of 150 mm in all calculations. The real measurement distance is of the order of the diameter of the heart or 110 mm.

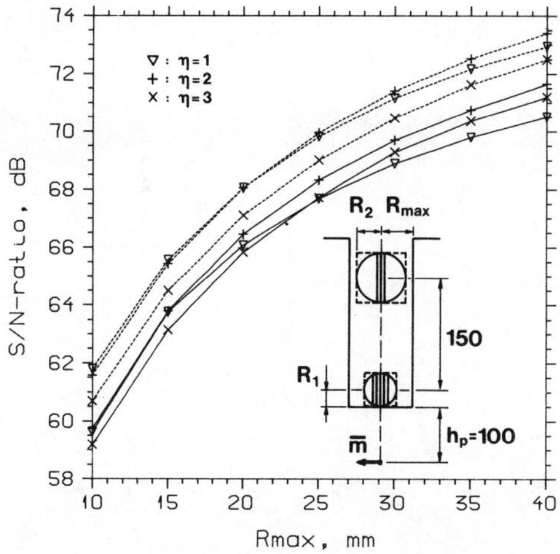

Fig. 3. Maximum S/N-ratio for y-gradiometers as a function of cryostat's inner radius when the cross-section of the gradiometer coils is circular (solid lines) and rectangular (dashed lines). The distance h_p from the periphery of the pick-up coil to the source is fixed to 100 mm. In the gradiometers with rectangular coils the coil area is equal to the area of the corresponding circular coil.

RESULTS

The optimization was made for the x-gradiometer with circular coils in three cases when the distance from the source to 1) the center of the pick-up coil, 2) the lower end of the pick-up coil and 3) the periphery of the pick-up coil of the y-gradiometer was fixed. In the first case the symmetric and the asymmetric gradiometer with $\eta = 2$ up to the maximum radius of 32 mm show nearly the same optimum S/N-ratios the symmetric one being only 0.5 dB better. The S/N-ratio of the asymmetric structure with $\eta = 3$ is about 2 dB lower than that of the symmetric gradiometer.

If the distance from the lower end of the pick-up coil to the source is fixed (this is the normal measurement situation with a single coaxial gradiometer) the highest signal is obtained also in this case with the symmetric gradiometer with the maximum coil radius R_{max} up to 40 mm. It is nearly 2 dB better than the asymmetric one with $\eta =$ 2 and 3 dB better than that with $\eta = 3$ in the range R_{max} = 10 - 35 mm, Fig. 2. It should be noted that in this case the structure for optimum sensitivity is not the same as the structure for maximum S/N-ratio. In this case the S/N-ratio of all three gradiometer types increases when the length of the second coil increases. For the symmetric case the length of the pick-up coil should be wound tightly, whereas for both asymmetric cases an optimum length can be found.

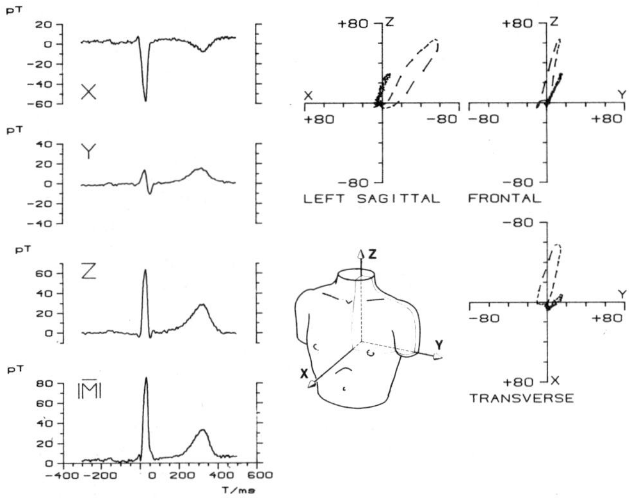

Fig. 4. Real time recording of the magnetic heart vector from a 30-year-old male. The measurement is performed with an optimized vector gradiometer using the unipositional lead system. The measurement distance is 106 mm.

TABLE 1 Dimensions and Data of the Optimized Vector Gradiometer

	Pick-up coil	Second coil
X-GRADIOMETER:		
Radius, mm	16.20	22.91
Length, mm	29.0	20.0
Number of turns	22	11
Y- AND Z-GRADIOMETER:		
Radius, mm	16.20	22.91
Length, mm	22.0	20.0
Number of turns	22	11

Baseline = 150 mm
Sensitivity, X: 20.3 $fT_{rms}(Hz)^{-1/2}$, Y,Z: 20.5 $fT_{rms}(Hz)^{-1/2}$
S/N-ratio, X: 75.2 dB, Y,Z: 69.2 dB
Inductance, X: 17.7 μH, Y,Z: 19.9 μH

According to this, a symmetric gradiometer is recommended for a single channel measurement if the maximum S/N-ratio is wanted.

In the vector gradiometer the distance from the source to the center of the pick-up coil depends on the dimensions of the y- and z-coils. If the y- and z-coils are circular and similar to the x-coil the optimum structure of the x-gradiometer is asymmetric with $\eta = 2$ when R_{max} is below 40 mm.

The optimization was made in the x-direction also with a circular current loop and a current ball surface, 50 mm in diameter. The only difference in the optimized structure with a current loop when compared to a dipole was that the length of the pick-up coil in asymmetric gradiometers tended to increase of the order of ten percent for the optimum result. The current ball surface source gave in all cases equivalent results with a dipole.

In the y- and z-gradiometers the optimization was made for both circular and rectangular coils when the distance from the source to the periphery of their pick-up coil was fixed. Among the y- and z-gradiometers with circular coils the asymmetric gradiometer $\eta = 2$ shows the best response on the whole examined range $R_{max} = 10 - 40$ mm, Fig. 3. The symmetric structure gives a similar S/N-ratio for small radii but on higher radii it is worst. When the y- and z-coils are rectangular the highest S/N-ratios are obtained with the structure where the pick-up and second coils are of the same width. In asymmetric cases the height of the pick-up coil is reduced. The S/N-ratios of the symmetric and the asymmetric structure with $\eta = 2$ are quite close to each other, Fig. 3. With rectangular coils the S/N-ratios are over 2 dB higher than with circular coils. The dimensions and properties of the optimized vector gradiometer are shown in Table 1. Figure 4 shows a real time vector MCG recording made with a multiplexed SQUID vector gradiometer (Lekkala and Malmivuo, 1984).

REFERENCES

Malmivuo, J. (1976). On the detection of the magnetic heart vector - An application of the reciprocity theorem. Acta Polytechnica Scandinavica, EL39, Helsinki.

Lekkala, J., and Malmivuo, J. (1984). Multiplexed SQUID-vector-magnetometer for biomagnetic research. J. Phys. E: Sci. Instrum., 17, 504-512.

STATISTICAL ANALYSIS OF THE RELATIONSHIP BETWEEN VMCG AND VECG

J. Nousiainen, J. Lekkala and J. Malmivuo

Biomedical Engineering Laboratory, Tampere University of Technology, P.O. BOX 527, SF-33101 Tampere, Finland

ABSTRACT

The statistical distribution, the angle and the magnitude ratio between the normal magnetic heart vector and electric heart vector are studied. A low sensitivity of the MHV to individual variations and detector displacement are observed. The angle and the magnitude ratio between the normal MHV and EHV are in accordance with results predicted theoretically.

KEYWORDS

Magnetic heart vector, electric heart vector, unipositional lead system.

INTRODUCTION

The theory of MCG and ECG shows that the relation between them depends on the electro-physiological restrictions in the cardiac source. A simple, widely used model for the cardiac activation is the cupshaped wavefront, which forms a uniform double-layer source. If the MCG and ECG were produced by this kind of source in a spherical conductor, then the magnetic heart vector (MHV), $\bar{m}$ and the electric heart vector (EHV), $\bar{p}$ should be related by (Wikswo, 1979) $\bar{m} = 1/2(\bar{r} \times \bar{p})$ where $\bar{r}$ is the vector from the origin to the center of the circular double-layer rim. This relation predicts, at least for a uniform double-layer source with a circular rim, that the MHV and EHV should be perpendicular. It shows also that the MHV is most sensitive to the current sources which produce a tangential EHV, and insensitive to the sources with radial EHV. The sensitivity of MCG will increase with radial distance from the origin. This is consistent with a discussion of the sensitivity lines of vector magnetocardiography (VMCG) and vector electrocardiography (VECG) in the lead field theoretical approach (Malmivuo, 1976).

METHOD

In this study, the relationship between VMCG and VECG is approached experimentally. The MHV and EHV is analysed from 18 normal persons without any history in cardiac diseases (Nousiainen, 1984). The MHV was determined from simultaneous recording of the three perpendicular components of the magnetic field by using the multiplexed SQUID vector magnetometer (Lekkala, 1984). The effect of inhomogeneities of the thorax was corrected using the unipositional lead system and multiplying the postero-anterior, right-to-left and infero-superior components of the recorded magnetic field with the coefficients 1,-1,-1, respectively (Eskola, 1984). In this lead system, the location of the detector is at the fourth intercostal space above the point V2 in the standard ECG. With these arrangements the most optimal lead field of the unipositional lead system in the heart's region was reached in the studies with the real torso model. The EHV was measured with the Frank lead system.

RESULTS

1. Normal MHV. The mean magnitude and orientation of the MHV and EHV at the maximum R and T waves and their standard deviations are given in Table 1.

TABLE 1 Direction and Magnitude of the Maximum R and T Waves of the MHV and EHV and their Standard Deviations in Spherical Coordinates.

		Magnitude	Horizontal angle	Vertical angle
QRS	MHV	47 pT ± 40 %	$160^{\circ} \pm 15^{\circ}$	$41^{\circ} \pm 14^{\circ}$
	EHV	1.5 mV ± 40 %	$102^{\circ} \pm 30^{\circ}$	$-38^{\circ} \pm 17^{\circ}$
T	MHV	13 pT ± 60 %	$127^{\circ} \pm 18^{\circ}$	$49^{\circ} \pm 11^{\circ}$
	EHV	0.6 mV ± 40 %	$53^{\circ} \pm 13^{\circ}$	$-32^{\circ} \pm 10^{\circ}$

The direction of the maximum R wave in the MHV varies less than in the EHV. The deviation ranges of the vector magnitudes are approximately similar.

2. Angle between the MHV and EHV. The mean angle between the MHV and EHV during the QRS complex is depicted in Fig. 1 (solid line). At the R wave the mean angle is about $98^{\circ} \pm 8^{\circ}$ (SD). The rapid change in the angle occurs about 60 ms after the onset of the QRS complex at the instant, when there is seen a notch in the magnitude tracings of the MHV and EHV in Fig. 2.

The angle at the maximum magnetic T wave was found to be about $102^{\circ} \pm 13^{\circ}$ (SD). The angle increses very slowly from the beginning of the T wave and varies less than 10° during a period of 100 ms around the maximum T vector.

3. Effect of the detector displacement. The effect of the detector location on the MHV was studied measuring the MHV at 25 locations. The orientation of the measured MHVs was compared to the direction

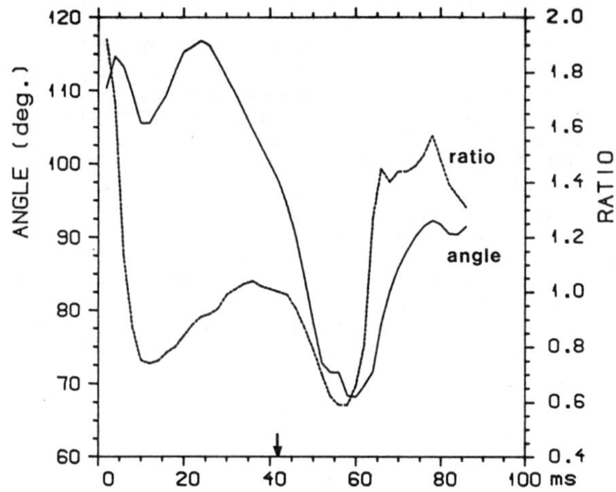

Fig. 1. Mean angle (solid line) and magnitude ratio (dashed line) between the MHV and EHV during the normalized duration of the QRS complex. The arrow indicates the instant of the maximum MHV.

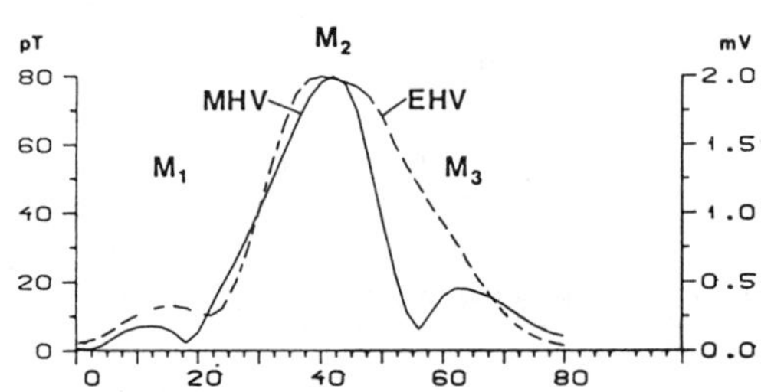

Fig. 2. Simultaneous plots of the magnitude curves of the MHV and EHV during the QRS complex for a normal subject.

of the MHV detected at the original measurement point. The deviations in the orientation of the MHVs are depicted as isogonal curves(solid) in Fig. 3. A 10 mm displacement of the detector in the frontal plane above the heart gives rise to a mean error of about 5° in the orientation and about 6 % in the magnitude of the maximum MHV.

The effect of the detector location on the angle between the MHV and EHV was also tested. The mean angle at the R wave was measured at 25 locations and is shown as isogonal curves (dashed) in Fig. 3. The angle of 90° is detected about 20 mm inferiorly from the detector location of the corrected unipositional lead at the fourth intercostal space.

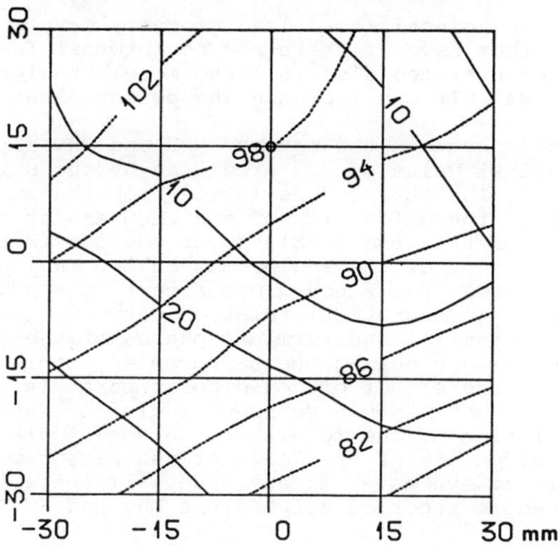

Fig. 3. Effect of the detector displacement on the orientation (solid isogonal curves) of the MHV in degrees when compared to the MHV measured at the original location (denoted by a circle), and on the angle between the MHV and EHV (dashed isogonal curves).

4. Ratio between the magnitudes of the MHV and EHV. The ratio between the magnitude plots of the MHV and EHV during the QRS complex can be used to describe the differences in relative sensitivities of the VMCG and VECG leads to the electric activity of the heart. The magnitude ratio between the MHV and EHV during the QRS complex, when the sensitivities are normalized to be equal at the maximum, is depicted in Fig. 1 (dashed curve). At the initial phase of the QRS complex the magnitude ratio is almost on every normal subjects (95 %) less than one indicating that the EHV is more sensitive than the MHV to the excitation in radial direction in both ventricles. The two minima in the ratio plot occur again at the instants, when there are seen notches in the magnitude tracings in Fig. 2. About 16-38 ms after the maximum MHV the mean sensitivity ratio is greater than one (for 60 % of the normal subjects) indicating greater sensitivity for the MHV, when the advance of the wavefronts is mostly tangential.

DISCUSSION

Our comparative VMCG and VECG study on the same normal material demonstrates that the interindividual variability of the mean characteristics of the MHV and EHV is approximataly similar. The observed low variation range in the direction of the QRS loop of the normal MHV as well as the small effect of detector displacement can be due to the self-centering effect suggested by Baule and McFee (1970). The great difference in conductivity between the well-conducting heart muscle and lowconducting lungs ensures that slight differences in the unipositional lead geometry have a smaller effect on the MCG than they have if there are no internal inhomogeneities. The effect is demonstrated also in torso model studies (Malmivuo, 1976, Eskola, 1984). It can increase the potential usefulness of MCG.

The observed angle between the maximum MHV and EHV differed slightly from the theoretical value of 90° predicted by the uniform double-layer in a sphere. The observed deviation from the expected relationship may be due to the unideality of the VMCG and/or VECG leads and/or a non-dipolar part in the double-layer sources (Wikswo, 1983). The temporal variation of the angle between the MHV and EHV during the QRS complex shows that a non-dipolar part in the double-layer sources distorts the theoretical relation between the MHV and EHV especially at the initial and terminal phases of the QRS complex. Because the angle changes during the QRS complex, it is very difficult to evaluate, when the effect of non-dipolar parts is at the minimum. These non-dipolar components have been studied by Karp et al. (1980). Whether the deviation is due to either the non-dipolarity of the source or the unideality of the leads or the measurement location, is yet difficult to evaluate. In any case, all these factors have a clear on the angle recorded between the MHV and EHV.

The ratio between the magnitudes of the MHV and EHV support the theoretically well known difference in sensitivities of the MHV and EHV: EHV seems to be more sensitive to radial part of the excitation and MHV more sensitive to the final phase of the ventricular depolarization , when the activation appears in a more tangential direction.

REFERENCES

Baule, G. and McFee, R. (1970). The Magnetic Heart Vector. *Amer. Heart J.*, 79, 223-236.

Eskola, H., Malmivuo, J., Nousiainen, J. and Lekkala, J. (1984). The corrected unipositional lead system for vector magnetocardiography. *IEEE Trans. Biomad. Eng.* (accepted for publication).

Karp, P., Katila, T., Saarinen, M., Siltanen, P. and Varpula, T. (1980). The normal human magnetocardiogram II: A multipole analysis. *Circ. Res.*, 47, 117-130.

Lekkala, J. and Malmivuo, J. (1984). Multiplexed SQUID-vector magnetometer for Biomagnetic Research. *J. Phys E: Sci. Instrum.*, 17, 504-512.

Malmivuo, J.A.V. (1976). On the Detection of the Magnetic Heart Vector - An application of the reciprocity theorem. Acta Polytechnica Scandinavica, Helsinki.

Nousiainen, J., Lekkala, J. and Malmivuo, J. (1984). Comparative Study of the Normal Vector Magnetocardiogram and Vector Electrocardiogram. Research Report, Electronics Laboratory, Tampere University of Technology, Tampere, Finland.

Wikswo, J.P. (1983). Theoretical aspects of the ECG-MCG relationship. In Biomagnetism, an interdisciplinary approach, S.A. Williamson, G-L. Romani, L. Kaufman and I. Modena (Ed.), Plenum Press, New York, pp. 311-326.

Wikswo, J.P., Malmivuo, J.A.V., Barry, W.H., Leifer, M.C. and Fairbank, W.M. (1979). The theory and application of magnetocardiography. *Adv. Cardiov. Phys.*, 2, 1-67.

INVESTIGATION OF MULTIPOLE REPRESENTATION OF THE CARDIAC BIOMAGNETIC FIELD WITH RESPECT TO THE REAL TORSO GEOMETRY

P. Tekel', K. Vicenik, I. Misiková, M. Maco
M. Kricfalusi, J. Kuzmová, P. Kneppo

Institute of Measurement and Measuring Technique,
Electro-Physical Research Centre of the Slovak
Academy of Sciences, Bratislava, Czechoslovakia

ABSTRACT

In this experiment the influence of the real torso geometry and the conductive boundary of the torso surface upon parameters of the multipole equivalent magnetic generator of heart is investigated. We assume that the torso is homogeneous and has a conductivity σ.

KEYWORDS

Cardiac electric and magnetic field, multipole equivalent generator, conductive boundary, superconducting quantum magnetometer.

INTRODUCTION

Theoretical simulations of the volume conductor current contribution to the total MCG signal have been done earlier (Horacek, 1973; Cuffin, 1974) based on computer simulation of heart-torso models. In these calculations the contribution of the volume currents to the total magnetic signal normal to the chest was calculated. An interesting experimental approach to this problem was presented by Eghrari and co-workers (1983) and Peters and co-workers (1983). The purpose of our experiment was the numerical determination of the influence of the torso geometry and conductive boundary upon parameters of the multipole equivalent cardiac magnetic generator.

METHODS

The theoretical base of this experiment was given by Geselowitz (1970), who determined the contribution of primary and secondary sources to the magnetic field outside an inhomogeneous volume conductor.

For our theoretical configuration the physical medium consists of two main regions: the homogeneous isotropic volume conductor with the conductivity $\sigma = 0.002$ S/cm bounded by a closed surface S, and an infinite external dielectric region outside the conductor. As to magnetic properties the whole space is homogeneous and isotropic. Under these conditions the fundamental equation has obtained the following form:

$$\bar{B}_p = \bar{B}_m - (\sigma/4\pi\mu_o) \int_S \phi\nabla(1/R) \times d\bar{S} = \bar{B}_m - \bar{B}_b,$$

where:

$\bar{B}_m$ - vector of the total magnetic induction measured above the torso

$\bar{B}_p$ - vector of the magnetic induction set up by primary sources usually characterized by impressed current density

$\bar{B}_b$ - vector of the magnetic induction representing the conductive boundary effect contribution

ϕ - electric potential on the torso surface

R - distance from an element of volume dV in the volume conductor to an external point, where the magnetic field is observed.

In order to describe the biomagnetic activity of the heart by means of its equivalent multipole magnetic generator it is necessary to determine both terms on the right side of the given equation. Because we use for the calculation of the multipole coefficients of the equivalent generator the values of z-component of $\bar{B}_p$, it is necessary to determine first the z-components of both terms on the right side of the equation.

The z-component of $\bar{B}_m$ was measured by means of a superconducting quantum magnetometer with a symmetric 2nd-order gradiometer at points given by the standard measuring MCG grid on the chest. This component was measured 2 cm above the thoracic wall in the frequency range 0 - 100 Hz. Filtering was used to remove the frequencies 50 Hz and 100 Hz and signal averaging for each lead (15 beats, sampling rate 250 Hz) to improve the signal to noise ratio. To sychronize the averaging process and to provide a common time reference for the MCGs recorded at different points of the grid, a trigger pulse was obtained from the simultaneously recorded ECG standard lead II. The R-wave maximum from this lead represents the time reference.

The z-component of $\bar{B}_b$ for each point of the standard measuring MCG grid was found by using a special automated measuring system EKKG-80 (Kneppo and co-workers, 1979) developed for cardioelectric research purposes in our institute. The EKKG-80 system enables sufficiently accurate approximation of the torso surface with a set of elementary triangles between the neighbour points of ECG measurements. The system finds the coordinates of these points. For a person of standard growth the ECGs were recorded with the person in a sitting position at 8 horizontal planes shifted 3 cm from another along the torso. In each plane the ECGs were recorded from 19 points located around the torso. For each of the so constructed elementary triangles a value of the electric potential is given to its centre of gravity, that equals 1/3 of the sum of potentials from the three measuring points creating the triangle. The z-component of $\bar{B}_b$ was calculated for each MCG measuring point as a sum of the contributions for all the above mentioned elementary triangles approximating the torso surface. The distance R is then the distance between the MCG measuring point and the centre of gravity of the elementary triangle with the potential as defined. As in the case of MCG recordings, we also used in the case of ECG recordings a sampling rate of 250 Hz and the common time reference as given by the R-wave maximum of the ECG lead II. The parameters of the equivalent magnetic multipole generator were calculated at the origin of the standard coordinate system for electrocardiology up to octupole order. These results were compared with values of the equivalent magnetic multipole generator parameters calculated under the assumption that the effect of the conductive boundary of the torso surface is neglected. Using this assumption, $\bar{B}_b$ is zero. The comparison was carried out in numerical and graphical form during the QRS interval (-40 ms up to +40 ms according to the R wave maximum of ECG lead II) of the heart action for a group of 7 healthy young men.

RESULTS

Figure 1 shows typical quality of MCG signals measured in our laboratory.

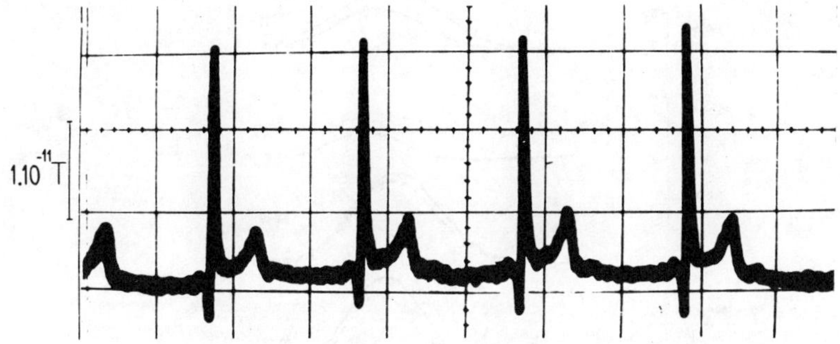

Fig. 1. Typical quality of MCG signals (direct oscillographic magnetometer output checking at the position D4, frequency range 0 - 100 Hz).

Typical multipole coefficients during the QRS complex of a healthy man are given in Fig. 2. The results obtained from all 7 investigated persons show a large influence of the conductive boundary of the torso surface upon the values of multipole parameters. Relatively the largest influence appears in parameters A_{10}, A_{11}, A_{20}, A_{22}, A_{30} and A_{32}. It was also investigated how the boundary effect influenced measured MCG data. This investigation was done in numerical form and graphically in the form of computer isomagnetic-field component maps. The results obtained are similar to those of Peters and co-workers (1983). The secondary sources on the torso boundary give a large contribution to the magnetic field (large differences in particular values of the z-component of the magnetic induction are present), but do not change the general features. The work described here represents a contribution to a more accurate evaluation of the reciprocal relation between the cardiac electric and magnetic multipole equivalent generator parameters (Titomir and Kneppo, 1983).

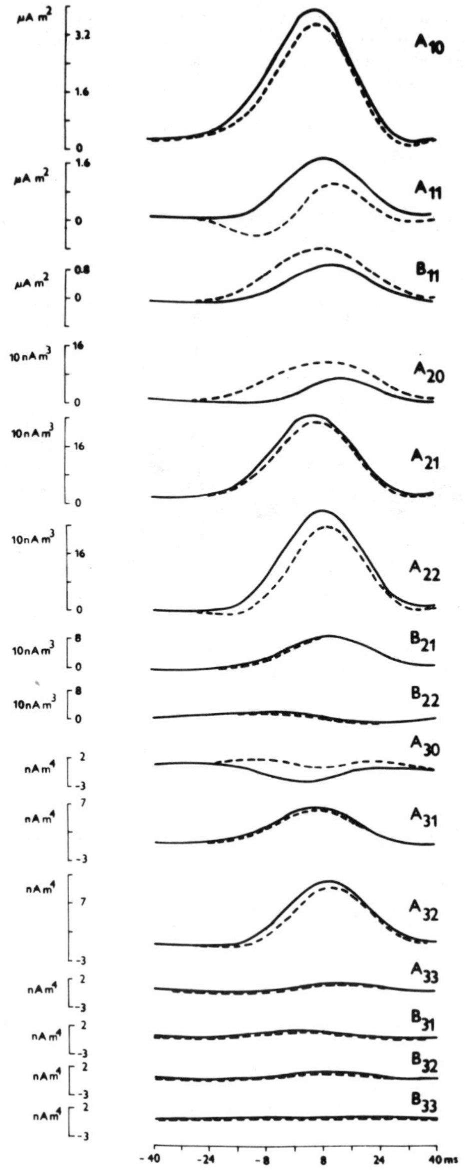

Fig. 2. Typical multipole coefficients during the QRS complex of a healthy person (man, 30 y.). (Full line - boundary effect neglected, dashed line - boundary effect accepted.)

REFERENCES

Cuffin, B. N. (1974). Ph.D. Thesis. Department of Electrical Engineering. The Pennsylvania State University.

Eghrari, I. R., Costa Monteiro, E., Costa Ribeiro, P. and von der Weid, J. P. (1983). On the influence of the volume conductor in magnetocardiography: an experimental approach. Il Nuovo Cimento, 2D, 346-355.

Geselowitz, D. B. (1970). On the magnetic field generated outside an inhomogeneous volume conductor by internal current sources. IEEE Trans. Magn., MAG-6. 346-347.

Horacek, B. M. (1973). Digital model for studies in magnetocardiography. IEEE Trans. Magn. MAG-9, 440-444.

Kneppo, P. and Titomir, L. I. (1979). Integral characteristics of human cardiac electric generator from electric field measurements by means of automatic coordinator. IEEE Trans. Bio-med. Engng, BME-26, 21-28.

Peters, M. J., Swenenhuis, M. J. M., van Oosterom and Wevers-Henke, J. J. (1983). The influence of inhomogeneities on the cardiac magnetic field distribution. Il Nuovo Cimento, 2D, 324-339.

Titomir, L. I. and Kneppo, P. (1983). On the possibility to determine integral characteristics of the cardiac electric generator from extracardiac electric and magnetic measurements. IEEE Trans. Bio-med. Engng, BME-30, 222-226.

IN VIVO STUDY OF THE EFFECT OF THE SECONDARY CURRENTS ON THE MCG

T. Varpula, T. Katila, T. Poutanen and M. Seppänen

Department of Technical Physics, Helsinki University of Technology, SF-02150 Espoo, Finland

ABSTRACT

The effect of the secondary currents on the MCG was studied by immersion of the normal subject in an electrically conducting bath. Further experiments were performed at different respiration stages of the subject. It was found that the secondary currents on the torso boundary have a significant effect on the MCG field, but they do not fully explain the fact that the zero-crossing line in the z-component of the MCG field distribution does not go over the center of the heart on the frontal plane.

KEYWORDS

Magnetocardiography, secondary and primary currents, volume currents, effect of the boundary

INTRODUCTION

The current-dipole model gives a satisfactory first-order approximation for the body surface potential distribution of the electrocardiogram. It was noted, however, from the early recordings of the magnetocardiogram that the current-dipole model is insufficient to give a full account of the cardiac magnetic field (Cohen and Chandler, 1969). Specifically, the prediction for the position of the zero-crossing line in the z-component of the MCG (perpendicular to the frontal plane) was not in accordance with experimental results. Moreover, even computer models aimed at describing more precisely the cardiac primary and secondary currents predict contrary to experimental results that the zero-crossing line goes approximately above the center of the heart (Horacek, 1973; Geselowitz, 1980; Peters and others, 1983), where the measured normal MCGs reach their maxima (Saarinen and others, 1978).

Two obvious reasons can be found for the discrepancy between theory and experiment. First, the primary cardiac currents are improperly described by the current dipole and the present computer models. One explanation would be that the primary current distribution in the cardiac muscle contains circular currents (vortices) (Wikswo and Barach, 1982) causing the maximum of the z-component of the MCG above the heart. These circular currents could be modelled in the first order either by a magnetic dipole or by a current quadrupole. Considering potential applicability of the magnetocardiography it is important to verify whether such circular currents exist. Theoretically, electrocardiography is incapable of detecting circular currents, whereas magnetocardiography can do it. Second, the secondary currents in nonisotropic tissues might not be accurately enough taken into account by the existing torso models. Computer models incorporating the nonisotropic muscles have yet to be implemented.

In the present paper, this problem was studied experimentally by changing cardiac secondary currents of normal subjects in vivo. In one series of experiments MCG field distributions were recorded at various stages of respiration. In these experiments not only the geometry of the volume conductor (the torso) changes but also the position and orientation of the heart varies thus altering the primary currents. This makes the interpretation of the results difficult. Therefore, a method was devised in which the effect of the torso boundary only is changed by immersing the subject in a conducting bath. The primary currents and torso geometry remain in this case unaltered. To our knowledge no previous reports on a succesful attempt to study this effect in vivo has been published although it has been considered before (Wilson and Johnston, 1938).

MATERIAL AND METHODS

The magnetocardiographic data were recorded from three voluntary male subjects (ages 23, 25 and 28 years) with no history of cardiac disease. The MCGs were mapped with a three-axial SQUID gradiometer with a baseline of 12 cm and sensitivity of 30 fT/$\sqrt{Hz}$ (Varpula and others, 1982). The measurement locations were chosen according to the standard MCG grid (Saarinen and others, 1978). In addition, a seventh column was added to the left side of the grid. In only one subject, whose results will be discussed in more detail here, extensive field mappings were performed. In two other subjects the recordings were less extensive, just to verify that the results can be generalized. The measurements were performed in the magnetically shielded room of the Helsinki University of Technology (Kelhä, 1981).

Two series of measurements were done. One, MCG mappings were taken at three respiration stages: normal, full inspiration and full expiration. Two, the subject was immersed in a large tank (length 1.6 m, width 0.6 m, height 0.4 m) approximating an infinite slab. For an infinite slab the boundaries have no effect on the normal component of the magnetic field (Cuffin and Cohen, 1977). The tank contained either pure water or electrolyte (NaCl) with an average electric conductivity of the tissues (0.2 A/Vm). The conductivity of the fluid was monitored with a four-electrode measurement. Proper preparation of the skin turned out to be essential for lowering its resistance and thus allowing the current to flow through the skin into the surrounding conducting fluid. Therefore, before the measurement the skin of the entire torso was carefully washed and rubbed with sand paper until it turned reddish. The temperature of the bath was kept around 32 degrees Celsius.

During the measurement the signals (MCG and ECG lead II) were stored on a magnetic tape with an instrumation recorder. At a later date the MCGs were averaged (10 - 20 repetitions) with a minicomputer using the R peak of the ECG II as a trigger. The measurement band was 0-100 Hz with a narrow 50 Hz filter.

RESULTS AND DISCUSSION

Results of the experiments are shown in Figs. 1 to 4.

The conducting bath caused substantial changes in the cardiac magnetic field. The z-component of the MCG was least affected when detected near the heart, whereas the bath had a substantial effect in the regions with a high curvature of the torso boundary, e.g. at B7 the change was 40 % of the maximum value at D4 (Figs. 2 and 3). The relative changes in some locations were still considerably higher (Figs. 2

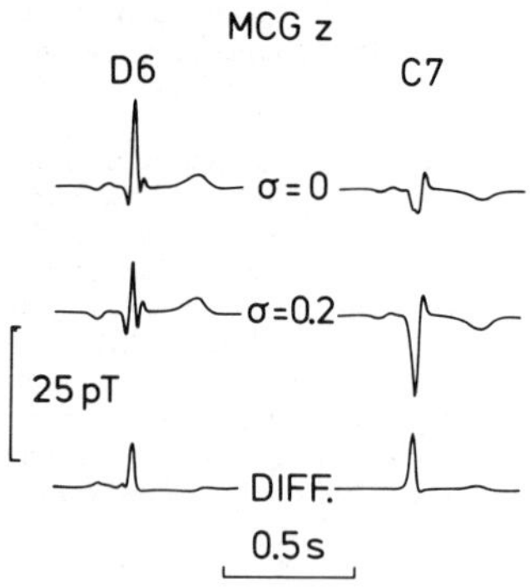

Fig. 1. The z-components of the MCG recorded at two locations (D6 and C7) of the MCG grid. The uppermost curves are the MCGs when the conductivity of the bath was zero. The MCGs in the middle are for the bath conductivity of 0.2 A/Vm. The lowest curves are computer-calculated differences between the two above MCGs and represent thus the approximate surface contributions in these locations.

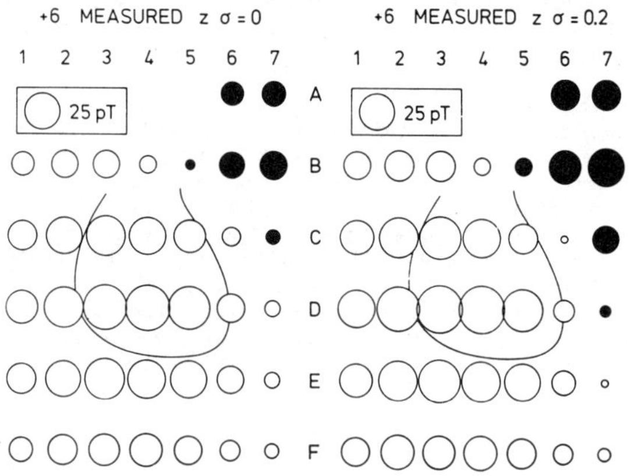

Fig. 2. Measured distributions of the z-component of the MCG for a subject immersed in a bath with zero-conductivity (left) and with the conductivity of 0.2 A/Vm (right) at the time instant of the maximum strength of the MCG field (6 ms before the maximum of the R peak in the ECG lead II). The area of the circle is proportional to the field strength at the respective location. The open circles indicate the magnetic field pointing toward the subject (positive direction of the coordinate axis). The outline of the cardiac silhouette is also shown.

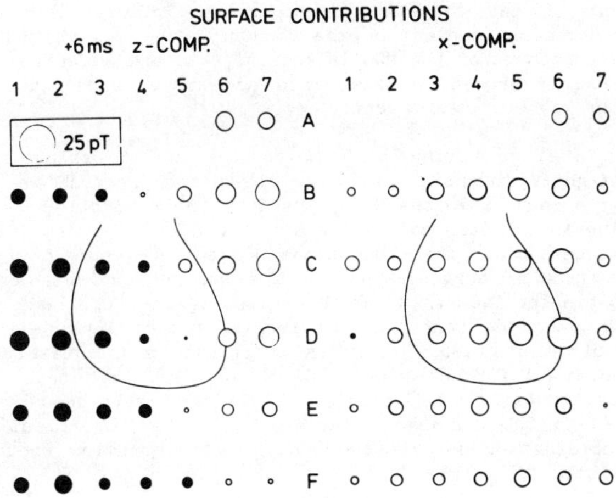

Fig. 3. Measured surface contributions to the z- (left) and x- (right) components of the MCGs at 6 ms before the maximum in the ECG II. The field values are obtained from the differences as shown in Fig. 1. A linear baseline correction of the MCGs was used.

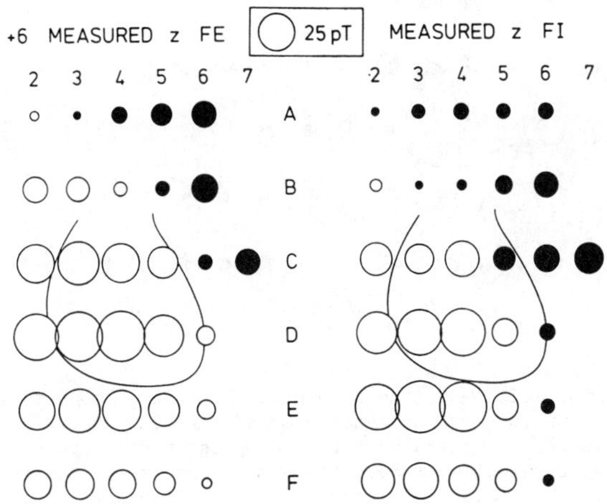

Fig. 4. Measured distributions of the z-component of the MCG for full expiration (left) and for full inspiration (right). The time instant is the maximum of the MCG. The conductivity of the bath is zero.

and 3). On the other hand, the x-component (pointing to the left of the subject) was affected by the bath considerably also near the heart (Fig. 3). Although the tangential component of the magnetic field in an infinite-slab geometry is contributed by the secondary current on the boundary, the result shows, however, that the tangential component of the MCG is sensitive to and affected by the torso boundary. This conclusion should be taken into account when considering possible methods to detect the magnetic heart vector.

Fig. 2 shows that the conducting bath moved the zero-crossing line of the z-component only slightly toward center over the heart. Because we are able to detect, however, large changes in the MCGs caused by the conducting bath and these changes are according to computer models (Geselowitz, 1980), we can infer that the conducting bath removes a major fraction of the MCG surface contribution (see Fig. 5). Therefore, the results suggest that the secondary currents on the torso boundary do not explain the fact that the zero-crossing line of the z-component in the MCG does not go over the heart on the frontal plane. Thus, either internal inhomogeneities or circular primary currents contribute substantially to the MCG field. This conclusion is further supported by the fact that both in the full expiration and particularly full inspiration the zero-crossing line is significantly closer to the cardiac center (Fig. 4) than in normal respiration (Fig. 2). These recordings do not, however, settle the question whether circular primary currents exist.

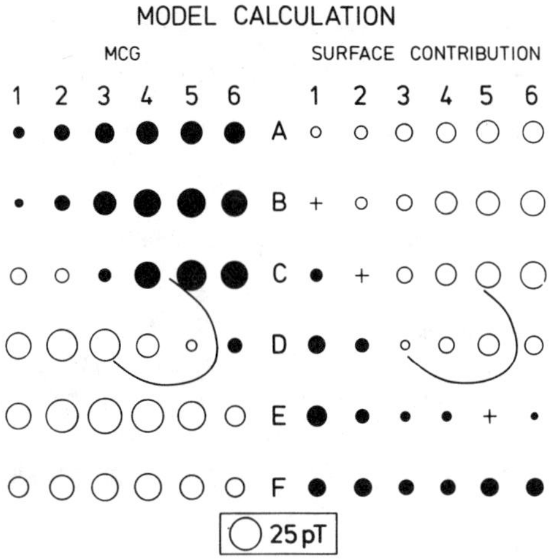

Fig. 5. Calculated distribution of the z-component of the MCG (left) and the z-component surface contribution (right) at a maximum of the MCG. The field values are calculated by a computer model developed by Geselowitz (1980). Note that the zero-crossing line in this calculated MCG field distribution goes over the heart.

ACKNOWLEDGEMENTS

The authors express their gratitude to Dr. R. Hari for her suggestions for the experiments. The aid of Mr. T. Tuomisto is gratefully acknowledged.

REFERENCES

Cohen, D., and L. Chandler (1969). Circulation, 39, 395-402.

Cuffin, B. N., and D. Cohen (1977). IEEE Trans. Biomed. Eng., BME-24, 372.

Geselowitz, D. B. (1980). IEEE Trans. Magn., MAG-16, 812-817.

Horacek, B. M. (1973). IEEE Trans. Magn., MAG-9, 3.

Kelhä, V. O. (1981). Proc. 3rd Interntl. Workshop on Biomagnetism, Berlin (West), 33-50.

Peters, M. J., M. J. M. Swennenhuis, A. van Oosterom, and J. J. Wevers-Henke (1983). Il Nuovo Cimento, 2D, 324-339.

Saarinen, M., P. Siltanen, P. J. Karp and T. E. Katila (1978). Ann. Clin. Res., 10, S 21, 1-43.

Varpula, T., P. Karp, T. Katila, T. Poutanen, and M. Seppänen (1982). Proc. XVI Annual Conference of the Finnish Physical Society, Report TKK-F-A466, abstr. 8:23.

Wikswo, J.P., Jr. and J.P. Barach (1982). J. Theoretical Biol., 95, 721-729.

Wilson, F., and F. D. Johnston (1938). Amer. Heart J., 16, 14.

EXPERIMENTAL STUDY ON CARDIAC RELATED MAGNETIC SUSCEPTIBILITY SIGNALS

R. Maniewski*, T. Katila, T. Poutanen, T. Varpula and J. P. Wikswo, Jr.**

Department of Technical Physics, Helsinki University of Technology, SF-02150 Espoo 15, Finland
*Institute of Biocybernetics and Biomedical Engineering, Polish Academy of Sciences, Warsaw, Poland. (Perm. address)
**Department of Physics and Astronomy, Vanderbilt University, Nashville, Tennessee, U.S.A.

ABSTRACT

Magnetic susceptibility signals caused by the mechanical activity of the heart were studied. Three experimental methods were used in order to seperate the different components in these signals. The results suggest that usually about 60 % of the signal measured near the heart originates directly from cardiac volume changes.

KEYWORDS

Biomagnetic fields, magnetic susceptibility measurements, cardiac mechanical activity, magnetic susceptibility plethysmogram.

INTRODUCTION

Magnetic susceptibility measurements offer new possibilities for noninvasive study of cardiac mechanical activity, e.g. in recordings of cardiac volume changes. This method was proposed by Wikswo, Opfer and Fairbank (1974, 1980) and then further studied by other authors (Boismier, Hlatky, and Tepley, 1979; Katila and others, 1981; Peters, Dunajski and van der Marel, 1981). However, the magnetic susceptibility signal has contributions from other cardiac related signals mainly arising from the movement of skin and chest-lung boundary.

Three regions with significant changes of the magnetic susceptibility are shown in Fig. 1 (Varpula, 1983). The signal related directly to the volume changes of the heart arises from the heart-lung boundary where the change in the magnetic susceptibility $\Delta\chi$ during the venticular ejection is of the order of $+5 \times 10^{-6}$ (SI units are used throughout the paper). Approximately the same susceptibility change occurs at the chestwall-lung boundary, but the polarity of the signal is opposite in this case ($\Delta\chi < 0$). The third source of magnetic signal indirectly related to the cardiac mechanical activity is skin displacement. This signal has the same polarity as the first one ($\Delta\chi > 0$) and its contribution is significant because the susceptibility change at skin-air boundary is higher ($\Delta\chi = +9 \times 10^{-6}$). In addition, the skin displacement region is situated close to the magnetometer.

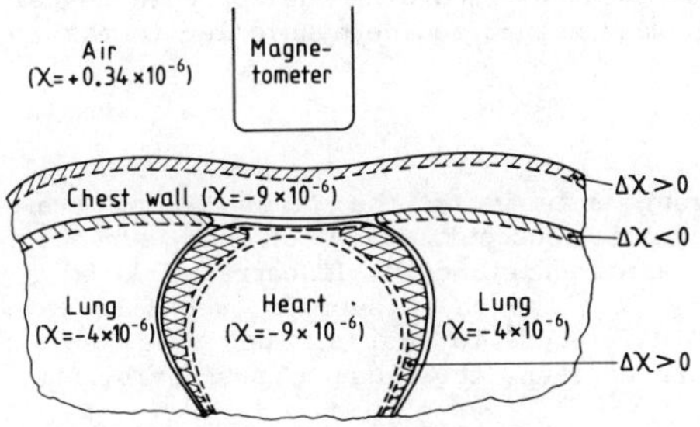

Fig. 1. Schematic representation of the transversal cross-section of the chest at the 8th thoracic vertebra. The regions with susceptibility changes during venticular ejection are indicated by shadowing.

It is possible to compensate for these signals by changing the magnetic susceptibility of the medium surrounding the thorax of the subject and in this way to separate the magnetic susceptibility plethysmogram. In the present study this theoretical possibility is experimentally confirmed.

METHODS

Measurements of magnetic signals were performed as in our previous study (Katila and others, 1982) in a laboratory room by applying a homogeneous magnetic field with a magnitude of 135 A/m. The subject in the field was in the supine position. The changes in magnetic field perpendicular to the frontal plane were recorded with two opposite directions of the applied field. A second-order SQUID gradiometer (baseline 10 cm, pick-up coil ϕ 2.5 cm) was employed. To suppress contributions unrelated to the cardiac activity, time-averaging was performed using the ECG lead II for timing. The magnetic signals corresponding to the opposite directions of the applied field were subtracted from each other, thereby removing the contribution of the MCG. Thus, the susceptibility signals obtained correspond to an external field of 270 A/m. The final frequency band of the signal processing was 0.03 to 20 Hz.

Three independent experimental methods were used. First, measurements were carried out when the thorax of the subject was surrounded by liquid (water solution of $MnCl_2$) with different values of magnetic susceptibility: close to that of blood ($\chi=-9 \times 10^{-6}$), of lung ($\chi=-4 \times 10^{-6}$) or that of air ($\chi=+0.34 \times 10^{-6}$). During the experiments the supine subject was surrounded by this water solution in a nonmagnetic vat made of plywood. To keep the surface of the liquid stationary and to eliminate cardiac related rippling, a plexiglass screen was placed on the liquid surface. The magnetic susceptibility signals were measured from two normal male subjects at the locations D5 and G3 of the standard MCG grid.

A second, similar experiment was performed using a technique introduced by Farrell and others (1980). A plastic bag containing the same solution as above with various values of magnetic susceptibility was placed on the chest of the subject. The upper surface of the bag was kept stationary with the plexiglass screen. The signals were recorded from one male subject at locations D4 and G3.

Third, magnetic susceptibility signals were measured at the locations D4 and G3 when the chest wall of the subject was either free to move in air or was fixed using a special vacuum casting. The immobilization of the chest wall was controlled in an additional experiment by recording the magnetic signal from the skin covered with ferromagnetic powder magnetized perpendicular to the chest wall.

RESULTS

The results of measurements using the first method are shown in Fig. 2. The decrease of the magnetic susceptibility of the surrounding medium caused the magnetic signal to diminish distinctly. It corresponds to a gradual cancellation of the component arising from the chest surface. When the magnetic susceptibility of the surrounding liquid is equal to that of lung (Fig. 2b) the motion of the chest wall is largely compensated and the measured susceptibility signal reflects mainly

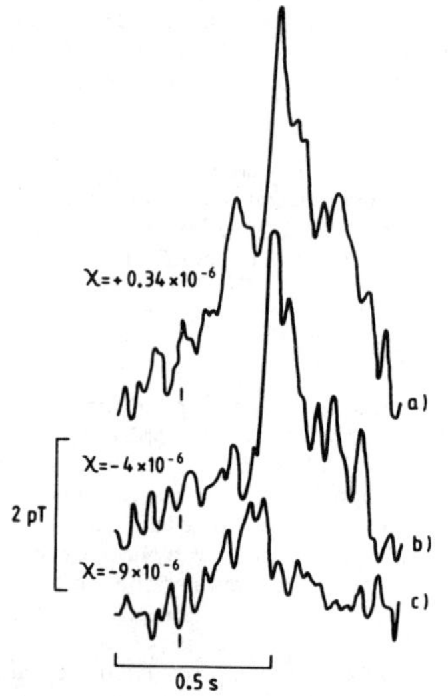

Fig. 2. The magnetic susceptibility signals measured at the location D5 from a subject immersed in water solution of $MnCl_2$ with various values of magnetic susceptibility. The R peak of the ECG is marked with vertical bars.

the changes in cardiac volume. The magnitude of the signal is in this case about 60% of that obtained in the case of liquid with a susceptibility equivalent to air (Fig. 2a). The skin displacement contribution can be eliminated from the magnetic susceptibility signal when the thorax of the subject was surrounded with water (Fig. 2c). In this case, the magnitude of the signal is about 60% smaller than that shown in Fig. 2a.

Similar results were obtained also by using the second method (Fig. 3). Replacement of air by a liquid with the same magnetic susceptibility (Fig. 3a) allows the elimination of some effects caused by the change of the density of the surrounding medium. The magnitude of the signal is in this case smaller than in air (Fig. 3d). Similar phenomena were also observed at the measurement location G3. A gradual compensation of chest wall movement by diminishing the magnetic susceptibility of

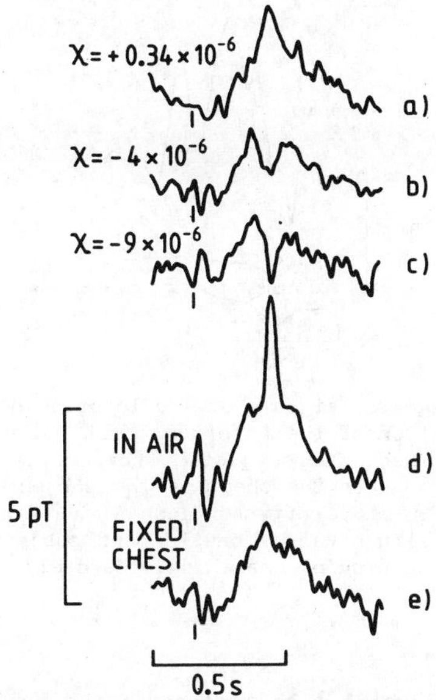

Fig. 3. The magnetic susceptibility signal measured at the location D4. Curves a, b and c were recorded when a plastic bag containing water solution of $MnCl_2$ with various values of magnetic susceptibility was placed on the chest of the subject. Curves d and e were obtained when the chest wall of the subject was free to move in the air (d) or fixed with a vacuum casting (e). The R peak of the ECG is marked with vertical bars.

the solution decreases the magnetic signal. The magnitudes of the signals when the susceptibility of the liquid is close to that of lung (Fig. 3b) and that of blood (Fig. 3c) are about 65% and 50%, respectively, of those obtained in the case when the susceptibility of the liquid was equivalent to air (Fig. 3a).

Further experiments performed according to the third method, i.e. with freely movable chest wall in air (Fig. 3d) and with fixed chest wall (Fig. 3e) show decreasing of magnetic susceptibility signal by about 30% when the chest was fixed. If the sharp peaks are ignored the signals look almost the same and the difference in there magnitudes is only about 15%. Results of controlling of the mobility of the chest wall during a fixation with vacuum casting are shown in Fig. 4. The magnetic signals from the layer of ferromagnetic powder spread on the chest are suppressed by more than 80 % when the chest is fixed (Fig. 4b,c) as compared to the signal obtained when the chest was free to move in air (Fig. 4a). Taking into account that the magnetic susceptibility signal caused by the motion of the chest contributes about 40 % of the total signal, one can state that the displacement of the chest during the measurement with fixed chest wall will produce not more than 10 % of the total susceptibility signal. Thus, when the chest was fixed, at least 90 % of the total susceptibility signal was produced directly by the mechanical activity of the heart.

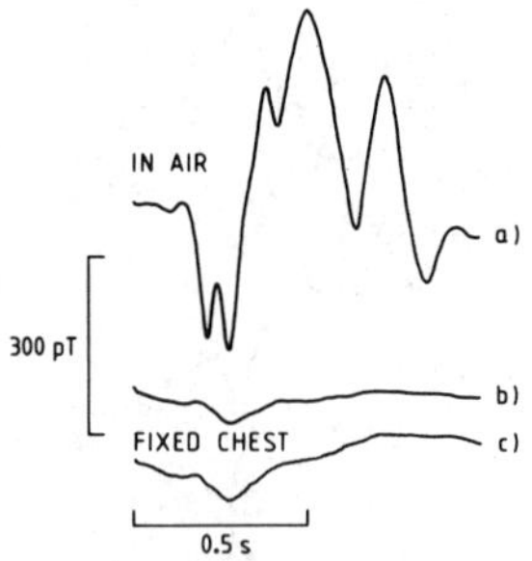

Fig. 4. The magnetic signals at the location D4 related to the motion of the anterior chest wall covered with magnetized ferromagnetic powder. Curve a was recorded when the chest of the subject was free to move in air; curves b and c when the chest was fixed with a vacuum casting and subject was either not breathing or breathing moderately.

CONCLUSIONS

Magnetic susceptibility signal reflects directly the cardiac mechanical activity when the chest of the subject is fixed or when the motion of the chest is compensated magnetically by surroundig it with a medium whose susceptibility equals that of lung. In these cases three independent experiments gave the same results: the contribution of the signal directly related to the cardiac volume change is about 60% of the total magnetic susceptibility signal. When the chest of the subject is surrounded by water the signal is additionally reduced by a component arising from the chest-wall-lung boundary (Fig. 1). It clearly decreased the signal to about 40% of the total susceptibility signal. These results confirm the theoretical considerations concerning changes in the magnetic susceptibility caused by the cardiac mechanical activity and especially the component arising directly from cardiac volume changes.

REFERENCES

Boismier, G., L. Hlatky, and N. Tepley (1979). J. Appl. Phys., 50, 2490 - 2493.

Farrell, D.E., J.H. Tripp, J.W. Zanzucchi, J.W. Harris, G.M. Brittenham and W.A.Muir (1980). IEEE Trans. Magn., MAG-16, 818-823.

Katila, T., R. Maniewski, T. Poutanen, T. Varpula, and P. Siltanen (1981). Proc. 5th Nordic Meeting on Medical and Biological Engineering, Linköping (Sweden), 245-247.

Katila, T., R. Maniewski, T. Tuomisto, T. Varpula, and P. Siltanen (1982). IEEE Trans. Biomed. Eng., BME-29, 16-25.

Peters, M. J., Z. Dunajski, and L. C. van der Marel (1981). Proc. 3rd Workshop on Biomagnetism. Walter de Gruyter, Berlin, 327-335.

Varpula, T. (1983). Il Nuovo Cimento, 2D, 624-641.

Wikswo, J. P., J. E. Opfer, and W. M. Fairbank (1974). Proc. AIP Conference, 18, 1335-1339.

Wikswo, J. P. (1980). Med. Phys., 7, 297-306.

Wikswo, J. P., J. E. Opfer, and W. M. Fairbank (1980). Med. Phys., 7, 307-314.

SOURCE IDENTIFICATION USING EVOKED POTENTIAL MEASUREMENTS

Charles C. Wood

Neuropsychology Laboratory, VA Medical Center
West Haven, Connecticut 06516
and
Departments of Neurology and Psychology
Yale University
New Haven, Connecticut 06510

ABSTRACT

Methods for identifying the anatomical sources of Evoked Potentials (EPs) recorded from the human scalp are reviewed with an emphasis on their implications for source identification using evoked magnetic fields (EFs). In attempting to identify sources using either potential or magnetic measurements one must consider: (a) the distinction between the neural currents active at a given instant and those that contribute to the potential or magnetic measurements in question; (b) the spatial and temporal extent of the potential or magnetic activity whose sources are being sought; (c) the distinction between source identification and source localization; and (d) most important, the lack of unique solutions for the "inverse problem" (i.e., the problem of deriving source configurations in the brain from surface potential or magnetic data). Because source configurations are underdetermined by surface data, additional types of evidence are necessary to test alternative source hypotheses. An example of such evidence is presented for the 20- and 30-msec somatosensory EPs evoked by median nerve stimulation, including the use of quantitative volume conduction models, cortical surface and intracerebral recordings, the effects of surgical excisions, and combined EP and EF recordings.

KEYWORDS

Evoked potentials; evoked magnetic fields; inverse problem; volume conduction models; source identification.

INTRODUCTION

My purpose in this paper is to review the problem of identifying electrical sources in the human brain using surface EPs and to consider its implications for source identification using EFs. Studies of human brain activity using EP data have a longer history than those using EFs, and a review of issues associated with source identification using potential measurements may be useful for magnetic studies. My comments should be interpreted as one investigator's view of the source identification problem; other EP workers may differ both on matters of substance and emphasis.

First, I consider some issues regarding the nature of the source identification problem and how it should be formulated. Although the issues to be considered are generally known, they are too frequently overlooked in empirical EP and EF studies. Second, I describe in some detail an example of how converging evidence from different types of data can be used to test alternative source hypotheses. The example is based on the 20- and 30-msec somatosensory EPs to median nerve stimulation which have been the subject of a continuing controversy over the last decade. Disagreement has focused on two hypotheses: (a) that these EPs are generated by a single tangential source in somatosensory cortex; and (b) that they are generated by separate radial sources in somatosensory and motor cortex. Results bearing on these and related hypotheses have come from cortical surface and intracerebral EP recordings, the effects of surgical brain excisions on surface EPs, and the use of quantitative volume conduction models. Although no one of these results is totally unambiguous, the pattern of results taken together provides strong support for the tangential source hypothesis. Third, I describe a comparison of EP and EF recordings of the 20- and 30-msec somatosensory activity that supports the tangential source hypothesis and illustrates how direct comparison of EP and EF recordings permits stronger source inferences than are possible using either technique alone.

SOURCE IDENTIFICATION: FOUR PRELIMINARY CONSIDERATIONS

1. At the outset it is important to distinguish between attempting to identify the the neural currents active at a given instant and and attempting to identify the currents that are responsible for a particular set of surface EPs or EFs. Because both EPs and EFs are due to ionic current flows in active neurons, both measures depend upon: (a) the spatial configuration of active cells and cell groups; (b) the spatial and temporal patterns of synaptic activation (spatial and temporal synchronization); (c) the location and orientation of the active currents relative to the measurement sites; and (d) the properties of electric and magnetic fields in conductive media, most importantly the principle of superposition. Consequently, EPs and EFs should be regarded as aggregate, biased, and incomplete measures reflecting whatever subsets of the neural currents occurring at a given instant superimpose to produce nonzero EP or EF activity at the locations being studied. Identifying the sources of a particular set of surface EPs or EFs is not equivalent to identifying the neural currents active at that time.

An important corollary of this point is that because potential and magnetic measurements are differentially sensitive to neuronal currents of different orientations and depths, the sources of EPs and EFs obtained under the same conditions are not necessarily the same. Theoretical calculations have demonstrated: (a) that EFs are sensitive only to the component of a source current tangential to the surface of the scalp, whereas EPs are sensitive to both tangential and radial source currents (for discussion, see Williamson and Kaufman, 1981); and (b) that EFs fall off more rapidly than EPs with increasing source depth, reaching zero for a source at the center of a sphere (Grynszpan and Geselowitz, 1973; Cuffin and Cohen, 1979). Thus, the sources of EPs and EFs will be the same only when they are due exclusively to relatively superficial, tangential sources and will become increasingly dissimilar with increasing involvement of radial or deep sources. As illustrated in the final section below, EPs and EFs may therefore be used in combination to place stronger constraints on source hypotheses than is possible with either measure alone.

2. It is important to specify the spatial and temporal extent of the EP or EF activity to which a given source hypothesis is intended to apply. Source hypotheses are generally formulated in terms of EP or EF "components", which are implicitly defined as the peaks in major waveform deflections. Hypotheses derived from data at the peak are presumed to apply to the entire duration of the deflection.

Although formulating hypotheses for data at the peaks of waveform deflections is a reasonable starting point, Fig. 1 illustrates how lack of attention to the temporal behavior of EPs or EFs can lead to seriously misleading source hypotheses. Figure 1 presents auditory EP distributions at three different latencies during what has been traditionally regarded as the auditory "N100" component.

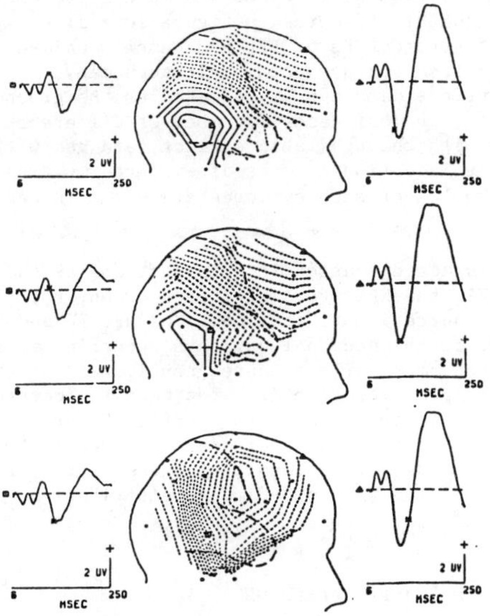

Fig. 1. Auditory EP distributions at latencies on the rising (top), peak (middle), and falling (bottom) phases of the "N100" component. Waveforms are from temporal (square) and frontocentral (triangle) locations; latencies at which distributions are plotted are indicated by asterisks; positive and negative voltage relative to a noncephalic reference are indicated by solid and dotted contours. From (Wood et al., 1984).

Note that there are systematic differences in the surface distributions at different latencies of the "same component", requiring a different source configuration at each latency. In fact, when the distributions are examined millisecond by millisecond over the entire duration of "N100", there is a continuous and systematic change over time. Such a change could reflect changes in the pattern of synaptic activation of a single anatomical structure, the summation of activity from multiple structures activated asynchronously, or both. Only when EP or EF distributions are stable over the entire duration of a major deflection can source hypotheses formulated for the peak be generalized to the entire deflection. In cases with changing distributions such as Fig. 1, the distribution at the peak and corresponding source hypotheses are no more representative than those for any other latency. (One might predict that EFs may have more stable distributions over time than EPs because the former are sensitive to a more restricted range of sources; too few direct comparisons of EPs and EFs in the same subjects and conditions are now available to test this prediction).

3. Although is it common to speak of source <u>localization</u> using EP or EF

measurements, I prefer the term source identification because it emphasizes that the configuration of source currents responsible for a given set of surface EPs or EFs may be quite complex and that devising an adequate source model is an essential first step in specifying the sources. Current dipoles are an important class of source models that have been useful for a wide variety of potential and magnetic phenomena (cf., Wood, 1982; Williamson and Kaufman, 1981). However, viewing the problem as one of source localization makes it easy to oversimplify the range of possible source configurations and to view the problem as simply one of finding the location of the best-fitting current dipole.

4. As a theoretical result, the lack of unique solutions for the inverse problem is well-known and needs no elaboration here beyond a simple restatement: Because of superposition, an infinite number of source configurations are capable of generating any given set of surface EPs or EFs and hence a unique source configuration cannot be derived from surface data. Although many possible sources for a given set of surface data can be rejected on the grounds of parsimony or anatomical plausibility, in most cases a number of different source hypotheses are consistent both with the available surface data and with know anatomical constraints. Therefore, additional types of evidence must be found to discriminate among them. Examples of such evidence are given in the next section.

To summarize, underlying each of the issues discussed in this section are the implications of the superposition of EPs and EFs from multiple sources, both in space and in time, for the problem of source identification. Surface EP and EF data provide measures of only a subset of the neuronal currents occurring at a given time and, because of superposition, may reflect source configurations that are quite complex. Unfortunately, superposition also dictates that the degree of source complexity need not be evident in the form of the surface EP or EF distributions. This problem has been recognized as a serious challenge in the EP literature for some time, and is becoming a more important issue in the EF literature as more examples of surface fields with non-dipolar distributions are reported.

TESTING SOURCE HYPOTHESES: AN EXAMPLE BASED ON SOMATOSENSORY EPs

In this section I attempt to illustrate how additional types of data can be used to evaluate alternative source hypotheses. The potentials to be considered are somatosensory EPs occurring about 20 and 30 msec following electrical stimulation of the human median nerve at the wrist. These potentials can be recorded from scalp electrodes contralateral to the nerve stimulated. At parietal locations, the EPs are negative-positive in morphology relative to a noncephalic reference and are termed N20-P30. EPs at frontal locations have peaks at similar latencies but are opposite in polarity (P20-N30). Figure 2 shows that the distributions of these potentials are generally dipolar in shape with extrema located over the parietal and frontocentral scalp.

Figure 3 shows that the surface distributions of these EPs are well fit by a current dipole source located in sensorimotor cortex near the central sulcus (Rolandic fissure), with a predominant orientation tangential to the surface of the scalp.

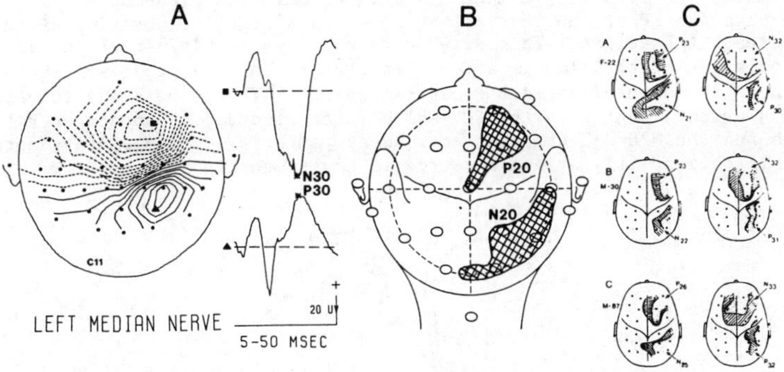

Fig. 2. Scalp distributions of the 20-msec (B and C) and 30-msec (A and C) somatosensory EPs from three different experiments (A: Wood, 1982; B: Goff et al., 1977; C: Desmedt and Cheron, 1981). Waveforms in A depict EPs at frontocentral (square) and parietal (triangle) locations; asterisks indicate the 30-msec peaks corresponding to the distribution in A.

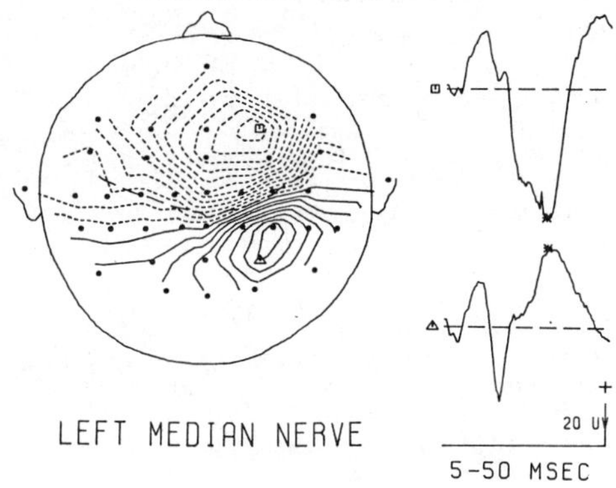

Fig. 3. Location and orientation of the best-fitting current dipole source for the 30-msec distribution. Dipole orientation and magnitude are shown in lateral, back, and top views of the head. From Wood (1982).

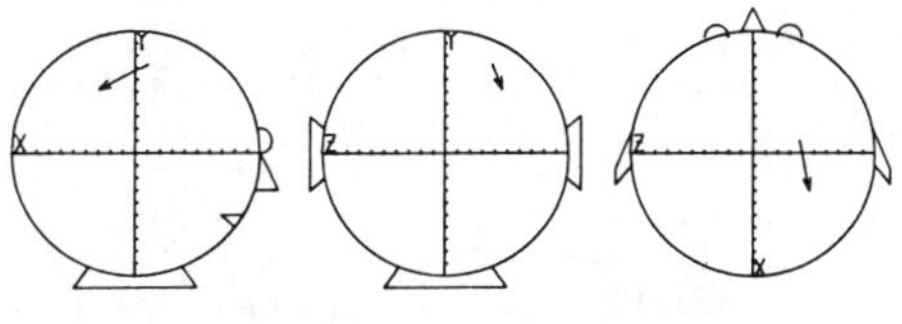

Figure 4 illustrates the two major source hypotheses that have been proposed for the 20- and 30-msec EPs. One hypothesis was originated by Broughton (1969, Broughton et al., 1981) and has been supported by Allison and his colleagues (Allison et al., 1980; Allison, 1982). They hypothesize that the tangentially oriented best-fitting dipole corresponds to a source located in Area 3b of somatosensory cortex which in humans is buried on the posterior bank of the central sulcus (solid arrows in Fig. 4). Area 3b receives heavy afferent somatosensory input from the thalamus and contains axially symmetric pyramidal cells that are capable of generating an anterior-posterior dipolar field of the type obtained. In animals, peripheral nerve stimulation produces a positive-negative EP at the cortical surface and a negative-positive EP in white matter beneath the cortex. This same pattern is hypothesized to occur in humans, only the orientation is anterior-posterior instead of dorsal-ventral because Area 3b is enfolded in the central sulcus. Thus, the Broughton-Allison hypothesis assumes that both the P20-N30 complex recorded at frontocentral scalp locations and the N20-P30 complex recorded at parietal locations are generated by a single tangentially oriented source in somatosensory cortex.

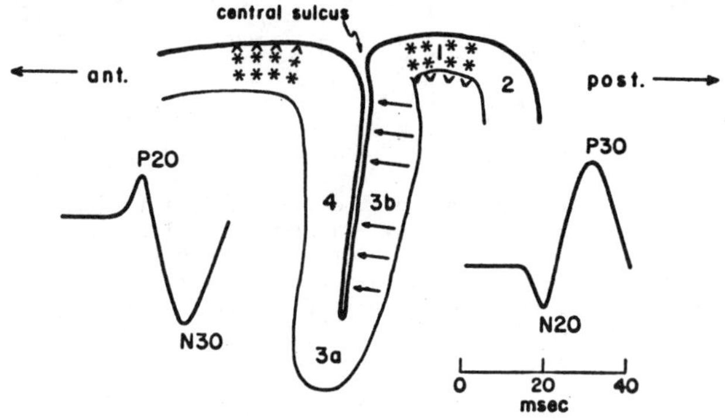

Fig. 4. Saggital section through left sensorimotor cortex illustrating the two source hypotheses for P20-N30 and N20-P30. The tangential hypothesis proposes that both sets of potentials are generated in somatosensory area 3b located on the posterior bank of the central sulcus (solid arrows), whereas the dual radial hypothesis (asterisks) proposes that P20 and N30 are generated by motor cortex in the crown of the precentral gyrus and that N20 and P30 are generated separately in the crown of the postcentral gyrus (somatosensory area 1). The direction of the arrows for each hypothesized source corresponds to the orientation of the positive side of the equivalent dipole source at 20 msec. Dipole orientation is reversed for the 30-msec peak.

Other investigators (e.g., Papakostopoulos and Crow, 1980; Desmedt and Cheron, 1980; Maccabee et al., 1983) have proposed that P20 and N30 are generated in motor cortex on the crown of the precentral gyrus and that N20 and P30 are generated in the crown of the postcentral gyrus (asterisks in Fig. 4). Thus, the tangential and radial source hypotheses differ both in the number of presumed sources (one versus two) and their predominant orientations (tangential versus radial). These source hypotheses provide an excellent illustration of the lack of a unique solution to the inverse problem: Both hypotheses provide a reasonably good account of the obtained surface EPs and both are consistent to a first approximation with the known anatomy of sensorimotor cortex. Let us now consider how additional types of data can distinguish between the tangential and dual radial hypotheses.

Figure 5 presents recordings of P20-N30 and N20-P30 activity from the exposed cortical surface of a human left hemisphere during neurosurgery. The EP waveforms are similar in morphology to those at the scalp (P20-N30 anterior to the central sulcus and N20-P30 posterior), but are considerably larger in amplitude and more focally distributed. Distributions at the 20-msec peak shown in Fig. 5 are maximal on the precentral and postcentral gyri; corresponding distributions at the 30-msec peak are similar in shape and location but opposite in polarity (negative frontocentrally and positive posteriorly).

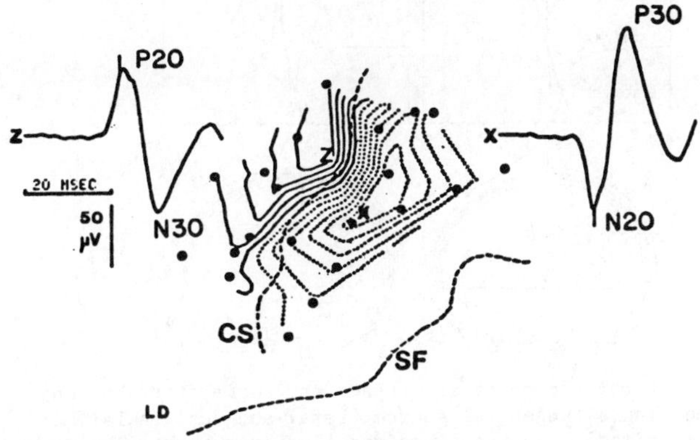

Fig. 5. Cortical surface recordings of the 20- and 30-msec EPs during neurosurgery. Waveforms are from locations indicated by Z and X, respectively, and other recording sites are indicated by dots. The topographic map is plotted at the latency of the 20-msec positive and negative peaks. Positive and negative voltage are indicated by solid and dotted contours. The central sulcus (CS) and Sylvian fissure (SF) are indicated by dashed lines.

The existence of large, focal EPs on the cortical surface with similar morphologies and spatial distributions to P20-N30 and N20-P30 on the scalp are consistent with a cortical origin for both sets of potentials. However, the cortical surface recordings alone do not provide a clear basis for distinguishing between the tangential and radial source hypotheses. That is, the cortical surface data can be explained either by a single source located tangentially to the cortical surface or by separate radial sources in exactly the same manner as the scalp data. However, when considered together the cortical surface and scalp data can help to distinguish between hypotheses by indicating whether either hypothesis can account for both sets of data with the same source configuration.

Figure 6 compares predicted cortical surface and scalp distributions derived from volume conduction models for the tangential and dual radial hypotheses. The cortical surface data were modelled by a homogeneous spherical volume conductor to approximate the conditions of the craniotomy, whereas the scalp data were modelled by a three-sphere head model with differing conductivities to approximate the brain, skull, and scalp (Darcey, 1979). The locations of the tangential and radial sources were selected to produce the extremely focal cortical surface distributions characteristic of the empirical data (10 degree separation of extrema in the model, corresponding to about 16 mm on the cortical surface). Having fixed the source locations to fit the cortical surface data, we can then ask whether those same sources produce distributions for the three-sphere data that are consistent with empirical scalp recordings. Figure 6 shows that the tangential source model produces scalp distributions with extrema separated by about 60 degrees whereas the extrema for the dual radial model are separated by only about 35 degrees. Since the empirical separation between extrema is 60-90 degrees (cf., Fig. 2), the tangential hypothesis provides

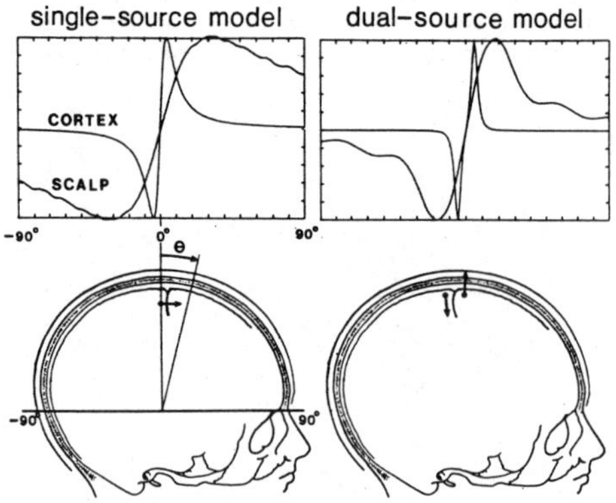

Fig. 6. Predicted cortical surface and scalp distributions for the single tangential source (left) and dual radial sources (right). Sources were modelled by current dipoles at the locations and orientations shown. A homogeneous one-sphere model was used for the cortical surface data and a three-sphere model for the scalp data. From Allison (1982).

a much better explanation of the relationship between cortical surface and scalp data. These results also illustrate how volume conduction models can be useful in distinguishing between source hypotheses in ways other than deriving solutions to the inverse problem for particular sets of source and head models.

Another source of evidence bearing on the tangential and dual radial models is based on human intracerebral recordings. In addition to their differential predictions about EP distributions on the scalp and cortical surface just described, the tangential and dual radial models make very different predictions about the morphology of EPs recorded in and around sensorimotor cortex. Specifically, the tangential hypothesis predicts that EPs posterior to area 3b should have the same negative-positive morphology regardless of whether they are recorded on the cortical surface or from depth locations (cf., Fig. 4). In contrast, the dual radial hypothesis predicts that EPs of opposite polarity to those on the scalp and cortical surface should be obtained from locations beneath somatosensory and motor cortex.

The most direct test of these predictions would be EP recordings from closely-spaced penetrations through the precentral and postcentral gyri; unfortunately there is little clinical justification for such data and only a few transcortical recordings have been reported (e.g., Stohr and Goldring, 1969; Allison et al., in preparation). However, data relevant to these predictions are available from human depth recordings obtained in conjunction with chronic implantation procedures performed to evaluate possible surgery for intractable epilepsy (Goff et al., 1980). Figure 7 depicts EPs recorded from selected depth locations anterior and posterior to the central sulcus. At locations 1-8 (anterior to the central sulcus), the same P20-N30 waveform seen in cortical surface and scalp recordings is evident. At locations 10-13 (posterior to the central sulcus), the N20-P30 waveform typical of posterior surface recordings is seen. Thus, there is no evidence of activity opposite in polarity to surface P20-N30 and N20-P30 as predicted by the dual radial hypothesis. Rather, the polarities of the depth recorded data conform to the predictions of the tangential hypothesis. Although this point would be even stronger if it were based on locations closer to the precentral and postcentral gyri (ideally on penetrations through those gyri), they nevertheless provide strong support for the tangential hypothesis.

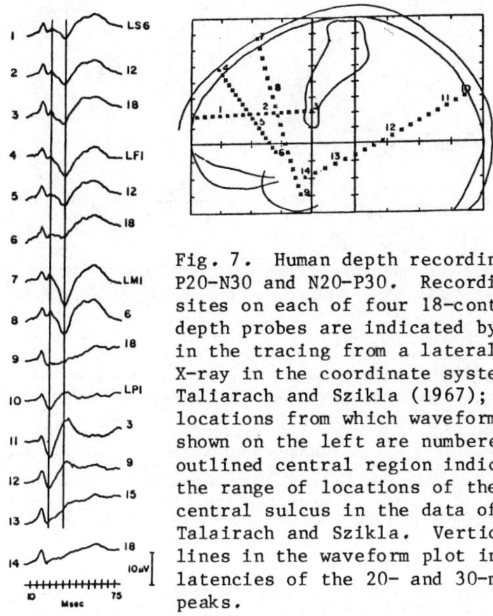

Fig. 7. Human depth recordings of P20-N30 and N20-P30. Recording sites on each of four 18-contact depth probes are indicated by dots in the tracing from a lateral skull X-ray in the coordinate system of Taliarach and Szikla (1967); locations from which waveforms are shown on the left are numbered. The outlined central region indicates the range of locations of the human central sulcus in the data of Talairach and Szikla. Vertical lines in the waveform plot indicate latencies of the 20- and 30-msec peaks.

Yet another type of evidence relevant to source hypotheses is based on the effects of brain damage, either naturally occurring (due to stroke, etc.) or due to surgical excisions (for removal of tumors, etc.). Both types of data are obviously less than ideal due to lack of control over the location and extent of damage, lack of replicability, and in the case of surgical excisions the presence of pathological tissue before excision. Despite these limitations, however, study of EPs following brain damage have proven useful as another means of testing source hypotheses.

Figures 8 and 9 present the effects of surgical excision of the hand representation area of somatosensory cortex on P20-N30 and N20-P30. This patient underwent an exploratory craniotomy because of rapidly worsening tonic and clonic seizures involving the left hand. Sensorimotor cortex was identified using cortical surface recordings obtained from a 64-channel array. Representative waveforms from the precentral and postcentral gyri are shown in Fig. 8. The median nerve stimulus elicited large paroxysmal discharges in the hand area at about 100 msec, but EPs at shorter latencies appeared to be within the normal range. Electocorticography revealed spontaneous and evoked paroxysmal discharges that were largest in the somatosensory hand area, and the postcentral gyrus was resected to the fundus of the central sulcus (Fig. 9, shaded area). Following the excision, recordings from the precentral gyrus and from parietal cortex anterior and posterior to the excised region showed no residual EP activity. Scalp recordings obtained five weeks later revealed only subcortical EPs and no evidence of the 20- and 30-msec cortical potentials. Finally, because the inital surgery failed to eliminate the seizures, a second craniotomy was performed six weeks later. Figure 9 shows that cortical surface recordings anterior to, posterior to, and directly over the excised tissue again reveal no 20- and 30-msec activity.

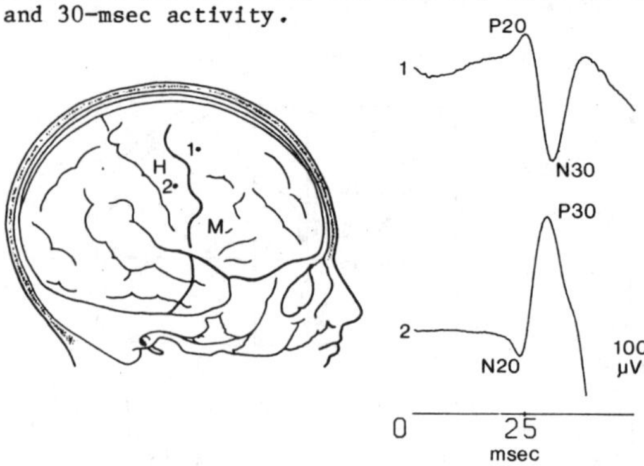

Fig. 8. Cortical surface EP recordings from the precentral (1) and postcentral (2) gyri immediately preceding the surgical excision of the hand representation area of somatosensory cortex. See text for details. From Allison et al. (in preparation).

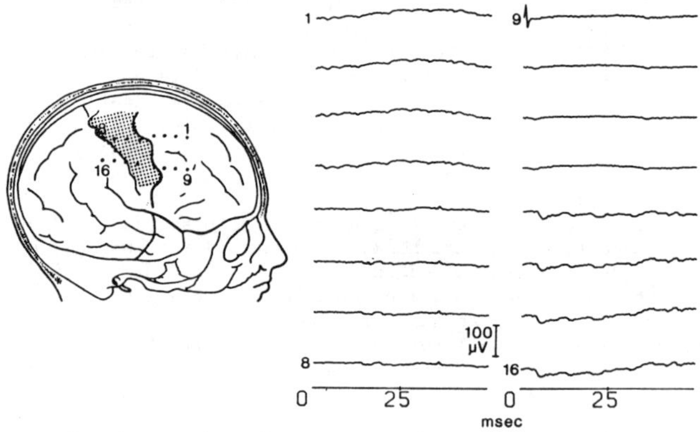

Fig. 9. Cortical surface EP recordings three months following excision of the hand area of somatosensory cortex (shaded). See text for details. From Allison et al. (in preparation).

Although the absence of cortical EPs immediately following the inital excision could be due to surgical trauma involving the precentral gyrus, their absence in scalp and cortical surface recordings weeks later is difficult to reconcile with the dual radial hypothesis. In contrast, the tangential hypothesis provides a straightforward explanation since it holds that both precentral and postcentral EPs are generated in the postcentral gyrus. One explanation that would be consistent with the radial hypothesis is that although cells in motor cortex in fact generate P20-N30, their afferent inputs project via somatosensory cortex.

For additional discussion of this point, see Allison et al. (in preparation).

SOURCE IDENTIFICATION USING COMBINED POTENTIAL AND MAGNETIC RECORDINGS

Theory suggests that EF and EP measurements differ in a number of respects that can be exploited in testing source hypotheses (Kaufman et al., 1980; Williamson and Kaufman, 1981; Cuffin and Cohen, 1979; Cohen and Cuffin, 1983). Two such differences were mentioned above: (a) surface EPs are sensitive both to radial and tangential sources whereas EFs are sensitive only to the tangential component of a source current; and (b) EPs have greater relative sensitivity to deep sources than EFs. Additional important differences are: (c) the inhomogeneous conductivities of the dura, skull, and scalp have little influence on surface EFs, whereas they decrease the spatial resolution of surface EPs; and (d) the spatial distributions of EF and EP activity on the surface differ in orientation by 90 degrees. In collaboration with David Cohen, Neil Cuffin, and Masaru Yarita of MIT, Truett Allison and I have recently attempted to exploit these differences as a further test of the tangential and dual radial source hypotheses (Wood et al., in press).

Median nerve EPs and EFs were recorded from 50-103 locations over the left hemisphere as shown in Figure 10. The EP waveforms (A) and spatial distributions (B) are similar to those shown in Fig. 1 and discussed above. The EPs at frontocentral locations (A-C and a-c) were positive-negative in morphology, with the positive peak at about 21 msec and the negative peak at about 29 msec. EPs at parietal locations (E-G and e-g) were negative-positive in morphology, with the major peaks at similar latencies. The isovoltage topographic maps show that the distributions at both latencies were approximately dipolar in shape and centered over sensorimotor cortex.

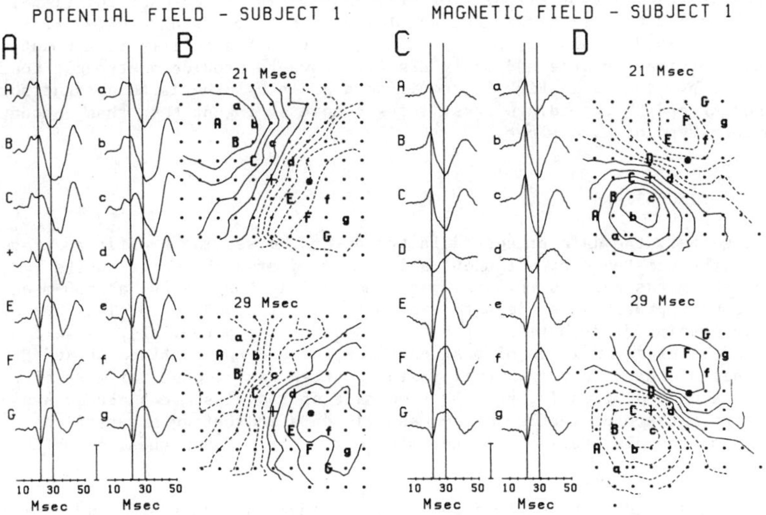

Fig. 10. EPs and EFs recorded over the left hemisphere following median nerve stimulation. A: EP waveforms from electrodes along the major axis of the EP distributions. Positive voltage relative to a linked-ear reference is

upward; calibration 5 uV. B: Isopotential distributions at the P20-N20 and N30-P30 peaks (21 and 29 msec, respectively). Solid and dotted contours indicate positive and negative voltage; electrode locations are indicated by dots or letters, with letters indicating locations corresponding to the waveforms shown in A. Scalp location C3, approximately over sensorimotor cortex, is indicated by the cross. The nose is to the left and the left ear at the bottom. C: EF waveforms from sites along the major axis of the EF distributions. Magnetic flux out of the head is upward; calibration 1 pT. D: Isofield distributions plotted at the same latencies shown in B and indicated by the vertical lines in A and C. Solid and dotted contours indicate magnetic flux flowing out of and into the head, respectively. From Wood et al. (in press).

Corresponding EF data are shown in Fig. 1 C and D. The EF waveforms were similar in morphology to the EP waveforms (although somewhat simpler - see below), with major peaks at virtually identical latencies. The EF distributions were also approximately dipolar in shape and were located over the same region of sensorimotor cortex as the EP distributions. However, the EF and EP distributions differed both in orientation (by approximately 90 degrees) and in extent (EF extrema were separated by about 60% the separation of the EP extrema).

The ability of the tangential and dual radial source hypotheses to account for these results was assessed by comparing the obtained EF and EP data to predicted results calculated from a four-shell volume conduction model of the head and either one or two dipole sources. This comparison demonstrated: (a) that the same tangential source can adequately account for both the EF and EP data; and (b) that a given pair of slightly off-radial dipoles can account for either the EF or EP data taken separately but not for both sets of data together (for details, see Wood et al., in press). Thus, just as the ability to account for combined cortical surface and scalp EPs was shown to provide a stronger test of the source hypotheses in Fig. 6, requiring a given hypothesis to account for combined EF and EP recordings constitutes a more stringent test than accounting for either type of data alone.

CONCLUSIONS

Taken together, the data presented in the preceding section provide a strong case for the existence of a tangentially oriented area 3b source on the postcentral gyrus that is in large part responsible for the 20- and 30-msec EPs. Such a source provides an adequate explanation of the scalp, cortical surface, and intracranial distributions of the 20- and 30-msec potentials, as well as the effect of surgical excision of postcentral gyrus and the combined EP and EF recordings of that activity. Dual radial sources can account for many of these results taken individually, but the same pair of radial sources cannot explain the entire pattern of results. The magnetic data indicating a superficial cortical source also rule out a thalamic source for N20 hypothesized by Chiappa (1983).

However, it should be noted that there is evidence in both scalp and cortical surface EP recordings for a contribution of radial sources, either in somatosensory cortex, motor cortex, or both, in addition to the tangential area 3b source. A radial source in somatosensory area 1 on the crown of the postcentral gyrus is thought to generate a P25-N35 complex that superimposes with the 20- and 30-msec area 3b potentials, particularly in the immediately vicinity of the central sulcus. In EP recordings, P25 is sometimes evident as a distinct peak between N20 and P30 in postcentral waveforms (Allison et al., 1980; Allison et al., in preparation). In other cases, however, there is no clear P25 peak and the combined effect of the tangential and radial sources is evident as a continuous anterior-to-posterior movement of the positive maximum in the spatial distribution during the 20 to 30 msec latency range. This movement reflects the initial activation of area 3b in the depths of the central sulcus followed in close succession by activation of more superficial regions of area 3b and ultimately area 1.

Because of the relative insensitivity of EF recordings to deep and to radial sources, systematic differences between EP and EF waveforms can be exploited to distinguish activity due to such sources from that due to superficial, predominantly tangential sources that contribute to both EP and EF recordings. For example, a diffuse positive peak is evident at 13-16 msec in the EP waveforms of Fig. 10 but is not evident in the EF data. This peak is thought to refect subcortical activity that would be expected to contribute negligibly to EF recordings. Differences between the EP and EF recordings are also evident in the latency region between the 20- and 30-msec peaks, probably reflecting the involvement of radial cortical sources in somatosensory cortex, motor cortex, or both. A quantitative approach to the evaluation of differences between EF and EP data and their implications for source identification is currently under investigation (Cuffin and Cohen, in preparation).

In summary, I have attempted both to provide some degree of perspective on the source identification problem and to illustrate how a reasonably convincing case can be made for a particular source hypothesis based on converging evidence from a number of different types of data. Together with parallel experiments in non-human animals, particularly primates, data of the sort I have described provide a workable approach to circumventing the lack of unique solutions to the inverse problem. Direct comparison of EP and EF recordings from the same subjects and conditions offers a means of making stronger inferences concerning source configurations than are possible using either technique alone.

ACKNOWLEDGMENT

I am grateful to T. Allison, D. Cohen, B.N. Cuffin, G. McCarthy, T.M. Darcey, and W.R. Goff for collaboration in various aspects of the research described in this paper.

REFERENCES

Allison, T. (1982). Scalp and cortical recordings of initial somatosensory cortex activity to median nerve stimulation in man. Ann. N. Y. Acad. Sci., 388, 671-677.

Allison, T., W. R. Goff, P. D. Williamson, and J. C. VanGilder (1980). On the neural origin of early components of the human somatosensory evoked potential. In J. E. Desmedt (Ed.), Clinical Uses of Cerebral, Brainstem and Spinal Somatosensory Evoked Potentials. Progress in Clinical Neurophysiology, Volume 7. Karger, Basel, pp. 51-68.

Allison, T., G. McCarthy, C. C. Wood, W. R. Goff, D. D. Spencer, and P. D. Williamson (in preparation). Somatosensory evoked potentials recorded from the human cortical surface and scalp to stimulation of median nerve. II. Cytoarchitectonic areas generating short-latency potentials.

Broughton, R. (1969). Discussion. In E. Donchin and D. B. Lindsley (Eds.),

Average Evoked Potentials: Methods, Results and Evaluations. Washington, NASA SP-191, pp. 79-84.

Broughton, T., T. Rasmussen, and C. Branch (1981). Scalp and direct cortical recordings of somatosensory evoked potentials in man (circa 1967). Canad. J. Psychol. 35, 136-158.

Chiappa, K. H. (1983). Evoked Potentials in Clinical Medicine. Raven Press, New York.

Cohen, D., and B. N. Cuffin (1983). Demonstration of useful differences between magnetoencephalogram and electroencephalogram. Electroenceph. clin. Neurophsiol., 56, 38-51.

Cuffin, B. N., and D. Cohen (1979). Comparison of the magnetoencephalogram and electroencephalogram. Electroenceph. clin. Neurophsiol., 47, 131-146.

Darcey, T. M. (1979). Methods for localization of electrical sources in the human brain and application to the visual system. Unpublished Ph.D. Dissertation, California Institute of Technology.

Desmedt, J. E., and G. Cheron (1980). Somatosensory evoked potentials to finger stimulation in healthy octogenarians and in young adults: Waveforms, scalp topography and transit times of parietal and frontal components. Electroenceph. clin. Neurophysiol., 50, 404-425.

Goff, G. D., Y. Matsumiya, T. Allison, and W. R. Goff (1977). The scalp topography of human somatosensory and auditory evoked potentials. Electroenceph. clin. Neurophysiol., 42, 57-76.

Goff, W. R., P. D. Williamson, J. C. VanGilder, T. Allison, and T. C. Fisher (1980). Neural origins of long latency evoked potentials recorded from the depth and from the cortical surface of the brain in man. In J. E. Desmedt (Ed.), Clinical Uses of Cerebral, Brainstem and Spinal Somatosensory Evoked Potentials. Progress in Clinical Neurophysiology, Volume 7. Karger, Basel, pp. 126-145.

Grynszpan, F., and D. B. Geselowitz (1973). Model studies of the magnetocardiogram. Biophys. J., 13, 911-925.

Kaufman, L., and S. J. Williamson (1980). The evoked magnetic field of the human brain. Ann. N. Y. Acad. Sci., 340, 45-65.

Maccabee, P.J., E. I. Pinkhasov, and R. Q. Cracco (1983). Short latency somatosensory evoked potentials to median nerve stimulation: Effect of low frequency filter. Electroenceph. clin. Neurophysiol., 55, 34-44.

Papakostopoulos, D., and H. J. Crow (1980). Direct recording of the somatosensory evoked potentials from the cerebral cortex of man and the difference between precentral and postcentral potentials. In J. E. Desmedt (Ed.), Clinical Uses of Cerebral, Brainstem and Spinal Somatosensory Evoked Potentials. Progress in Clinical Neurophysiology, Volume 7. Karger, Basel, pp. 15-26.

Stohr, P. E., and S. Goldring (1969). Origin of somatosensory evoked scalp responses in man. J. Neurosurg., 31, 117-127.

Talairach, J., and G. Szikla (1967). Atlas of Stereotaxic Anatomy of the Telencephalon. Masson et Cie, Paris.

Williamson, S. J., and L. Kaufman (1981). Biomagnetism. J. Magn. Magn. Mater., 22, 129-202.

Wood, C. C. (1982). Application of dipole localization methods to source identification of human evoked potentials. Ann. N. Y. Acad. Sci., 388, 139-155.

Wood, C. C., G. McCarthy, N. K. Squires, H. G. Vaughan, Jr., D. L. Woods, and W. C. McCallum (1984). Anatomical and physiological substrates of event-related potentials: Two case studies. Ann. N. Y. Acad. Sci., 425, 681-721.

Wood, C.C., D. Cohen, B. N. Cuffin, M. Yarita, and T. Allison (in press). Electrical sources in human somatosensory cortex: Identification by combined magnetic and potential recordings. Science.

LOCALIZATION OF CEREBRAL SOURCES BY NEUROMAGNETIC MEASUREMENTS*

Gian Luca Romani and Roberto Leoni

Istituto di Elettronica dello Stato Solido - C.N.R.
Via Cineto Romano 42, 00156 Roma, ITALY

ABSTRACT

The typical procedure for neuromagnetic localization of cerebral sources is described. Different methods proposed for analyzing data of spontaneous brain activity are presented and their respective possibilities are compared. Finally, the validity of the theoretical model used for fitting experimental data distribution and the reliability of three-dimensional localization are discussed.

KEYWORDS

Neuromagnetism; evoked potentials and fields; Relative Covariance; current dipole.

INTRODUCTION

The possibility of three-dimensional localization of sources of normal or pathological cerebral activity represents one of the most exciting features provided by the neuromagnetic approach to the investigation of the human brain. Important results in source localization were recently achieved in the study of primary and secondary sensory areas (Hari and colleagues, 1984; Okada and colleagues, 1984; Romani, Williamson and Kaufman, 1982a), in that of higher levels of brain functions (Fiumara and colleagues, 1984; Okada, 1984a; Weinberg and colleagues, 1984), and in the investigation of normal (Chapman and colleagues, 1984) and pathological brain activity (Barth and colleagues, 1982; Chapman and colleagues, 1983). In all these studies a scalp distribution of magnetic fields associated with neural activity was suitably detected and interpreted in terms of a relatively simple model. Identification of the equivalent source location was eventually achieved in almost the totality of the cited researches. The accuracy of localization was in some cases extremely good as compared with that provided by corresponding electrical potentials measurements (Wood, 1982).

---

*Also supported by Progetto Finalizzato Tecnologie Biomediche e Sanitarie, CNR.

These results, and others which are being collected, are particularly promising also from a clinical point of view. Nevertheless, the effective validity of the procedures used for data analysis should be checked and the consequent localization reliability should be verified as compared with that available from other investigation techniques.

LOCALIZATION PROCEDURES

The typical procedure of neuromagnetic localization consists of the following steps:

1) Detection of magnetic fields over specific areas of the scalp, generally accompanied by simultaneous electric potentials recording.
2) First data processing to get iso-field contour maps illustrating the spatial distribution of magnetic fields.
3) If the experimental pattern shows similarity with that generated by a simple model source (henceforth a current dipole in a homogeneously conducting sphere, unless differently specified), this source is assumed to be the "equivalent" generator of the measured signals. The best sphere fitting the subject's head is evaluated and the three-dimensional localization of the equivalent source which best accounts for the experimental data distribution is achieved by means of iterative numerical procedures.

It should be remarked that the concept of equivalent dipole is a pure mathematical abstraction, i.e. a simple and feasible tool to interpret electric potential and magnetic field distributions and yet sufficiently realistic. A localization of an equivalent dipole in a specific region of the brain does not permit, by all means, any further speculation on the physiological characteristics of the investigated source.

Each step will be discussed in the appropriate section, with specific emphasis on points which our experience identifies as crucial ones. Basic concepts concerning instrumentation, modeling and statistical data analysis will not be described here; the interested reader is referred to appropriate references (Burington and May, 1958; Romani, Williamson and Kaufman, 1982b; Williamson and Kaufman, 1981).

Experimental methods

It is commonly agreed that a measurement of the component of the magnetic field perpendicular to the scalp provides information independent of volume currents if the underlying source is representable as a current dipole and the head as a sphere with homogeneous conductivity or with a spherically symmetric conductivity distribution (Grinzspan and Geselowitz, 1973). The measurement is typically performed positioning the magnetic detector at different sites of the scalp according to a pre-determined grid. As is well known, the sensor geometry can significantly affect the characteristics of the measured pattern (Romani, Williamson and Kaufman, 1982b; Williamson and Kaufman, 1981). The example illustrated in Fig.1 concerns the simple case of a current dipole located 4 cm below the surface of a semi-infinite medium with homogeneous conductivity and

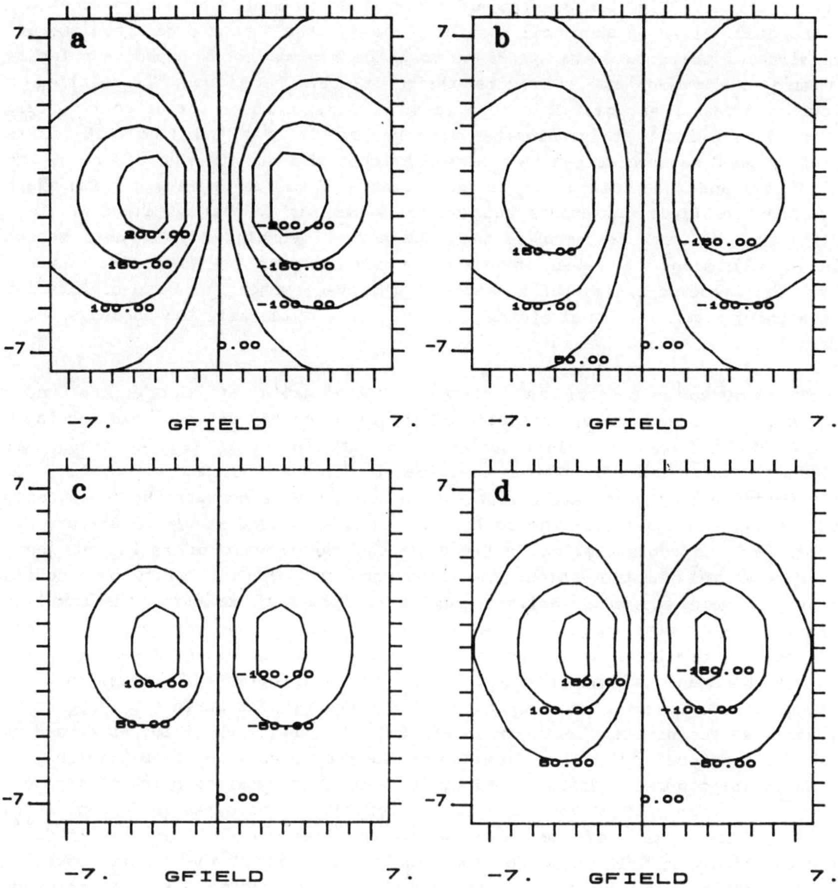

Fig. 1. Maps illustrating the effect of the pickup coil geometry on the magnetic field generated by a current dipole located 4 cm below the surface of a semi-infinite medium with homogeneous conductivity. a) Iso-field contour map showing the spatial distribution of the component of the field perpendicular to the surface B_n. b-d) same field as mesured by a magnetometer, 5 cm in diameter (b); by a 2nd-order gradiometer with neglegible area and 3 cm baseline (c); by a 2nd-order gradiometer 1.5 cm in diameter and 5 cm baseline. Frame units are expressed in cm. Field amplitudes in arbitrary units.

oriented parallel to it. Fig.1a shows the spatial distribution of the component of the magnetic field perpendicular to the surface. The use of a pickup coil with finite area produces a spreading of the pattern and a modest weakening of the measured amplitude. Fig.1b shows the effective field as sensed by a magnetometer 5 cm in diameter. If the pickup coil has the geometry of a second-order gradiometer, the baseline of the gradiometer tends to move the two maxima of opposite polarity one toward the other and to reduce their relative amplitude, simulating a shallower and weaker source. This is evident in Fig.1c, where the effective field as measured by a 2nd-order gradiometer with neglegible area and 3 cm baseline is reported. A good compromise can be achieved by choosing the diameter of the pickup coil (1.5 cm) and the baseline of the gradiometer (5 cm) as in Fig.1d. The shape of the effective field pattern is substantially similar to that of the field (a). Nevertheless, it must be remarked that the actual geometry of the used sensor should by all means be taken into account when trying to solve the inverse problem. The values defining the geometry of the gradiometer in d) are of interest when designing a multi-channel system: however they match only the chosen source, but do not have a general validity.

The choice of grid spacing is related to the depth of the source under investigation. Indeed, a rough estimate of the relationship between grid spacing G and source depth from the scalp D suggests that the former is made at most equal to 0.7(D+d), where d is the distance between the pickup coil and the scalp. For example in order to investigate a source at a depth D=2 cm beneath the scalp, with a pickup coil 1 cm apart from the scalp, the grid spacing G should be at most 2.1 cm. The previous consideration is based on the requirement of having at least three experimental positions along the line connecting the two maxima of opposite polarity. A more accurate analysis can be performed by applying the sampling theorem for the spatial frequency (Lehmann, 1983). Let us consider the simple case of a current dipole located at a depth D=2 cm below the surface of a semi-infinite medium with homogeneous conductivity. If the dipole is oriented parallel to the y direction, the component of the magnetic field normal to the surface B_n depends on the transverse coordinate x as shown in Fig.2a. The effective field, as sensed by a 2nd-order gradiometer, 1.5 cm diameter of the pickup coil and 5 cm baseline, is reported in the figure. A distance d=1 cm is assumed as separating the pickup coil and the plane. The spatial Fourier transform of $B_n(x)$, reported in Fig.2b, shows that the greatest part of the spectrum is confined below, say, 0.25 cm^{-1}. Therefore, choosing this value for the Nyquist frequency, the spatial sampling frequency must be twice as large. Consequently a grid spacing G=2 cm is required to get an appropriate representation of the field distribution.

The overall size of the grid to be used for investigating a dipolar source is also related to its depth D. Fig.3 illustrates this dependence when the overall grid dimensions are 14x14 cm and a dipolar source is located at increasing depths below the surface of a homogeneously conducting half-space. The effective field is displayed as measured by a second order gradiometer, 1.5 cm pickup coil diameter and 5 cm baseline. The figure shows that the chosen dimensions are inadequate already for a source 6 cm far from the pickup coil. As a typical value of the distance d is 1.5 cm, the mentioned figure corresponds to a source located at 4.5 cm beneath the scalp, case not infrequent in neuromagnetic studies.

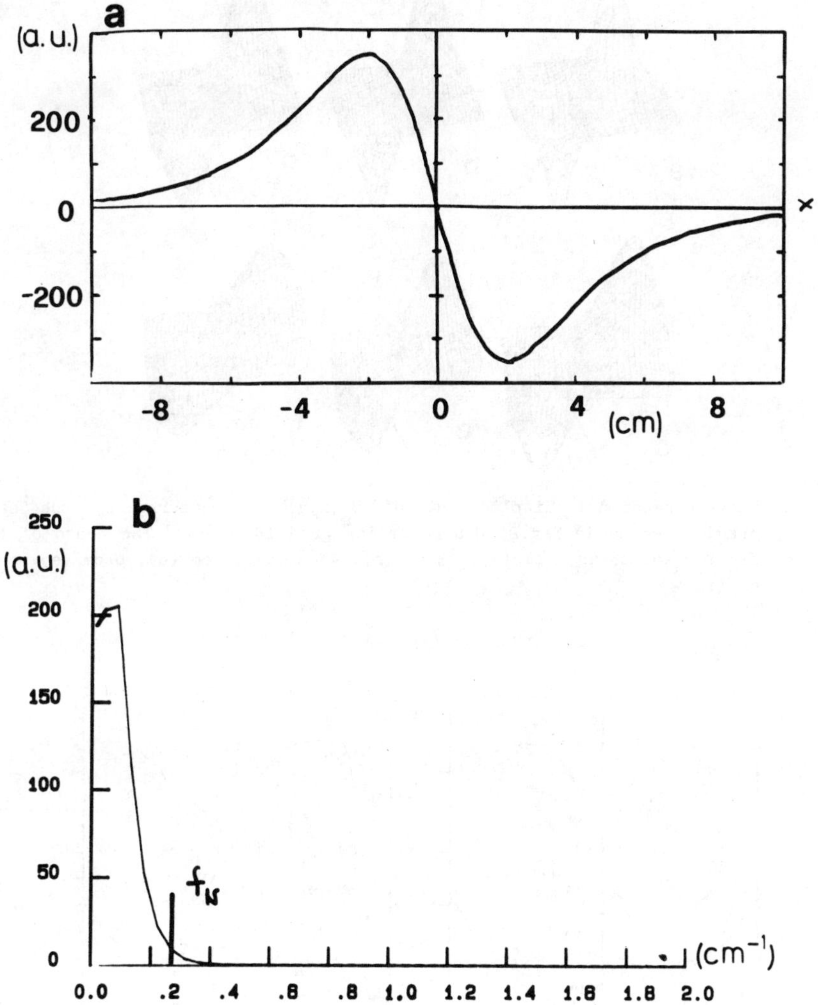

Fig. 2. a) $B_n(x)$ as generated by a dipole 3 cm deep below the surface of the half space and sensed by a 2nd-order gradiometer 1.5 cm in diameter and 5 cm baseline. b) Spatial Fourier transform of $B_n(x)$. The vertical bar identify a possible value for the Nyquist spatial frequency.

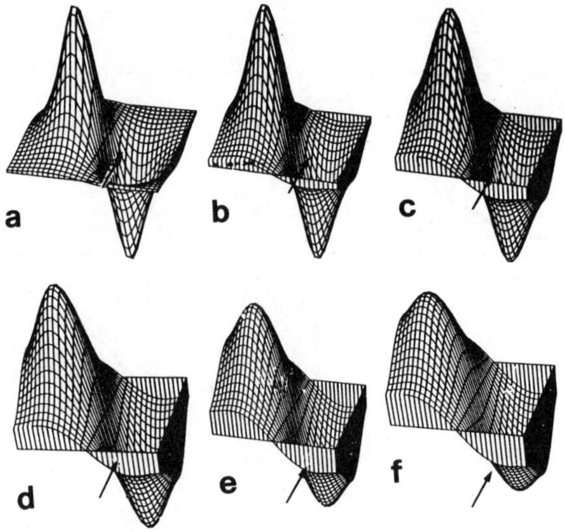

Fig. 3. Three-dimensional displacement of $B_n(x,y)$ as measured by the same gradiometer as in Fig.2 in a recording grid 14x14 cm. The depth of the dipolar source is 2 cm (a), 3 cm (b), 4 cm (c), 5 cm (d), 6 cm (e) and 7 cm (f).

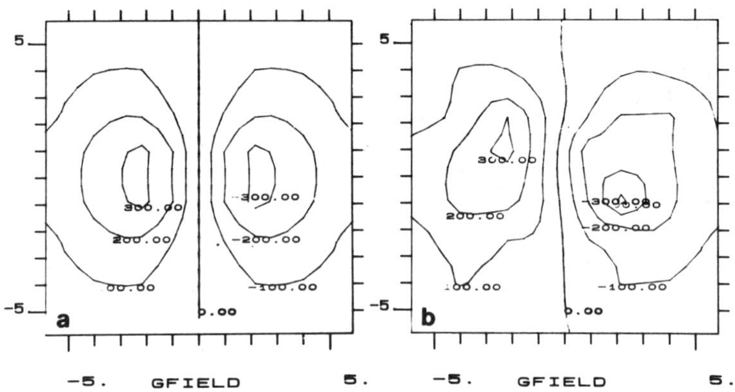

Fig. 4. a) Iso-field contour map illustrating the spatial distribution of B_n as measured by the same 2nd-order gradiometer as in Fig. 2 with the pickup coil at a distance d=1 cm from the surface of the half space. The dipolar source is 2 cm below it. b) "Perturbed" distribution obtained adding to d an error with average peak-to-peak fluctuation 0.5 cm.

Another important remark concerns the distance d. Variations of the order of few mm with respect to a nominal value can significantly affect the detected field amplitude and consequently artificially modify the measured pattern. This effect proves to be dramatic when maximum sensitivity is required, as it becomes comparable with the average noise (Okada and colleagues, 1984). Fig.4a shows the theoretical distribution of B_n as calculated onto a plane below which a current dipole is located at a depth D=2 cm. The field is calculated as previously, in correspondance of a distance d=1 cm between the pickup coil and the plane. If the value of d is perturbed by adding an error with gaussian distribution having a peak to peak average fluctuation σ=0.5 cm and mean value zero, the spatial distribution is significantly modified, as shown in Fig.4b. Consequently, the localization of the equivalent source becomes less reliable.

The simultaneous detection of electric potentials generally does not present difficulties, if silver-silver cloride or gold electrodes are used. The choice of the electric reference, however, which is crucial in the case of electric mapping (Nunez, 1981), must be carefully made if electric and magnetic scalp distributions are to be compared or a correlation between simultaneous MEG and EEG signals is required. A usual choice is the contralateral ear lobe reference. Alternatively the average reference, i.e. the istantaneous mean value of all the measured electric signals, is commonly used by several experimentalists (Darcey, Ary and Fender, 1980; Nunez, 1981).

Data analysis

In the case of evoked field measurements the non-simultaneity of magnetic recordings constitutes only a minor problem, in that signals measured in correspondance of each recording site are averaged with respect to the stimulus onset. The experimental conditions are identically reproduced each time but for unavoidable stress of the subject and consequent weakening of concentration in case of overlengthy sessions. Iso-field contour maps illustrating the distribution of the magnetic fields at different latencies with respect to the stimulus onset are derived from the averaged signals and can be inspected as a first check on the measured pattern.

The problem is much more complicated in the case of spontaneous brain activity. Although it has been recently demonstrated (Barth and colleagues, 1982) that an averaging procedure can be successfully used in the study of magnetic signals originating from neural activity in some cases of focal epilepsy, in the majority of pathological cases and in the investigation of normal brain rhythms the non-simultaneity of magnetic recordings can prove to be an insurmountable problem. Barth and colleagues (1982) have studied several cases of focal epilepsy which showed high-amplitude spike-like electric signals well reproduced during interictal activity. These signals were used as a trigger for averaging simultaneously recorded magnetic signals in correspondance of different sites of the grid. Only signals time-locked with the electric trigger could be extracted from the noise, as related to the same source. An example resulting from this kind of analysis is shown in Fig.5, where averaged magnetic tracings, obtained from a patient with a clinical history of right temporal lobe epilepsy, are reported in

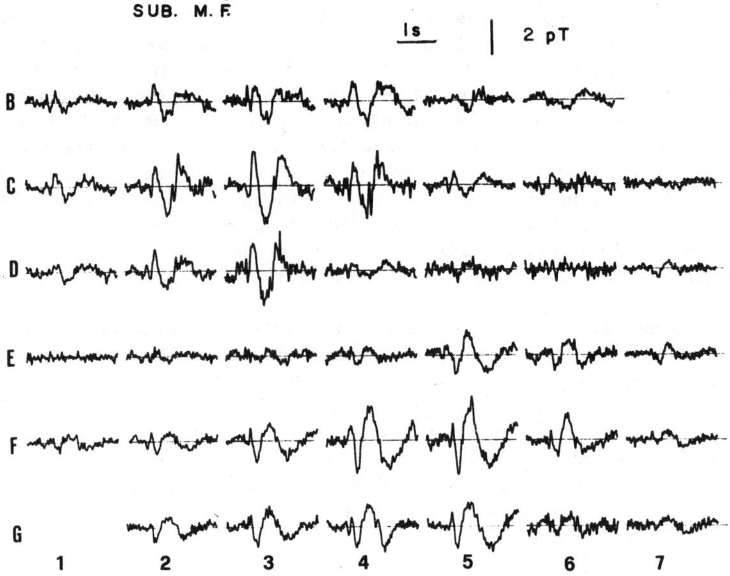

Fig. 5. Averaged magnetic tracing obtained for the subject MF displayed in the recording bandwidth 0.1÷32 Hz. For a description of the reference grid see the text.

correspondance of their experimental recording site. Each signal was obtained on the basis of 4 to 12 samples, depending on the number of interictal artifact-free spikes occurring during the recording time. The reference grid was positioned over the right temporal bone of the subject, with the horizontal axis tilted about 30° with respect to the line connecting the inion to the nasion. The ear canal was approximately below E1. Signals are displayed in the recording bandwidth 0.1÷32 Hz. The map reported in the figure shows an evident polarity reversal in the measured field and suggests that a dipolar-like source can be successfully used to interpret the data. This first hint will be confirmed by the successive analysis.

Nevertheless, our experience in the study of focal epilepsy (Chapman and colleagues, 1983; Carelli and colleagues 1983; Modena and colleagues, 1982; Ricci and colleagues, 1981; Ricci, 1983) showed that quite often the electric pathologic activity does not permit averaging, in that only low-level spikes or quasi-periodic rhythms are detected. Sometimes the electric activity is poor as compared with a sustained magnetic one. Two examples illustrating what is stated above are reported in Fig.6. The figure shows typical real-time MEG and EEG tracings as recorded from two patients affected by focal epilepsy. The first patient RM presents repetitive spike-like signals both in the MEG and in the EEG: the electric pathological signals can be successfully used as a trigger for averaging the corresponding MEG spikes. By contrast, the second patient EI shows a quasi-rhythmic activity at about 2 Hz in both the EEG and the MEG, which cannot be suitably analyzed by the illustrated procedure.

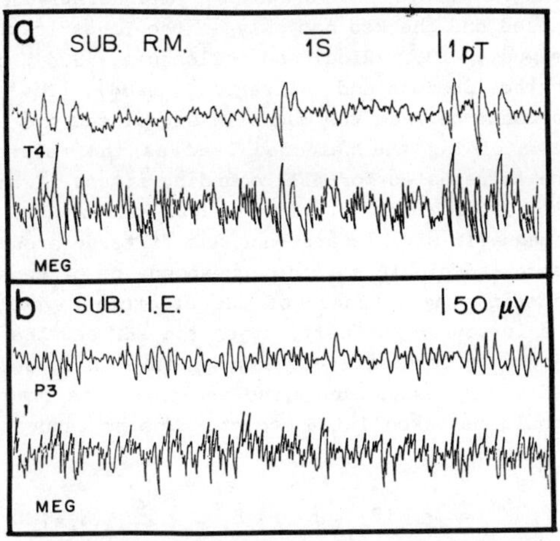

Fig. 6. Typical real-time MEG and EEG tracings as recorded from two patients (RM and EI) affected by focal epilepsy.

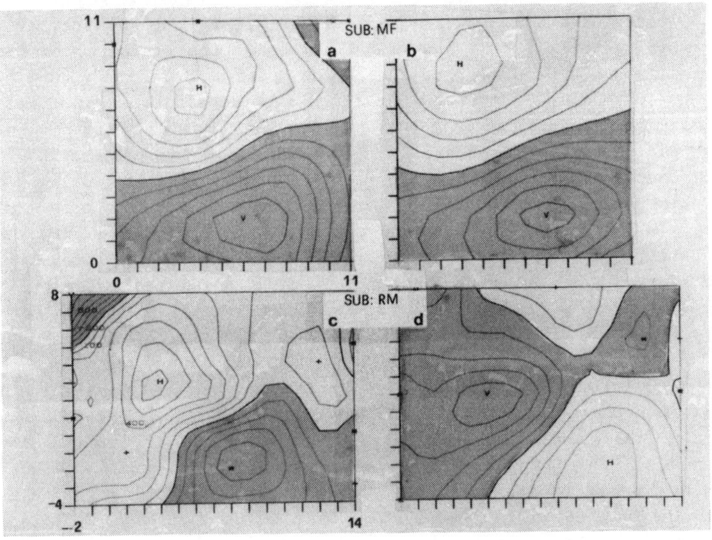

Fig. 7. Comparison between the iso-field (a and c) and the iso-RC (b and d) contour maps for subjects MF and RM. Frame units are expressed in cm.

In order to partially overcome this drawback, the Relative Covariance method has been recently proposed and checked not only in the investigation of focal epilepsy (Chapman and colleagues, 1983; Ricci and colleagues, 1984), but also in the study of normal alpha-rhythm (Chapman and colleagues, 1984). The basic principle is to calculate the covariance between the magnetic signal and one of the simultaneously recorded EEGs. Maintaining the selected lead as the reference electrode, the described procedure is repeated for all recording sites. If the FFT analysis shows that the normal or pathological activity is confined around specific frequencies both electric and magnetic signals are bandpass filtered around those frequencies, before the covariance is calculated. In correspondance of each recording site, the covariance is divided by the variance of the electric signal, to compensate for possible variations in source intensity. When the MEG and the EEG are in phase the covariance is expected to be positive, and negative in an out of phase condition. Randomic phase shifts are neglected by the analysis, as only signals coming from the same source should be taken into account. The procedure can be summarized by the following expression:

$$RC_n = \sum_t (B_{t,n} - B_{m,n})(E_t - E_m) / \sum_t (E_t - E_m)^2$$

where $B_{t,n}$ and E_t are the amplitudes of simultaneous magnetic and electric signals at times $t=1,\ldots N$, $B_{m,n}$ and E_m their respective mean values, and RC_n the Relative Covariance at the n-th recording site. For each measured position the computed value of RC is a linear function of the magnetic field. Its distribution with respect to the position on the scalp reflects that of the magnetic field.

Figure 7 illustrates a comparison between the iso-field contour maps derived from averaged experimental distributions and the iso-RC maps evaluated by processing the same data with the latter procedure. The upper part of the figure refers to the same subject as Fig.5, and the iso-field map is calculated in correspondance of the first strong component of the tracings. The RC is calculated with respect to the T4 lead of the 10-20 system, contralateral ear reference. The lower part of the figure concerns magnetic signals detected over the right temporal and frontal scalp of the patient RM, affected by right frontal lobe epilepsy (Ricci and colleagues, 1984), and whose real-time tracings are reported in Fig.6a. Once more the iso-field map on the left is derived from averaged signals, while the map on the right depicts the RC distribution. In both cases the evident similarity between respective distribution confirms the substantial congruity of the two procedures.

As previously mentioned, the Relative Covariance method was recently applied to the study of normal brain rhythms (Chapman and colleagues, 1984). In their investigation these authors concentrated the analysis on the alpha rhythm which was measured at recording sites spread over approximately the whole scalp. The spatial distribution of the RC values for one subject (RI), as calculated with respect to the O_z electrode (10-20 system, left ear lobe reference) is reported in Fig.8. In the reference frame used in the figure the subject's inion is positioned at (0,0), the horizontal axis is parallel to the inion-nasion line and 5 cm below it, and all vertical axes represent lines pointing to the vertex. We want to remark that only a multi-sensor system featuring a high number of magnetic channels would provide a correspondingly complete iso-field map. The iso-RC

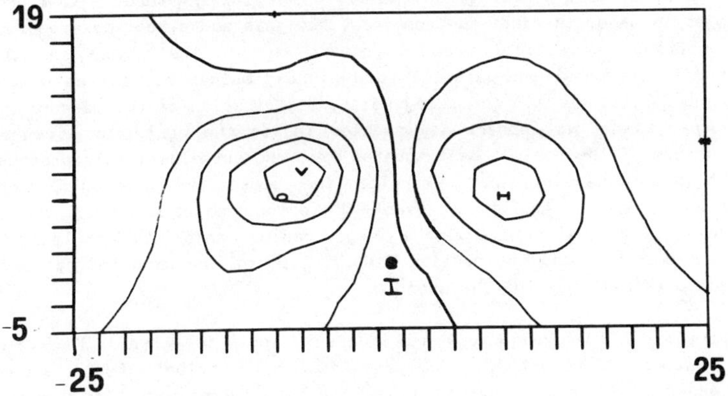

Fig. 8. Spatial distribution of the RC values for subject RI obtained in the study of the alpha rhythm. Frame units are expressed in cm. For a description of the reference grid see the text.

pattern shows a spatial distribution which can be interpreted in terms of an equivalent dipolar source, at least in a first approximation.

Source Localization

One of the most crucial problems to be faced when attempting source localization is the identification of appropriate parameters which can be used to check the applicability of a specific model. The consideration that significant results in the relatively young history of biomagnetism were collected on the basis of simple eye-inspection of field distributions and localization procedures that nowadays could be considered as very crude, is somewhat conforting and supports the validity of the method - even if it has been regarded as questionable by some people. Certainly, an eye-inspection of the iso-field or iso-RC maps provides a first check to figure out whether or not a specific model is useable to interpret data distributions. For example, the maps shown in Fig.7 show evident similarities with the simple pattern generated by a current dipole in a homogeneously conducting sphere.

A more reliable procedure should, however, evaluate the parameters relative to the "equivalent" source which was chosen to account for the measured pattern. This can be achieved by minimizing the differences between experimental and theoretical distributions as compared with the experimental error. In practice, if the model used for interpretation is the above mentioned one, five parameters must be estimated: three for the dipole position, one for the dipole strength and one for its direction, as only its tangential component is involved. Their best value can be obtained minimizing the χ^2 function, through numerical iterative procedures. The minimized χ^2 value also provides a statistical significance level which permits an estimate of the appropriateness of the used model.

An important point concerns the choice of the sphere to fit the head. The procedure typically used by our group consists in digitizing three actual profiles of the subject's head, as derived from x-ray analysis and direct craniometry: the sagittal profile, a vertical profile passing through the vertex and the ear canals, and a horizontal one parallel to a plane passing through the nasion and the inion and 2 cm above it. According to the distribution of the measuring sites over the scalp, the best sphere is inserted inside the posterior, central or anterior portion of the scalp (least squares fit). Nevertheless, this procedure is only partially satisfactory, as quite often the projections of several recording points onto the sphere poorly represent the actual positions over the scalp. Alternatively, a "local" sphere with a larger radius can be used at least in the temporal region, provided the spatial distribution of the experimental sites is constrained to a small area of the scalp.

A first example of localization of an equivalent dipole is shown in Fig. 9. The iso-field contour map reported in d) represents the distribution of the magnetic field as measured in the bandwidth 0.1÷100 Hz over the right hemisphere of a normal subject. These fields were recorded in correspondance of stimulation of the auditory system by 1 KHz tone bursts, 50 ms duration, 70 dB SPL. The map illustrates the spatial distribution of the N100 component of the field. The positions of the recording sites as projected onto the fitting sphere are identified by crosses in a-c). The solid circles represent the three-dimensional location of the equivalent generator as obtained by means of the χ^2 fit - which provides a significance level larger than 15%. The uncertainty region is contained within the circle. The adequacy of the model used for the localization is further confirmed by the quite good agreement between the experimental iso-field map and the theoretical one (e) as obtained from the fitted source.

Somewhat less satisfactory results are achieved in case of spontaneous activity. Fig. 10 shows the three-dimensional localization of the equivalent source for the epileptic subject MF (see Fig.5 and 7). The iso-field map in d) reproduces the distribution already shown in Fig.7a. The filled squares identify the location of the source as obtained from this map, whereas the open circles represent the position of the equivalent generator as derived from the RC distribution. For both localizations the significance level was less than 0.001. However a satisfactory agreement between the experimental and theoretical (e) distributions is appreciable. Probably this less satisfactory result should be related to the features of the actual source: indeed the electrocorticographic analysis performed on the exposed cortex before lobectomy revealed that a relatively large area of cortex - a few cm^2 - was pathologically active (Ricci and colleagues, 1984).

Finally we want to discuss the localization achieved in the study of the magnetic alpha rhythm. Fig.11 shows the three-dimensional localization obtained for subject RI (see Fig.8), when fitting the iso-RC distribution using a single equivalent source. The generator is placed at about 6cm below the scalp. The significance level provided by the fit is 1%. Nevertheless, the authors of this investigation (Chapman and colleagues, 1984) preferred a different hypothesis on the source configuration, i.e. two equivalent generators synchronously acting and located at a smaller depth. This was based on several argumentations out of which we mention only one: the intensity of one single generator would have been too high as

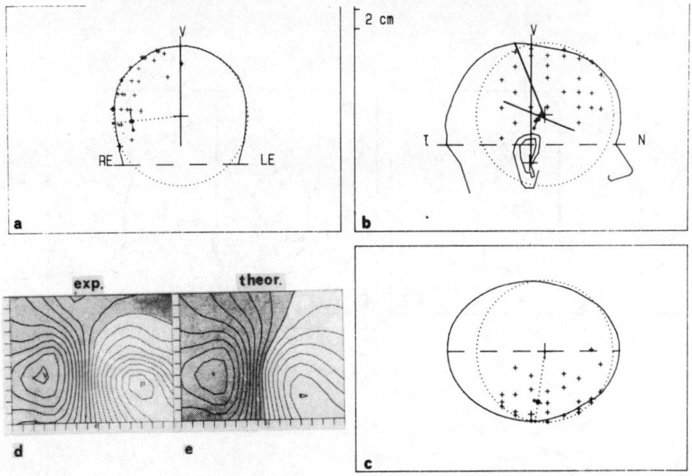

Fig. 9. a-c) 3-dimensional localization of the equivalent source of the magnetic field distribution obtained stimulating the auditory system. d) Experimental iso-field map, depicting the distribution of the N100 component, used for localization. e) Theoretical distribution of the magnetic field as obtained from the fitted source. Crosses represent the recording sites as projected onto the sphere.

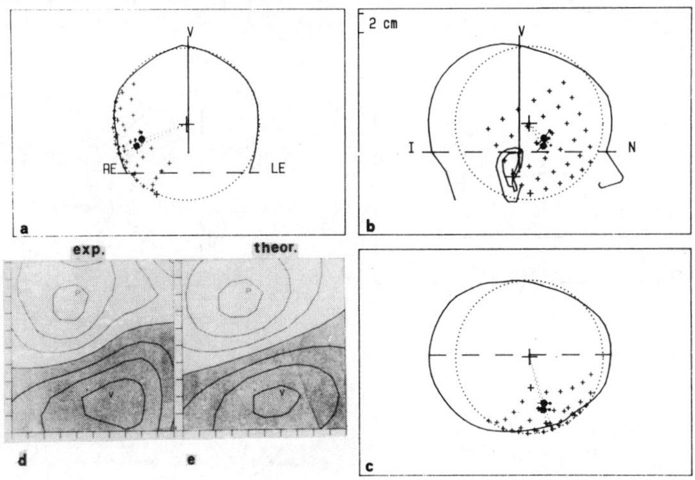

Fig.10. a-c) 3-dimensional localization of the equivalent source for the epileptic subject MF. The filled squares represent the localization using the iso-field distribution whereas the open circles represent the equivalent generator as derived from the RC distribution. The iso-field map used for localization is illustrated in d). e) Theoretical distribution of the magnetic field as obtained from the fitted source.

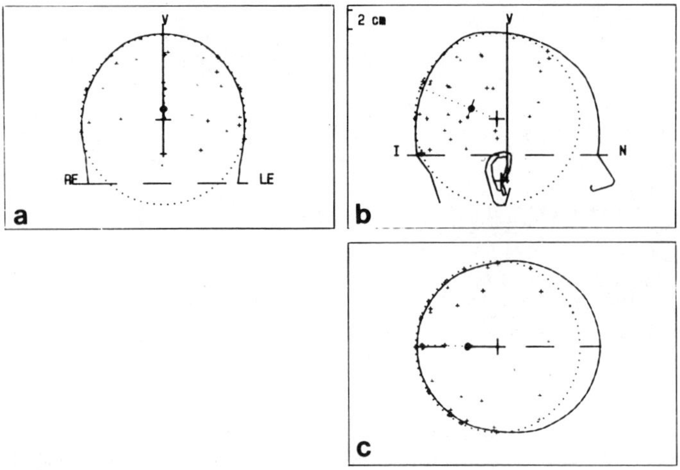

Fig.11. a-c) 3-dimensional localization of a single equivalent source based on the RC distribution shown in Fig.8.

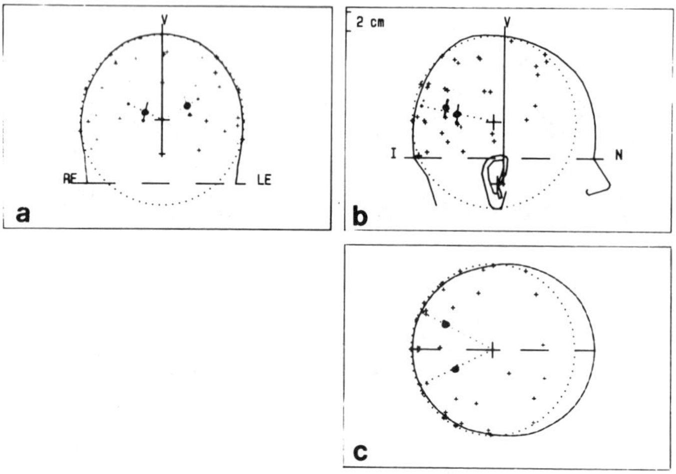

Fig.12. a-c) 3-dimensional localizatiun of two equivalent generators accounting for the same RC distribution of Fig.8.

compared with typical values of current density in the cortex. A fitting procedure successively performed using the two-dipole model confirmed this hypothesis. Fig.12 illustrates the two-dipoles localization as obtained from the same iso-RC distribution by means of a 10 parameters fit. The two equivalent generators are located at an average depth of 3.7 cm and the actual sources could be spread along the terminal portion of the calcarine fissure. The level of significance provided by the fit is 7%. Although no statistical comparison has been made between the two models, this last finding provides additional support to the idea that a two-dipole model is moderately preferable with respect to the single-dipole one.

CONCLUSIONS

The present state of art of neuromagnetic localization is relatively satisfactory. When the experimental field distribution resembles that of the simple dipolar model quite good results can be achieved: the localization provides a high significance level and a small uncertainty region.

One of the aims of this paper was, however, at illustrating that several experimental parameters may significantly affect the shape of the measured field distribution and therefore must be carefully controlled: the geometry of the magnetic sensor, the spacing of the grid used for detecting fields at different sites of the scalp and, finally, the distance between the probe and the scalp are some of them.

When these experimental ambiguities have been supposedly eliminated, the field distribution can be compared with that of a simple theoretical model. We want to remark again that the localization of an equivalent dipolar source is a procedure certainly meaningful from a physical-mathematical point of view. However one should not conclude that this source has a real physiological correspondance. In fact the same equivalent generator could equally well account for either a concentrated population of neurons or for one spread inside a cup-shaped region of cortex (Okada, 1984b).

Nevertheless, the measured field distribution is often more complex, as in some of the examples described here and in others which was not possible to mention for space limitations, like for instance those observed in the study of higher levels of brain functions (Fiumara and colleagues, 1984). Consequently, a more complex structure of the actual source(s) is expectable and more complicated model sources should be developed. Despite that, the results so far achieved are certainly encouraging in that the identification of the equivalent source location can be obtained with an accuracy at least comparable with, if not better than that provided by other techniques of "functional" localization.

ACKOWLEDGEMENTS

The authors are indebted to all the members of the biomagnetic group of Rome and in particular to Prof. I. Modena for helpful discussions and comments. Special thanks are due Dr. S.N. Ernè for suggestions concerning the use of the sampling theorem for the spatial frequency, and him and Dr. S. Barbanera for a critical revision of the manuscript. Finally, the authors want to express their gratitude to Prof. A. Paoletti for continuos encouragements and help.

REFERENCES

Barth D.S., W.Sutherling, J.Engel and J.Beatty (1982). Science, 218, 891-894.

Burington R.S. and D.C. May (1958). Probability and Statistics with Tables. Handbook Publishers, Inc., Sandusky, Ohio.

Carelli P., I. Modena, G.B. Ricci and G.L. Romani (1983). Biomagnetism: an interdisciplinary approach. S.J. Williamson, G.L. Romani, L. Kaufman and I. Modena Eds., Plenum Press, New York-London, 469-482.

Chapman R.M., G.L.Romani, S.Barbanera, R.Leoni, I.Modena, G.B.Ricci and F.Campitelli (1983). Lett. Nuovo Cimento, 38, 549-554.

Chapman R.M., R.J.Ilmoniemi, S.Barbanera and G.L.Romani (1984). Electroencephalogr. Clin. Neurophysiol., in press.

Darcey T.M., J.P. Ary and D.H. Fender (1980). Prog. Brain Res. 54, 128-134.

Fiumara R., F. Campitelli, G.L. Romani, R. Leoni, M. Caporali, M. Zanasi, A. Cappiello, G. Fioriti and I. Modena (1984). This volume.

Grynzspan F. and D.B. Geselowitz (1973). Biophys. J. 13, 911-925.

Hari R., K. Reinikainen, E. Kaukoranta, M. Hamalainen, R. Ilmoniemi, A. Penttinen, J. Salminen and D. Teszner (1984). Electroencephalogr. Clin. Neurophysiol., 57, 254-263.

Lehmann H.P. (1983). Biomagnetism: an interdisciplinary approach. S.J. Williamson, G.L. Romani, L. Kaufman and I. Modena Eds., Plenum Press, New York London, 591-624.

Modena I., G.B. Ricci, S. Barbanera, R. Leoni, G.L. Romani and P. Carelli (1982). Electroencephalogr. Clin. Neurophysiol., 54, 622-628.

Nunez P.L. (1981). Electric fields of the brain. The neurophysics of EEG. Oxford University Press, New York-Oxford.

Okada Y. C. (1984a). Approaches to defining the intracranial generators of event-related electrical and magnetic fields. Proc. EPIC VII Meetings, Florence, Italy, in press.

Okada Y. C. (1984b). This volume.

Okada Y. C., R. Tanenbaum, S.J. Williamson and L. Kaufman (1984). Electroencephalogr. Clin. Neurophysiol., in press.

Ricci G.B. (1983). Il Nuovo Cimento, 2D, 517-537.

Ricci G.B., G.L. Romani, I. Modena, S. Barbanera, R. Leoni, and P. Carelli (1981). Electroencephal. Clin. Neurophysiol., 52, 95.

Ricci G.B., R. Leoni, G.L. Romani, F. Campitelli, S. Buonomo and I. Modena (1984). This volume.

Romani G.L., S.J. Williamson and L. Kaufman (1982a). Science, 216, 1339-1340.

Romani G.L., S.J. Williamson and L. Kaufman (1982b). Rev. Sci. Instrum. 53, 1815-1845.

Weinberg H., P.A. Brickett, J. Vrba, A.A. Fife and M.B. Burbank (1984). Annals of New York Academy of Sciences, in press.

Williamson S.J. and L. Kaufman (1981). Biomagnetism. S.N. Ernè, H.D. Hahlbohm and H. Lubbig Eds., Walter de Gruyter, Berlin-New York, 353-402.

Wood C. C. (1982). Annals of New York Academy of Sciences, 139-155.

BIOMAGNETISM IN THE STUDY OF BRAIN FUNCTIONS

R. Hari[1], D. Cohen[2], L. Kaufman[3], G.L. Romani[4], and H. Weinberg[5]

[1]Low Temperature Laboratory, Helsinki University of Technology, Espoo, Finland; [2]Francis Bitter National Magnet Laboratory, MIT, Cambridge, Massachusetts, USA; [3]Neuromagnetism Laboratory, New York University, New York, USA; [4]Istituto di Elettronica della Stato Solido, CNR, Rome, Italy; [5]Department of Psychology, Simon Fraser University, Burnaby, British Columbia, Canada

KEYWORDS

Magnetic fields; evoked responses; human brain.

The panel 'Biomagnetism in the study of brain functions' at the 5th World Conference on Biomagnetism (Vancouver, Canada, August 1984) was chartered to discuss the issues and new developments of neuromagnetism. This integrated paper is based on preconference correspondence. The panel members were asked to express their opinions on some central questions of interpretation and applications of neuromagnetic recordings in brain research. Some authors concentrated on a few questions only. The paper has been edited by R. Hari (chairperson of the panel) and the contributions of different authors are given in alphaphetical order.

Question 1: What do you consider the most important results of your group? Have the neuromagnetic recordings provided such information about the function (psychophysiology, neurophysiology) of the human brain that could not have been obtained by other means?

COHEN: Our group has been involved with MEG for some time, both experimentally and theoretically. Experimentally (aside from the early demonstrations that the brain produces measurable magnetic field) perhaps a significant result has been the clear demonstration of the three MEG-EEG map differences, using the somatosensory N20 dipole source (Cohen and Cuffin, 1983). Theoretically, perhaps a useful result has been that the component of magnetic field which is perpendicular to the skull does not see the volume current, only the source current (Cohen and Hosaka, 1976).

Our neuromagnetic recordings have produced two types of functional information, difficult or perhaps impossible to obtain with the EEG alone; one type concerns the sources of some evoked responses while the other concerns sources of the alpha rhythm. Of the evoked sources we found that the sources of N20 and P30 are each primarily a tangential dipole in or near area 3b of the central sulcus (Wood and coworkers, 1985). There has been disagreement among neurophysiologists as to whether this is indeed a tangential dipole in area 3b, or is instead either two primarily radial dipoles located more superficially or a deeper source in the thalamic region. We also found, for evoked response, that the source of N100 due to auditory click (in at least one subject) cannot be a single source in primary auditory cortex, as had occasionally been supposed, but is at least two sources located some distance apart (Wood and Wolpaw, 1982). The first source is indeed in primary auditory cortex at the Sylvian fissure, but the second source is quite far away, in the temporal lobe. We found this by careful modeling of MEG and EEG maps, where the information was combined (in what we call the four-step method). This new information concerning these sources could probably not have been obtained with the EEG alone, mostly because of the large modeling errors in the inverse solution of EEG maps; these errors involve layer thickness and conductivity, to which the EEG inverse solution is most sensitive.

Of the alpha-rhythm source, the new information concerns the geometric orientation of these sources in the cortex. We have seen an unusual phenomenon which we call the polarization of the alpha rhythm (Cohen and Cuffin, 1979). This unique phenomenon (seen only with the MEG) is explained if the alpha rhythm sources are oriented perpendicurlarly to the cortical surface. This orientation has only been surmised from potential depth recordings (change in polarity in going through various layers) but never proven. The explanation of polarization therefore is a confirmation of this orientation. For example, if the alpha generators were randomly oriented with respect to the cortical surface, then it appears that magnetic polarization would not be possible.

HARI: The first neuromagnetic evoked responses at the Helsinki University of Technology were recorded in prof. T. Katila's group using auditory stimuli (Hari and colleagues 1980; 1982). It turned out that the magnetic counterpart of the auditory N100 deflection was most probably generated at the supratemporal plane, at or near the primary auditory cortex. Earlier the opinions about the generation of N100 had been controversial. The later comparison of the effects of stimulus repetition rate on simultaneously measured electric and magnetic evoked responses indicated, however, that the auditory vertex potential can not be totally explained by activation of the auditory cortices.

Since 1982, the group at the Low Temperature Laboratory at the Helsinki University of Technology has worked mainly with the somatosensory system and pain (Hari and colleagues 1983, 1984; Kaukoranta and colleagues, this volume). We were able to distinguish between activities at the primary and secondary somatosensory cortices. Recently the experiments have become more painful: electric stimulation of an intact tooth elicited neuromagnetic responses whose field pattern suggests activation of the frontal operculum. The results indicate that tooth pulp has cortical representation at this area in man.

One important achievement of the group is the design and construction of a four channel SQUID magnetometer (Ilmoniemi and colleagues, 1984), which has been in routine use since summer 1983.

KAUFMAN: Our earliest work in the Neuromagnetism Laboratory of NY University was on the field evoked by changing visual patterns, and we went on to study somatosensory, auditory and motor fields. The work of Romani, Williamson and

Kaufman (1982) on the organization of the human auditory cortex was a landmark study since it gave us information about the brain that we could not have gotten by other means. The study by Pelizzone, Williamson and Kaufman (this volume) shows that the auditory cortex may be subdivided still further, with different areas having what appears to be different functions. This too could not have been learned using any other technique. The work of Okada and colleagues (1983) on endogeneous components of the event related potential may well have gone a long way towards clearing up problems in the area of cognitive psychophysiology, and this too depended upon the neuromagnetic method. This work, as well as the work that went on at nearly the same time in other laboratories, established at least the possibility that it is easier to resolve the sources of evoked and event related fields than it is to resolve the sources of comparable electrical measurements.

ROMANI: One of the most promising features provided by the neuromagnetic method is the possibility of localizing sources of physiological and pathological cerebral activity. This property, which has been significantly pointed out during the last few years, permitted identifying sources related to activity elicited in primary (Williamson and Kaufman, 1983) as well as in secondary areas (Hari and colleagues, 1984) by stimulation of peripheral sensory systems. More recently it has been shown that also signals related to higher levels of brain functions can be successfully measured and the location of their equivalent generator at least approximately identified. As a consequence of the localization capability, the functional organization of the brain can be investigated. One example is the recent verification of the tonotopic organization of the human auditory cortex (Romani, Williamson and Kaufman, 1982).

The most important results achieved by my group concern the investigation of spontaneous brain activity in normal and pathological subjects. We started this study in 1980 and so far we have performed a systematic investigation on more than 20 normal subjects and about 100 epileptic patients. The first results (Ricci and colleagues, 1981; Modena and colleagues, 1982) obtained in the clinical side demonstrated that magnetic signals associated with focal epilepsies were quite often sharply localized over specific areas of the scalp, whereas the simultaneously recorded electric activity was either widely spread over the scalp or much weaker than the magnetic one. In some cases even no electric counterpart of pathological magnetic signals was observed. This property was also confirmed by applying the Fourier analysis method to both magnetic and electric channels (Barbanera and colleagues, 1983). However, only with the development of the Relative Covariance method it was possible to achieve a full three-dimensional localization of the equivalent generator corresponding to the active area of the cortex (Chapman and colleagues, 1983; Ricci and colleagues, this volume). An example of the state of art of the localization procedure presently performed in my laboratory was presented in this meeting by Ricci and colleagues. The same correlation technique was applied to the magnetic and electric signals related to the normal alpha rhythm and provided also in this case the three-dimensional identification of the location of the equivalent generators responsible for the major portion of the occipital alpha activity (Chapman and colleagues, 1984).

Question 2: What would be the most suitable way of presenting the data?

COHEN: If one is recording a scalar quantity, such as the component of magnetic field perpendicular to the skull, then I believe there is no choice but to present the data in isoline form. However, in recording the gradient of this field component, say due to bipolar recordings (two side-by-side coils in the MEG or electrodes in the EEG), then the quantity is no longer scalar. The quantity to be displayed is then a vector. In that case, the map must be some form of vector or arrow map. For example, we use bipolar maps (our 2-D system) along with the usual monopolar Br map, because they nicely complement each other; the 2-D arrow map can pin down the location of the extrema due to a dipole very well, in comparison to the monopolar map. Stated otherwise, maps of higher order than those of a simple scalar quantity must be presented in higher order form, that is, in vector form (or even more complex forms for higher orders yet).

HARI: Up to now we have used isocontour maps to represent field patterns. The equivalent current dipoles were then calculated using a spherical model of the head. The assumption of concentric inhomogeneities does not seem to be good enough at the frontal areas of the head, and more complex volume conductor models will be studied. Attempts have been made to use alternative presentations of the data, i.e., minimum norm current estimates which can be calculated without source models (Ilmoniemi and colleagues, this volume).

ROMANI: We prefer to identify the location of the equivalent source directly in three sections of the subject's head as derived from x-ray analyses or craniometry. The best sphere fitting the skull (on the basis of the three profiles) is inserted in the appopriate portion of the skull and the experimental points are projected onto the sphere (Ricci and colleagues, this volume).

Question 3: What are the most serious problems in interpreting the data? What is the physiological/ theoretical relevance of the dipole model?

COHEN: In my opinion, the most serious problem is performing accurate modeling on combined MEG and EEG maps, in order to obtain meaningful source information. This is not an easy thing to do, to combine the very best low-error techniques of the experimentalist in obtaining the map, and that of the theoretician in obtaining the solutions to inverse prolems. In both cases the MEG community has made only a beginning; there is a long way to go to solve the problem, but it can and should be done.

Concerning the modeling, what other choices do we have for the interpretation of MEG data except with dipoles? I believe that it is too complicated to use multipole expansion in the case of the brain, when the source is so large compared to the source detector spacing. Much good analysis can be done with multi-dipole models, and in many ways it is a natural thing to do.

HARI: The first task in interpreting the neuromagnetic recordings is to be sure that the responses are of cerebral origin. Therefore the field patterns produced by all possible artefacts, for example by eye blinks (Antervo and colleagues, this volume) should be carefully studied. The possibility of contamination of MEG by muscular activity has not yet been sufficiently discussed.

In interpreting the data we must keep in mind that with MEG we are restricted to study the mass activity of neuronal populations oriented and organized in a specific way. Location of these populations is important and interesting, but many aspects of relevant brain activity, for example electrotonic communication between adjacent cortical dendrites, remains beyond our reach.

In many cases the dipole model seems to explain surprisingly well the main features of the measured field patterns. At neuronal level the dipole approximation is applicable for measurement points at distances larger than a few lengths of the neuron (Nunez, 1981). Current source density analyses (cf., Mitzdorf, 1985) indicate that even mass activity of cortical cells can be represented by dipoles, at least at some moments of time and far away from the source. At the usual measurement distance of at least two centimeters also dipole layers can be reasonably well be described by equivalent single dipoles when the extension of the layer does not exceed 20 degrees (Okada, this volume). When more complicated field patterns are present, multidipole models are preferable to multipole expansions which are difficult to interpret physiologically. Further, the terms of the magnetic multipole expansion are electrically silent and cannot explain the electric potential distributions.

KAUFMAN: Thus far most of the data seem to be accounted for in terms of a simple dipole-inside-a-sphere type of model. At any rate, the data are often so noisy that one cannot perform an experiment with sufficient precision to reject this type of model. The view of our group is that the important issue of the adequacy of the prototypical current dipole as the putative source of observed fields is a matter that should not cause us to suspend the recording of data. Nor, for that matter, need this issue cause us to suspend attempts to predict the locations of sources based on the dipole model until a better model is in hand. In fact, most of us believe that the types of models we will use in the future will involve multipole sources enclosed within conducting volumes having a variety of shapes. These head shapes may well belong to different anthropometric classes. They vary across people, and also with age. In fact, predicting the shape of the head of the adult from that of the child is an enterprise that actually involves the development of a descriptive geometry and of growth functions uniquely suited to describing the shapes of heads. However, this type of work will not become useful in neuromagnetism until our measurements are sensitive enough to permit discrimination among predictions of diverse models. As yet we do not have such data in hand, although there are signs that we will be led to recognize the limited applicability of the equivalent current dipole model. Some of these signs are in the form of data, while others are in the form of advances in existing instrumentation.

The kinds of data I refer to are quite different from, say, those portions of the evoked field that correspond to the N20 - P30 complex of the somatosensory evoked potential. As you will hear from others at this conference, an equivalent current dipole underlying this complex is located with equal precision by both electrical and magnetic measures. However, this result cannot be generalized to other more complicated phenomena, e.g., P300. We are all aware that even in simple situations, e.g., N20 - P30, treating the simple equivalent current dipole as a hypothetical source is fraught with problems (see Desmedt, 1980, for alternative interpretations). Nevertheless, there is something satisfying about such simple solutions. Also, even when there is no obvious correspondence between electrical and magnetic data, the resulting solutions based on the assumption of an equivalent current dipole source seem both elegant and reasonable. However, other data are appearing with which we can deal only tentatively in such simplistic terms.

One example is that of the data obtained by Lounasmaa, Williamson, Kaufman and

Tanenbaum (this volume). We expected the change in speed of a moving grating to activate cells in both the primary visual cortex and in the so-called "extra-visual areas" of the brain as well. Work with monkeys supported this conjecture. However, we came up with a bumper crop of responses from many parts of the head, and they do indeed suggest that multiple sources in widely separated parts of the brain respond to the change in speed of a moving pattern. These results are tentatively interpreted in terms of overlapping fields which are associated with several different equivalent current dipole sources. However, we have no way of knowing if other models might not give us more precise predictions of the complicated field pattern we observed. As things stand now, the fit to elementary current dipole sources leaves something to be desired. We are beginning to feel the need for more sophisticated levels of analysis, but are not yet sure of the form it will take.

If the head can be modelled by the standard sphere (a big "if" in many contexts), then the component of the field normal to the scalp is unaffected by the presence of layers of different conductivity. That is, the spherical head is "transparent" to fields that arise from small segments of current in the brain. These small segments of current are usually modelled as current dipoles. It is worth noting, however, that even extended sources encompassing a large number of neurons will produce a field that is indistinguishable from that attributed to a current dipole, provided that the distance of the source from the surface is large relative to its spatial extent. Also radially oriented current dipole sources will not produce any detectable field normal to the "scalp" of the sphere model, since this field is due solely to the tangential component of the current dipole. This prediction from a sphere model is consistent with empirical findings which show that the stronger evoked fields can be attributed to sources in the walls of sulci of the brain. Since elongated neurons are largely normal to the surface of the cortex, such neurons in the walls of sulci will be tangential to the scalp. Thus far the data are consistent with the theoretical conclusion that the field normal to the scalp is due to intracellular currents rather than widespread volume ("backflow") currents. For if the field were attributable to volume currents associated with depolarization or hyperpolarization of membranes of neurons, then the sources could not be so sharply localized and the neuromagnetic data would probably not appear to differ from electrical data, which does exhibit a contribution from radially oriented sources. In addition, it may be worth noting that if the source were a magnetic dipole due to a circulating current, then sources would be as likely to be located in gyri as in sulci, and a strange variety of field patterns could be encountered. Thus far, none such have been reported. Finally, the strength of the field falls off more rapidly than predicted by the inverse square law as the depth of the source increases within a sphere. This rapid fall-off of field strength with distance to the source works in favor of sharper spatial resolution of different though simultaneously active sources. This may well be one of the major reasons for the large difference in morphology of the waveforms of the evoked potential and field, e.g., as typified by the somatic responses which, in the case of the field, resembles the potential recorded from the exposed surface of the brain while the same potential recorded from the scalp shows signs of significant contributions from very different sources.

WEINBERG: It is interesting to note that the new science of biomagnetism has landed itself smack into the middle of one of the most pervasive controversies in the history of the brain sciences; the issue of what is meant by localisation of function in the brain. Based on our knowledge of brain physiology to what extent can we reasonably expect complex function to be localised anatomically? The raison d'etre of biomagnetism as applied to the analysis of the brain, as some see it, is to localise sources. But to what extent does localisation exist, and if it exists what is meant by localisation?

The issue has had a long history and began in earnest with Gall in the beginning of the 18th century, with the assertion that Phrenology could localise complex behaviour patterns. By the end of that century Florens was writing "..the nervous system is not a homogenous system. The cerebral lobes do not act like the cerebellum, nor the cerebellum like the spinal cord, nor the spinal cord like the nerves. But it is a single system. All its parts conspire, consent. What distinguishes them is some particular a determinate mode of action; what unites them is a reciprocal action on their common energy."

After that the psychologists arrived: Karl Lashley concluded, after extensive studies of cortical lesions in rats in relation to visual discriminations, that "the maze habit is not intefered with by any purely sensory or motor defect, and that the formation of a sensory habit is not retarded by absence of the corresponding cortical sensory area, and that the deterioration following lesions in different cortical sensory fields is qualitatively the same for all fields" (Lashley, 1929).

After that the brain surgeons arrived. Penfield observed that "Stimulation of different points on the lateral aspect of the temporal lobe ... produced in the patient differed portions of her dream ... when the electrodes were held in place for a longer period, the hallucination progressed like a story or memory unfolding" (Penfield and Rasmussen, 1950).

In the same period of 1940-1960 there developed a strong movement to localise sites of hunger, sex, thirst and basic "motivations" in the brain stem and temporal lobe, (e.g., Hess won a Nobel Prize for this and Elliot Stellar was the foremost proponent of motivational centers of the brain), until it was discovered by Valenstein and others that different behaviour could be elicited from the same stimulation. The behaviour elicited depended on the immediate environmental stimulus context. No one talks about strict localisation of motivational centres these days. In 1973 Luria talked about a revision of the concept of 'localization' and said ... "Our examination of the structure of functional systems in general and of the higher psychological functions in particular has led us to take a complete fresh look at the classical ideas of localization of mental function in the human cortex. Whereas elementary functions of a tissue can, by definition, have a precise localization in particular cell groups. There can of course be no question of the localization of complex functional systems in limited areas of the brain or of its cortex."

Of course this issue has been plaguing those interested in the analysis of electrical fields since development of the EEG. There has been a great investment of resources in the field of evoked potentials today, both human and otherwise, in the attempt to localise components of evoked potentials. This issue is subtly reflected in the now more popular name for evoked potentials, event related potentials. Electrical potentials are said to be not <u>evoked</u> by stimuli in the sense that there is a fixed brain responses defined by only the stimulus characteristics. Rather the response observed is the result of the processing of the stimulus input in the context of memories (previous stimulus experience) and their interaction with a neurophysiology which is in the constant process of update. Mary Brazier said the brain can never be considered a static system; that each repetition of a physically identical stimulus impacts on a different brain simply by virtue of that stimulus having occurred previously. Therefore, insofar as the brain is concerned, each stimulus is a different stimulus.

Fifty years of work has centered on the description of how different components of evoked potentials are related to information processing, storage and retrieval (retrieval in its broadest sense, that of emergent behaviour). What everyone wants to know is: from where are they coming? Because to know from whence they arise is to know the sites of the brain active in processing, storage and retrieval.

I believe there are two important issues about which students of MEG must seriously think. The first is: is the question being asked important, i.e., is knowledge about the localisation of systems in the brain important to an understanding of the organisation of the brain? The second is: what is meant by localisation? Of course these are not independent issues since the answer to each implies an interpetations of the other.

A good example of those who think the question of where components of EPs are localized is unimportant is Emanuel Donchin. His argument is that evoked potentials index the means through which the brain analyses, stores and retrieves information; i.e., the types of programmes it uses and its critical subroutines. The analogy is to the computer. Who cares where precisely the computer stores information? What is more important are the programmes used for analysis, and for achieving the desired or observed output. In fact, it does not matter at all where the information is stored, as long as it can be accessed by the appropriate software. It is the software which is responsible for functional organisation of the hardware.

In my view the answer to whether localization is important lies in the answer to what is meant by localisation of function in the brain. The great promise of MEG is that it can be used to better localise the "sources" of components of electrical changes. Until the last few years it was tacitly assumed that the source could be most usefully described as a single dipole. Except for some simple cases, in which localisation is of the first or second stage of sensory projection systems, there are two reasons why the use of a single dipole to describe sources will not work. Firstly, no neurophysiologist believes that a single dipole is responsible for complex electrical events in the brain, especially those events which have a time course over many, sometimes hundreds, of milliseconds. Secondly, if some set of dipoles which are actually widely distributed are described as a single equivalent dipole, that description adds very little information about the nature of that dipole distribution.

Localisation of function in the brain does not mean localisation of a single site but rather the identification of a spatio-temporal system. For some functions of the brain this system may be spatially and temporally "tight", i.e., it may be within some small space and time. Other systems may be spatially distributed and be active over longer periods. In all likelihood, the more complex the function the more distributed will be the system. For this reason MEG research must turn its attention to the means through which research can help to identify the spatial and temporal characteristics of distributed systems. If MEG is to be used for the localisation of complex sources, it is necessary to apply multiple source models in combination with anatomical and physiological knowledge about probable sources. Although a major advantage of the MEG is that it is undistorted by the tissues of the head, there is no unique inverse solution for the intracerebral current flow responsible for an observed MEG distribution. Therefore it is necessary to use existing knowledge about neurophysiology in order to limit the solutions to those which are reasonable.

I believe that everyone recognises the need for new and imaginative approaches to the "inverse solution". Are there methods which can be combined with what we already know about neurophysiology? Certainly we know that if a cardiac source is localised outside the heart, something is wrong. The same principal can be applied to "source-systems" in the brain. If it is known from past studies of complex functions of the brain that several parts of the brain are contributing to a function, or that it is likely that many areas must be involved, then we should be looking for source systems which are distributed. Furthermore, it is necessary to be able to distinguish whether dipoles are changing their spatial locations over time (e.g., during an evoked potential) or whether their location remains the same whilst their strengths wax and wane.

Clearly what is needed is a method for interfacing known neurophysiology with mathematical estimates of the spatial distribution of constellations of dipoles which are active at some sequence of instants in time. Whether such spatial and temporal configurations can best be modeled by current or magnetic dipoles is a question which has been raised and which should be examined.

ROMANI: Major poblems in interpretation of data concern the choice of the sphere to fit the head, its location inside the skull and, above all the adequacy of the sphere model to fit the head in specific regions (Romani and Leoni, this volume).

It should be pointed out that the current dipole model is a simple tool, mathematically accessible and yet sufficiently realistic. However, the localization of an equivalent source in a specific area of the brain does not imply, by all means, a physiological correspondence of the actual source to the model. It is a task for neurophysiologists to speculate on this problem and, if necessary, to set the motivations for more complex models.

Question 5: How should the present experience of neuromagnetic recordings guide the future design of magnetometers?

COHEN: First, I believe it is necessary to use a multilocation MEG detector in order to eliminate placement and other errors which take place with the single-location detector. Second, I believe that gradient detectors (for example, two side-by-side coils as our 2-D system) will be used to give greater accuracy in finding the extrema in a map.

KAUFMAN: Recent advances in instrumentation also make it important to consider more elaborate models. The introduction by the Helsinki group of a four-SQUID array in a single dewar makes it possible to measure the field at three places at once, and also the field gradient along the main axis of the dewar. Our own group is now using a 9 SQUID system (Williamson and colleagues, this volume). Five dc-SQUIDs and second order gradiometers are used to measure the field at five different places (2 cm apart), and 4 rf-SQUIDs are used for electronic subtraction of low-frequency noise. Ultimately such systems will provide data that are less noisy, if only because they reduce positional errors (the detection coils are at fixed distances from each other), and because variations in the strength of the response due to habituation, adaptation or long-term physiological changes of the subject can be discriminated from differences in field strength due to position alone. In essence, multi-sensor systems make it possible to avoid some of the potential pitfalls associated with the assumption of stationarity, which underlies all mapping studies employing a single sensor. In addition, due to the remarkable reduction in time needed for mapping the field, the new systems will make it possible to obtain more detailed maps of the field pattern than has been possible thus far. We look forward to the advent of the 30 SQUID array that is now being seriously considered by the group at Los Alamos, and to still more intricate systems that are in the dream stage of development. The reason is simple. We need to have snapshots of the field all about the head at each instant of time, and this pattern will not only tell us that we need more sophisticated models, it may well give us hints about the properties that such models must have. This, however, is one or two years away. Meanwhile, we should use the simplest models available to derive a data base for our area of work. I think that it was von Neumann and Morgenstern who, in the preface to their work on the theory of games, said that there is no need for precise methods when our basic concepts are unclear. The opposite side to this coin is that there is little need for precise and complicated theories when our data are so crude that they can be handled equally well by simple theories. We simply do not yet know which conceptual tools will be needed to intepret the data of the future. Hence, the data base I am calling for should not be considered as immutable since it is limited by the nature of existing instrumentation, procedures, and by where we look.

Where should we look and what should we look for? A single sensor system allows us to look at one place at a time. By the time we move it elsewhere, the state of the subject could change, and such changes in his or her state may well confound the systematic changes with position that are to be expected on purely physical grounds. Therefore, repeated measures are needed, and careful adherence to the principle of counterbalanced experimental design must be part of the routine. Better still, with several simultaneously active sensors we can repeat one measurement while taking another measurement at another time and at a somewhat different position. It is my view that multisensor systems must replace all of the existing single sensor systems in the very near future, perhaps for this reason alone.

ROMANI: The answer to this question is contained in the reference Romani and Leoni (this volume). Certainly the experience so far achieved in neuromagnetic recordings will yield the development of a new generation of sensors, with a multi-channel configuration, specifically designed for the investigation of various sources.

Question 6: What about the future?

COHEN: We have begun to do animal MEG experiments, mainly to verify the orientation of electrical generators in the cortex. Specifically, we are trying to see if the MEG is supressed compared to the EEG, in the rabbit brain compared to the human brain. If the MEG indeed sees only tangential generators in the fissures, then there should be a supressed MEG in the rabbit, because there are almost no fissures. Preliminary results are that the MEG is indeed suppressed, although verification and modeling is needed to see the degree of suppression.

We will continue our work with Wood and Allison in clarifying the sources of some evoked responses in the brain, for example the sources due to auditory click (say in the latency range of 30-250 ms). Also, we are beginning a project to use the MEG to localize epileptic foci. In this case we would use both the ac and the dc capability of the MEG, where the dc is subdivided into two kinds: "on-off" dc which accompanies bursts of ac events such as a spike, and persistent dc which can last for many hours. Several years ago we began to look for the dc from the human brain, and got side-tracked by analyzing the dc which is due to touching the hair follicles of the scalp; now we want to go through the scalp and into the brain.

Concerning purely theoretical work Cuffin plans to continue using the moving-dipole method to find the inverse solution; he is investigating the properties of this type of solution, especially for complex sources. He also plans to continue his investigations on the effects of fissures on the MEG and EEG, and other inhomogeneities.

HARI: Most people working in biomagnetism are highly trained in the experimental method and trying to find relevant applications. Although this kind of approach has been fruitful even in brain research, I feel that our attempts should be more problem-oriented. In the future we should try to develop methodological and neurophysiological knowledge side-by-side and the methods of data analysis should be inspired by relevant problems of brain research.

We are planning to continue to study the auditory and the somato-sensory systems as well as pain mechanisms. Theoretical work will be done to find out alternative approaches to interpret the measurements for example by means of current maps. A large effort will be put to the multichannel magnetometer project started a few months ago.

KAUFMAN: I am in full agreement with those of our colleagues in the area of evoked potentials who decry the widespead practice of trying to deduce the functions of the brain from measurements made with one or two electrodes on the scalp. Our own work progressed from studying visual evoked fields detected at a very few places over the occipital region of the head to careful mapping of fields at many different places about the head. The reason is simple, if you don't look, then you may well miss detecting activity of parts of the brain that are quite distant from the primary projection areas. These parts of the brain are discussed in the physiological literature, but the species to which they apply are monkey and cat. It is interesting to observe that both the visual and auditory areas have multiple representations in these species. However, the multiple sensory areas of old world monkeys differ dramatically from those of new world monkeys. Therefore, we cannot generalize from work on any species of monkey to humans. The only way to gain access to these areas in humans is through the methods of neuromagnetism, because it has the requisite spatial and temporal resolution for regions that contain differentially active neurons. This puts us in a unique position, because we can begin to describe uniquely human functions.

Let me conclude on a note of caution about these "uniquely human functions". Thus far most of us have adopted a rather primitive research strategy. We stimulate some sense organ, for example, and find an area of the brain in which lies a "source" for a magnetic field normal to the scalp that changes in step with the stimulus. We then alter the properties of the stimulus and see if the source changes in position, strength, latency of response, or whatever. In a few cases the source is relatively sharply tuned to a particular type of stimulus. Now, there is a danger in this strategy since it entails the hidden assumption that functions are identified with places. Speech, which is a uniquely human function, has been closely identified with particular brain structures - Broca's area for example. Although we magneticians have yet to "see" Broca's area, it would be a mistake to assume that if we did we would have any very profound insight into how the brain actually works. We would know nothing of the intricate and truly widespread processes that must underlie an ongoing monologue, for example. In fact, if we were to stop our work only after identifying particular structures with particular functions, then we would be rediscovering an old science - somewhat like reinventing the wheel. I would be tempted to call this new version of an old science "magnetophrenology". However, the true test of the value of our work will be to prove that we can tease apart stages of cognitive functions which may well share structures with other functions. This network of stages and the deciphering of their modes of operation and interactions is the proper application of neuromagetism to psychology. Let this be one of the problems we shall address in the future.

WEINBERG: The thrust of our own laboratory has been threefold. Firstly, we are engaged in a systematic attempt to explore methods through which MEG can be used to assess human capabilities. I believe much more of this needs to be done. We are working on the assumption that the MEG is better able, than is EEG, to directly access those systems of current sources which are active during visual and auditory discriminations. Secondly, we are interested in following the movement of patterns of current sources during the course of an interictal seizure. We are also continuing to model multiple dipole sources, using multiple dipoles of known configuration and spatial location in a human skull, for the purpose of developing inverse solutions which can establish probabilities of different combinations of dipoles which could theoretically result in the empirical data.

ROMANI: The activity of my group will be concentrated on the investigation of normal and pathological subjects, with the aim at verifying the localization procedure with different diagnostic tools, including NMR, PET and stereo-electroencephalography. The use of a recently set up four-channel system will permit a significant reduction of the recording time and more sophisticated procedures of data analysis. Finally, we are going to use the evoked field technique to investigate pathological subjects and get further evidence of functional localization.

REFERENCES

Antervo, A., R. Hari, T. Katila, T. Ryhänen, and M. Seppänen (1985). Magnetic field pattern produced by eye blinking. This volume.

Barbanera, S., G.B. Ricci, and I. Modena (1983). Localization of epileptic foci by means of biomagnetic measurements. Il Nuovo Cimento, 2D, 547-557.

Chapman, R.M., R.J. Ilmoniemi, S. Barbanera, and G.L. Romani (1984). Selective localization of alpha activity with neuromagnetic measurements. Electroenceph. clin. Neurophysiol., in press.

Chapman, R.M., G.L. Romani, S. Barbanera, R. Leoni, I. Modena, G.B. Ricci, and F. Campitelli (1983). SQUID instrumentation and the relative covariance method for magnetic 3D localization of pathological cerebral sources. Lett. Nuovo Cimento, 38, 549-554.

Cohen, D., and B.N. Cuffin (1979). Magnetic measurement and display of current generators in the brain. Parts I and II. Proceedings of the 12th International Conference on Med. and Bio. Engineer., Israel.

Cohen, D., and B.N. Cuffin (1983). Demonstration of useful differences between

MEG and EEG. Electroenceph. clin. Neurophysiol., 56, 38-51.
Cohen, D., and H. Hosaka (1976). Magnetic field produced by a current dipole J. Electrocardiol., 9, 409-417.
Desmedt, J.E., and E. Brunko (1980). Functional organization of far-field and cortical components of somatosensory evoked potentials in normal adults. In J.E. Desmedt (Ed.), Clinical Uses of Cerebral, Brainstem and Spinal Somatosensory Evoked Potentials. Karger, Basel, pp. 27-50.
Florens, P. (1842). Recherches Experimentales sur les Proprietes et les Fonctions du System Nerveaux dans les Animaux Vertebres. J.B. Bailliere, Paris.
Hari, R., K. Aittoniemi, M.-L. Järvinen, T. Katila, and T. Varpula (1980). Auditory evoked transient and sustained magnetic fields of the human brain. Localization of neural generators. Exp. Brain Res., 40, 237-240.
Hari, R., K. Kaila, T. Katila, T. Tuomisto, and T. Varpula (1982). Interstimulus-interval dependence of the auditory vertex response and its magnetic counterpart: Implications for their neural generation. Electroenceph. clin. Neurophysiol., 54, 561-569.
Hari, R., E. Kaukoranta, K. Reinikainen, T. Huopaniemi, and J. Mauno (1983). Neuromagnetic localization of cortical activity evoked by painful dental stimulation in man. Neurosci. Lett., 42, 77-82.
Hari, R., K. Reinikainen, E. Kaukoranta, M. Hämäläinen, R. Ilmoniemi, A. Penttinen, J. Salminen, and D. Teszner (1984). Somatosensory evoked cerebral magnetic fields from SI and SII in man. Electroenceph. clin. Neurophysiol., 57, 254-263.
Ilmoniemi, R.J., M.S. Hämäläinen, and J. Knuutila (1985). The forward and inverse problems in the spherical model. This volume.
Ilmoniemi, R., R. Hari, and K. Reinikainen (1984). A four-channel SQUID magnetometer for brain research. Electroenceph. clin. Neurophysiol., 58, 467-473.
Kaukoranta, E., R. Hari, and K. Reinikainen (1985). Magnetic fields evoked by peroneal nerve stimulation. This volume.
Lashley, K. (1929). Brain Mechanisms and Intelligence. University of Chicago Press, Chicago.
Luria, A.R. (1973). The Working Brain. Penguin Press, London.
Lounasmaa, O.V., S.J. Williamson, L. Kaufman, and R. Tanenbaum (1985). Signs of neuromagnetic sources in extrastriate areas of human cortex. This volume.
Mitzdorf, U. (1985). The current source density method and its applications in the cat's cerebral cortex: investigation of evoked potentials and EEG-phenomena. Physiol. Rev., in press.
Modena, I., G.B. Ricci, S. Barbanera, R. Leoni, G.L. Romani and P. Carelli (1982). Biomagnetic measurements of spontaneous brain activity in epileptic patients. Electroenceph. clin. Neurophysiol., 54, 622-628.
Nunez, P.L. (1981). Electric Fields of the Brain. Neurophysics of EEG. Oxford University Press, New York-Oxford.
Okada, Y. (1985). Discrimination of localized and extended current dipole sources and localized single and multiple sources. This volume.
Okada, Y.C., L. Kaufman, and S.J. Williamson (1983). The hippocampal formation as a source of the slow endogeneous potentials. Electroenceph. clin. Neurophysiol., 33, 417-426.
Penfield, W., and T. Rasmussen (1950). The Cerebral Cortex of Man. The MacMillan Company, New York.
Pelizzone, M., S.J. Williamson, and L. Kaufman (1985). Evidence for multiple areas in the human auditory cortex. This volume.
Ricci, G.B., G.L. Romani, I. Modena, S. Barbanera, R. Leoni, and P. Carelli (1981). Magnetic recording (MEG) in epilepsy. Electroenceph. clin. Neurophysiol., 52, 95.
Ricci, G.B., R. Leoni, G.L. Romani, F. Campitelli, S. Buonomo, and I. Modena

(1985). 3-D neuromagnetic localization of sources of interictal activity in cases of focl epilepsy. This volume.

Romani, G.L., S.J. Williamson, and L. Kaufman (1982). Tonotopic organization of the human auditory cortex. Science, 216, 1339-1340.

Romani, G.L., and R. Leoni (1985). Localization of cerebral sources by neuromagnetic measurements. This volume.

Williamson, S.J., and L. Kaufman (1983). Application of SQUID sensors to the investigation of neural activity in the human brain. IEEE Trans. Magn., MAG-19, 835-844.

Williamson, S.J., L. Kaufman, Y. Okada, and M. Pelizzone (1985). Nine channel SQUID installatíon for unshielded neuromagnetic measurements. This volume.

Wood, C., D. Cohen, N.B. Cuffin, M. Yarita, and T. Allison (1985). Electrical sources in human somatosensory cortex: Identification by combined magnetic and potential field recordings. Science, in press.

Wood, C.C., and Wolpaw, J.R. (1982). Scalp distribution of human auditory evoked potentials. II. Evidence for overlapping sopurces and involvement of auditory cortex. Electroenceph. clin. Neurophysiol., 54, 25-38.

NEUROMAGNETIC LOCALIZATION OF INTRACRANIAL CURRENT SOURCES IN THE EPILEPSIES: INTRODUCTION

Jackson Beatty*, Daniel S. Barth* **, William Sutherling**

Departments of Psychology* and Neurology**
University of California, Los Angeles 90024

Susumu Sato*, Douglas Rose**, and Roger Porter***

Epilepsy Branch*, Clinical Neuroscience Branch**,
and Medical Neurology Branch***
National Institute of Neurological and Communicative Disorders and Stroke
Bethesda, Maryland 20205

Giovanni B. Ricci

Instituto di Elettronica dello Stato Solido, Rome

David Cohen

Massachusetts Institute of Technology
Cambridge, Massachusetts 02139

ABSTRACT

The purpose of this panel was to summarize the present state of neuromagnetic research in the epilepsies. Contributions were made by representatives of the various laboratories that are working on this problem.

KEY WORDS

Epilepsy; Neuromagnetometry; SQUID, Magnetic Field Mapping.

INTRODUCTION TO THE PANEL

The epilepsies are a related family of disorders that are characterized by an abnormal and excessive neuronal discharge within the central nervous system. Epilepsy is not uncommon. With an incidence of approximately 0.5 percent, over 1 million persons in the United States suffer from the disorder. But epilepsy is not only a clinical problem; it also a rich subject for scientific inquiry. The epilepsies are among the most exquisite experiments of nature and the study of epilepsy has provided fundamental insights into the workings of the human brain. That frontier of brain research is now being extended, as neuromagnetometric techniques are being applied to the study of epilepsy. This is a young endeavor, but already the results are exciting.

The present panel summarizes recent advances in the neuromagnetic study of epilepsy. The panel consisted of two lengthy presentations, the first by Barth and the second by Sutherling, and three brief reports by Sato, Ricci, and Cohen.

Barth presented both the theoretical justification and the first experimental results from a subdiscipline that might be called animal neuromagnetometry. Theoretical calculations concerning the effects of coil size and baseline upon magnetic fields produced by the brain of a small laboratory animal demonstrated that the instrument presently used from human recording is also suitable for animal work. Using a penicillin model of epilepsy in the rat, Barth obtained the first magnetic recording of an electrocortically verified seizure and also provided evidence of slow magnetic field shifts that accompany ictal discharge.

Sutherling next summarized human epilepsy studies to date. He explained one approach to field interpretation and raised several problems of field distortion. Initial validation studies were also discussed.

The remaining portion of the panel was devoted to three brief presentations. Sato described preliminary work at the National Institutes of Health and displayed a field map for hand-selected epileptic spikes. He also discussed a range of other possible methods of MEG and EEG analysis.

Ricci reviewed the work of the Italian group. He summarized ongoing validation studies and future plans for increasing the accuracy for clinical MEG localizations.

Finally, Cohen discussed slow magnetic phenomena, a topic of long standing interest. He stressed the range of artifacts that can contaminate the measurement of slow and standing magnetic fields.

It was the feeling of all panelists that clinical neuromagnetometry has had a vigorous beginning. The neuromagnetic study of epilepsy has proven to be of both basic and clinical interest. Human epilepsy, nature's exquisite experiment, is again revealing basic insights into human brain function in both health and disease.

ANIMAL NEUROMAGNETOMETRY AND ITS SPECIFIC APPLICATION TO THE STUDY OF FOCAL EPILEPTIC PHENOMENA

Daniel S. Barth* **, William Sutherling**, and Jackson Beatty*

Departments of Psychology* and Neurology**
University of California
Los Angeles, California, 90024, U.S.A.

ABSTRACT

A new area of biomagnetism, animal neuromagnetometry, is introduced. Both theoretical and empirical results are presented indicating that the extracranial magnetic fields produced by synchronized neuronal discharge in animal cortex may be measured and mapped using large coil gradiometers designed for human neuromagnetic studies.

KEY WORDS

Animal neuromagnetometry; focal epilepsy; computer simulation; SQUID

INTRODUCTION

Recent investigations of the human brain have demonstrated that the magnetoencephalogram (MEG) permits the non-invasive exploration of normal and pathological neuronal currents within the three-dimensional cranial space (Williamson and Kaufman, 1981a; Beatty et al., 1983; Barth et al., 1984a). This results from several properties of neuromagnetic fields that are not shared with measures of the brain's electrical potential in the scalp electroencephalogram (EEG). First, the interpretation of magnetic fields measured about a focal region of discharging cortex is simplified because the magnetic field is not distorted by uneven conductive properties of the skull. Second, the pattern of the normal component of an extracranial magnetic field (that component measured with present MEG sensors) is not changed by the orientation of a source with respect to the skull, only the strength of the field is affected. Third, unlike electrical currents, neuromagnetic fields are not attenuated by the highly resistive properties of the skull. Thus, the MEG may record brain activity not detected in the scalp EEG. Fourth, electromagnetic theory indicates that neuromagnetic fields mapped in the MEG are predominantly the product of intracellular currents and not complementary volume currents distributed throughout the extracellular space. Finally, neuromagnetic measures are able to record slowly changing or steady cellular currents because there is no electrode-tissue interface to serve as a capacitive barrier at low frequencies (Cohen, 1972).

The sensitivity and spatial resolution of neuromagnetic measures have led to the experimental application of MEG to localizations of cortical current sources underlying evoked sensory response (for reviews see Williamson and Kaufman, 1981a; Beatty et al., 1983). These same properties have also permitted the practical localization of epileptic foci in patients with partial complex (focal) seizure disorders (Barth et al., 1982; Barth et al., 1984a; Barth, 1984), and the ability to discriminate spatiotemporal discharge patterns between multiple adjacent epileptic foci in the same patient (Barth et al., 1984a; Barth et al., 1984b; Barth, 1984).

Yet, the neuromagnetic exploration of brain activity promises much more than this. Given the unique potential of neuromagnetic measures to chart massed intracellular brain currents, and the ability to record both rapid and extremely slow changes in these cellular currents, MEG measurement may be reasonably applied to the study of cellular mechanisms underlying localized electrical brain activity. However, the application of magnetoencephalographic methods to the study of cellular mechanisms underlying localized brain activity requires a detailed knowledge of both brain magnetic fields and intracellular currents. Without an empirical foundation, the interpretation of neuromagnetic fields and assumptions about their neurogenesis must be based on electromagnetic theory and indirect evidence in humans. In human experimentation, a precise knowledge of intracellular currents is not possible. In animal experimentation, however, matters are very different. By using precisely controlled and well-studied animal models, knowledge of cellular activities becomes possible. For these reasons, we introduce here a new area of biomagnetic research, animal neuromagnetometry.

ANIMAL NEUROMAGNETOMETRY

The neuromagnetic study of an animal model of central nervous events introduces a number of scientific advantages over similar studies in humans. First, activation of focal brain regions may be precisely controlled and verified. Activation may be introduced through sensory stimulation, but may also be produced by direct cortical stimulation or the focal application of epileptogenic agents. The location and size of the activated cortical region may be easily verified using deoxyglucose tracing (Collins et al., 1976), and its electrical properties may be accurately determined with invasive surface and laminar electrode arrays when appropriate (Goldensohn et al., 1977; Vollmer et al., 1978; Pockberger et al., 1983). Even more detailed information about the type and orientation of cells responsible for a given neuromagnetic field may be obtained through standard histological analysis. A second major advantage of animal magnetometry is that highly stereotyped brain activity may be produced, without the variability seen in most human studies. This is of critical importance to work performed with present single channel MEG sensors in which magnetic field maps must be reconstructed from a number of sequential measurements, often requiring hours of recording time. Animal studies also provide a control of artifacts not possible in human work. Perhaps the most important example of this is in studies of epilepsy in which movements generated by pathological brain discharge may be eliminated by the use of paralytic agents (Barth, 1984; Barth et al., 1984c). Finally, most animal models of both normal and pathological brain activity bring with them a wealth of knowledge from other electrophysiological studies with which neuromagnetic data may be compared.

Neuromagnetic studies of animals have not been previously attempted, primarily because of doubts about its feasability. Foremost amoung these concerns were that magnetic signals produced by animal cortex would be too small to detect using present MEG sensors, and that if detected, could not be usefully mapped with large coil instruments designed for human use. However, such concerns are without foundation. Computer simulations of instrumental effects produced by mapping magnetic fields about a small animal cranium using a large coil gradiometer demonstrate the theoretical feasability of field mapping in animal magnetometery. Further, at least one form of brain electrical discharge in the animal, that produced by interictal spiking using a penicillin model of focal epilepsy in the rat, generates substantial magnetic fields that may be measured in the MEG. These calculations and results are as follows.

SIMULATION OF INSTRUMENTAL EFFECTS IN MAGNETIC FIELD MAPPING

To determine the feasibilty of neuromagnetic measurement on the small cranial size of the rodent using a neuromagnetometer designed for the human brain, we carried out a computer simulation to detect possible measurement distortions introduced by using a large coil MEG sensor to map magnetic fields about a small sphere (Barth, 1984).

An ideal one channel sensor for measuring extracranial magnetic fields in both animal and man would be a single very small pick-up coil capable of sampling the radial component of flux at precise extracranial points. In practical measurements, this has not been realized due to both the low sensitivity of small coils and the presence of large noise fields in the environment. The present generation of instruments designed for human magnetometry typically employ a large diameter coil (2.4 cm) to average flux over a wide area, thereby improving sensitivity. The pick-up coil is also wound coaxially with coils further removed on the Z axis, forming a second spatial derivative gradiometer with the capacity to substantially cancel the uniform magnetic fields of distant noise sources. Both of these deviations from the ideal may introduce systematic distortions in the field pattern sensed by the magnetometer (Williamson and Kaufman, 1981a; Romani et al., 1982). Here, a computer model of a magnetometer and gradiometer is employed to evaluate instrumental effects when mapping fields from two relevant cranial sizes approximating those of the human and the rodent.

The radial component of the magnetic field (Br) produced by a dipole source (Q) in a conducting sphere of a given radius (R) was computed from an adaptation of the law of Biot-Savart (Williamson and Kaufman, 1981a):

$$B_r = \frac{\mu_o Q}{4\pi d^2} \frac{b/d \cdot \cos\varphi \sin\theta}{[1 + 2bR/d^2(1-\cos\theta)]^{3/2}} \qquad (1)$$

In equation 1, d represents the radial distance of the source from the measurement surface and b is the radial distance of the source from the center of the sphere. Theta denotes the angle of declination toward the measurement position along the X axis (line connecting the field extrema) with respect to the surface location of the source. Phi represents the azimuthal angle with respect to the X axis (Fig. 1). This equation yields values of the field strength measured by an ideal sensor. To compute the spatially averaged response of coils with finite radii, the radial field component was integrated over the coil area. This was approximated by dividing the coil into a series of closely spaced segments, calculating the radial field component passing through the center of each segment, multiplying the area of the segment times this field, and summing the products for all segments. To simulate the response of a gradiometer, this calculation was then repeated for each of the back-coils, taking into account coil polarity (direction of winding) and added distance from the measurement surface.

Assuming a single dipole as the underlying current source, the magnetic field produced by such a source may be completely determined through measurements along the X axis of the coordinate system, a line orthogonal to the dipole connecting the two magnetic field extrema. Successive recordings along the X axis will produce a plot of the two symmetrical field extrema, reversing in polarity directly above the location of the dipole. Because the magnetic field is symmetrical about the axis of the dipole, only one field extrema need be calculated for computer modeling purposes.

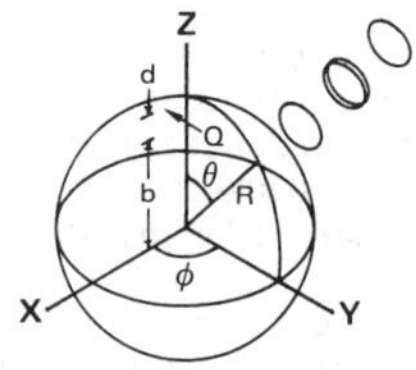

Fig. 1. Schematic diagram of the coordinate system used for modeling magnetic fields measured about a conducting sphere. Q represents the current dipole oriented along the Y axis and positioned on the Z axis at a depth d below the surface of the mapping sphere. The axis of the depicted gradiometer coils is along a given radius at an angle of declination (theta) from the Z axis and angle of rotation (phi) from the X axis.

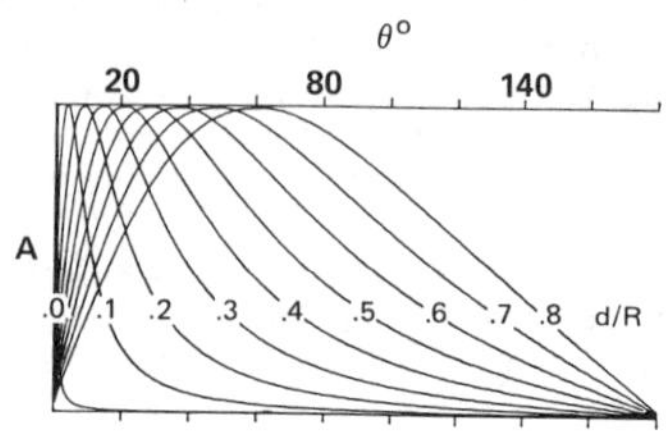

Fig. 2. The normalized amplitude (A) of the magnetic field sensed by a point magnetometer for measurement points passing along the X axis through one of the field extrema. Seperate plots are generated for dipoles of different depths.

Figure 2 depicts the computer generated plots for magnetic field extrema of dipoles at various depths as measured using an ideal sensor with no distortion. For a dipole of a given depth, the normalized amplitude of the magnetic field is plotted as a function of measurement position along the X axis. Because the measurements are computed for a spherical mapping surface, the position along the X axis is recorded in terms of the angle of declination (theta) from the surface location of the dipole to the measurement point. These calculations indicate that as the depth of the dipole increases, or as the ratio of the depth to the radius of a given sphere increases, the location of the extrema of the radial magnetic field component progressively shifts along the X axis away from the location of the source. Because of the symmetry of the magnetic field extrema about the dipole axis, the total angle between the field extrema for a dipole of a given depth may be determined from these plots by multiplying theta by two. Thus, as previously reported (Williamson and Kaufman, 1981b), the angle between the magnetic field extrema increases systematically for dipoles of increasing depth, approaching 180 degrees as the source approaches the center of the sphere (Fig. 3).

These plots reflect the response of an ideal recording system or point magnetometer. Because superconducting sensing coils must be operated inside of a large dewar, the first constraint to be imposed on the ideal system is that of a fixed distance between the actual measurement coils and the outside of the dewar. In presently used equipment this is 10 mm. The dotted curves in Fig. 4 reflect the response of a point magnetometer adjusted for this constant distance. The effect of the constant will differ depending on the radius of the mapping surface. Plots were therefore computed for two separate spheres approximating the radii of a human and a rodent cranium. Both of these plots begin with an initial angle between the field extrema reflecting the distance of the sensing coils from the measurement surface.

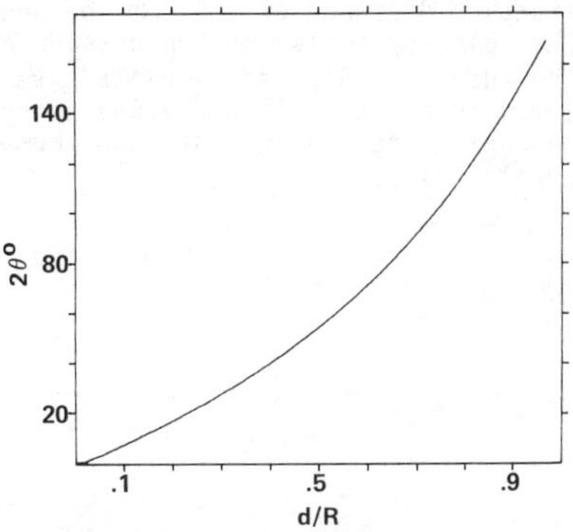

Fig. 3. The total angle between the field extrema (2*theta) measured by a point magnetometer plotted as a function of source depth.

A second constraint added to the ideal system is the requirement of back-coils to cancel the uniform field gradients of noise in the environment. The coil configuration of the presently employed second derivative gradiometer is shown schematically in Fig. 1. For computer simulations, a coil separation of 32 mm was used which equals that of most commercially available systems. To isolate the effect of the back-coils on recorded magnetic fields, each coil was first modeled as a single recording point, thus forming a point gradiometer. The solid curves in Fig. 4 indicate the computed response of the point gradiometer when measuring magnetic fields from the large and small sphere. In both cases, the angle between measured field extrema increases less rapidly with source depth than would occur for the point magnetometer. This result is expected since the second spatial derivative of the magnetic field falls off more rapidly with distance. Furthermore, the effect of the gradiometer coils is greater for measurements on a large sphere. As the depth of the source is increased on the large sphere, the absolute distance of the dipole from the sensing coils is greater than for measurements on the small sphere. At these greater distances the gradient of the magnetic field is not as steep and thus the relative effect of the back-coils may be expected to be more pronounced.

A third constraint on the ideal recording system is the requirement of large sensing coils to average the magnetic flux over a finite area and improve sensitivity. To isolate this effect, calculations were performed for an area magnetometer. Thus, the effect of the back-coils was removed and the effect of spatial averaging introduced. In contrast to the gradiometer, computed fields for the area magnetometer demonstrate a slightly more rapid increase in the angle between the field extrema with source depth (solid curves, Fig. 5) than calculated for the point magnetometer (dotted curves, Fig. 5). The effect of spatial averaging is very small for measurements performed on a large sphere but becomes more significant for the smaller sphere were the area of the sensing coils (24 mm) is a large fraction of the total mapping area.

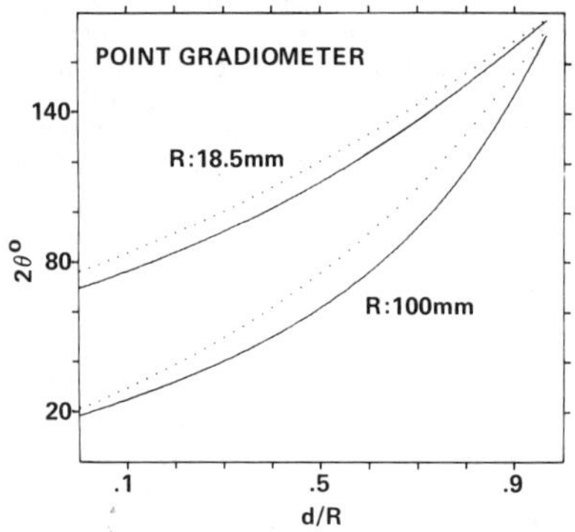

Fig. 4. The total angle between the field extrema (2*theta) measured by a point magnetometer (dotted curves) and point gradiometer (solid curves) plotted as a function of source depth. These plots assume a fixed starting distance of 10 mm between the pick-up coils and the outside of the dewar. Seperate plots were computed for a small and large sphere approximating the rodent and human cranial size.

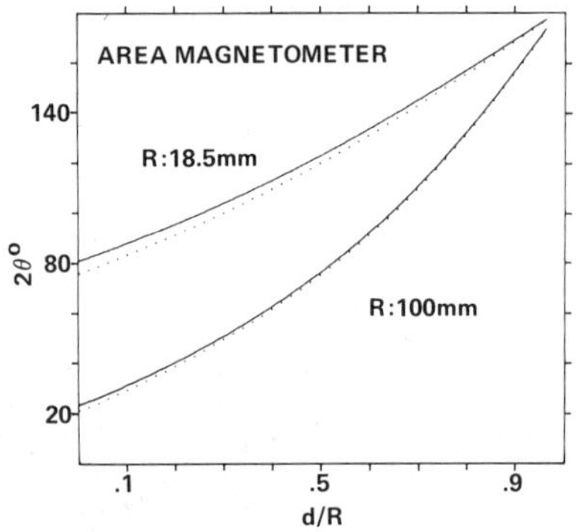

Fig. 5. Plots are similar to Fig. 4 except the response of an area magnetometer is depicted.

Figure 6 displays the combined response of spatial averaging and back-coils in the area gradiometer. This is the most accurate model of the SHE Biomagnetic Probe that we employ for neuromagnetic measurements. Although the effects of these two constraints are somewhat self cancelling, the gradiometer effect is larger and therefore remains as a significant source of localization error. This is especially true for measurements in large spheres (solid curve) when compared to the response of the point magnetometer (dotted curve). This same effect may be seen in Fig. 7 where the morphology of the field extrema computed for an area gradiometer are plotted for sources of various depths in the large and small sphere. Although the angle between the field extrema is affected by instrumental distortion, thus influencing derived source depth, the shape and polarity reversal point of the measured magnetic field remains undistorted. These calculations suggest that irrespective of the instrument effects and the size of the measurement sphere, the surface location of the source determined from the point were the magnetic field reverses sense should remain quite accurate.

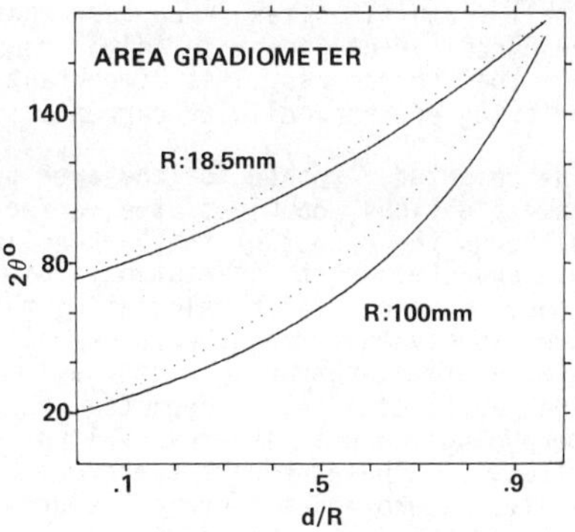

Fig. 6. Plots are similar to Fig. 4 except the response of an area gradiometer is depicted.

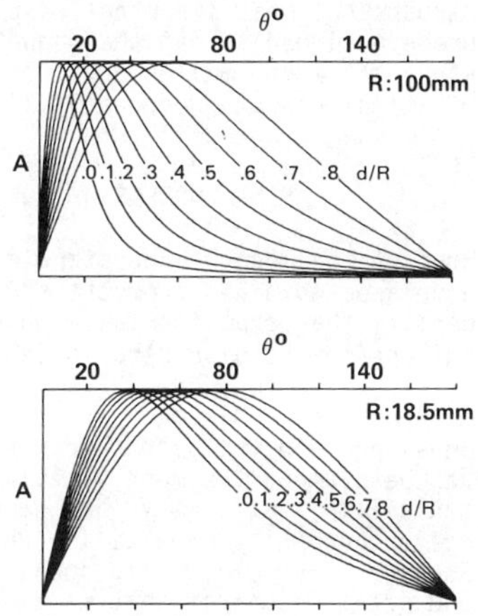

Fig. 7. Plots are similar to Fig. 2 except the response of an area gradiometer is depicted for measurements on a large and small sphere.

These calculations indicate that instrumental distortions may produce significant errors in the localization of sources underlying neuromagnetic fields. The effects modeled here for the point gradiometer and the area magnetometer both involve errors in the measured angle between the magnetic field extrema. The largest effect is produced by the back-coils of the gradiometer, causing the angle between the extrema to increase less rapidly with increasing source depth, resulting in an underestimate of source depth. This effect is most pronounced in recordings computed for a large sphere, were the field gradients of distant sources are less steep and therefore impose a greater influence on the back-coils relative to the pick-up coil.

The effect of spatial averaging computed for the area magnetometer is less pronounced and becomes a significant source of error only for computed measurements on a small sphere where the area of the coil is a large fraction of the total mapping area. Here, the effect is opposite to that of the gradiometer, causing a more rapid increase in the distance between the field extrema with increasing source depth. Therefore, the effect of spatial averaging could cause small overestimates of source depths when measuring from small cranial sizes such as that of the rodent. However, this error may be rendered insignificant in animal magnetometery where determination of the depth of the source is less important than the surface location, orientation, and polarity of intracellular currents.

The computed response for the area gradiometer presently used for neuromagnetic investigations combines the effects of the back-coils and spatial averaging, although the effect of the back-coils dominates the results. Therefore, it may be expected that in routine neuromagnetic measurements, the greatest possible source of error is in calculating the depth of underlying current sources due to the involvement of the back-coils and that if these results are not used for compensation, the error would be in underestimating the depth of the source. The computed fields reported here do not indicate any distortions in the morphology or polarity reversal points between the extrema due to instrumental effects. These results therefore strongly suggest that the surface location of a given neuromagnetic current source recorded from a small sphere such as the animal cranium, should be quite accurate if it is derived from the polarity reversal of the emerging and reentering magnetic flux.

Perhaps the most important result of these calculations is that a large coil probe designed to map the magnetic fields of the human brain may also be used to study the much smaller brain of the rodent. This finding opens the possibility of animal neuromagnetometry using presently available sensing devices.

ANIMAL MODELS OF FOCAL EPILEPSY

Most of the modern understanding of the intracellular events underlying epilepsy have been obtained from the study of animal models of epilepsy. Such procedures permit the experimental production, verification, and replication of epileptic-like disorders in laboratory animals.

Although there has been a large variety of animal models of focal epilepsy developed in the past thirty years (Purpura et al., 1972), the penicillin model has been among the most useful. There are several reasons for this. First, the penicillin focus may be carefully controlled both in size and degree of epileptogenicity by altering the quantity and concentration of injected drug. Second, the penicillin focus is characterized by highly stereotyped interictal and ictal discharge that remains stable for the several hours required for magnetic measurement. Third, the penicillin focus is one of the most intensively investigated models of focal epilepsy, providing a wealth of knowledge with which to compare our findings. Finally, there are a number of detailed electrophysiological studies of the spatial distribution of sources within the penicillin focus producing both interictal (Tharp, 1971; Goldensohn et al., 1977; Elger and Speckmann, 1983; Pockberger et al., 1982) and ictal (Vollmer et al., 1978; Gumnit, 1979; Petsche et al., 1978; Pockberger et al., 1983) discharge, permitting investigation of the neurogenesis of measured magnetic fields.

The spatial organization of epileptiform activity in the penicillin focus, the reliability with which it may be produced, and its similarity in many ways to human focal epilepsy, make the penicillin model ideally suited to the investigation of magnetic phenomena associated with focal epilepsy. We have used this model to obtain the first measurements of magnetic epileptic phenomena in an animal recorded both between and during seizures.

Penicillin induced focal epilepsy was measured in four paralyzed laboratory rats

(Barth, 1984; Barth et al., 1984c). Following craniectomy, penicillin was injected unilaterally into the left cingulate area. Interictal electrical spikes were monitored in the electrocorticogram (ECoG) using a solder-plated copper wire placed directly above the injection site. As in our human studies, electrical spikes were later used as time markers to average epileptiform events in the MEG recorded from two extracranial sites on opposite sides of the focus.

Figure 8 shows a typical averaged interictal spike recorded in both the ECoG (A) and two MEG recording sites (B, solid and dotted traces). The averaged magnetic spike reverses polarity between the two MEG recording sites indicating substantial (.5 pT) magnetic fields emerging from and reentering the cranium. Interictal magnetic fields of similar amplitude and polarity were recorded in all of the animals.

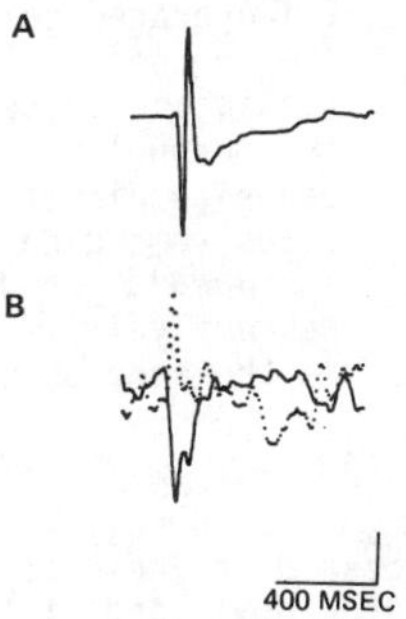

Fig. 8. (A) Averaged biphasic interictal spike in the ECoG (negative down). (B) Averaged interictal spikes in the MEG recorded from bilateral sites (solid and dotted traces) differ from the ECoG spike in morphology and are of opposite direction between sites, reflecting magnetic flux simultaneously emerging from (upward) and reenetering (downward) the cranium. Vertical calibration, MEG = 240 femto Tesla (fT), ECoG = 500 V.

Figure 9 shows electrical and magnetic recordings of two typical focal seizures in one of the animals (B and C). The onset of both seizures was characterized by the appearance of afterdischarge in the ECoG (middle trace) followed by rapid seizure spiking (arrows) and terminated abruptly with postictal depression. The most apparent response in the MEG (top traces) recorded from both points was a steady magnetic field shift. In all seizures recorded from these animals, a steady field shift began at the appearance of afterdischarge in the ECoG, peaked at the onset of rapid seizure spiking, and returned abruptly to baseline at the termination of the seizure. Like the magnetic fields of interictal penicillin spikes, the steady magnetic field shift was also of opposite polarity on oppsite sides of the focus.

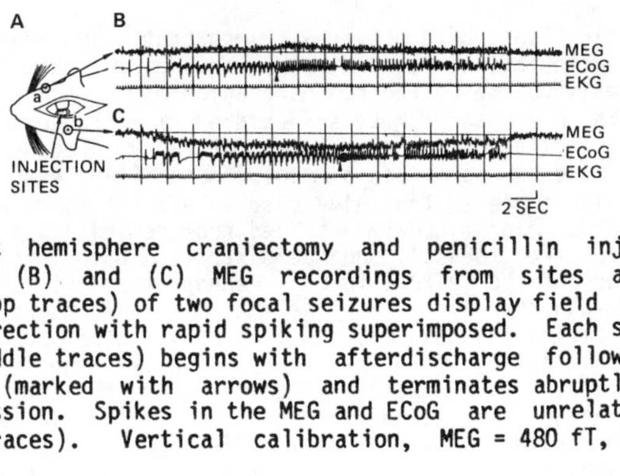

Fig. 9. (A) Left hemisphere craniectomy and penicillin injection sites (dots). (B) and (C) MEG recordings from sites a and b respectively (top traces) of two focal seizures display field shifts of opposite direction with rapid spiking superimposed. Each seizure in the ECoG (middle traces) begins with afterdischarge followed by rapid spiking (marked with arrows) and terminates abruptly with postictal depression. Spikes in the MEG and ECoG are unrelated to EKG (bottom traces). Vertical calibration, MEG = 480 fT, ECoG = 800 V.

Superimposed on the large ictal magnetic field shifts were rapid seizure spikes. These are more easily visualized after averaging, using the first 20-30 rapid spikes in the ECoG as time-locking points. The time locking points for rapid seizure spikes were located in the midpoint of the 1-second averaging window, resulting in an averaged spike in the center of the trace in both the ECoG (Fig. 10A) and MEG recorded from the bilateral sites (Fig. 10B). These spikes were surrounded by smaller deflections at the approximate period of the repetitive discharge. This fast activity also reversed direction between the two MEG recording sites.

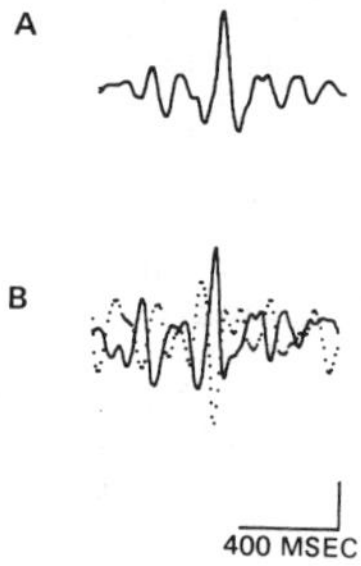

Fig. 10. (A) Averaged rapid seizure spike in the ECoG (negative down). (B) Averaged seizure spikes in the MEG recorded from bilateral sites (solid and dotted traces) are similar in morphology to the ECoG and are of opposite direction between recording sites (emerging and reentering fields = upward and downward traces respectively). Vertical calibration is the same as Fig. 8.

These data indicate that large extracranial magnetic fields accompany both interictal and ictal discharge in the rat penicillin focus. Although mapping was not performed, the measured fields in all animals were of similar polarity, suggesting an organized field pattern. Perhaps most important, the largest magnetic signal recorded from the penicillin focus was a steady field shift accompanying seizures. These slow magnetic phenomena are similar to d-c electrical shifts recorded from the penicillin focus using nonpolarizable electrodes. Such d-c electrical changes are believed to be an electrophysiological characteristic of epileptic seizures in general but have not been widely studied, in large part because of technical difficulties associated with d-c electrical recording.

CONCLUSION

These investigations have several important implications. First, the theoretical calculations demonstrate convincingly that brain events occurring within the small volume of the rodent cranium may be accurately and effectively measured with neuromagnetometers designed for human subjects; this finding opens the possibility of animal neuromagnetometry. Second, the animal model of epilepsy studied here provides a means of specifying exactly the type and locus of intracellular currents that give rise to epileptiform events of the brain magnetic field. Magnetic data obtained from penicillin-induced epilepsy in the rodent cortex are robust, and demonstrate both ictal and interictal manifestations of the seizure focus. Perhaps most interesting are steady field shifts observed accompanying the fast electrical and magnetic seizure phenomena; these slow field changes may in time prove to be among the most interesting aspects of brain epileptiform activity. Finally, the results reported here demonstrate that neuromagnetic fields may be usefully measured in a controlled animal preparation, permitting for the first time empirical investigations of the neurogenesis of brain magnetic fields, and the neuromagnetic study of cellular mechanisms underlying both normal and abnormal brain discharge.

ACKNOWLEDGEMENTS

This research was supported in part by the Environmental Physiology Program of the Office of Naval Research on contract N00014-76-C-0616 and by NINCDS grant 1-R01-NS20806-01.

REFERENCES

Barth, D.S. (1984). Neuromagnetic measurements of epilepsy in animal and man. Doctoral Thesis, UCLA.

Barth, D.S., Sutherling, W., Engel, J. Jr. and Beatty, J. (1982). Neuromagnetic localization of epileptiform activity in the human brain. Science, 218:891-894.

Barth, D.S., Sutherling, W., Engel, J. Jr. and Beatty, J. (1984a). Neuromagnetic localization of single and multiple sources underlying epileptiform spikes in the human brain. In R.J. Porter et al. (Eds.), Advances in Epileptology: XVth Epilepsy International Symposium. New York: Raven Press, pp. 379-384.

Barth, D.S., Sutherling, W., Engel, J. Jr. and Beatty, J. (1984b). Neuromagnetic evidence of spatially distributed sources underlying epileptiform spikes in the human brain. Science, 223:293-296.

Barth, D.S., Sutherling, W. and Beatty, J. (1984c). Fast and slow magnetic phenomena in focal epileptic seizures. Science, 226:855-857.

Beatty, J., Richer, F. and Barth, D.S. (1983). Magnetoencephalography. In M.G.H. Coles, S.W. Porges and E. Donchin (Eds.), Psychophysiology: Systems, processes and applications. New York: Guilford Press, (in press).

Cohen, D. (1972). Magnetoencephalography: Detection of the brain's electrical activity with a superconducting magnetometer. Science, 175:664-666.

Collins, R.C., Kennedy, C., Sokoloff, L. and Plum, F. (1976). Metabolic anatomy of focal motor seizures. Arch. Neurol., 33:536-542.

Elger, C.E. and Speckmann, E.-J. (1983). Vertical inhibition in motor cortical epileptic foci and its consequences for descending neuronal activity to the spinal cord. In E.-J. Speckmann and C.E. Elger (Eds.), Epilepsy and motor system. Baltimore: Urban and Schwarzenberg, pp. 152-160.

Goldensohn, E.S., Zablow, L. and Salazar, A. (1977). The penicillin focus. I. Distribution of potential at the cortical surface. Electroencephal. Clin. Neurophysiol., 42:480-492.

Gumnit, R.J. (1979). Field potentials in partial seizures - A reappraisal. In E.-J. Speckmann and H. Caspers (Eds.), Origin of cerebral field potentials. Massachusetts: Georg Thieme Publishers, pp. 183-194.

Petsche, H., Muller-Paschinger, I.B., Pockberger, H., Prohaska, O., Rappelsberger, P. and Vollmer, R. (1978). Depth profiles of electrocortical activities and cortical architectonics. In M.A.B. Brazier (Ed.), Architectonics of the cerebral cortex. New York: Raven Press, pp. 257-280.

Pockberger, H., Petsche, H. and Rappelsberger, P. (1983). Intracortical aspects of penicillin-induced seizure patterns in the rabbit's motor cortex. In E.-J. Speckmann and C.E. Elger (Eds.), Epilepsy and motor system. Baltimore: Urban and Schwarzenberg, pp. 161-178.

Purpura, D.P. (chief Ed.), (1972). Experimental Models of Epilepsy New York: Raven Press.

Romani, G.L., Williamson, S. and Kaufman, L. (1982). Biomagnetic instrumentation. Rev. Sci. Instrument., 53:1815-1845.

Tharp, B.R. (1971). The penicillin focus: A study of field characteristics using cross-correlation analysis. Electroencephal. Clin. Neurophysiol., 31:45-55.

Vollmer, R., Petsche, H., Pockberger, H., Prohaska, O. and Rappelsberger, P. (1978). Spatiotemporal analysis of cortical seizure activities in a homogeneous cytoarchitectonic region. In M.A.B. Brazier (Ed.), Cytoarchitectonics of the cerebral cortex. New York: Raven Press, pp. 281-305.

Williamson, S.J. and Kaufman, L. (1981a). Magnetic fields of the cerebral cortex. In S.N. Erne, H.D. Hahlbhm and H. Lubbig (Eds.), Biomagnetism. Berlin: Walter de Gruyter, pp. 353-402.

Williamson, S.J. and Kaufman, L. (1981b). Biomagnetism. J. Magnetism and Magnetic Materials, 22:129-202.

MAGNETIC FIELDS OF EPILEPTIC SPIKE FOCI: EQUIVALENT LOCALIZATION AND PROPAGATION

William Sutherling*, Daniel S. Barth* ** and Jackson Beatty**

Departments of Neurology* and Psychology**
UCLA, Los Angeles, California 90024, U.S.A.

ABSTRACT

Magnetic fields of the human focal epileptic interictal spike are robust and well-organized, allowing predictions of localization and propagation of equivalent sources. Some distortions of field patterns may be due to skull anatomy. Predictions are consistent with validation studies to date.

KEYWORDS

Partial Epilepsy; Neuromagnetism; Field Map Distortions; SQUIDs

INTRODUCTION

The study of human partial (focal) epilepsy is likely to be of basic scientific value to biomagnetism. Robust phenomena occur in epilepsy which are the largest electrical signals observed in the human nervous system (Penfield and Jasper, 1954; Daly, 1980). In partial epilepsy, these signals are the interictal spike and the usually rhythmic ictal electrographic discharge (Ajmone-Marsan and Ralston 1957; Walter, 1973). They have, by definition, a large signal-to-noise ratio (Chatrian and colleagues, 1974; Celesia and Chen, 1976) which markedly facilitates investigations into localization. Large relative distances between electrically active structures in the human brain with spatial extension in both the surface and depth dimensions allow accurate testing of spatiotemporal relations (Talairach and colleagues, 1958). There is substantial experience in epilepsy with interpretation of electrophysiological recordings both extracranially and intracranially (Penfield and Jasper, 1954; Engel and colleagues, 1981). Thus magnetic techniques can be investigated in an area about which much is already known. Furthermore, medically intractable focal epilepsy is a rare situation where it appears ethically justified to perform invasive intracranial electrical recordings in humans to localize the origin of seizures for focal excisional surgery, since the results of this therapy are excellent (Engel and colleagues, 1981; Cahan and colleagues, 1984; Ward, 1983; Van Buren and colleagues, 1975). This allows the unique opportunity to test predictions with techniques that do not have the inverse problem (Helmholtz, 1853), since the recording radius of intracranial contacts is less than 5 millimeters under most routine recording conditions for electrocorticography or stereotaxic depth electroencephalography (Cooper and colleagues, 1965; Ajmone-Marsan, 1973; Babb and Crandall, 1976; Babb and colleagues, 1973).

Conversely, neuromagnetic study, the magnetoencephalogram (MEG), is likely to be of value to epileptology. It theoretically allows the non-invasive study of the epileptic electrical currents of the cerebral cortex including the limbic system. Most of the deeper cortical structures involved in epilepsy including the majority of hippocampus are 2 centimeters or more from the center of the head (Talairach and colleagues, 1958) and therefore accessible to neuromagnetic measurement, as are fissures and sulci of lateral cortex (Williamson and Kaufman, 1981a; 1981b; Cohen and Hosaka, 1976). Furthermore, the MEG can be studied over long periods of time *in vivo* non-invasively, under different conditions. Due to the ability to repeatedly study the same patient with different magnetic sensor placements, sampling error for cortical structures is much reduced for tangential components of current sources. In many institutions, invasive intracranial electrodes are only placed in areas which are resectable (Crandall, 1973) and it is not feasible or ethical to sample all brain structures or to replace the electrodes once they have been positioned (Talairach and colleagues, 1958; Ajmone-Marsan, 1980). In this way neuromagnetic recording could give information complementary to invasive intracranial recordings due to less sampling error. From a practical point of view, the non-invasive sampling of multiple structures in the depths of the brain over prolonged periods is important in focal epilepsy surgery, where the academic exercise of precise localization of intracranial electrical currents is essential for effective treatment (Ward, 1983; Walter, 1973). Today, there are at least 60,000 patients who are conservatively considered to be good candidates for focal epilepsy surgery in the United States (Ward, 1983).

Magnetic recording of spontaneous epileptic signals has been performed by several laboratories (Cohen, 1972; Hughes and colleagues, 1977; Modena and colleagues, 1982). Recently, mapping of averaged and spontaneous spikes and precise predictions of localization of interictal activity have been done (Barth and colleagues, 1982; 1984). Validations and comparisons with the electric potential field (EEG) have been started at this symposium.

Two questions were asked in this series of investigations. First, are the magnetic fields of interictal epileptic currents sufficiently well-organized to allow prediction of localization of the epileptic focus for verification? In these experiments, the important simplification allowed clinically was that the patients studied did, in fact, have focal epilepsy in which the originating brain structure could be assumed to be spatially restricted rather than widely extended (Commission CTILAE, 1981). This has major implications for the specific methodology and analysis approach which has evolved and is discussed below. Second, can non-invasive magnetic field recording detect interictal epileptic propagation patterns, already reported in intracranial electrophysiological recordings (Buser and colleagues, 1973)? This is important for localization, since epileptic currents are known to propagate and, therefore, it is essential to identify the initial component or origin of the activity and to exclude propagated activity (Brazier, 1973). Furthermore, propagation patterns can, in certain cases, help localize the primary epileptic focus in the frequent case of confounding epileptiform patterns at other sites (Buser and Bancaud, 1983). Finally, propagation patterns are of interest in the investigation of the basic mechanisms of the epilepsies and their functional anatomies (Brazier, 1973).

METHODOLOGY AND APPROACH TO INTERPRETATION

Our approach in neuromagnetic studies was context-specific and a detailed explanation is therefore given. The phenomenon chosen for investigation was a prototype well-suited to a localization exercise: the focal interictal epileptic spike in patients with documented partial (focal) seizure disorders (Commission CTILAE, 1981). The interictal epileptic spike measured in the scalp electric potential field is believed to be localized to a restricted region of cortex, usually about 6 square centimeters (Abraham and Ajmone-Marsan, 1958; DeLucchi and colleagues, 1962; Cooper and colleagues, 1965). For such restricted source configuration, the use of a single equivalent source model is initially justified by the levels of noise in the system, since this corresponds to an angular extent superficially of approximately 17-20 degrees, at the limit of the ability to distinguish a focal from a more extended source (Okada, this symposium) and within the apparent limits of equivalent source accuracy (Cuffin, this symposium). The interictal spike at the epileptic focus is also a stereotyped, repetitive signal which is approximately stable over time in its equivalent localization (Rossi and colleagues, 1984). Using morphology and spatial extension in the scalp electric potential recording as a guide to select and trigger similar magnetic events for analysis, a large number of samples can be obtained in a certain percentage of patients to allow complete mapping of the field. Finally, the interictal focal epileptic spike is similar in many of its characteristics to the initial components of the primary specific evoked response, which has been accurately modeled in many situations as a restricted dipolar sheet. This is important for modeling and is discussed below.

Two dimensional mapping of the instantaneous magnetic field of averaged spikes was done for several reasons. Mapping reduces data for visual analysis. It allows more detailed interpretation of the field and easier recognition of configuration of equivalent current sources. As a first step in data interpretation, it is customary in clinical neurophysiology to determine first if a measurement is artifactual or cerebral. To make this determination the interpreter usually proves a believable cerebral field of an event or accepts it as artifact. An important finding in the electric potential field literature is that artifact recognition is very difficult by automated procedures and visual analysis is still usually required (Gotman and colleagues, 1979). A field map facilitates this initial, critical step in interpretation. If there are non-physiological or unexpected distortions in the field in certain regions, investigations can be made for possible artifacts in that area. After eliminating the possibility of artifact, one can accept a field pattern as cerebral in origin. As a second step, the configuration of the source can be inferred if obvious or can be subjected to computer analysis so that the appropriate model can be used for localization and characterization of the field. In addition to exclusion of artifact and selection of appropriate models, mapping enables easier comparison of studies between or among individuals and easier comparison of regions within the sampled area (Darcey, 1979). This is especially important in the newly evolving studies of brain epileptic fields, where specific patterns could correspond to a certain anatomy and where statistical reliability in large patient studies is essential.

Averaging is useful in spike mapping because low or zero amplitude data contain important information, for example at the "null point" directly over a tangential current source (Cohen and Cuffin, 1983). Averaging enough events to an EEG trigger increases the signal-to-noise ratio and allows reliable field mapping at these points. Furthermore, although spikes in the electric potential field are chosen to be similar by morphology and distribution to allow accurate comparison of different points in the map, similarity in the routine spontaneous EEG does not necessarily indicate similarity in the magnetic field and averaging provides statistical reliability. Although the comparison of highly selected, individual spikes can be powerful and give orderly field maps (Sato and colleagues, this symposium), in some cases which we have reviewed in our laboratory, single spikes even with good signal-to-noise ratios do not give orderly field patterns, whereas averaged spikes do. The two techniques appear to be complementary and we tend to use averaging to increase statistical reliability of the signal and to determine the null point as well as the maxima in the map. The combination of detailed magnetic and electric potential field recording simultaneously may help detect spike variability and the use of multichannel sensors may allow more general interpretations of individual spontaneous spikes.

The first approximation model of an equivalent dipole in a sphere was used because of its simplicity, power and apparent appropriateness in the specific questions asked (Williamson and Kaufman, 1981). It is important to be careful in the application of this model. Pitfalls in interpretation usually arise from the inverse problem (Helmholtz, 1853) and from approaching the brain as a "black box". Thus, in the case of generalized phenomena such as the 3 Hz spike-and-wave of "petit mal" or "absence" seizures and the alpha rhythm (Daly, 1979; Gloor, 1979; Perez-Borja and colleagues, 1968; Committee CTILAE, 1981), previous investigators have shown that the single equivalent dipole model can give non-physiological localizations deep in the white matter close to a ventricle or in the midline where there is no cerebral cortex (Schneider and Gerin, 1965; Cohen, 1968). Errors from the inverse problem can frequently be avoided in the particular case by using other, independent converging lines of evidence (Darcey, 1979; Lehmann and colleagues, 1982). In our investigations, patients were studied whose detailed electro-clinical studies had already confirmed a focal epilepsy (Committee CTILAE, 1981; Engel and colleagues, 1981). These patients also had additional studies of brain structure and function including xray or magnetic resonance imaging and positron emission computed tomography, which usually adds information about focality (Engel and colleagues, 1981). Our belief is that such independent knowledge is important for appropriate selection of a model since the anatomy and physiology of the human brain are so complex that most models in source localization are only approximations. Therefore, in our study designs, we try to determine which approximation is most useful and appropriate to answer the specific scientific question asked, within the testable limits of validation studies. This does not appear to be an approach unique to our laboratory (Freeman, 1975; Darcey and colleagues, 1979; 1980).

The equivalent dipole model appears to be a good first approximation for the localization of the initial component of the epileptic interictal spike. There is much evidence to support this model in analysis of the earlier components of primary specific evoked responses and there are major similarities of the initial components of the evoked response and of the spike (Freeman, 1975; Vaughn, 1982; Gloor and colleagues, 1963; Creutzfeldt and colleagues, 1966). The single equivalent dipole model is simple to use for localization, in determining both depth and surface location from the distance between the maxima and from the "null point" of polarity reversal. This can be done visually when the field pattern is symmetrically dipolar (Williamson and Kaufman, 1981) or by least squares fitting of models to the data in more complex situations (Romani and colleagues, this symposium; Darcey, 1979). Finally, the equivalent dipole model forces an unambiguous localization which can then be subjected to scientific testing. This forced localization is appropriate and essential to determine the localization of the initial component of the epileptic interictal spike which is expected to be restricted in extent from intracranial recordings. The use of a spherical head model has been used for spike foci which are both deep and more superficial (Williamson and Kaufman, 1981).

REVIEW OF RESULTS

Magnetic spikes were robust. Their predicted localizations were reproducible and consistent with initial validation studies. The magnetic fields of the averaged interictal spike focus showed high amplitude signals with an excellent signal-to-noise ratio in our series of 25 patients to date. This is illustrated by the patient in Fig. 1 who had a left temporal interictal spike focus (Barth and colleagues, 1982). Both the averaged spikes in the matrix (1B) and the isofield maps (1C and 1D) were well-organized and the patterns observed above noise. The isofield map for the initial most localizing component of the spike (1C) shows an approximately dipolar pattern, but there is a distortion from what would be expected for an underlying equivalent dipole, most prominently in the right lower quadrant of the map, discussed below. Using the equivalent dipole model, the localization of the first spike component is in the left anterior temporal lobe. This localization is consistent with the localization of the spike from electric potential field measurements and with other converging evidence from the patient's neurological evaluation and observation of her seizure. Preliminary evidence from this laboratory indicates that the magnetic localization of the interictal epileptic spike focus correlates well with intracranial electrical recording techniques in six patients who have undergone stereotaxic depth electroencephalography, electrocorticography, positron emission tomography, computerized axial xray tomography and post-operative follow-up.

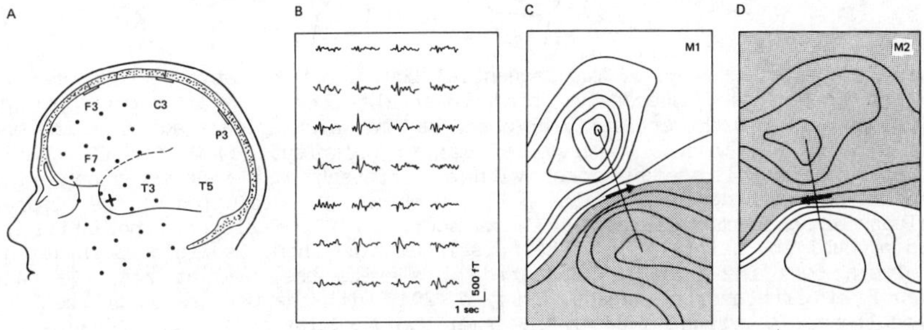

Figure 1. Magnetic Fields of a Focal Temporal Lobe Interictal Spike. (A) Sampling Matrix over left hemisphere. (B) Averaged magnetic spikes using EEG trigger. Polarity inversion for initial spike component (M1), indicating exiting and entering magnetic flux. (C and D) Isofield maps of first and second components, respectively. Arrows indicate location of predicted equivalent current dipolar sources of the spike and the presumed direction of intracellular currents, in the lateral cortex of the temporal lobe. Isofield calibration 150 femtoTesla per line. Flux entering the head is in shaded region. Matrix point spacing is 2 centimeters. (Taken from Barth and colleagues, 1982, p. 893, with permission of the authors and publisher. Copyright 1982 by the American Association for the Advancement of Science.)

Magnetic propagation patterns of the equivalent epileptic spike were clear and discrete. This is illustrated in the patient in Fig. 2 (Barth and colleagues, 1984). This patient had a right temporal interictal spike focus. The magnetic field of the initial component of the spike (2H) shows an orientation of maxima and a field morphology that is different from a second, later component of the spike (2I). The equivalent source localization in two dimensions of these two components is different, suggesting two current sources which have a different net configuration and net localization. Considering the depth of the currents, the two sources are separated spatially by approximately 1 centimeter. The subsequent field patterns at 2J and 2K resemble the patterns of the second and first components, respectively, although there are distortions, discussed below. In one-half to two-thirds of patients studied in our laboratory, such multiple equivalent magnetic sources have been observed.

Two additional studies are ongoing in our laboratory. Preliminary findings indicate that the magnetic and electric fields of the interictal spike behave similar to the fields which have been previously reported for the somatosensory evoked response (Cohen and Cuffin, 1983). There is also preliminary evidence that electrical currents during spontaneous typical behavioral and electrographic seizures can be recorded magnetically.

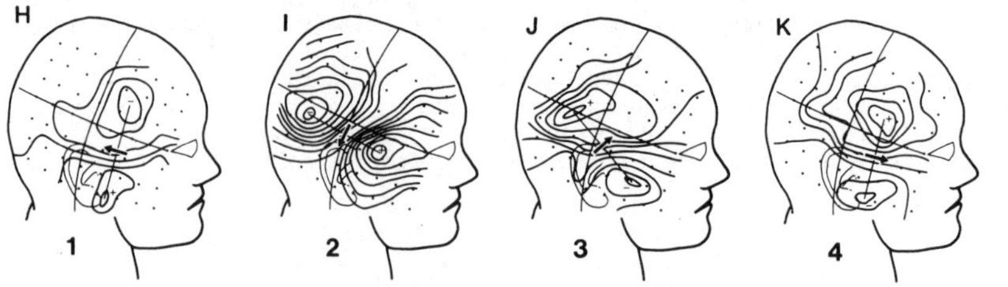

Figure 2. Magnetic Fields of Two Sequential Equivalent Magnetic Sources of a Focal Temporal Lobe Spike. Successive time intervals from left to right. (H) First equivalent source with vertically oriented maxima which appears again at the end of the discharge (K) with a reversal of maxima polarity. (I) A second, spatially discrete source follows the first with a different maxima orientation and also reappears (J) with opposite maxima polarity during the evolution of the discharge. Simultaneous characterization of these sources in the electric potential field would probably be helpful in further distinguishing them. Isofield calibration 100 femtoTesla per line for "I". Magnetic flux exits the head at positive values. (Taken from Barth and colleagues, 1984, p. 294, with the permission of the authors and publisher. Copyright 1984 by the American Association for the Advancement of Science.)

DISCUSSION

The predicted localizations of the interictal spike are testable and this has major implications for the non-invasive investigation of the electrical activity of the brain. The localizations using the equivalent dipole model are easily obtainable from the well-organized field patterns. Predictions are unambiguous for intracranial validation techniques which do not have the inverse problem. This demonstrates the important and non-trivial fact that, despite the great complexity present in the human epileptic interictal focus (Ajmone-Marsan, 1980; Gloor, 1975; Ward, 1983; Lieb and colleagues, 1981), equivalent source localizations in epilepsy are amenable to more detailed scientific analysis. This implies that neuromagnetic accuracy can be definitively tested where it is still ethically justified to make direct recordings from the human brain, in human focal epilepsy considered for surgery. The implications of a non-invasive technique for determining brain physiology *in vivo* is of general interest to neuroscience. Clinical relevance of non-invasive evaluation of patients is substantial.

The propagation patterns observed are testable and important for accuracy of localization, further diagnostic specificity and investigation of the basic mechanisms of the epilepsies. The magnetic propagation patterns appear robust and discrete, and with the equivalent dipole model give distinct, unambiguous localizations for apparent multiple sources. The ability to detect propagation patterns is important for localization of equivalent sources and diagnostic specificity, since epileptic activity is known to propagate (Jasper, 1969; Brazier, 1973). The most important signal for determining the nature of the epilepsy and its treatment is usually the earliest detectable paroxysmal electrical current (Walter, 1973; Crandall, 1973). Certain propagation patterns of interictal activity are useful in identifying the epileptic focus (Buser and colleagues, 1973; Buser and Bancaud, 1983). Furthermore, propagation patterns are of importance in investigations of the basic mechanisms of epilepsy since there is evidence that anatomical pathways not usually used may become functional in the interictal or ictal epileptic state (Buser and colleagues, 1983; Lange and colleagues, 1984). Electrically-induced changes in the use of neural pathways are of general interest in investigation of brain function (Bliss and Lomo, 1973). The phenomenon of epileptic propagation is also relevant to certain questions about interictal-to-ictal transition (Jasper, 1969; Prince and Wilder, 1967) and kindling (Goddard and colleagues, 1969).

A major finding in these studies is the presence in the field maps of distortions from patterns expected for a simple equivalent dipolar source. There appear to be two types of distortion, those for single sources in which the lower part of the map looks less dipolar (Fig. 1C) and those for multiple sources in which distortions from a dipolar pattern appear in other, frequently more central parts of the map (Fig. 2). The first type of distortion could be interpreted in two ways. It could be: (1) artifactual due to non-radial measurement over the ear, mastoid process and jaw below the inferior surface of and at a greater distance from the brain or (2) cerebral in origin, indicating a more complex, possibly extended and irregular source configuration and inappropriate for an equivalent dipole model. Since the general appearance of the field in Fig. 1C is approximately dipolar and since the upper part, in which the radial magnetic field was measured over the more spherical surface of the scalp, is what would be expected for an equivalent dipolar source, it is not unreasonable to assume that the field pattern could be generated by a restricted dipole layer, whose center can be approximately modeled by an equivalent dipole and there are, in addition, artifactual distortions of the field. This interpretation implies that in mapping the upper and lower parts of the matrix one is comparing the radial component of the magnetic field from the upper part, presumably generated by impressed currents, to a combination of radial and tangential components in the lower part which may include contributions from volume currents (Williamson and Kaufman, 1981b; Tripp, 1983). Such contributions could arise from the irregular boundaries of the middle and anterior fossae of the skull which are less spherical than its lateral convexity (Gray, 1977; Carpenter and Sutin, 1983). The modeling of such complicated boundaries appears to be a difficult problem and work on this has been begun in this symposium. Recording over the cheek and jaw also reduces the absolute amount of the radial component of the magnetic field detected due to different recording angle and larger distance from the source. This could result in the smaller signal usually found for the lower maxima. A complementary, direct approach to modeling such complexities is the empirical testing of alternative models with validation studies to test the equivalent dipole model. While the validation studies are in progress, if one is conservative and assumes worst case, the distortions are suspect to be in part artifactual and not necessarily representative of underlying cerebral activity. This approach is similar to that usually taken in clinical neurophysiology. Although obviously biased as a general rule, this approach has usually resulted in greater specificity without sacrificing significant sensitivity in electrophysiological measurements.

The second type of distortion in the multiple magnetic source fields may be different in origin. In the case of the patient depicted in Fig. 2H, there is distortion of the field from an approximate dipolar pattern, and the distortions appear to be in the parts of the map which subsequently show the maxima of an apparent second source (2I). Other distortions occur for subsequent maps (2J and 2K) and suggest a similar persistence of one field pattern during the maxima of another. These data suggest that part of these distortions from approximate dipolar patterns could be due to their overlapping in time. Another interpretation is that there is only one extended dipolar sheet which is simultaneously active and that the multiple apparent sources are due to changing relative orientations of the underlying electrical currents within the extended source. Although this is possible and the electric potential field needs to be mapped simultaneously to test this possibility with certainty, the large time differences between the sources are similar to those observed in the electric potential field for apparent different sources and correspond to time delays expected between cortical structures (Darcey and colleagues, 1979), with a stable pattern in between times of rapid change. These interpretations can be tested directly with intracranial techniques.

Future directions of neuromagnetic investigation in human epilepsy are likely to depend on important equipment developments. The limitations of single channel sensors to long sampling periods with biological variability, the resulting fewer numbers of patients studied with less powerful statistics, the restriction to magnetic recording biased by the electric potential field trigger used for averaging or selecting spikes and the difficulty in mapping single spontaneous events are a problem. This presently is most restrictive in monitoring ictal currents during spontaneous seizures. Multiple sensor arrays and new distributions of coils or coil geometries (Karp and Duret, 1980) would allow several centers of clinical neuromagnetometry to validate unbiased magnetic predictions and make comparisons with the electric potential field.

ACKNOWLEDGEMENTS

This research has been supported by grants NS00678 and NS20806 from the National Institute of Neurological and Communicative Disorders and Stroke and by contract N00014-76-C-0616 from the Office of Naval Research.

REFERENCES

Ajmone-Marsan, C. (1973). Electrocorticography. In A. Remond (Ed.), Handbook of Electroencephalography and Clinical Neurophysiology. Vol. 10C. Elesevier, Amsterdam. pp. 1-49.

Ajmone-Marsan, C. (1980). Depth electrography and electrocorticography. In M. J. Aminoff (Ed.), Electrodiagnosis in Clinical Neurology. Churchill Livingstone, New York. pp. 167-189.

Ajmone-Marsan, C. and B. L. Ralston. (1957). The Epileptic Seizure: Its Functional Morphology and Diagnostic Significance. A Clinical-electrographic Analysis of Metrazol-induced Attacks. Charles C. Thomas, Springfield, Illinois. 255 pp.

Abraham, K. and C. Ajmone-Marsan. (1958). Patterns of cortical discharges and their relation to routine scalp electroencephalography. Electroenceph. clin. Neurophysiol., 10, 447-461.

Babb, T. L., E. Carr and P. H. Crandall. (1973). Analysis of extracellular firing patterns of deep temporal lobe structures in man. Electroenceph. clin. Neurophysiol., 34, 247-257.

Babb, T. L. and P. H. Crandall. (1976). Epileptogenesis of human limbic neurons in psychomotor epileptics. Electroenceph. clin. Neurophysiol., 40, 225-243.

Barth, D. S., W. Sutherling, J. Engel and J. Beatty. (1982). Neuromagnetic localization of epileptiform spike activity in the human brain. Science, 218, 891-894.

Barth, D. S., W. Sutherling, J. Engel and J. Beatty. (1984). Neuromagnetic evidence of spatially distributed sources underlying epileptiform spikes in the human brain. Science, 223, 293-296.

Bliss, T. V. P. and T. Lomo. (1973). Long-lasting potentiation of synaptic transmission in the dentate area of the anesthetized rabbit following stimulation of the perforant path. J. Physiol, 232, 331-356.

Brazier, M. A. B. (1973). Electrical seizure discharges within the human brain: The problem of spread. In M. A. B. Brazier (Ed.), Epilepsy: its Phenomena in Man. Academic Press, New York. pp. 155-171.

Buser, P., J. Bancaud and J. Talairach. (1973). Depth recordings in man in temporal lobe epilepsy. In M. A. B. Brazier (Ed.), Epilepsy: Its Phenomena in Man. Academic Press, New York. pp. 67-97.

Buser, P. and J. Bancaud. (1983). Unilateral connections between amygdala and hippocampus in man. A study of epileptic patients with depth electrodes. Electroenceph. clin. Neurophysiol., 55, 1-12.

Cahan, L. D., W. Sutherling, M. A. McCullough, R. Rausch, J. Engel and P. H. Crandall. (1984). Review of the 20-year UCLA experience with surgery for epilepsy. Cleveland Clinic Quarterly, 51, 313-318.

Carpenter, M. B. and J. Sutin. (1983). Human Neuroanatomy. William and Wilkins, Baltimore, 8th Ed. 872 pp.

Celesia, G. G. and Chen R. C. (1976). Parameters of spikes in human epilepsy. Diseases Nervous System, 37, 277-281.

Chatrian, G. E., Bergamini L., Dondey M., Klass D. W., Lennox-Buchthal M. and Petersen I. (1974). A glossary of terms most commonly used by clinical electroencephalographers. Electroenceph. clin. Neurophysiol., 37, 549-553.

Cohen, D. (1968). Magnetoencephalography: Evidence of magnetic fields produced by alpha rhythm currents. Science, 161, 784-786.

Cohen, D. (1972). Magnetoencephalography: Detection of the brain's electrical activity with a superconducting magnetometer. Science, 175, 664-666.

Cohen, D. and J. Hosaka. (1976). Magnetic field produced by a current dipole. J. Electrocardiology, 9, 409-417.

Cohen, D. and B. N. Cuffin. (1983). Demonstration of useful differences between the magnetoencephalogram and electroencephalogram. Electroenceph. clin. Neurophysiol., 56, 38-51.

Commission on Classification and Terminology of the International League Against Epilepsy. (1981). Proposal for revised clinical and electroencephalographic classification of epileptic seizures. Epilepsia, 22, 489-501.

Cooper, R., A. L. Winter, H. J. Crow and W. G. Walter. (1965). Comparison of subcortical, cortical and scalp activity using chronically indwelling electrodes in man. Electroenceph. clin. Neurophysiol., 18, 217-228.

Crandall, P. H. (1973). Developments in direct recordings from epileptogenic regions in the surgical treatment of partial epilepsies. In M. A. B. Brazier (Ed.), Epilepsy: Its Phenomena in Man. Academic Press, New York. pp. 287-310.

Creutzfeldt, O. D., S. Watanabe, L. D. Lux. (1966). Relation between EEG phenomena and potentials of single cortical cells. Electroenceph. clin. Neurophysiol., 20, 1-37.

Darcey, T. M. (1979). Methods for localization of electrical sources in the human brain with applications to the visual system. PhD thesis. California Institute of Technology.

Darcey, T. M., J. P. Ary and D. H. Fender. (1979). Spatio-temporal evoked potentials to partial field pattern stimulation. European Conference on Visual

Perception. Nordwijkerhout, Netherlands.

Darcey, T. M., J. P. Ary and D. H. Fender. (1980). Methods for the localization of electrical sources in the human brain. In H. H. Kornhuber and L. Deecke (Eds.), Motivation, Motor and Sensory Processes of the Brain. Prog. Brain Research, 54, 128-134.

DeLucchi, M. R., B. Garoutte, R. B. Aird. (1962). The scalp as an electroencephalographic averager. Electroenceph. clin. Neurophysiol., 14, 191-196.

Engel, J., R. Rausch, J. P. Lieb, D. E. Kuhl and P.H. Crandall. (1981). Correlation of criteria used for localizing epileptic foci in patients considered for surgical therapy of epilepsy. Annals of Neurology, 9, 215-224.

Freeman, W. J. (1975). Mass Action in the Nervous System. Academic Press, New York. pp 202-249.

Gastaut, H. and R. Broughton. (1972). Epileptic Seizures: Clinical and Electrographic Features, Diagnosis and Treatment. Charles C. Thomas, Springfield, Illinois. 286 pp.

Gloor, P., C. C. Vera, L. Sperti. (1963). Electrophysiological studies of hippocampal neurons. Electroenceph. clin. Neurophysiol., 15, 353-402.

Gloor, P. (1975). Contributions of electroencephalography and electrocorticography to the neurosurgical treatment of the epilepsies. In D. P. Purpura, J. K. Penry and R. D. Walter (Eds.), Advances in Neurology: Neurosurgical Management of the Epilepsies. Vol. 8. Raven Press, New York. pp. 59-105.

Gloor, P. (1979). Generalized epilepsy with spike-and-wave discharge: A reinterpretation of its electrographic and clinical manifestations. Epilepsia, 20, 571-588.

Goddard, G. V., D.C. McIntyre, C. K. Leech. (1969). A permanent change in brain function resulting from daily electrical stimulation. Exp. Neurology, 25, 295-330.

Gotman, J., J. R. Ives, P. Gloor. (1979). Automatic recognition of interictal epileptic activity in prolonged EEG recordings. Electroenceph. clin. Neurophysiol., 46, 510-520.

Gray, H. (1977). Anatomy, Descriptive and Surgical, 15th Ed. Bounty Books, New York. p. 104.

Helmholtz, H. (1853). Ueber einige Gesetze der Verteilung elektrischer Stroeme in koerperlichen Leitern mit Anwendung auf die thierisch-elektrischen Versuche. Ann. Physik u. Chemie, 165 (89), 211-233.

Hughes, J. R., J. Cohen, C. I. Mayman, M. L. Scholl, D. E. Hendrix. (1977). Relationship of the magnetoencephalogram to abnormal activity in the electroencephalogram. J. Neurology, 217, 79-93.

Jasper, H. H. (1969). Mechanisms of propagation: Extracellular studies. In H. H. Jasper, A. A. Ward and A. Pope (Eds.), Basic Mechanisms of the Epilepsies. Little, Brown and Company, Boston. pp 421-438.

Karp, P. and D. Duret. (1980). Unidirectional magnetic gradiometers. J. Applied Physics, 51, 1267-1272.

Lange, H. H., J. P. Lieb, J. Engel and P. H. Crandall. (1983). Temporo-spatial patterns of pre-ictal spike activity in human temporal lobe epilepsy. Electroenceph. clin. Neurophysiol., 1983, 56: 543-555

Lehmann, D., T. M. Darcey and W. Skrandies. (1982). Intracerebral and scalp fields evoked by hemiretinal checkerboard reversal and modeling of their dipole generators. In J. Courjon, F. Mauguiere and M. Revol (Eds.), Clinical Applications of Evoked Potentials in Neurology. Raven Press, New York. pp 41-48.

Lieb, J. P., J. Engel, A. B. Gevins and P. H. Crandall. (1981). Surface and deep correlates of surgical outcome in temporal lobe epilepsy. Epilepsia, 22, 515-538.

Modena, I., G. B. Ricci, R. Barbanera, R. Leoni, G. L. Romani and P. Carelli. (1982). Biomagnetic measurements of spontaneous brain activity in epileptic patients. Electroenceph. clin. Neurophysiol., 54, 622-628.

Perez-Borja, C., G. E. Chatrian, F. A. Tyre and M. H. Rivers. (1962). Electrographic patterns of the occipital lobe in man: A topographical study based on use of implanted electrodes. Electroenceph. clin. Neurophysiol., 14, 171-182.

Prince, D. A. and B. J. Wilder. (1967). Control mechanisms in cortical epileptogenic foci: "Surround" inhibition. Arch. Neurology, 16, 194-202.

Romani, G. L., S. J. Williamson and L. Kaufman. (1982). Biomagnetic Instrumentation. Rev. Sci. Instrument., 53, 1815-1845.

Rossi, G. F., G. Colicchio and P. Pola. (1984). Interictal epileptic activity during sleep: A stereo-EEG study in patients with partial epilepsy. Electroenceph. clin. Neurophysiol., 58, 97-106.

Schneider, M. and P. Gerin. (1970). Une methode de localization des dipoles cerebraux. Electroenceph. clin. Neurophysiol., 28, 69-78.

Talairach, J., M. David and P. Tournoux. (1958). L'Exploration Chirurgicale Stereotaxique du Lobe Temporal dans l'Epilepsie Temporale. Masson et Cie, Paris. 136 pp

Tripp, J. H. (1983). Physical concepts and mathematical models. In S. J. Williamson, G. L. Romani, L. Kaufman and I. Modena (Eds.), Biomagnetism. Plenum Press, New York. pp. 101-139.

Van Buren, J. M., C. Ajmone-Marsan, N. Mutsuga and D. Sadowsky. (1975). Surgery of temporal lobe epilepsy. In D. P. Purpura, J. K. Penry and R. D. Walter (Eds.), Advances in Neurology: Neurosurgical Management of the Epilepsies, Vol. 8. Raven Press, New York. pp. 155-196.

Vaughn, H. G. (1982). The neural origins of human event-related potentials. Ann. N. Y. Acad. Sci., 388, 125-138.

Walter, R. D. (1975). Principles of investigation of surgical candidates. In D. P. Purpura, J. K. Penry and R. D. Walter (Eds.), Advances in Neurology: Neurosurgical Management of the Epilepsies, Vol. 8. Raven Press, New York. pp. 49-58.

Ward, A. A. (1983). Perspectives for surgical therapy. In A. A. Ward, J. K. Penry and D. P. Purpura (Eds.), Epilepsy. Raven Press, New York. pp. 371-390.

Ward, A. A, H. H. Jasper, and A. Pope. (1969). Clinical and experimental challenges of the epilepsies. In Jasper, H. H., Ward, A. A. and Pope A. (Eds.), Basic Mechanisms of the Epilepsies. Little, Brown and Company, Boston. pp. 1-12.

Williamson, S. J. and L. Kaufman. (1981a). Magnetic fields of the cerebral cortex. In S. N. Erne, H. D. Hahlbohm and H. Lubbig (Eds.), Biomagnetism. Walter de Gruyter, Berlin. pp 353-402.

Williamson, S. J. and L. Kaufman. (1981b). Biomagnetism. J. Magnetism Magnetic Materials, 22, 129-202.

SINGLE MAGNETIC SPIKE MAPPING

Susumu Sato*, Douglas Rose**, and Roger Porter***

Branches of Epilepsy*, Clinical Neuroscience**, and Medical Neurology***
National Institute of Neurological
and Communicative Disorders and Stroke
Betheseda, Maryland, 20205, U.S.A.

ABSTRACT

Simultaneous EEG and MEG interictal spikes were recorded at 42 locations in a 22 year-old man with partial complex seizures. Representative MEG spikes were selected for each location and an isofield contour map was constructed from these data. The resulting map showed an asymmetrical configuration, which may be due to a difficulty in positioning the SQUID probe in the region of the cheek.

KEY WORDS

Partial complex epilepsy; neuromagnetism; SQUID.

INTRODUCTION

EEGs with focal discharges usually show a well-defined spike focus. The spike discharges in such a focus, however, may be of several different types as revealed by differences in morphology, amplitude, duration and maximal representation in adjacent channels. These different types of EEG spikes accompany different types of magnetic spikes. This finding raises the question of whether averaging of magnetic spikes triggered by EEG spikes is the optimal method for the mapping of magnetic spikes.

METHODS AND RESULTS

With the Data Percision Model 6000 Waveform Analyzer, we compared the phase relationship between identical EEG spikes and simultaneous magnetic spikes. Figure 1 shows six pairs of spikes at the same location. There is a small but significant "phase" difference between pairs. This finding led us to look at individual spikes and their distribution.

We have collected simultaneous EEG and MEG spikes at 42 (6 x 7, 2 cm apart) over the right temporal area in a 22 year-old man with complex partial seizures. The EEG was recorded with the international 10-20 system. Several different types of EEG spikes were found in the right anterior temporal area. At each location, 2 to 3 samples of EEG-MEG pairs of the same type were extracted and the most representative sample was used for further analysis. Figure 2 shows the representative magnetic spikes at each location. The maximaum amplitude of the upward deflection of the magnetic spikes was seen at location 21 and phase reversal at locations 24, 30, and 35.

Figure 3 shows an isocontour map of the magnetic spikes depicted in Fig. 2. The isocontour lines were drawn on the basis of the amplitude of upward deflection of the magnetic spikes. This mapping shows an asymmetrical configuration, which may be due to difficulty in positioning the SQUID probe radially to the brain surface in the lower temporal to cheek area. In this case, the distance between extrema is estimated at 5.5 cm. Similar treatment of other components of the same spike and other types of spikes is under way, and preliminary results suggest that they have slightly different locations of extrema and dipole orientation from the one shown here.

We are planning to compare these results with those of the averaging technique. With averaging, it may be difficult to document a moment-to-moment shift or spread of dipole (spike) or (spikes), but the approach described here may have such a capability, particularly with sequencing spike occurrence. The question, however, may not be answered until a multichannel system is available.

This method may have the potential to delineate the boundary of a discharging focus rather than the center of gravity (of a discharging focus) which is the most likely approximation by the averaging method.

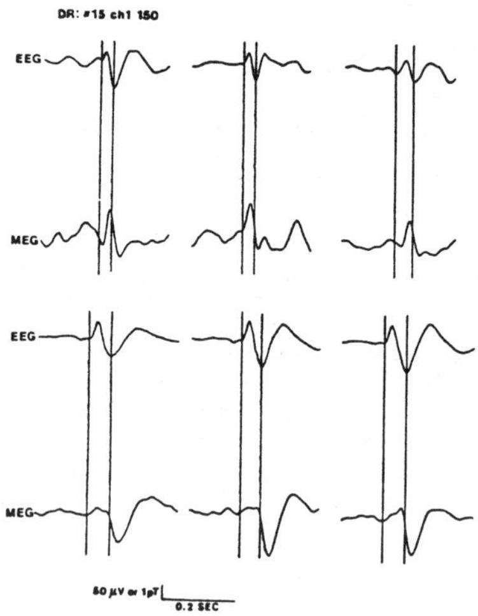

Figure 1. Six pairs of EEG and MEG spikes at the same location.

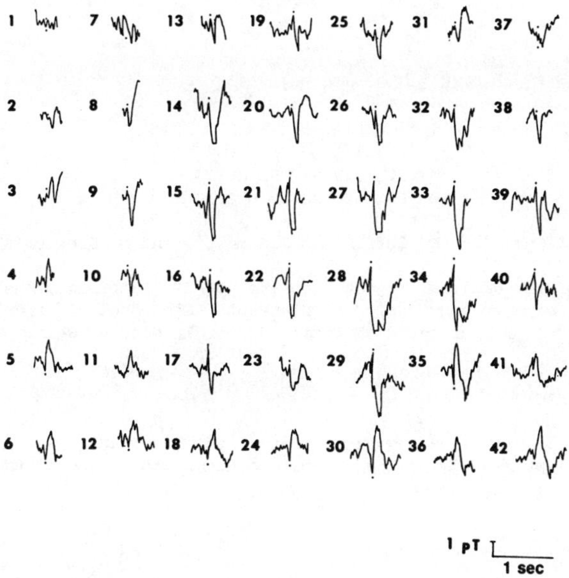

Figure 2. Representative single magnetic spikes at each location.

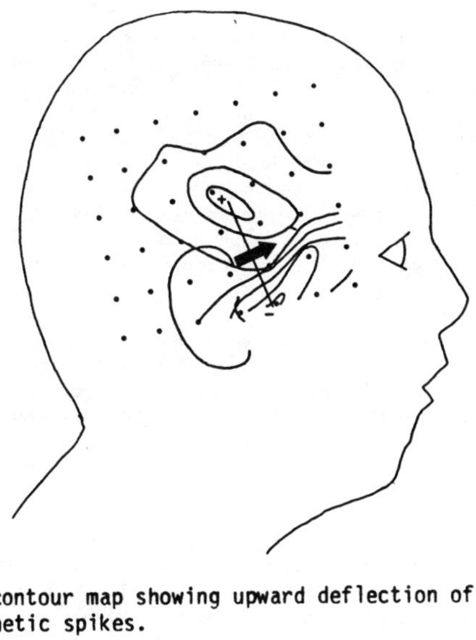

Figure 3. Isocontour map showing upward deflection of single magnetic spikes.

EVALUATING THE dcMEG FOR LOCALIZING EPILEPTIC FOCI

D. Cohen[1], B.N. Cuffin[1], R.J.Gumnit[2], and D. Schomer[3]

[1]Francis Bitter National Magnet Laboratory, MIT, Cambridge, Mass., USA
[2]Dept. of Neurology, Univ. of Minnesota, Minneapolis, Minn., USA
[3]Dept. of Neurology, Beth Israel Hospital, Boston, Mass., USA

ABSTRACT

The dcMEG (the MEG with low-frequency limit at dc) can readily see dc signals associated with epileptic activity; an evaluation has begun to see if these signals can help localize epileptic foci.

KEYWORDS

dcMEG, epileptic foci, on-off dc, persistent dc, injury current.

INTRODUCTION

As part of a program to see how well the MEG generally can localize epileptic foci in the human brain, an evaluation has begun of the dcMEG specifically; this is the MEG with lower frequency at dc. The dcMEG is less subject to the artifacts which make dcEEG measurements unreliable, hence offers a new way of measuring any dc produced at a focus, therefore of locating this focus. We divide dc into two types: on-off dc, and persistent dc, where each can be related to epileptic foci. On-off dc only appears for a period of seconds or minutes, and accompanies a burst of ac signal, say a single wave or a train of spike-and-wave. An example of dcMEG with human spike-and-wave is seen elsewhere in these proceedings (Cohen, 1985), and an experimental animal example has also been reported (Barth and colleagues, 1984).

In potential measurements within animal brains during experimental seizures it was found that the on-off dc accompanying sharp waves played an intimate role in the local excitability; also the distribution of this dc around the focus was more focal than the sharp waves (Gumnit and Takahashi, 1965; Gumnit and colleagues, 1970). However, on-off dcEEG (scalp) measurements have not been significantly used to find foci, mostly because of the galvanic skin problem. In contrast, although the dcMEG does receive an on-off dc from the scalp when it is touched (Cohen and colleagues, 1980), the level is relatively much smaller than in the dcEEG.

Persistent dc exists for hours or longer, and is due to long-term physiological processes; one example is the "current of injury". Because there is evidence of abnormal [K^+] at an epileptic focus (Somjen, 1983), and abnormal [K^+] is believed to cause a persistent dc (Cohen and colleagues, 1980), a persistent dc may exist at a focus; further, the spread of persistent current could influence the excitability of neighboring neurons, and could be a mechanism for seizure spread. The persistent dc potential has not been measured in the brain because of various slow dc drifts of the measuring system; these particular problems are not present with the dcMEG.

METHOD

The planned procedure for evaluating the dcMEG will be the following. First, a study will be made of reducing or eliminating the main artifact for on-off dc (touching the scalp). A study will also be made of sources of persistent dc which are non-neural, such as subcutaneous muscle; if these are significant in comparison to the expected neural dc levels, then efforts will be made to reduce them. What is needed is "a clear window" for viewing persistent dc originating in the brain itself. If the non-neural sources present too difficult a problem, then the persistent dc study will be discontinued.

Measurements will then be made in partial seizure patients with a single documented focal lesion due to tumor or recent stroke. The dcMEG maps (as well as acMEG and acEEG maps) will be measured of interictal events (and ictal events where possible) in these patients. Then, to determine the foci locations due to dcMEG measurements, accurate computer modeling will be performed. The locations determined from the dcMEG measurements will be compared with locations determined from combined acMEG-acEEG maps, and from the acEEG maps alone. This comparison will show if the foci determined by the dcMEG (on-off and persistent) can be more accurately located than with ac maps.

The measurements will all be made in the MIT shielded room. It is planned that two separate MEG detectors will simultaneously be used: Our present single location 2-D detector (three channels), and a multi-location system now being designed. This will yield a total of at least eight dcMEG channels for viewing the epileptic events so that complicated cross-referencing (as with a single channel) will be unnecessary. We believe that the program of evaluating the dcMEG for location of epileptic foci will involve three or four years of work.

REFERENCES

Barth, D.S., W. Sutherling, and J. Beatty (1984). Fast and slow magnetic phenomena in focal epileptic seizures. Science, 226, 855-857.

Cohen, D (1985). Introduction: A 1984 perspective on biomagnetism. This volume.

Cohen, D., Y. Palti, B.N. Cuffin, and S.J. Schmid (1980). Magnetic fields produced by steady currents in the body. Proc. Natl. Acad. Sci., USA 77, 1447.

Gumnit, R.J. and T. Takahashi (1965). Changes in direct current activity during experimental focal seizures. Electroenceph. clin. Neurophysiol., 19, 63.

Gumnit, R. J., H. Matsumoto, and C. Vasconetto (1970). DC activity in the depth of an experimental epileptic focus. Electroenceph. clin. Neurophysiol., 28, 333.

Somjen, G.G. (1983). Interstitial ion concentration and the role of neuroglia in seizures. In Electrophysiology of Epilepsy. Academic Press, London.

Discrimination of localized and distributed current dipole sources and localized single and multiple sources*

Yoshio Okada
Neuromagnetism Laboratory
Department of Psychology
New York University, New York, N. Y. 10003 U. S. A.

ABSTRACT

The magnetic field associated with a current dipole in a spherical conductor was compared with the field patterns associated with distributed generators and multiple localized generators. The dipole field was found to be statistically indistinguishable from those of the extended or multiple generators under a wide variety of conditions. Thus a single dipole may serve as an adequate model of the generator of the human magnetoencephalogram. In some conditions, however, the field patterns may clearly deviate from that of a dipole and could reveal the presence of a distributed generator or multiple generators.

KEYWORDS

magnetoencephalogram; current dipole; current generators; MEG

INTRODUCTION

The current generator of the human magnetoencephalogram (MEG) has been commonly modeled as a point source and represented by a current dipole (e.g., Brenner et al., 1978). This first-order approximation has been justified on the ground that the current dipole, which is the first term of the multipole representation of any current generator, dominates when the dimension of the actual generator is small relative to the distance between the generator and the field point. However, it is unclear as to how well the dipole term dominates the extracranial field pattern for generators of various sizes and depths, since there is no study that has provided quantitative answers to this question. The present study defines some of the boundary conditions within which one may justify the use of a localized single current dipole as a model of the generator of the MEG. This study also provides quantitative answers to another key theoretical issue, namely whether the MEG can be used to identify the presence of an extended generator or multiple localized generators.

The problems were analyzed by evaluating the goodness of fit of the single current dipole field to the fields of theoretical extended sources and multiple localized sources. This method of analysis was chosen because, on one hand, one may justify the current dipole as the most parsimonious model source when one can not statistically distinguish its field pattern from the actual field pattern obtained in an experiment and, on the other hand, a statistically significant deviation can suggest that a single point source model may need to be abandoned in favor of a multiple source model or an extended source model.

METHOD

The goodness of the fit was evaluated for three types of current generators - linear, sheet and 2-dipole sources - embedded in an 8-cm radius spherical conductor with concentric inhomogeneity in conductivity. The linear source was located either along an arc or along a radial line of the sphere (see the inset of Fig. 1b) and was represented by a set of current dipoles oriented parallel to each other, perpendicular to the axis of the source and tangential to the surface of the volume conductor. The continuous sheet source (Fig. 2) was represented by an ensemble of discrete current dipoles oriented parallel to each other, perpendicular to the midsagittal plane of the sphere in the case of a radial sheet or perpendicular to one side of the sheet in the case of a tangential sheet, and oriented tangential to the surface of the sphere. The two-dipole sources (inset of Fig. 3) were tangential to the spherical surface and located along the midline at a given depth. The spacings between adjacent dipoles were 1° for the linear sources and 2° for the sheets - they were sufficiently small for representing a continuous source by a set of discrete sources.

For each source, Biot-Savart law was used to calculate the normal component of its magnetic field that would be sensed by a second order gradiometer (2.4 cm coil diameter and 3.2 cm baseline) placed at 39 positions on the surface of a spherical conductor (see Fig. 1a). The normal component of the field for a distributed source was determined simply by summing the amplitudes of the field component of individual dipoles.

Once the field was calculated, an iterative algorithm (STEPIT; Chandler, 1965) was used to find the equivalent current dipole that gave the best fit in the sense of least squares to the field of a given source (Maclin, 1982; Maclin et al., 1983; Okada, et al., 1984). Two parameters of the model - strength and depth of the source - were free to vary, since the surface location and orientation of the dipole on the tangential plane were known from the symmetry of the sources investigated. These two parameters plus the sum of squared deviations in the fit were thus obtained from this analysis.

The goodness of fit of a single dipole field to the actual experimental data depends on the quality of data. Therefore, we evaluated the statistical significance of the sum of squared deviations we found above in terms of various levels of noise (expressed in terms of the percentage of the maximum field amplitude) that would be present in the actual experimental data. Specifically, we calculated F-ratio, that is:

$$F = (\Sigma d^2 + \Sigma e^2)/\Sigma e^2,$$

where Σd^2 is the sum of squared deviations in the fit of the model to the noiseless field of a theoretical source calculated from the above steps and Σe^2 is the sum of squared errors of data points that would be present in an actual experiment (Winer, 1971).

RESULTS

Goodness of fit

Figure 1a shows the isofield contour maps of the linearly extended sources along an arc at a depth of 2 cm from the surface of the sphere. The source lies below the origin of the map and is directed along the horizontal axis. As the angular extent of the source increases, the isofield becomes more elongated and distinguishable from the field of a single dipole.

Figure 1b presents the relation between F-ratio and angular extent of the linear source, separately for sources located at depths of 2, 4 and 6 cm below the surface of the sphere. There are two major conclusions that can be drawn from this figure. First, it shows that the field due to the linearly extended source is not statistically distinguishable from that due to a single dipole even for fairly large sources. F-ratio of the 95% confidence level is indicated by the horizontal line at F = 1.7. F-ratio is below the significance level for linearly extended sources with an angular extent of as much as 30°, or 3 cm in arc length, at the depth of 2 cm, when the variability of each data point is 5% of its mean. When the source is deeper, the field of the extended source becomes more and more indistinguishable from the field of a single dipole. The quality of data on human MEG today is at best about 5% of the maximum field amplitude. Therefore, we may conclude that the use of a single dipole as the model of the actual current source

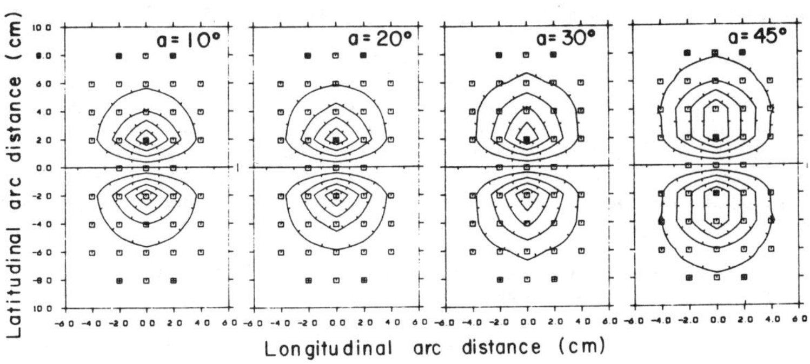

Fig. 1a. Isofield maps for four linear sources at a depth of 2 cm. a = angular extent in degrees.

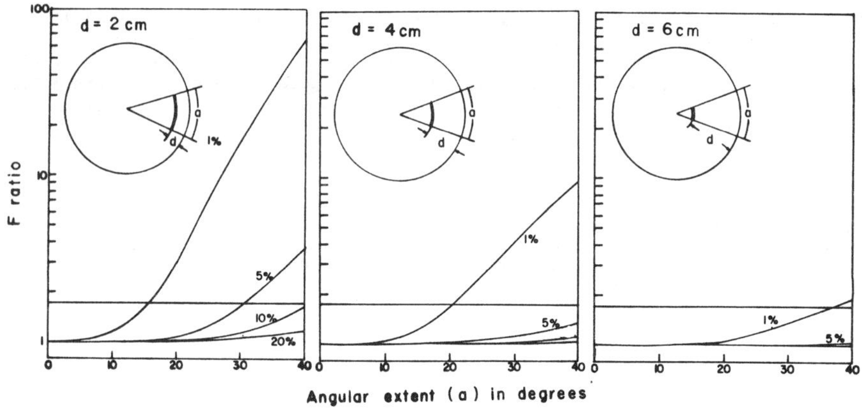

Fig. 1b. F-ratio as a function of angular extent (a) for a family of linear sources indicated in the inset.

is justified when we stimulate a narrow strip of active tissues that approximates a linear source of the type examined in Fig. 1, even when the angular extent of the source is as much as 20 - 30°.

The second conclusion is complementary to the first. The field pattern of a linearly extended source may be different both qualitatively as we saw in the right-most panel of Fig. 1a and statistically as shown in Fig. 1b, when the extent of the source is more than 30° for the data with a 5% variability or as small as 15° for the high quality data with a 1% variability. The field pattern of a line source is more discriminable from that of a point source as the source becomes shallower. Thus, it should be possible to infer under some conditions that the actual source is not a localized source based on MEG data.

The same analysis for other types of sources defines the boundary conditions within which the field of a given type of source is statistically indistinguishable from that of a single dipole. Fig. 2a shows F-ratios as a function of angular width of a sheet source located on the surface of a sphere at a depth of 2 cm below the conductor. F-ratio does not become significant for the data with a 5% variability until the source is about 18° wide or 18 mm in arc length. Fig. 2b indicates that the field pattern of a radial sheet is not apparently easily distinguishable from the single dipole field, even when its angular extent is as much as 20°. This is true in spite of the fact that the shallowest part of the source is only 1 cm beneath the surface of the conductor.

Fig. 3 presents the results for a family of two-dipole sources. The two dipoles are arranged as in the inset of Fig. 3b. They are symmetrically arranged so that the sum of the fields due to the individual dipoles is symmetrical about the origin of the coordinate system used in Fig. 3a. The isofield maps are for two-dipole sources separated by 2 cm in arc length and located at a depth of 2 cm below the surface of the spherical conductor. The field pattern is already

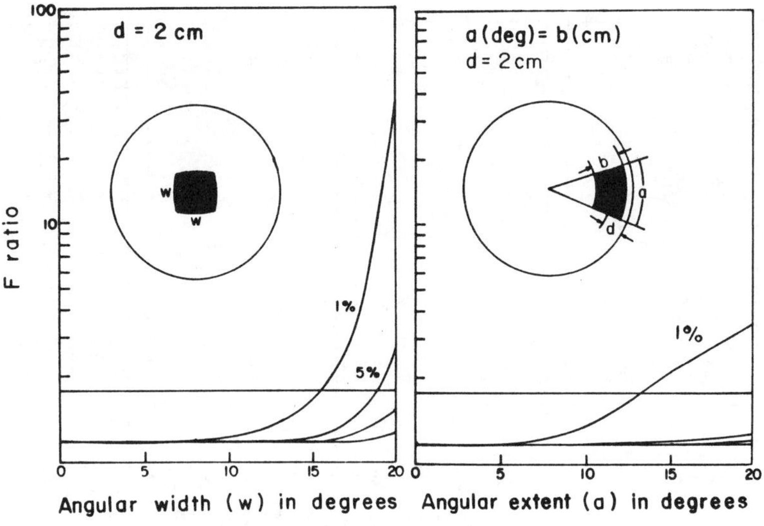

Fig. 2. F-ratio as a function of angular width of tangential and radial sheet sources at a depth of 2 cm.

different from that a single dipole for $\theta = 0°$, as clear from a comparison with the isofield map for the shortest linear source presented in Fig. 1a. As the angular orientation between the two dipoles increases, we see that the isofield map becomes more asymmetrical about the vertical axis.

F-ratios in Fig. 3b indicate that the two-dipole field is not easily statistically distinguishable from that of a single dipole with the quality of data obtainable today on human MEG when the sources are separated by as much as 0.5 - 1 cm. However, the result also shows that the field pattern can be dinstinguished when the separation is wider than about 1 - 2 cm. F-ratio is significant even for the worst case of 20% variability when the angular orientation is about 40°.

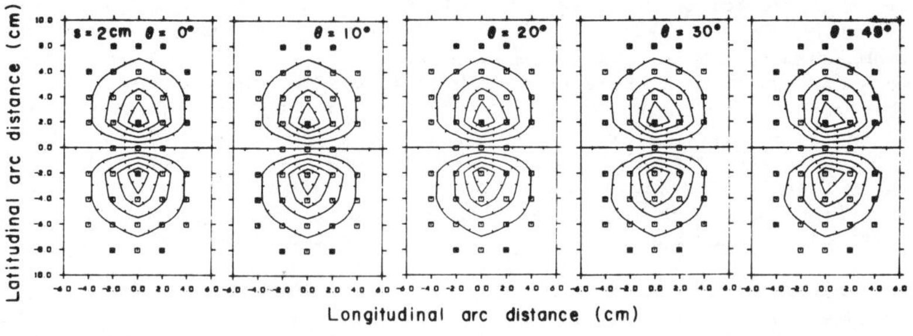

Fig. 3a. Isofield maps for two-dipole sources separated by an arc length of 2 cm at a depth of 2 cm. The sources are shown in the inset of Fig. 3b.

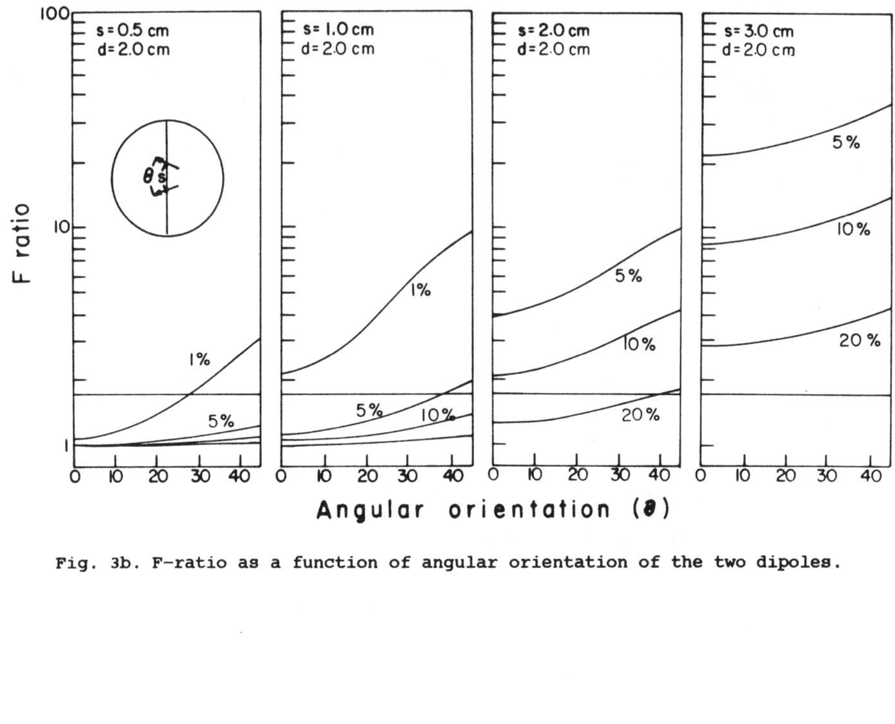

Fig. 3b. F-ratio as a function of angular orientation of the two dipoles.

Depth estimates

Table 1 presents the depth of the single dipole that gave the best fit in the sense of least-squares to the field patterns of the line sources embedded in the 8-cm radius spherical conductor and extending along an arc over the midline of the sphere. The equivalent sources are deeper but the errors are actually fractions of a centimeter for sources as wide as 20° of arc at a depth of 2 cm.

Table 1: Depth of the Equivalent Dipole for Tangential Line Source

Size	Depth (cm)		
	2	4	6
10°	2.05	4.04	6.02
20°	2.22	4.16	6.07
30°	2.44	4.36	6.13
40°	3.11	4.63	6.24

Table 2: Depth of the Equivalent Dipole for Sheet Sources

Size	Type of Source	
	Cup	Fan
4°	2.01	2.00
8°	2.05	1.99
12°	2.12	1.97
16°	2.18	1.95

Although the results are not presented here, the equivalent dipoles for the 2-cm deep radially oriented line sources are shallower than the actual sources; but the difference is again only 10% or 2 mm for such a line source extending as much as 2 cm - from 3 cm to 1 cm below the spherical surface. The results for the line sources agree with the earlier results obtained by Maclin (1982).

Table 2 presents the result for the tangential sheet sources (cup) and radial sheet sources (fan) whose geometrical centers are located at a depth of 2 cm below the sphere. Again the discrepancies between the actual and estimated depths are small (i.e. less than 10% of the actual depth for sources as wide as 16° x 16°).

Table 3 shows the results for the 2-dipole sources. The depth is affected almost solely by the separation between the two dipoles; the difference in the orientations of the two dipoles does not affect the depth, at least for the range of orientations studied here.

Table 3: Depth of the Equivalent Dipole for 2-Dipole Sources

Separation (cm)	Orientation (degrees)				
	0	10	20	30	45
0.5	2.0	2.0	2.0	2.0	2.0
1	2.1	2.1	2.1	2.1	2.2
2	2.9	2.7	2.7	2.7	2.7
3	3.8	4.1	4.1	4.1	4.1
4	4.9	5.4	5.4	4.6	5.3

DISCUSSION

The above results are relevant for the study of human MEG. Both the goodness of fit of the field patterns and the depth estimates indicate that the model is quite likely adequate for most of the sources studied so far. The field of a distributed source is not distinguishable from that of a point source when a radial sheet source is as extensive as 20° and 2 cm wide radially, which is probably the dimension of sources in the somatic or visual cortices that have been studied so far. The over- or under-estimation of the depth of the actual line or extended source is small, being only 4 mm even for such a shallow and large source as a line source extending 30° along an arc at a depth of 2 cm.

This conclusion, however, does not imply that the MEG can not be used to discriminate distributed sources or multiple sources. The field patterns are quite different from that of a single dipole when they are produced by a tangential sheet larger than 20° x 20° or by two-dipole sources with a separation greater than 1 - 2 cm. The depth estimates for the multiple sources are affected dramatically by the separation between the dipoles.

The boundary conditions established in the present study are useful not only for those theorests who apply a point source model to interpret human MEG, but also for those who might attempt to use or argue for the necessity of more complex models to better account for the data. For the former group, the present result indicates that a point source may be used as the model for an extended physiological current generator. However, one must carefully examine the experimentally obtained field pattern and determine whether the data suggest the need for a more complex model, namely an extended source or a set of multiple sources. For the latter group, the result indicates that one can not unconditionally expect to extract the information concerning the width and length of the actual generator or the number of simultaneously active generators from any MEG data one collects. The solution may be non-unique for relatively deep and small generators. Nevertheless, it should be possible to extract such an information under certain conditions such as when there are two discrete regions of active tissues separated by as little as 1 - 2 cm.

REFERENCES

*The preliminary results from this study were reported at Winter Conference on Brain Research, Steamboat Springs, Col. in January 1984. The study was supported in part by ONR grant N00014-76-C-0568 awarded to Dr. S. J. Williamson and Dr. L. Kaufman to whom the author wishes to exress his appreciations.

Brenner, D., J. Lipton, L. Kaufman, and S. J. Williamson (1978). Somatically evoked mangetic fields of the human brain. Science, 199, 81-83.

Chandler, J. P. (1965). STEPIT. Indiana University.

Maclin, E. L. (1982). Ph.D. thesis submitted to New York University.

Maclin, E. L., Y. C. Okada, L. Kaufman, and S. J. Williamson (1983). Retinotopic map on the visual cortex for eccentrically placed patterns: First non-invasive measurement. Il Nuovo Cimento, 2, 410-419.

Okada, Y. C., R. Tanenbaum, L. Kaufman, and S. J. Williamson (1984). Somatotopic organization of the human somatosensory cortex revealed by neuromagnetic measurements. Exp. Brain Res., in press.

Winer, B. J. (1971). Statistical Principles in Experimental Design, McGraw-HIl New York.

TOPOGRAPHY OF SIMULATED MEG AND EEG GENERATED BY MULTIPLE INTRACRANIAL DIPOLES

H. Weinberg, P. Brickett, F. Coolsma, and M. Baff

Brain Behaviour Lab., Dept. of Psyc., Simon Fraser Univ., Burnaby, British Columbia, Canada, V5A 1S6

ABSTRACT

Single and multiple current dipoles energised by isolated sources were located in known positions inside a human skull filled with an electrically conductive medium. Maps of the measured electrical and magnetic fields confirmed the predicted relationships between magnetic and electrical fields for both single and multiple dipoles. Values of the recorded magnetic fields were close to those calculated for a dipole in a homogeneous sphere. For multiple dipoles energised simultaneously the resulting maps are found to be equal to those made from the sum of measurements for the individual dipoles. Combinations of two dipoles were investigated for which the magnetic field maps appeared similar to those for a single dipole. Quantitative comparisons were then made of how well best-fitting single and double dipole models could account for the magnetic field.

KEYWORDS

Source localisation; brain; MEG; EEG; dipoles.

INTRODUCTION

Information from the measurement and mapping of magnetic fields of the brain may be combined with electroencephalographic (EEG) data and known neurophysiology of the brain to identify current "sources" associated with normal and pathological states. Cohen and Cuffin (1983) have described the theoretical relationship between the magnetic and electric fields produced by single dipoles. Okada, Kaufman, and Williamson (1983) have localised a single source for the P300 component of the visual evoked potential. However, a consideration of the neurophysiology of the brain would suggest that sources associated with complex processes, such as those which produce components of endogenous evoked potentials, are probably multiple and also widely distributed. The following study was designed to test localisation methods using single and multiple sources at known locations in a human skull. A similar physical model was used by Henderson, Butler, and Glass (1975) who localised sources based on only electrical recordings. Snyder (1982) investigated the magnetic field pattern measured in a plane for a current dipole and found good agreement with the theoretical predictions for a conducting half-space model.

Method

The model consisted of a human skull which had all openings except the foramen magnum sealed with wax. The skull was filled with EKG Sol electrode cream which had a resistivity of 650 Ohm·cm. Each current dipole was 22 mm in length, constructed from aligned ends of pairs of wires forming the top of a "T", insulated except at the ends which were capped with 1 mm spheres of solder. Three such dipoles, arranged to form an equilateral triangle and mounted on a plexiglass rod, were inserted into the skull through the foramen magnum and positioned 6 mm below the parietal bone. The posterior corner of the triangle was located at the vertex (Fig. 1 a,b). Each of the dipoles could be energised separately by an isolated constant current source which supplied a 45 Hz square wave current of 125 μA for EEG recording and 70 μA for MEG recording. Recordings were made for each dipole energised individually and also for dipoles I and II energised simultaneously from isolated sources. Electrical potentials were recorded from gold cup electrodes attached to the skull with dental cement at the 10/20 international system locations and additional locations near the dipoles (Fig. 1c). Linked reference electrodes were attached 15 mm above the external auditory meati; grounding electrodes were attached to the cheekbones.

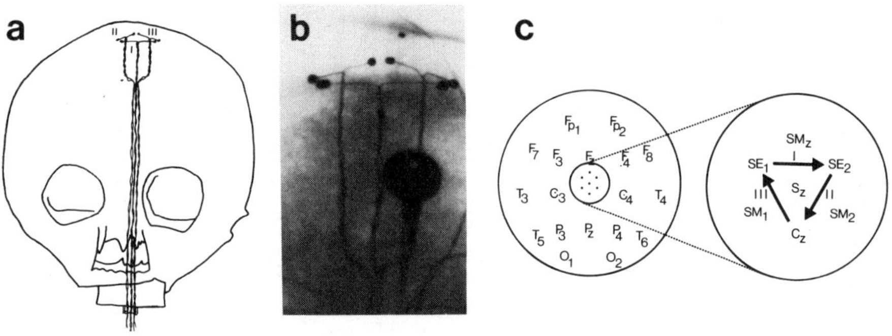

Fig. 1(a): Tracing of radiograph showing location of dipoles and connecting wires in the skull. (b): Enlarged section of the same radiograph. One EEG electrode is also shown. (c): Schematic diagram of the dorsal surface of the skull showing 10/20 system and the additional recording positions. The locations of dipoles I, II, and III are indicated by arrows in the five times enlargement of the central area.

Magnetic recordings from the same positions were obtained with a third order gradiometer manufactured by CTF Systems Inc. (Vrba and others, 1982). The diameter of the sensing coil was 38 mm and the inter-coil separation was 55 mm. Thirty-two 250 msec epochs of data were averaged, rectified about the mean, and integrated. The resulting values were used to construct isofield and isopotential contour maps of the whole scalp and a five times enlargement of the area near the dipoles. Details of the mapping techniques are included in another paper in this volume (Weinberg and others).

Results and Discussion

Figure 2 shows maps of the magnetic and electric data for dipoles I, II, III, and for combinations of II and III. Figure 2(a,b,d) demonstrates that a single dipole

Fig. 2. (Next page) Magnetic (M, upper row) and electric (E, lower row) isocontour maps for locations as indicated in Fig 1(c). For each pair of maps, the energised dipoles are indicated by arrows on the enlargement to the right. (a-f): Maps for single dipoles (I, II, III), simultaneously energised pairs of dipoles (II&III, II&-III), and arithmetically combined dipoles (II+III). Arrowheads pointing up indicate fields exiting the head or negative potentials. For magnetic maps the darkest shade corresponds to 60 picoTesla. For electric maps the darkest shade corresponds to 2.6 mV for the whole head maps and 5.6 mV for the enlargement.

within a real skull produces relatively symmetric fields. Consistent with the right hand rule, the line connecting the magnetic field extrema is perpendicular to the dipole axis. The electrical field shown in Fig. 1(a) is rotated 90 degrees with respect to the magnetic field. Maximum positivity and negativity of the electrical field is located over the positive and negative poles of the dipoles, respectively. These results are what would be theoretically predicted for a single dipole. Figure 2(c) shows a two dipole configuration (II&III) in which each dipole is polarised to approximate a system in which current flow is from one pathway to the other, i.e., the positive pole of one dipole is in close proximity to the negative pole of the second dipole. If it were assumed that the field in Fig. 2(c) was the result of a single equivalent dipole, that dipole would have been described as having an orientation pointed to the left. The arithmetic sum of the empirically measured fields from each of the dipoles energised separately (II+III) is shown in Fig. 2(e), and is virtually identical to the empirically measured field for the two dipoles activated simultaneously. As shown in Fig. 2(f), when the same two dipoles were polarised such that the positive poles of both were in close proximity (II&-III), a configuration which approximates the convergence of two current pathways, the magnetic field is rotated 90 degrees with respect to the field shown in Fig. 2(c) and an equivalent dipole would be located in the median plane. The electrical field in Fig. 2(f) shows a broader distribution of negativity consistent with the spatial separation of the negative poles of both dipoles, and a spatially confined positivity consistent with the summation of electrical potentials at the location of the positive poles. The magnetic and electric fields for this combination of dipoles was also virtually identical to that calculated arithmetically (II-III, not shown).

For a comparison of the recorded and theoretical field patterns, Fig. 3 shows the predicted field strength due to dipole I as a function of position along the median

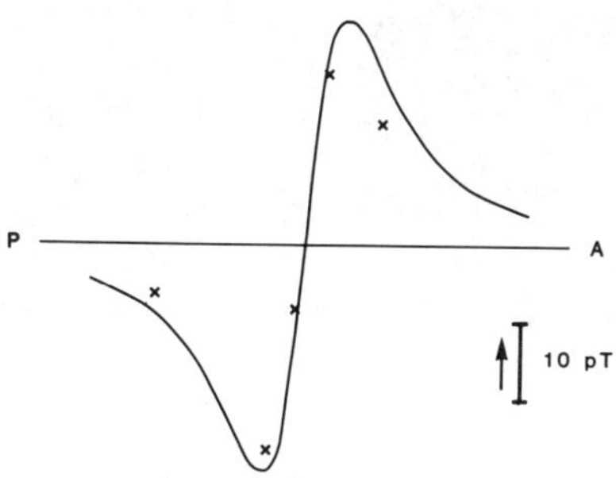

Fig. 3. Comparison of measured field values for dipole I (Xs), and predicted values along the midline in posterior (P) to anterior (A) direction.

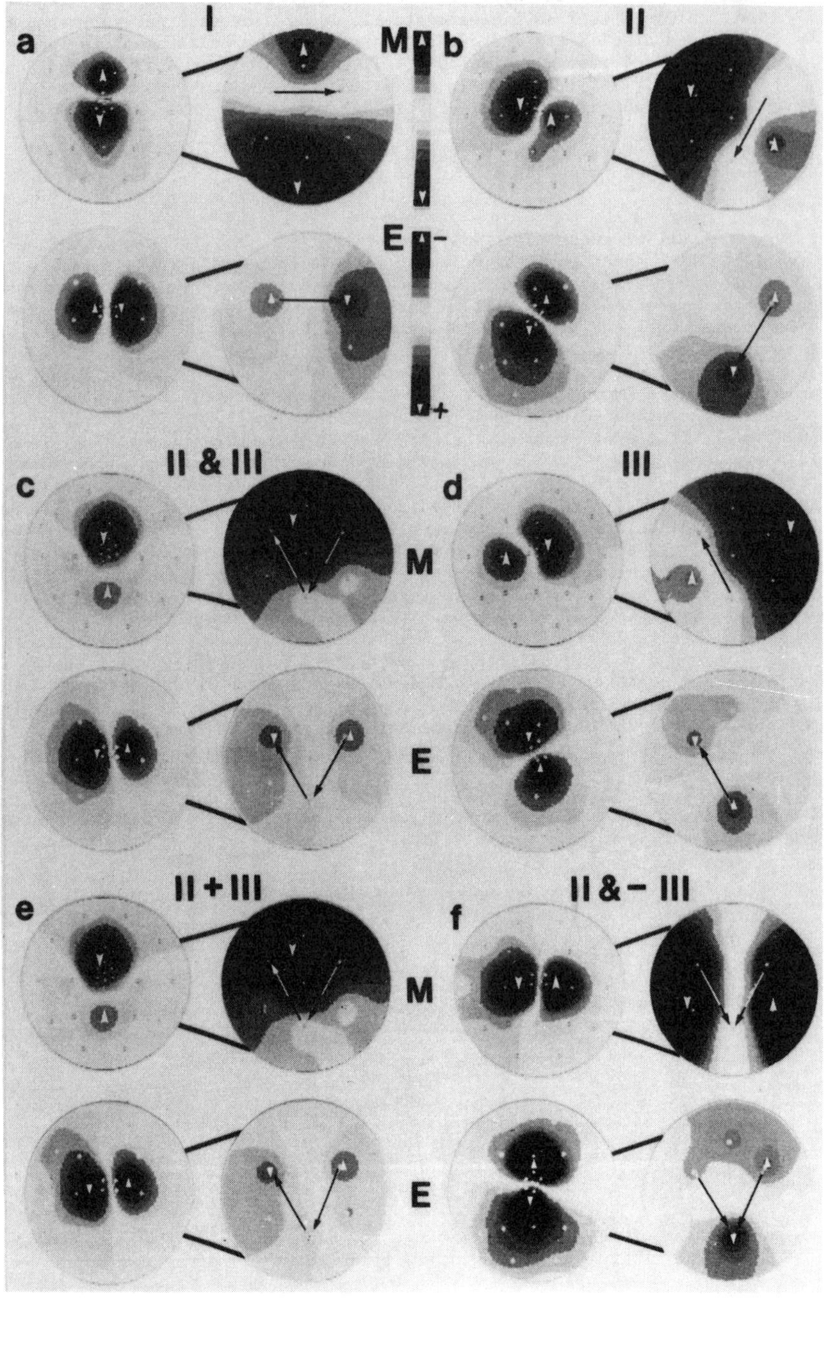
I
a
M
b
II
E
-
+
II & III
c
M
E
III
d
II + III
e
M
E
II & - III
f

line as calculated for a spherical model with equations from Williamson and Kaufman (1981). Five measured values, indicated by the Xs, show close agreement near the location of the dipole, while the more distant positions show smaller amplitudes than predicted. Possible sources of error which we are presently investigating include the gradiometer sensing coil diameter and coil separation distance, the difference between the normal and radial magnetic components, and also the length of the dipole.

Additional work with this model now being prepared for publication will be briefly summarized. Using the recorded field values, a least-squares estimate was computed for the five parameters which define the location, orientation, and magnitude of a tangential dipole. The head was modelled as a sphere with a fixed origin located halfway between Fz and Oz and with a variable radius equal to the measured distance to each recording position. Dipoles II and III were investigated in combination because this pair is orientated symmetrically with respect to the recording locations. For both dipoles this procedure accurately found the position and orientation within 2 mm and 4 degrees, respectively, although the dipole moment was underestimated by approximately 15%. In relation to how well the measured values correspond to theoretical predictions, 98% of the variance for both dipoles II and III was accounted for by the single dipole model. The same percentage of variance was also accounted for by a two dipole solution for the combinations II+III and II-III. However, a single dipole model fit to the data for II+III accounts for 92%, but this figure is 98% for II-III. This shows that when current flow is in a primarily posterior direction (see II&-III, Fig. 2(f)), the field pattern is virtually indistinguishable from that produced by a single dipole. The field for II+III however is significantly different from that for the best fitting single dipole ($F(5,20) = 10.2$, $p < .001$). Results from these and other simulations may hopefully lead to more sophisticated localisation methods and a better understanding of the field patterns recorded from the brain.

ACKNOWLEDGEMENT

Supported by N.S.E.R.C, N.R.C., Ministry of Transport, D.S.S of Canada, and the Science Council of B.C.

REFERENCES

Cohen, D. and B. N. Cuffin (1983). Demonstration of useful differences between magnetoencephalogram and electroencephalogram. Electroenceph. clin. Neurophysiol., 56, 38-51.

Henderson, C. J., S. R. Butler, and A. Glass (1975). The localization of the equivalent dipoles of EEG sources by the application of electric field theory. Electroenceph. clin. Neurophysiol., 39, 117-130.

Okada, Y. C., L. Kaufman, and S. J. Williamson (1983). The hippocampal formation as a source of the slow endogenous potentials. Electroenceph. clin. Neurophysiol., 55, 417-426.

Snyder, U. K. (1982). Neuromagnetism: A short theoretical and experimental study. M.A. thesis, Dalhousie University.

Vrba, J., A. A. Fife, M. B. Burbank, H. Weinberg, and P. A. Brickett (1982). Spatial discrimination in SQUID gradiometers and 3rd order gradiometer performance. Canadian J. Physics, 60, 1060-1073.

Williamson, S. J. and L. Kaufman (1981). Magnetic fields of the cerebral cortex. In S. N. Erne', H-D. Hahlbohm and H. Lubbig (Eds.), Biomagnetism. Walter de Gruyter, Berlin. pp. 353-402.

THE FORWARD AND INVERSE PROBLEMS IN THE SPHERICAL MODEL

R.J. Ilmoniemi, M.S. Hämäläinen, and J. Knuutila

Low Temperature Laboratory,
Helsinki University of Technology, 02150 Espoo, Finland

ABSTRACT

We analyze the forward problem in the case of the spherical model, introducing simple ways of dealing with the tangential components of the magnetic field. For the inverse problem, we present an alternative to the use of source models, namely the construction of best estimates which are based, in general, both on the results of measurements and on some prior knowledge.

KEYWORDS

Magnetoencephalography; spherical model; forward problem; inverse problem; best estimate; minimum norm estimate.

INTRODUCTION

Traditionally in magnetoencephalography, the magnetic field component normal to the scalp is measured. With single channel magnetometers, this may well be the best approach. However, the designer of multichannel instruments should seriously consider whether the measurement of tangential components at selected points would improve the precision of locating brain events. Another tradition, widely believed to be the only possibility in the determination of sources of biomagnetic fields, is the use of source models. The following discussion attempts to challenge these traditions. For more extensive treatments of pertinent concepts, see, e.g., Cuffin and Cohen (1977) and Tripp (1983).

THE FORWARD PROBLEM IN THE SPHERICAL MODEL

We assume that the conductivity of the head is spherically symmetric about some origin of coordinates: $\sigma(\vec{r}) = \sigma(r)$. Then the radial component of the magnetic field, produced by the current dipole $\vec{Q} = Q\vec{e}_y$ at $\vec{r}_Q = a\vec{e}_z$ (Fig. 1a), can be calculated directly from the dipole expression (Cuffin and Cohen, 1977):

$$B_r(\vec{r}) = \frac{\mu_0}{4\pi} |\vec{r}-a\vec{e}_z|^{-3} \vec{Q}\times(\vec{r}-a\vec{e}_z)\cdot\vec{e}_r = -AG^{-3}\sin\theta\cos\phi , \qquad (1)$$

where $A = \mu_o Q/4\pi a^2$, $G = (1-2\rho\cos\theta+\rho^2)^{1/2}$, and $\rho = r/a$. In the quasistatic approximation, $\nabla\times\vec{B}=0$ outside the head, which leads to $rB_\theta = \int_\infty^r (\partial B_r/\partial\theta)dr$ and $r\sin\theta B_\phi = \int_\infty^r (\partial B_r/\partial\phi)dr$. Inserting the derivatives, obtained from Eq. 1, and integrating, we get the tangential components of the magnetic field:

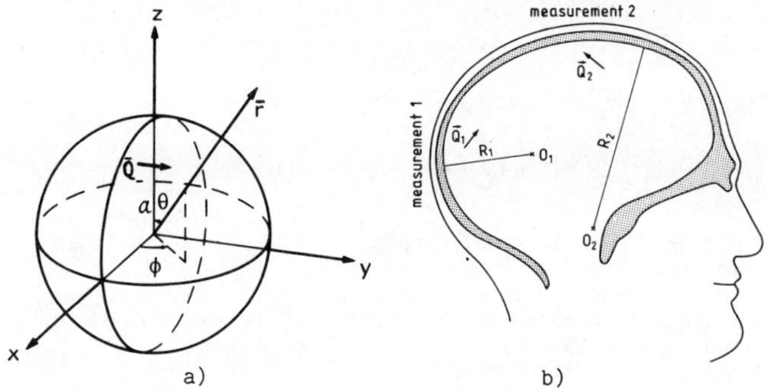

Fig. 1. (a) The coordinate system for the spherical model. (b) The choice of the origin should be based on the local curvature of the skull's inner surface.

$$B_\theta = \frac{-A\cos\phi}{\rho G^3 \sin^2\theta}\left[1+\rho\cos\theta(\rho\cos\theta-1-\cos^2\theta-G^2)+G^3\right] , \tag{2}$$

$$B_\phi = \frac{A\sin\phi}{\rho G \sin^2\theta}(\rho-\cos\theta-G) . \tag{3}$$

To apply the spherical model it suffices that $\sigma(\vec{r})=\sigma(r)$ in the region of appreciable current flow. Figure 1b illustrates this important point: the current flow patterns produced by the dipoles $\vec{Q}_1$ and $\vec{Q}_2$ are influenced mainly by the nearest parts of skull's inner surface with local radii of curvature R_1 and R_2, respectively. Thus, the appropriate origins for the analysis of magnetic field measurements 1 and 2 are O_1 and O_2. The assertion that the choice of origin should be based on the skull's inner surface relies on the assumption that the currents flowing in the poorly conducting skull and in the scalp are not important as far as magnetic fields are concerned.

It is often stated correctly that the effect of volume currents in a spherically symmetric head is zero on the radial field component and nonzero on the tangential part of the field. From this, one is tempted to draw the erroneous conclusion that volume currents complicate the analysis of tangential components. We emphasize that it is not necessary to consider the volume currents explicitly when calculating any of the magnetic field components in the spherical model. The situation is illustrated in Fig. 2. The radial field component B_r is produced solely by the source current distribution J^i, which also gives rise to the volume current J^v; both J^i and J^v contribute to the tangential part of the field B_t. B_t, however, can be determined from B_r, as shown above. How can it be that J^v affects the value of B_t but not B_r while B_t can be calculated from B_r? The answer is simple: the effect of J^v on B_t is independent of any details of conductivity as long as spherical symmetry is maintained. Therefore, although the volume currents are affected by the conductivity profile, their effect on B_t depends only on J^i.

Further insight to the spherical model can be obtained from the diagrammatic equations of Fig. 3. Each circle with a subscript B is a symbol for the magnetic field outside a spherically symmetric head. The arrows denote primary current elements. According to Eq. a, an inversion of current changes the sign of the magnetic field. Eq. b is an assertion of linearity, while Eq. c states that the magnetic field produced by a radial current source is zero. In Eq. d, two zeros are added to the magnetic field of a current element

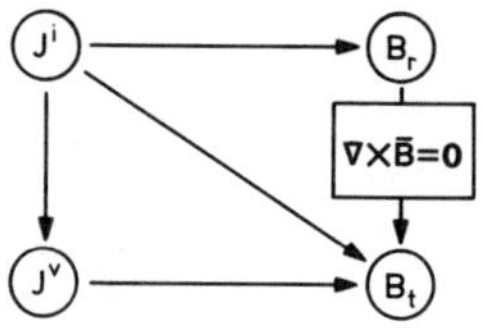

Fig. 2. The radial component of the magnetic field is produced by the source current alone, while both source and volume currents contribute to the tangential component. The tangential part of the field can be calculated from the radial component.

(arbitrary length). Applying Eqs. a and b, we see that the result is an equivalent current loop. The volume current has been replaced by the radial parts of this loop. We notice immediately that the source can roughly be approximated by a magnetic dipole located at the center of mass of the triangle, i.e., two thirds of the distance from the origin towards the current element. The calculation of higher multipole moments from the loop is a simple matter, too. $\vec{B}(\vec{r})$ produced by the loop can be calculated simply from the magnetic scalar potential ϕ_m, defined by $\vec{B} = -\nabla\phi_m$:

$$\phi_m(\vec{r}) = \frac{\mu_0}{4\pi} I \iint d\vec{S}'\cdot(\vec{r}-\vec{r}')|\vec{r}-\vec{r}'|^{-3} , \tag{4}$$

where I is the current circulating in the loop, and $d\vec{S}'$ is a surface element of the triangle. The resulting magnetic field for a short current element, i.e., for a current dipole $\vec{Q}$, located at $\vec{r}_Q$, can be expressed as a linear combination of the vectors $\vec{R} = \vec{r}-\vec{r}_Q$, $\vec{S} = \vec{r}+\vec{r}_Q$, and $\vec{V} = \vec{r}_Q\times\vec{Q}$:

$$\vec{B}(\vec{r}) = \frac{\mu_0}{2\pi}\nabla(tf/d) = \frac{\mu_0}{2\pi d}\left[f\vec{V} + t(g\vec{R} + h\vec{S})\right] \tag{5}$$

where $d = R^2S^2-(\vec{R}\cdot\vec{S})^2$, $t = \vec{r}\cdot\vec{V}$, $f = R-2r+(\vec{R}\cdot\vec{S})/R$, $h = 2(\vec{R}\cdot\vec{S}-R^2)f/d-(1/r)+1/R$, and $g = 2(\vec{R}\cdot\vec{S}-S^2)f/d-(1/r)+(2/R)-\vec{R}\cdot\vec{S}/R^3$. The effect of volume currents is implicit: the origin must lie at the center of symmetry for Eq. 5 to be correct. In applying Eqs. 1 - 3 to arbitrary dipoles one must compute trigonometric functions and transform between coordinate systems. As these tasks are avoided with Eq. 5, we have been able to speed field calculations by more than a factor of two.

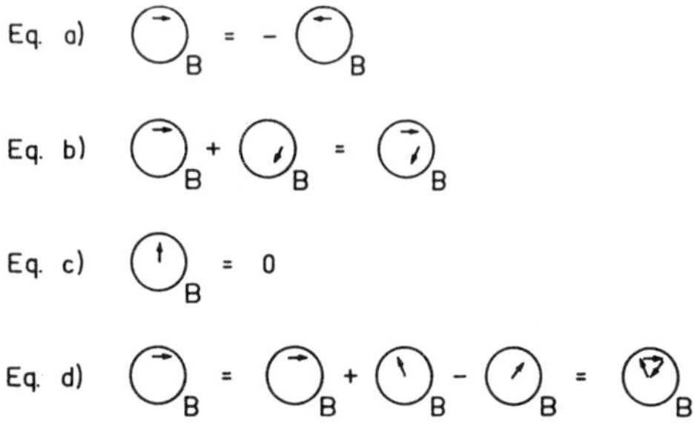

Fig. 3. Equations with symbols that denote the magtic field outside a spherically symmetric source area.

The vector notation of Eq. 5 shows that there is no profound complexity in the analysis of the tangential components of the magnetic field. This encourages us to consider possible advantages of measuring tangential components of the neuromagnetic field. First, restricting the possible analysis to the radial component would require one to orient the magnetometer normal to the inner surface of the skull (Fig. 1b). With a fixed multichannel magnetometer array this may be impossible for all the channels simultaneously. Second, for a multichannel magnetometer in a dewar of fixed size, the ability to obtain information about the magnetic field can be improved by making some channels tangential. We have demonstrated by computer simulations that dipole fitting from tangential field data is as accurate as from radial field data.

THE INVERSE PROBLEM - MINIMUM NORM CURRENT ESTIMATES

The inverse problem is easy to state: Calculate the primary current distribution in the brain from the magnetic field outside the head. The solution is not unique: an infinity of different current distributions can explain a given magnetic field. Which of the distributions should we choose? Or should the information be expressed in some other form, e.g., by multipole expansions (Katila and Karp, 1983) of the field or of the current? The answer depends on the purpose of the experiment. Multipole coefficients, except for the dipole term in the multipole expansion of the current, seem useful only as tools for mathematical analysis and perhaps for the classification of normal and abnormal heart activity. Our goal is to express the neuromagnetic data in the form of estimates of primary currents in the brain. The neuroscientist is not interested in multipole moments, equivalent current loops, not even in the magnetic field or the field gradient. These are mere tools in the determination of the location and other characteristics of brain activity.

We present the inverse problem in the framework of estimation theory: from measured magnetic field values and some prior knowledge, we want to construct the best estimate for the primary current distribution. It is necessary to define explicitly what is meant by "best". We have studied the application of best estimates to MEG in the hope that, eventually, current density maps will be routinely available for the neuroscientist. We have only taken the first step (Hämäläinen and Ilmoniemi, 1984), demonstrating by simulations and by analyses of measured data that brain activity can be located without using source models.

For our discussion, a few definitions are needed. The lead field of the ith magnetometer, designated by L_i, describes the sensitivity of the magnetometer to primary currents. It is convenient to consider L_i as a vector in an infinite-dimensional function space; the corresponding three-dimensional function, $\vec{L}_i(\vec{r})$, is defined by the equation $B_i = \int \vec{L}_i(\vec{r})\cdot\vec{J}^p(\vec{r})dV$, where B_i is the output of magnetometer i and $\vec{J}^p(\vec{r})$ is the primary current density at $\vec{r}$. By defining the inner product of L_i and L_j, $(L_i,L_j) = \int \vec{L}_i(\vec{r})\cdot\vec{L}_j(\vec{r})dV$, we obtain the norm of L_i: $||L_i|| = (L_i,L_i)^{1/2}$. Similarly, the norm of the current distribution J^p is $||J^p|| = (J^p,J^p)^{1/2}$, where $(J^p,J^p) = \int \vec{J}^p(\vec{r})\cdot\vec{J}^p(\vec{r})dV$. The norm of J^p is a measure for the amplitude of the current distribution.

Of all the current distributions that explain the measured magnetic field, the minimum norm estimate is the one with the smallest norm. If no prior knowledge of the primary current distribution is available, except that it is confined to a known volume of space, e.g. the head, the minimum norm estimate is the best in the sense that the expectation value of $||J^p_{est}-J^p||^2$ is minimized. Here, J^p_{est} is the estimate for the primary current J^p. Fig. 4 shows an example: the left median nerve was stimulated at the wrist, and the magnetic field was recorded successively from the locations indicated by crosses. The rectangles on the left contain isocontour maps for the magnetic field 30 and 150 ms after the stimulus. On the right, corresponding minimum norm estimates are shown 35 mm below the plane defined by the magnetometer locations. It was assumed, for convenience, that the primary current is confined to the rectangle. Of all the current distributions in this rectangle that can explain the measured signals, the one shown has the minimum norm. The measured signals provide no information about the difference between the original current and this estimate. Thus, to improve the estimate, additional information about the source current has to be used. Systematic methods to do this should be the goal of future research on the neuromagnetic inverse problem.

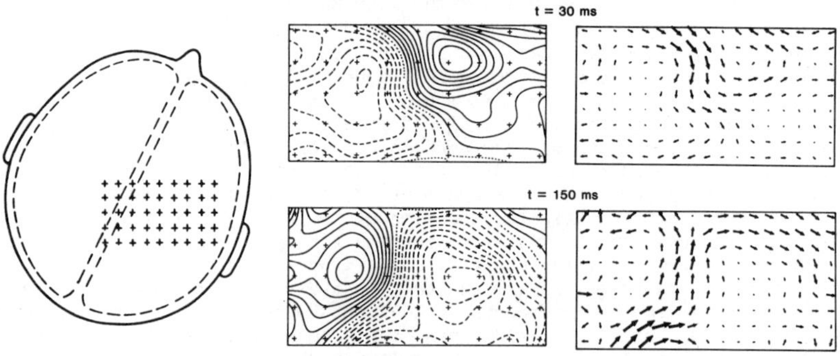

Fig. 4. Isofield maps and minimum norm current estimates. The continuous (positive) and dashed (negative) lines denote flux out of and into the skull, respectively. The crosses show the magnetometer locations, 20 mm apart from each other.

ACKNOWLEDGEMENT

We are grateful to Riitta Hari and Olli Lounasmaa for criticism and encouragement. This work was supported by the Academy of Finland.

REFERENCES

Cuffin, B.N., and D. Cohen (1977). Magnetic fields of a dipole in special volume conductor shapes. IEEE Trans. bio-med. Engng, BME-24, 372-381.

Hämäläinen, M.S., and R.J. Ilmoniemi. Interpreting magnetic fields of the brain - estimates for current distributions, in preparation.

Katila T., and P. Karp (1983). Magnetocardiography: Morphology and multipole presentations. In S.J. Williamson, G.L. Romani, L. Kaufman, and I. Modena (Eds.), Biomagnetism: An Interdisciplinary Approach. Plenum Press, New York. pp. 237-263.

Tripp, J.H. (1983). Physical concepts and mathematical models. In S.J. Williamson, G.L. Romani, L. Kaufman, and I. Modena (Eds.), Biomagnetism: An Interdisciplinary Approach. Plenum Press, New York. pp. 101-139.

SOURCE LOCALIZATION BASED ON EEGs AND MEGs

C.J. Stok*, W.J.J. Kouijzer**, Maria J. Peters*

* Dep. of Electr. Eng. and Dep. of Techn. Physics, Twente Univ. of Techn., Enschede, The Netherlands
** Dep. of Techn. Physics, Eindhoven Univ. of Techn., Eindhoven, The Netherlands

ABSTRACT

The work reported here is concerned with visually evoked electric and magnetic responses in man. We used a current dipole as a model of the source in the brain, and 4 concentric spheres as a model of the volume conductor, i.e. the head. We estimated and compared the positions and strengths of such sources based on electric and magnetic responses.

KEYWORDS

Visual evoked response; MEG; EEG; source localization; isofunction plot.

INTRODUCTION

The localization of sources in the visual cortex, based on a measured potential distribution at the scalp and a magnetic field distribution outside the scalp, depends on the model used. The purpose of this paper is to discuss the following question: Is it possible, using a simple model, to compute the position and strength of the source which could have generated the measured EEGs as well as the measured MEGs? If so, one has an affirmation that the model can be used for source localization procedures in vision research.

We modelled the source by a single current dipole and the head by taking 4 concentric spheres that bound respectively the brain, the cerebrospinal fluid, the skull and the scalp (Cohen, 1983). With the model used, there are two main differences to be expected between MEG and EEG (Williamson, 1981): (1) A radially oriented dipole gives rise to a scalp EEG but produces no MEG outside the head. (2) The MEG generated by a tangentially oriented dipole is deformed less by volume currents than the EEG.

When the experimental and computed results of both the EEGs and the MEGs are displayed as isofunction plots, we obtain maps which are suited for comparison.

MEASURING METHOD

The visually evoked electric and magnetic responses were recorded from two subjects with normal vision. The steady-state responses were obtained using a 4 Hz reversal stimulus: A checkerboard was presented on a TV screen during 120 ms. During the next 120 ms the reversed checkerboard (i.e. black and white squares interchanged) was shown. The transient responses were obtained by using a 1.25 Hz appearance-disappearance stimulus: A checkerboard was presented for 300 ms every 800 ms. Responses caused by light intensity flashes from both appearance and disappearance were avoided by keeping the mean luminance of the stimulus constant. The visual angle subtended by the TV screen was $3^{o}.3 \times 2^{o}.5$; the visual angle subtended by each check in the pattern was $0^{o}.25 \times 0^{o}.25$. The pattern had a contrast of 80 % and was displayed with a mean luminance of 60 cd m^{-2}. The subject was asked to focus on a red light-emitting diode which was attached to one of the edges of the monitor, in order to obtain half field stimulation.

During magnetic measurements the subject lay down on a wooden bed looking downward at a surface coated mirror that reflected the TV screen. The display was placed at a distance of 5.4 m in order to avoid disturbances in the magnetic field. The MEG was measured by a symmetric second-order gradiometer with a baseline of 60 mm and a coil diameter of 30 mm, coupled to a SQUID. The axis of the gradiometer was oriented vertially, with a distance of approximately 14 mm between the pick-up coil and the subject's scalp at the inion. The head of the subject was fixed with respect to the bed by supporting the forehead and using a mouthpiece which was made to fit the front teeth. The subject was moved with respect to the gradiometer by transverse and lateral movements of the bed with an accuracy of 2 mm. Thus all magnetic measurements were performed in a horizontal plane normal to the axis of the gradiometer. The measurement points (i.e. 25 points for subject KS and 30 for WK) were spaced 20 mm apart on a grid with the origin over the inion. The surfaces of a sphere around and a plane nearby the subjects head, both offer measuring positions which are reproducable and easy to define. A disadvantage of the plane measuring area is that the sensitivity in most points is decreased as compared to the sphere, because the distance to the source is larger. In our case however the sources are in the visual cortex which means that the plane does not need to be large and the loss of sensitivity does not pose serious problems. Furthermore the plane allows for a fast and convenient measuring procedure and when a more realistic model of the head is used, which is a subject of our current research, the concept of a normal field component on a sphere has no significance.

The MEG signal was prefiltered (0.3 - 200 Hz) and amplified before it was applied to a digital filter, with a passband of 0.3 - 35 Hz for reversals and of 0.3 - 95 Hz for app.-disapp. measurements. In each position 200 responses were averaged. Two EEGs were recorded during some of the magnetic measurements in order to control the stability of the response. The reproducability was tested by repeating some of the recordings, especially when after a pause the measurements were continued.

Under conditions comparable to those mentioned above, 16 unipolar EEGs were recorded of which 8 simultaneously. Electrodes were spaced 40 mm apart. The ground electrode was placed at the vertex and the reference electrode was placed on the forehead. The EEG was filtered from 0.3 - 100 Hz. Again 200 responses were averaged on-line.

Isofunction plots were computed off-line, using a computer program that increased the number of data points by interpolation (Akima, 1978) and drew contours according to this extended data set (Snijder, 1978). In the isomagnetic field component maps, lines were drawn that connect those points having an equal value of the effective magnetic induction, which is defined as the total magnetic flux through the gradiometer divided by the area of the pick-up loop. In the isopotential maps lines were drawn to connect those points with equal potential measured with respect to the reference electrode. The mutual distances between the measuring electrodes on the scalp were conserved in the plane in which the map was constructed.

MODEL

We calculated the distribution of the electric potential and the magnetic field which is generated by a current dipole located in a set of 4 concentric spheres bounding regions with different conductivities. For the magnetic field this problem is equivalent to that of a current dipole in a single homogeneous sphere. The fundamental mathematical expressions describing the magnetic field were derived by Grynzpan and Geselowitz (1973). The formulas for the electric potential distribution are given by Cuffin and Cohen (1979). The conductivities and radii were taken from the same article, but the radii for subject KS were adapted according to transverse NMR scans of his head. A computer program, based on the expressions mentioned above, was used to calculate (1) the magnetic field distribution in a plane outside the outermost sphere and (2) the electric potential on the surface of this sphere, caused by a current dipole inside the innermost sphere. From these results we calculated the effective magnetic induction for the gradiometer used in our experiments, and the electric potential with respect to a fixed point on the outermost sphere (i.e. the reference electrode). The isofunction contourplots of the calculated magnetic and electric data were computed in the same way as those of the measured data. Figure 1 shows the model of the head and the relative position of the plane (M) where the MEG was calculated.

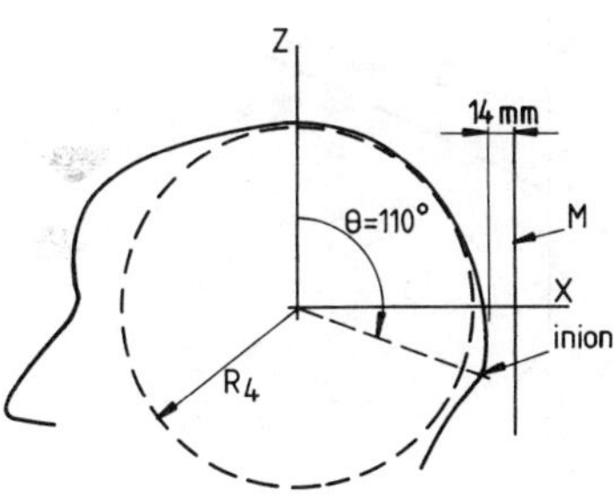

RADII	SUBJECTS WK	KS
R_1 =	79 MM	63 MM
R_2 =	81	65
R_3 =	85	71
R_4 =	88	75

CONDUCTIVITIES

$\sigma_1 = 3.3 * 10^{-3}$ $(\Omega.\mathrm{CM})^{-1}$
$\sigma_2 = 10.0 * 10^{-3}$
$\sigma_3 = 4.2 * 10^{-5}$
$\sigma_4 = 3.3 * 10^{-3}$

Fig. 1. A sphere can be fitted to the back part of the head. The inion and the plane in which the MEG was calculated (M) are indicated. Regions 1 through 4 model respectively the brain, the cerebrospinal fluid, the skull and the scalp.

RESULTS AND DISCUSSION

For instants of time during which the responses showed a maximal amplitude, we found MEG and EEG patterns which in some cases could be described by our model. Two such cases will be discussed.

Figure 2 shows measured and computed MEG and EEG data for subject KS. Details concerning the stimulus conditions are given in the legends. The contourstep in the measured data is for MEG 70 fT and for EEG 0.5 uV. The cross indicates the position of the projection of the inion, the star indicates the projection of the center of the spheres and the arrow the projection of the dipole. Compare the locations of the inion in the contourplots of the EEGs. The best fit for EEG and MEG simultaneously is shown in the plots and was obtained by using a dipole that was oriented in such a way that the ratio of the tangential and the radial component was

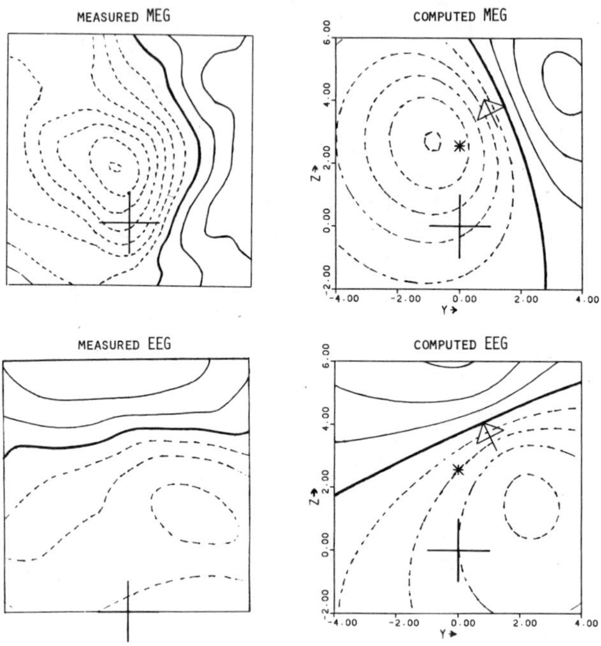

Fig. 2. Experimental and computed contourplots of MEG and EEG of subject KS. Averaged response to app.-disapp. stimulus. Left visual half field stimulation. Time: 110 ms after onset. Positive (negative) potential differences and magnetic flux in negative (positive) x-direction are indicated by solid (dashed) contourlines. Axis measures are in cm.
The position of the dipole in spherical coordinates is R=60 mm, $\theta=80^0$, $\phi=10^0$. The dipole direction is given by $\theta=51^0$, $\phi=-135^0$.

3:2. A good fit could be obtained for the MEG, whereas the EEG seems to be shifted 2 cm to the top of the head. The source is located in the region of the visual cortex, in the right hemisphere as might be expected because of the half field stimulation that was used.

Figure 3 shows data for subject WK. Details concerning the stimulus conditions are given in the legends. The contourstep in the measured data is for MEG 70 fT and for EEG 2 uV. The cross, the star and the arrow have the same meaning as in figure 2 (resp. inion, center of sphere, dipole). In this case the EEG seems to be mainly generated by the radial component of a dipole. However, the MEG shows that the dipole also has a tangential

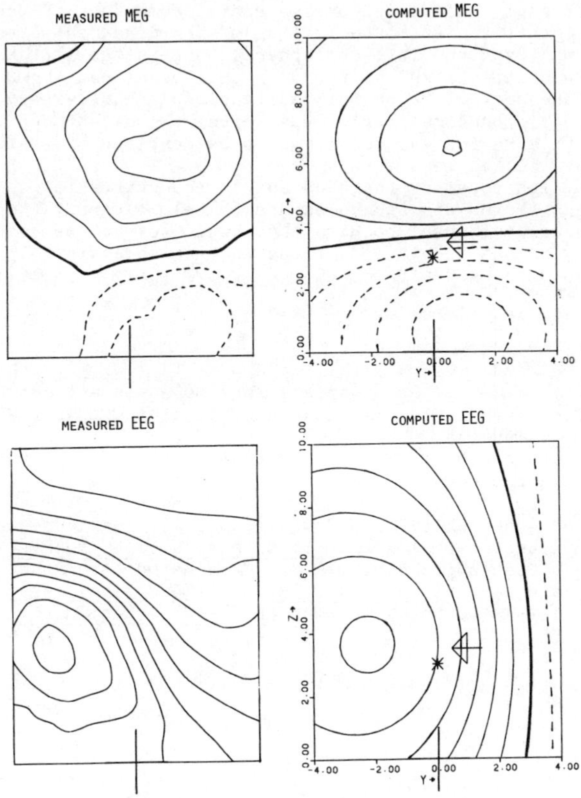

Fig. 3. Experimental and computed contourplots of MEG and EEG of subject WK. Averaged response to reversal stimulus. Left visual half field stimulation. Time: 100 ms after reversal. Solid and dashed lines: see legends fig. 2. The position of the dipole in spherical coordinates is R=55 mm, $\theta=85^0$, $\phi=10^0$. The dipole direction is given by $\theta=90^0$, $\phi=-60^0$.

component. The position of the dipole was estimated according to these magnetic results. After fitting the maps by choosing the component values, the ratio of the tangential and radial components was found to be 3:1, indicating that the dipole is mainly tangential after all. Note the depth of the source and the position of the extreme in the potential map.

We did not use extensive procedures to fit the experimental and theoretical results. Nevertheless, in the results shown, we could not move a dipole more than 2^{0} in θ- or ϕ-direction, nor 5 mm in R-direction, without reducing the overall fit considerably.

Figure 3 shows that the MEG was useful in estimating the location of the source in a case were this could not be done easily using the EEG, because the EEG was strongly influenced by the radial component of the dipole. We noticed before that the absolute maximum of the measured potential distribution was over the left hemisphere, whereas the left half of the visual field was stimulated. We found in this case that the dipole lies very deep in the cortex. Further investigations are necessary to determine whether the active cortical region was actually located deep in the brain, or else that this region was too large to be desribed by a single current dipole and was located less deep.

A more detailed model of the head and/or the source is necessary to attenuate the observed discrepancies and to help clearify those MEGs and EEGs among our experimental results that could not yet be described satisfactorily by the model used. Although this model is helpfull in some cases, it should be used with care.

REFERENCES

Akima H. (1978). A method of bivariate interpolation and smooth surface fitting for irregularly distributed data points [E1]. ACM Trans. Math. Software, 4, 160-164

Cohen D. and Cuffin B.N. (1983). Demonstration of useful differences between magnetoencephalogram and electroencephalogram. Electroenc. & clin. Neur. 56, 38-51

Cuffin B.N. and Cohen D. (1979). Comparison of the magnetoencephalogram and the electroencephalogram. Electroenc. & clin. Neur. 47, 132-146

Grynzpan F. and Geselowitz D.B. (1973). Model studies of the magnetocardiogram. Biophys. J., 13, 911-925

Snijder W.V. (1978). Contour plotting [J6]. ACM Trans. Math. Software, 4, 290-294

FLUX REVERSAL PHENOMENA IN SPATIAL DISTRIBUTIONS OF THE MAGNETOENCEPHALOGRAMS

Shoogo Ueno, Hitoshi Wakisako, and Koosuke Harada

Department of Electronics, Kyushu University
Fukuoka, 812, Japan

ABSTRACT

The head consists of different materials with different electrical conductivities such as brain tissues, skull, ventricles and brain lesions. The influence of these inhomogeneities on the spatial distributions of electrical potentials and magnetic fields over the surface of the head is studied by means of a computer simulation. It is found that the spatial patterns of EEGs and MEGs are influenced by the inhomogeneities in the head. The computer simulation suggests that flux reversal phenomena can appear in spatial MEG patterns when pathological situations of brain lesions are changed from edema to calcification.

KEYWORDS

Biomagnetism; MEG imaging; source localization; isofield contour map; computer simulation; flux reversal; inhomogeneity; current dipole; brain lesion; ventricle.

INTRODUCTION

Ionic currents of neuroactivities in the brain produce weak magnetic fields around the head as well as the electroencephalogram (EEG) at the scalp. These magnetic fields, called magnetoencephalograms (MEGs) were first measured by Cohen (1968, 1972, 1975). Since the successful measurements of the MEG, both spontaneous and evoked magnetic fields of the human brain have been studied well to clarify the origin of the brain electrical activities. During the past few years, evoked magnetic fields have been measured mostly to study source localization, and rather substantial advances in understanding the functional organization of the human brain have been made through a technique of isofield contour maps of evoked magnetic fields (Hari et al., 1980; Williamson and Kaufman 1981; Romani et al., 1982; Okada et al., 1982; Okada, 1983; Williamson and Kaufman, 1983). Most of these investigations were concerned with estimation of position, orientation and depth of the current sources which were modeled as single current dipoles in homogeneous volume conductors. However, the head consists of different materials with different electrical conductivities such as brain tissues, skull, ventricles, brain lesions, etc.. Brain lesions themselves can be also expected to have different conductivities according to the various pathological situations such as edema and calcification. In the present paper, the influence of these inhomogeneities on the isofield contour maps is studied by means of a computer simulation. Theoretical models of the MEG have been reported by Cuffin and Cohen (1977) and others. However, no investigations on the effects of the inhomogeneities of the ventricles and brain lesions on the MEG patterns have been reported previously.

INHOMOGENEOUS MODEL OF THE HEAD

The head is assumed to be a spherical volume conductor with an electrical conductivity σ_1, in which an inhomogeneous area with different conductivity σ_2 is located as shown in Fig. 1. The hatched area represents inhomogeneous region with a conductivity σ_2. The upper part shows a hemispherical model, the middle part shows a brain lesion model, and the lower part shows a ventricular model. An internal current source in the brain is assumed to be a current dipole which is marked with an arrow in the sphere. We assume that the current dipole represents any types of current sources of neural activities in the brain. The dipole represents either a source of spontaneous EEG and MEG activities, or evoked potentials and magnetic fields, or endogenous potentials and magnetic fields.

(a) Hemispherical model
The brain is covered by the skull and located in the upper part of the head. The lower part of the head is occupied by other organic systems such as the nasal and oral cavities. Let us divide the head into two hemispheres; the upper hemisphere of the cerebral cortex and the lower hemisphere occupied by the other. The conductivity of the lower hemisphere may be lower than the upper. A current dipole is located at (X,Y,Z)=(0.25,0.25,0.3535) in the left frontal region of the unit sphere and is oriented in the direction parallel to the Inion-Nasion axis, or X axis.

(b) Brain lesion model
The electrical conductivity of brain lesions varies with pathological situations such as edema and calcification. Edema gives a high conductivity, whereas calcification is expected to be a low conductivity. A brain lesion is assumed to

a)

b)

c)

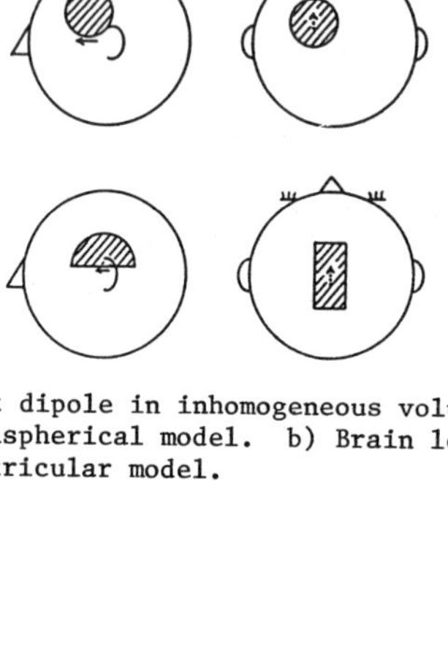

Fig. 1. Current dipole in inhomogeneous volume conductor.
a) Hemispherical model. b) Brain lesion model.
c) Ventricular model.

be a small sphere with a radius 0.3 and positioned at (X,Y,Z)=(0.25,0.25,0.3535). A current dipole is located at (X,Y,Z)=(0.25,0.25,0.05), parallel to X axis.

(c) Ventricular model

The cerebral ventricles are located in the center of the head, and the inside of ventricles is filled with the cerebrospinal fluid. The wall of the ventricle is made with bone material which gives a low conductivity, whereas the cerebrospinal fluid inside the ventricles gives a high conductivity. Let us assume that the ventricles are represented by a simple shape as shown in Fig.1 c), although the ventricles consist of plural chambers with very complex patterns. The model ventricle, with a radius 0.4 and bottom size 0.4 x 0.8, is positioned at (X,Y,Z)=(0.0,0.0,0.1). A current dipole is located at (X,Y,Z)=(0.0,0.0,0.05), parallel to X axis.

METHODS

Electrical potentials

It is assumed that an inhomogeneous volume conductor can be divided into a number of regions in each of which the conductivity is uniform and isotropic. If current sources J^i exist in the conductor, the electrical potential V at any point in a region of conductivity σ is given by

$$V = -\frac{1}{4\pi\sigma}\int \frac{1}{r}\nabla\cdot J^i\, dv - \frac{1}{4\pi\sigma}\int (\sigma'-\sigma'')\, V\, \nabla(\frac{1}{r})\cdot dS_j \tag{1}$$

where r is the distance from an arbitrary point to the element of volume or area, σ' and σ'' are the conductivities of two different adjacent regions separated by the surfaces S_j, and the vector dS_j is directed from the primed region to the double primed one.

Barr et al. (1966) have introduced a simultaneous equation to solve Eq.(1) on a computer as follows:

$$A\cdot V = B \tag{2}$$

The elements of the matrices A and B are given by

$$[A]_{ij} = (\sigma'_j - \sigma''_j)\,\Delta\Omega_{ij} - 4\pi\sigma'_i\delta_{ij} \tag{3}$$

$$[B]_i = \int \frac{\nabla\cdot J^i}{r_i}\, dv \tag{4}$$

where $\Delta\Omega_{ij}$ is the solid angle of the j-th segment as seen from the observation point just inside the i-th segment, δ_{ij} is Kronecker's delta, and r_i is the distance from the source to the i-th segment.

In our brain model, potentials on the surface of the sphere can be calculated by Eq.(2).

Magnetic fields

The expression of magnetic fields H generated outside an inhomogeneous volume conductor by internal current sources has been derived by Geselowitz (1970) as follows:

$$H = \frac{1}{4\pi}\int J^i \times \nabla(\frac{1}{r})\, dv + \frac{1}{4\pi}\sum_j \int (\sigma' - \sigma'')\, V\, \nabla(\frac{1}{r}) \times dS_j \tag{5}$$

where r is the distance from an external point where H is evaluated to an element of volume dv inside the body. The first term gives the field that arises from the sources J^i, whereas the second term gives the effect of inhomogeneities and external boundary on the magnetic field.

Let us rewrite Eq.(5) in summation form:

$$H = Hp + \frac{1}{4\pi}\sum_i (\sigma'_i - \sigma''_i)\, V_i\, \nabla(\frac{1}{r_i}) \times \Delta S_i \tag{6}$$

where H_p is the field determined by the Biot-Savart law, V_i is the electrical potential at the i-th segment and is calculated by Eq. (2), r_i is the distance from the i-th segment to the field point and ΔS_i is the area vector of the i-th segment directed from the region with conductivity σ_i' to the region with conductivity σ_i''. In the numerical calculation, magnetic fields perpendicular to the surface of the sphere were obtained.

Mapping

Electrical potentials and magnetic fields over the surface of the sphere were calculated by changing the conductivity ratio σ_2/σ_1. In the field mapping, the surface of the upper hemisphere was projected into the unit circle on the two dimensional plane. The values of electrical potentials were quantized to 11 levels, 0 - 9 plus X instead of 10, and magnetic fields were quantized to 21 levels, from the North pole to the South pole (N,9,8,$\cdots$,2,1,0,a,b,$\cdots$,h,i,S).

RESULTS AND DISCUSSION

In the numerical calculation, the conductivity σ_1 of the sphere always equalled 1.0, and the conductivity of the hatched area σ_2 was varied.
Spatial distributions of electrical potentials (EEG) and magnetic fields (MEG) in the case of the hemispherical model were mapped as shown in Fig. 2. The upper part shows potential distributions and the lower part shows the magnetic fields. In each map, the top of the circle corresponds to Nasion and the bottom is Inion. Compared with the homogeneous case (σ_2/σ_1=1), the maximum value of the potential increases with the decrease of σ_2, whereas the isofield contour maps of the magnetic fields are not so influenced by the inhomogeneity.

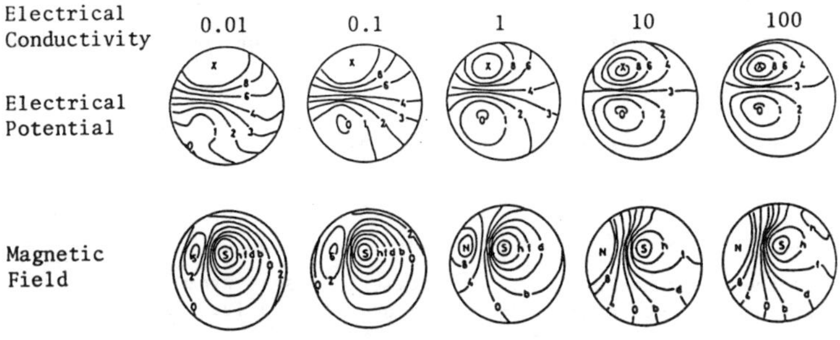

Fig. 2. Hemispherical model.

Figure 3 shows calculated results in the brain lesion model. The upper part shows potential distributions and the lower part shows the magnetic fields. Maps in cases of σ_2/σ_1=10,100 correspond to edema, and maps in cases of σ_2/σ_1= 0.1,0.01 correspond to calcification. The patterns of potential distributions are analogous to the homogeneous case, although the amplitudes decrease with the increase of σ_2. In contrast with the potential distributions, the isofield contour maps of magnetic fields are influenced dramatically by the inhomogeneity. The poles of the North and South are reversed in the range between σ_2/σ_1=10 and σ_2/σ_1= 0.1 . The results suggest that flux reversal phenomena can appear in isofield contour maps of evoked magnetic fields when pathological situations of brain lesions are changed from edema to calcification.
Figure 4 shows calculated results in the ventricular model. The upper part shows potential distributions and the lower part shows the magnetic fields. The spatial patterns of potential distributions are analogous to the homogeneous case, whereas the magnetic fields are reversed in the range between σ_2/σ_1=10 and σ_2/σ_1=1 .

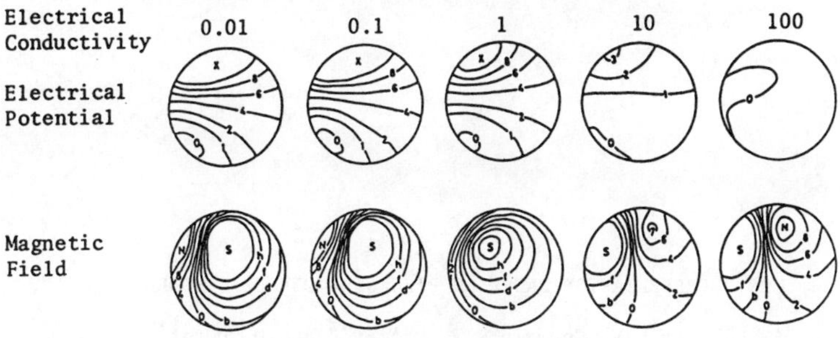

Fig. 3. Brain lesion model.

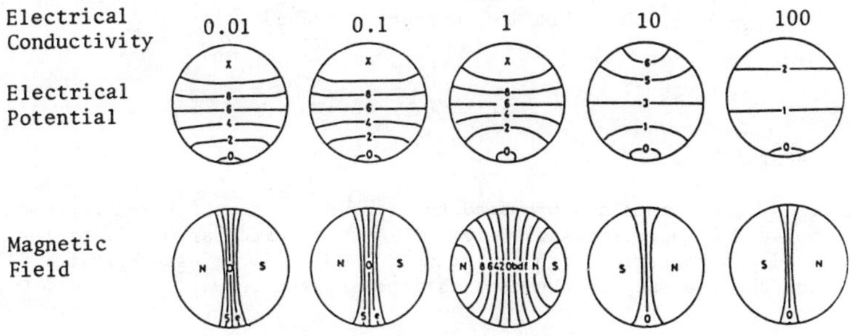

Fig. 4. Ventricular model.

The change in spatial patterns of the MEGs simulated by our brain model can be explained by the behavior of ionic currents flowing through the areas with different conductivities in the head. The pathways of volume currents are strongly affected by the inhomogeneous medium which results in the change of spatial patterns of isofield contour maps. The degree of variation is dependent on the inhomogeneities and also dependent on the geometrical relationship between the current sources and the boundary of the inhomogeneous medium.

REFERENCES

Barr, R.C., T.C.Pilkington, J.P.Boineau, and M.S.Spach (1966). IEEE Trans. Biomed. Eng., BME-13, 88-92.

Cohen, D. (1968). Science, 161, 784-786.

Cohen, D. (1972). Science, 175, 664-666.

Cohen, D. (1975). IEEE Trans. Magn., MAG-11, 694-700.

Cuffin, B.N. and D. Cohen (1977). IEEE Trans. Biomed. Eng., BME-24, 372-381.

Geselowitz, D.B. (1970). IEEE Trans. Magn., MAG-6, 346-347.

Hari, R., K.Aittoniemi, M.L.Jarvinen, T.Katila, and T.Varpula (1980). Exp. Brain Res., 40, 237-240.

Okada, Y., S.J.Williamson, and L.Kaufman (1982). Int. J. Neurosci., 17, 33-38.

Okada, Y. (1983). Il Nuovo Cimento, G.L.Romani and S.J.Williamson (Eds), 2D, 379-409.

Romani, G.L., S.J.Williamson, and L.Kaufman (1982). Science, 216, 1339-1340.

Williamson, S.J. and L.Kaufman (1981). Biomagnetism, S.N.Erne, H-D.Hahlbohm and H.Lubbig (Eds), Walter de Gruyter, Berlin, 353-402.

Williamson, S.J. and L.Kaufman (1983). IEEE Trans. Magn., MAG-19, 835-844.

MEG REPRESENTATION BASED ON PIECEWISE CIRCULAR MODELS

P. Nicolas*, V. Germain*, T. Thomas** and D. Duret**

*Unité INSERM U. 84, Hôpital de la Salpêtrière, Paris, France

**CENG-LETI/MEM, Grenoble, France

ABSTRACT

New geometrical algorithms are proposed to improve the computerized representation of dipolar MEG sources distributions. Piecewise circular models of the skull are used, in connexion with the standard EEG 10-20 reference system. The stability and the accuracy of the computations are studied.

KEYWORDS

MEG ; dipolar sources ; computing geometrical algorithms ; piecewise circular approximation ; 3-D imaging.

INTRODUCTION

It is generally admitted that one of the main features of the MEG measurements is to be a good means for approximating the intracranial locations of certain neuroelectrical activities (Kaufman and Williamson, 1982 ; Romani, Williamson and Kaufman, 1982 ; Barth and co-workers, 1982, 1984). More precisely, if the neuromagnetic fields are well fitted to dipolar brain current models, which is often the case, then we are able to compute "approximately" the location of the barycentre of the sources involved. Although the reliability of such computations remains controversial (Cuffin, 1982), we are accustomed to do them, in using classical representations of dipolar fields arising, either from a conducting sphere, or from a conducting half space (Williamson and Kaufman, 1981).

The most critical problem then comes from the structure of the brain itself, which is not a sphere, nor a half-space. Yet, the architecture of the brain is so complex that any small error in the source localization may cause our physiological interpretation to fall completely wrong. Let us recall only that the grey matter is a few millimeters thick. Therefore, any error of the same order in computing the location of a given focused activity should lead us to localize it either inside the white matter or outside the brain ...

Such dramatic errors may occur at each step of the measurement sessions and at each level of further data processing. That was the reason for which we felt useful to search for new computing algorithms intended to improve the accuracy of the MEG imaging and the reliability of our MEG based interpretation.

PIECEWISE CIRCULAR REPRESENTATION OF THE SKULL

With few exceptions, the localization of the neuromagnetic fields dipolar sources is based on the so-called "spherical model" of the skull.

Plane models (Williamson and Kaufman, 1981) are seldom taken as reference, except for superficial sources, whereas Cuffin and Cohen (1977) proposed spheroïdal systems that are not seemingly in use to date. In the spherical model, either the head or part of it are assumed to be globally spherical, with average radius estimated from measures of their curvature (Kaufman and Williamson, 1982). This allows to know approximately the angular span between two opposite extrema of the field, from which the depth of the expected dipole is deduced.

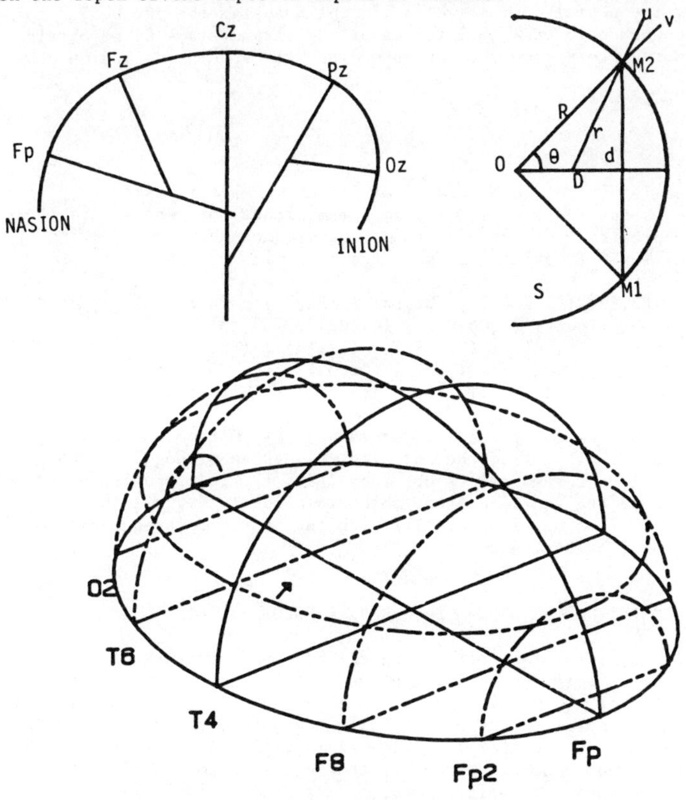

Fig. 1. Piecewise circular representations. Top left : cross section through the midline, showing different arcs measured and their computed radii of curvature. Top right : geometrical data and parameters. At the bottom : 3-D view of the skull, showing the 3 orthogonal sections (solid lines) from which other sections (dotted lines) may be interpolated.

As already pointed out, the accuracy of the result closely depends on the errors coming from both model and computation. It may be improved in using a piecewise circular matching of the 10-20 EEG electrode system (Jasper, 1958), which is commonly used in both EEG and MEG studies. This system is based on standard landmarks that are distributed along different lines which join several anatomical references : frontal, central, parietal, temporal, etc. Each of them are plane, in order to allow representations of different cross sections of the skull, providing an adequate coverage of the head. A 3D-approximated representation may be obtained from the above system as follows. Three orthogonal such cross sections are choosen, on which the arc and the chord between every standard consecutive locations of the electrode line are measured. In this way, we obtain several pairs of numbers (C,A). Let us assume now that the line is circular between every 10-20 consecutive locations. Then, all the arcs are circular. Moreover, they are continuously and tangentially connected each other (there are no vortices on the head : only the normal derivative is discontinuous at the connecting points). Hence, from each pair of actual values of the parameters A,C, we obtain the local radius of curvature R from the corresponding half angular deflexion θ by :

$$\theta = 2\,(A/C)\,\sin\,(\theta/2) \qquad (1)$$

$$R = C/2\theta \qquad (2)$$

The equation (1) is solved by classical iterative algorithms.
This method is applied successively to the midline, the central and the temporal lines. More precisely for each of them, the arc and the chord joining the following pairs of 10-20 locations are measured :

- Midline : (Fp,Fz) ; (Fz,Cz) ; (Cz,Pz) ; (Pz,Oz) ; Oz,inion).
- Central line : (T4,C4) ; (C4,Cz) ; (Cz,C3) ; (C3,T3).
- Temporal line : (Fp,Fp1) ; (Fp1,F7) ; (F7,T3) ; (T3,T5) ;
  (T5,O1) ; (O1,Oz) ; (Oz,O2) ; (O2,T6) ;
  (T6,T2) ; (T2,F6) ; (F6,Fp2) ; (Fp2,Fp).

Intermediate points may be used, if necessary, to improve the accuracy of the approximation. Finally, from the so calculated three orthogonal cross sections, a 3D representation of the head is obtained (Fig. 1), in computing a three-linear interpolation of the skull'surface, which is uniquely determined. After this last operation, any cross section parallel with one of the reference plane may be easily displayed by the computer.

COMPUTING THE DEPTH OF CURRENT DIPOLES

In what follows, it will be still assumed that the curvature of the arcs may be different between every two points on the scalp. Hence, let S be the sphere centered on the origin O, with radius R. A current dipole is assumed to be located inside S, parallel to the z-axis, with its barycentre D situated at a distance d from the surface of S (Fig. 1). It is well-known that the magnetic isofield lines are core-shaped. Moreover, the radial component of the dipolar field vanishes everywhere on the plane (Oz,D) ; its maximum isofield energy line intersects the surface of S in two points M1,M2, that are located symmetrically with regard to this zero plane (Williamson and Kaufman, 1981). Let be B_R this radial component of the dipolar induction. From the Biot-et-Savart law, the following representation for B_R is at our disposal :

$$B_R = C\,\frac{k \times u}{r^2}\,.\,v \qquad (3),$$

where C>0 is constant, being the vectors and scalars defined as follows (Fig. 1)

$$k = (0,0,1) \ ; \ a = 1 - \frac{d}{R} \ ; \ r = R\ (1 + a^2 - 2a\cos\theta)^{1/2} \ ;$$

$$u = \frac{R}{r}\ (\cos\theta - a,\ \sin\theta,\ 0) \ ; \ v = (\cos\theta,\ \sin\theta,\ 0).$$

The problem under consideration is to find d from the only measured arc and chord between M1,M2. For that purpose, we start from the product :

$$k \times u = r^{-1}\ (-\sin\theta,\ \cos\theta - a,\ 0);$$

and let us consider the function f such as :

$$f\ (\theta) = r^{-2}\ k \times u.v$$

$$= -\ aR^{-2}\ (1 + a^2 - 2a\cos\theta)^{-3/2}\ \sin\theta.$$

We must find the extrema of f over the interval $(0, \pi/2)$, i.e. the values of θ, between 0 and $\pi/2$, for which the derivative $df/d\theta$ vanishes. They correspond to the second order algebraic equation:

$$a^2\cos\theta + a\ (\cos^2\theta - 3) + \cos\theta = 0, \qquad (4)$$

which is equivalent to that given by Williamson and Kaufman (1981).
As $\cos\theta$ may be known from the equation (1), the expected solution is given by :

$$a = [3 - \cos^2\theta - [(\cos^2\theta - 3)^2 - 4\cos^2\theta]^{1/2}]\ /\ 2\cos\theta \qquad (5)$$

It is the only convenient root of the equation (4), which satisfies the necessary condition : $a \le 1$.

So we finally obtain :

$$d = R\ (1 - a), \qquad (6)$$

which solves the problem. The dipolar source may be then adequately inserted in the above piecewise circular 3-D representation of the skull, which leads to the 3-D imaging expected.

STABILITY AND ACCURACY

To apply the above algorithms, we must know the domains on which they are stable with regard to the measurement errors. More precisely, the error Δ on the computed results have to be bounded, on some interval as a function of the error ε which is made in measuring the arc A and the chord C.

In practice, for each pair (A,C) and each ε, we must find the maximal error on the results $\Delta(A,C,\varepsilon)$ so that, for example, $\Delta \le 2\varepsilon$. This allows to evaluate the accuracy of the method.

Fig. 2 shows the confidence domain of the depth d, for A = 7 cm as a function of θ (A,C). It can be seen that there is θ_o for which the algorithm (1), (2), (3), .. (6) falls dramatically unstable when θ is smaller than θ_o. In this case, the plane model (PM) provides better accuracy than the piecewise circular model (PCM), because the measurement errors then may give the chord greater than the arc, which explains the notch points (θ_o, $d - \Delta$) in Fig. 2. From both theoretical and numerical calculations, the following conclusion may be drawn : Being the curvature strong enough, the PCM algorithm is the best, if the arc and the chord A, C are known with small error with regard to that which can be accepted on the results (in Fig. 2, d is obtained with accuracy better than 0.1 cm for θ greater than 20°, if the errors on A is less than 0.01 cm). For weak curvatures and poor measurements accuracy, the plane model PM giving $d = 2^{-1/2}A$ must be preferred.

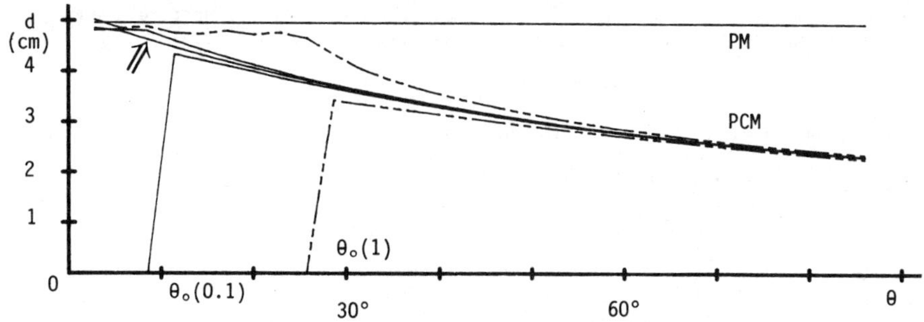

Fig. 2. Errors on the computed depth of dipolar MEG sources in using the PCM as a function of θ. True values of d (arrow) are framed by the upper and lower bounds $d(\theta) \pm \Delta(A,C,\varepsilon)$ for maximal measurement error $\varepsilon = \pm$ 0.01 cm (solid curves) and $\varepsilon = \pm$ 0.1 cm (dotted curves).

CONCLUSION

In the above study, we limited ourselves to the only geometrical problems dealing with 3-D representation of the skull and the source location ; to be quite satisfactory, any algorithm should take into account the overestimates that come from the dimension and other features of the flux tranformer (Williamson and Kaufman, 1981). However, the actual reliability of the MEG 3-D imaging appears to be closely dependent on the accuracy of the geometrical measurement techniques. An attempt should be made for improving them, as well as to standardize the experimental methods and the different data processing in use. This is the only means for doing valid comparisons of results taken in several laboratories.

REFERENCES

Barth, D.S., W. Sutherling, J. Engel Jr., and J. Beatty (1982). Science, 218, 891-894.

Barth, D.S., W. Sutherling, J. Engel Jr., and J. Beatty (1984). Science, 233, 294-297.

Cuffin, B.N. (1982). Abs. 4th. Int. Workshop Biomagnetism (Rome), 78-79.

Cuffin, B.N., and D. Cohen (1977). IEE Trans. Biomed. Engr. BME 24 N° 4, 372-381.

Jasper, H.H. (1958). EEG Clin. Neurophysiol., 10, 371-375.

Kaufman, L., and S.J. Williamson (1982). Ann. N. Y. Acad. Sci. 388, 197-213.

Romani, G.L., S.J. Williamson, and L. Kaufman (1982). Science, 216, 1339-1341.

Williamson, S.J., and L. Kaufman (1981). In Erné, S.N. Hahlbom, and H. Lübbigs (Eds), Biomagnetism. Walter de Gruyter, Berlin, 353-402.

ANALYSIS OF MAGNETOENCEPHALOGRAPHIC DATA OBTAINED WITH A FOUR-CHANNEL SQUID MAGNETOMETER

M.S. Hämäläinen, R.J. Ilmoniemi, J. Knuutila, and K. Reinikainen

Low Temperature Laboratory,
Helsinki University of Technology, 02150 Espoo, Finland

ABSTRACT

We compare three methods of locating brain activity: 1) least-squares fitting of a dipole to the measured field maps, 2) determination of the 'equivalent dipole' from the maxima of isocontour magnetic field maps, and 3) the method of minimum norm estimates for the current distribution.

KEYWORDS

Magnetoencephalography; multichannel magnetometer; equivalent dipole; least-squares fitting; minimum norm estimate; subject noise.

INTRODUCTION

Improvements in the experimental technology for neuromagnetic measurements make it increasingly important to consider data analysis in detail; here we compare three methods for determining the locations of cerebral activity. All the measurements were done with our four-channel SQUID magnetometer, subsequently referred to as 4M (Ilmoniemi, Hari, and Reinikainen, 1984).

THE FOUR-CHANNEL MAGNETOMETER

The four pickup coils of 4M are at the vertices of a regular tetrahedron (Fig. 1a): the lower three are at a distance of 12 mm from the head; the fourth, 12.5 mm higher, provides information about the variation of the magnetic field along the radial direction. Each flux transformer is in the form of a gradiometer (base length 60 mm). The field sensitivity is 22 $fT/\sqrt{Hz}$ for the lower channels and 15 $fT/\sqrt{Hz}$ for the upper one. Figures 1c and 1d show isocontour field maps that were obtained, simultaneously, with the three lower channels and with the upper one, respectively.

SIMULATIONS

A simple method to locate cerebral activity is to determine the tangential

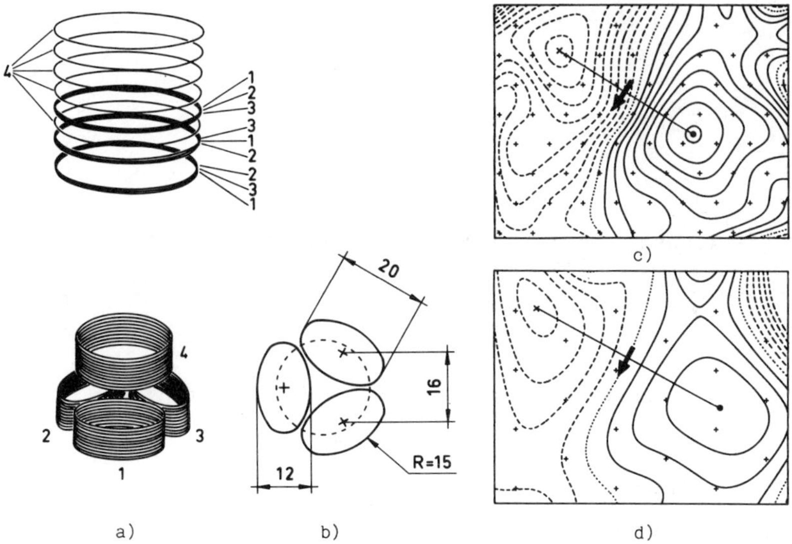

Fig. 1. (a) Pickup and compensating coils of 4M. The axes of the coils are parallel. (b) Pickup coil dimensions (top view). (c) Field map obtained with the lower channels at 80 ms after a 1000 Hz beep to the left ear. (d) The map measured with the upper channel. The contour separation is 40 fT.

current dipole that fits best to the measured magnetic field in the least-squares sense. By computer simulations of measurements with 4M, we investigated the effect of noise on the fitted dipole. Similar studies have been performed for electrical (Gaumond, Lin, and Geselowitz, 1983) and for both electrical and magnetic measurements in the case of 18 current dipoles of fixed location (Cuffin, 1978). We found the least-squares fit by using a constrained minimizing algorithm (Himmelblau, 1972).

In our first set of simulations, the grid of 21 positions of 4M (Fig. 2a) was kept fixed. The accuracy of determining the location, orientation, and strength of a test dipole in a spherically symmetric volume conductor was studied as a function of the location and orientation of the dipole, and as a function of noise. The test dipole was placed at five positions (Fig. 2a), each 30 mm below the scalp, and oriented 0°, 30°, 60°, and 90° with respect to the long side of the measurement area. The center of symmetry of the volume conductor was assumed to be 140 mm from the lower pickup coils.

Since the accuracy of the least-squares fit did not depend on dipole orientation, the results were averaged over the orientations. Figure 2a shows the location error at each dipole site as a function of noise, which is expressed as percentage from the maximum value of the radial field component in the measurement area. At 10 % noise level the location error is smaller than 2 mm. At this noise level, the error in orientation is less than 5°, and at 30 %, less than 10°. The amplitude of the dipole could be determined with an error less than 20 % in each case.

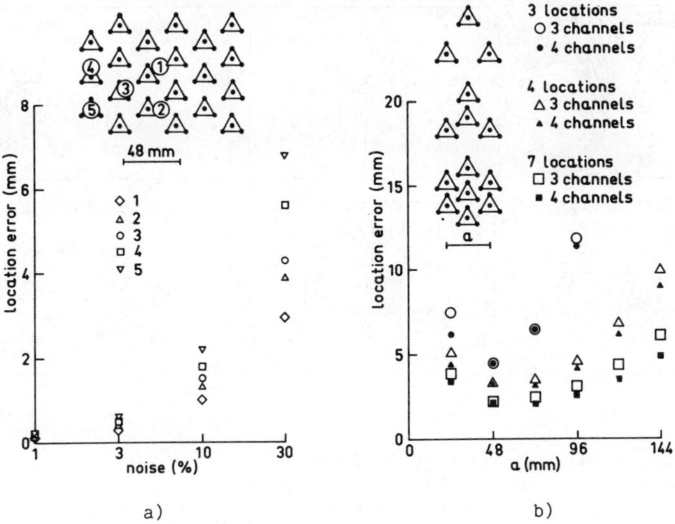

Fig. 2. (a) Location error as a function of noise. On top, the measurement grid (a triangle signifies 4M) and the test dipole locations 1 - 5 are shown. Each value is an average from 24 simulations. (b) Location error for different magnetometer configurations (shown on top) as a function of separation between 4M positions. Each value is an average from 40 simulations.

In the second set of simulations, the measurement grid was varied; the noise level was kept fixed at 10 %. Least-squares fits were made for 3, 4, and 7 locations of 4M (Fig. 2b), using only the lower channels or using all channels. The test dipole was kept fixed at 5 mm off the symmetry axis of the measurement grid. The location errors for all six magnetometer configurations are shown in Fig. 2b as a function of the grid constant a. The optimum separation is 50 - 60 mm. Near optimum, the accuracy with only three channels is the same as with four channels but off the optimum the four-channel measurement is superior.

FITTING A DIPOLE TO MEASURED DATA

In the experiments of Sams and co-workers (1984), a sequence of 1000 Hz beeps was delivered to the subject's left ear, interrupted at random intervals by tones of 1030 Hz. Data for the 1000 Hz tones measured from two subjects 120 ms after the beep are shown in Fig. 3a. The 'equivalent dipoles' were determined from the extrema (Williamson and Kaufman, 1981) and by least-squares fitting (shaded and black arrows, respectively). It is seen that the method of extrema can result to an error of 10 mm and consequently, to misinterpretations.

Figure 3b shows field patterns calculated from the equivalent dipoles; in Fig. 3c, the differences between the measured field patterns and those calculated from the equivalent dipoles are shown. No dipolar structures are evident in the difference maps. This shows that the 120 ms peak in the

S1

S2

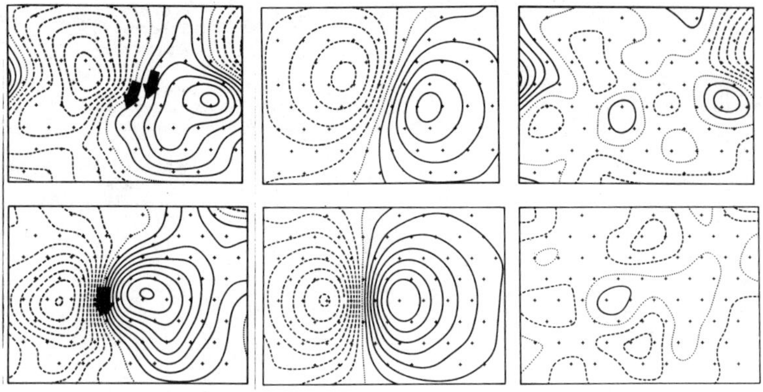

a) b) c)

Fig. 3. (a) Field patterns evoked by tone bursts. The contour separation is 50 fT. The 'equivalent dipole' determined from the extrema (shaded arrow) is seen to miss the equivalent dipole (black arrow) found by least-squares fitting. (b) Field patterns calculated from the equivalent dipoles (azimuthal equidistant projection, sphere radius 150 mm). (c) The differences between a and b. S1, S2 = Subject 1, 2.

auditory evoked response can be reasonably explained by a current dipole located in the primary auditory cortex. However, some peaks are left in the map. They tend to comprise values from triplets of simultaneous measurements, suggesting that an important contribution to noise comes from the variability of measured responses from session to session.

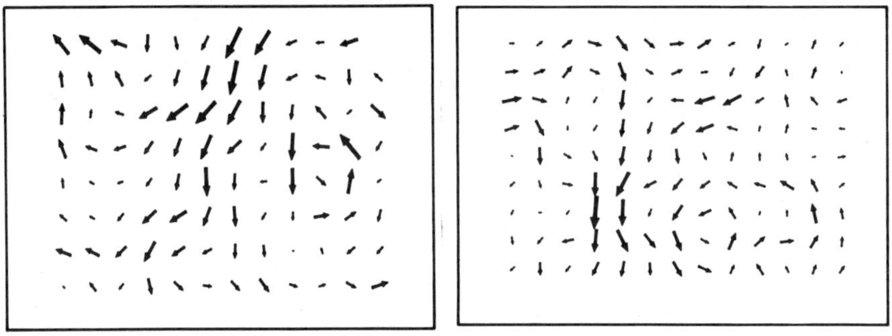

a) b)

Fig. 4. Minimum norm estimates. The primary current was assumed to lie in a plane 20 mm from the scalp. (a) The estimate constructed from the data of Fig. 3a, Subject 1. (b) The estimate for Subject 2.

MINIMUM NORM ESTIMATES

The minimum norm estimate for the current distribution (Ilmoniemi, Hämäläinen, and Knuutila, 1984) is obtained as a weighted sum of the magnetometer lead fields. The norm of a current distribution $\vec{J}(\vec{r})$ is $||J|| = \sqrt{\int \vec{J}(\vec{r}) \cdot \vec{J}(\vec{r})\, dV}$. Minimum norm estimates were calculated for the data of Fig. 3a. The primary current was assumed to lie in a plane at a depth of 20 mm from the scalp. The estimated primary current distributions are displayed in Fig. 4. The arrows show the direction and amplitude (length of arrow) of the estimate. The estimates are seen to be localized with a resolution of the order of 1 cm to the same area where the equivalent dipole is. The important difference to the current dipole is that no assumption of localized primary current was made. The minimum norm estimate is the best that can be constructed when the only a priori assumption is that the current is confined to some source area.

CONCLUSION

Our results indicate that the determination of the equivalent current dipole from the field extrema can lead to a large error. If the assumptions implicit in the dipole model are justified and the spherical model is correctly applied, one can approach accuracies of the order of a millimeter in locating the dipole. If the assumptions are poorly founded, biased results will be obtained. For these cases, minimum norm estimates are suggested.

ACKNOWLEDGEMENT

We acknowledge Olli Lounasmaa for criticism and support; Riitta Hari, Elina Kaukoranta, and Mikko Sams, for the measurements; and Jaripekka Salminen, for computer programs. This work was supported by the Academy of Finland.

REFERENCES

Cuffin, B.N. (1978). On the use of electric and magnetic data to determine electric sources in a volume conductor. Ann. Biomed. Eng., 6, 173-193.

Gaumond, R.P., J.-H. Lin, and D.B. Geselowitz (1983). Accuracy of dipole localization with a spherical homogeneous model. IEEE Trans. bio-med Engng, 30, 29-34.

Himmelblau, D.M. (1972). Applied nonlinear programming. McGraw-Hill, New York.

Ilmoniemi, R.J., M.S. Hämäläinen, and J. Knuutila (1984). The forward and inverse problems in the spherical model. This volume.

Ilmoniemi, R., R. Hari, and K. Reinikainen (1984). A four-channel SQUID magnetometer for brain research. Electroenceph. clin. Neurophysiol. In press.

Sams, M., R. Hari, E. Kaukoranta, K. Reinikainen, K. Alho, M.S. Hämäläinen, R.J. Ilmoniemi, R. Näätänen, and J. Salminen (1984). Magnetic responses to pitch changes in a sequence of auditory stimuli. This volume.

Williamson, S.J., and L. Kaufman (1981). Biomagnetism, J. Magn. Magn. Mat., 22, 129-201.

3-D NEUROMAGNETIC LOCALIZATION OF SOURCES OF INTERICTAL ACTIVITY IN CASES OF FOCAL EPILEPSY*

G.B. Ricci°, R. Leoni, G.L. Romani, F. Campitelli°°, S. Buonomo°°° and I. Modena°°°°

Istituto di Elettronica dello Stato Solido - C.N.R.
Via Cineto Romano 42 - 00156 Roma, ITALY
°Dipartimento Scienze Neurologiche, Università "La Sapienza", Roma
°°Elettronica S.p.A., Roma
°°°Dipartimento di Neurologia, Ospedale Generale, Formia
°°°°Dipartimento di Fisica, Università "Tor Vergata", Roma

ABSTRACT

Several cases of focal epilepsy have been extensively studied by the neuromagnetic method. In all cases but one the epileptic focus was 3-D localized analyzing data by the Relative Covariance method and using a simple equivalent dipolar fitting source. When possible, an averaging technique was also used, and a quite good agreement between the two procedures was observed. The comparison between the neuromagnetic localizations and those available from x-ray analyses and, in two cases, from surgical findings confirmed the adequacy of the model for foci located in the central, parietal and occipital regions. The model seems partially satisfactory for the frontal region, and is questionable for temporal lobe foci.

KEYWORDS

Focal epilepsy; neuromagnetism; current dipole; Relative Covariance.

INTRODUCTION

Previous studies have shown that magnetoencephalographic (MEG) measurements well identified areas of pathologic activity in epileptic patients (Carelli and colleagues, 1983; Modena and colleagues, 1982; Ricci, 1984; Ricci and colleagues, 1981). Recently, three-dimensional localization of sources of interictal activity

*Supported also by Progetto Finalizzato Tecnologie Biomediche e Sanitarie, CNR.

in cases of focal epilepsy was achieved, not only when simultaneous EEG spikes of high amplitude were available, thus permitting an averaging procedure (Barth and colleagues, 1982; Barth and colleagues, 1984; Ricci and colleagues, 1984), but also when low-voltage electric - not necessarily spikes - showed a good correlation with the simultaneous MEG. In the latter case the Relative Covariance method (Chapman and colleagues, 1983) proves to be successful in a wider population of clinical cases.

The investigation described in this paper was carried out with the purpose of verifying the validity of neuromagnetic localization as compared with evidence from other diagnostic techniques, surgical findings and, in two cases, electrocorticography (ECoG) before right temporal lobectomy. In particular, clinical cases with pathological activity arising from different cerebral regions were selected to verify whether or not the simple model used for data interpretation and source localization provided appropriate reliability.

METHODS AND MATERIALS

The typical experimental procedure consisted in measuring the magnetic field at different sites of the investigated scalp area according to a pre-determined grid (2x2 cm). Magnetic signals were detected by means of a second-order gradiometer, 3 cm pickup coil diameter and 5 cm baseline, coupled to a rfSQUID. The field noise spectral density measured inside the biomagnetic facility at the Istituto di Elettronica dello Stato Solido (wooden hut) was typically 40 $fT/\sqrt{Hz}$ above 0.5 Hz. Depending on the patient studied, the magnetic field at 30 to 80 recording sites was sequentially measured. The simultaneous standard 10-20 system EEG was recorded (ear lobe reference, contralateral to the side of the epileptic focus). The magnetic channel was filtered at the line frequency and harmonics using a comb filter. All channels were bandpass filtered in the frequency bandwidth 0.5-32 Hz, using a 48dB/octave filter, with flat delay characteristics. Signals were A/D converted (12 bits) at a sampling frequency of 128 Hz. For each magnetic recording site data were collected for 2 minutes and stored onto disc. The computer used for data acquisition was an HP1000F.

Data were analyzed using, when possible, the averaging procedure first proposed by Barth and colleagues (1982) and, in all cases, the Relative Covariance (RC) method (Chapman and colleagues, 1983). This procedure was described in detail in the mentioned reference and was also successfully used for investigating sources of spontaneous brain activity in normal subjects (Chapman and colleagues, 1984). We only recall here that the covariance between the magnetic channel and one of the simultaneous electric ones is evaluated in correspondance of each recording site, and divided by the variance of the electric channel, to compensate for variations in source activity. Clearly, all the RC values are calculated with respect to the same electric lead. Both electric and magnetic channels are bandpass filtered around specific frequencies identified by previous FFT analysis. In-phase and out-of-phase MEG and EEG signals should yield positive or negative RC values respectively, while the overall spatial distribution of the RC coefficient should reflect that of the magnetic field, provided the same source is producing the measured signals. Substantial agreement between the averaging procedure and the RC one was demostrated by Chapman and colleagues (1983) and will be confirmed in the present paper. Localization was performed when the iso-RC contour maps clearly showed a dipolar-like pattern. The three-dimensional location of the equivalent dipolar source was obtained by fitting the experimental data distribution through numerical iterative procedures. In all the fitted cases the simple model of the current dipole in a homogeneously conducting sphere was used. In each case the best sphere was fitted into the actual anatomical profiles of the head of the studied patient, as they were obtained by direct craniometry, and from x-ray slides and computerized tomography. The five parameters identifying the dipole position and strength (tangential component only) were iteratively adjusted minimizing the χ^2 function.

Thirteen cases of focal epilepsy were studied. Ten out of them clearly showed focal localization on the basis of anatomical-pathology and of x-ray findings. Four patients showed focal localization in the temporal region, two in the frontal one, four in the central area, one in the parietal one and the last one in the occipital region. Another patient (GD) did not show any clear localization on the basis of any diagnostic means, but presented a clinical history of paroxysmal paresthesia mainly involving the contralateral arm. Two subjects, both from the temporal epilepsy sample, were studied before surgical operation, during which their ECoGs were also recorded.

RESULTS AND DISCUSSION

In all cases but one a three-dimensional localization was achieved by means of the neuromagnetic investigation. We want to remark that in the non-localized case (NS), a clear clinical (right motor jacksonian seizures) and CT scan (left rolandic parasitical cyst) localization was available: the MEG analysis revealed rare spikes over the left rolandic region of the scalp, but the absence of EEG corresponding signals prevented any correlation study. In two cases the electrical signals were much more significant that the magnetic ones. Nevertheless thanks to the RC method , a three-dimensional localization of the equivalent source of the interictal rhythmic activities was achieved. The same result was obtained for two patients, GD (see above) and EI, where a correlation analysis of a theta focal rhythm permitted localization of the equivalent source.
Space does not permit describing all cases in detail, thus we will illustrate only the most interesting ones.

Central, Parietal and Occipital Localizations

The results obtained within this group of clinical cases (6 out of 7) are particularly satisfactory. In all the studied patients excellent agreement between the neuromagnetic localization and that available from previous analyses was found. As expected, the sphere model proved to be particularly appropriate for fitting experimental data covering the posterior and central portions of the head. Fig.1 shows the 3-d localization obtained in one case of rigth motor jacksonian

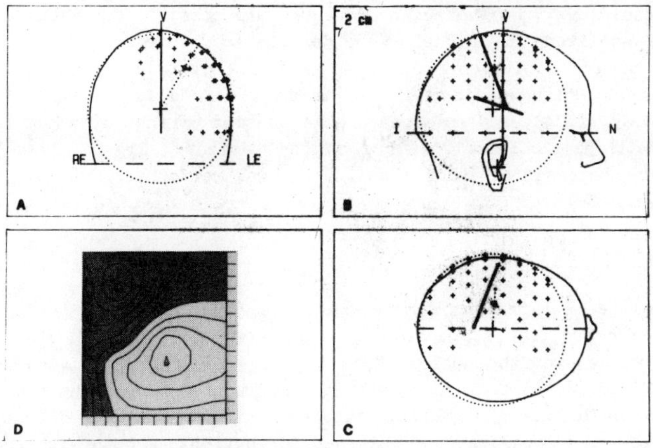

Fig.1. 3-D equivalent source localization for subject EI (right motor epilepsy); A) vertical section through the vertex, B) sagittal section, C) horizontal section; the sylvian and rolandic fissures are also shown; crosses represent experimental points as reported onto the fitting sphere. D) Constant contour map illustrating the RC values distribution; distance between adjacent tics 1 cm. The iso-RC lines are separated by 4 pT/mV.

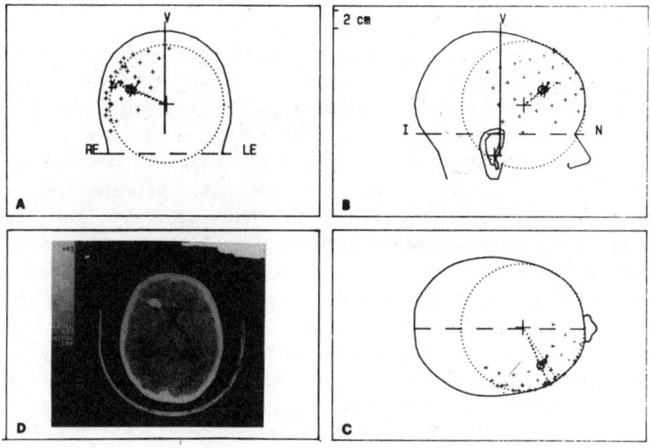

Fig.2. A-C) Localization in one case of frontal epilepsy (RM). Small and large circles identify locations of the equivalent sources as derived by the averaging procedure and the RC method respectively. D) CT scan of the same subject at the appropiate level.

epilepsy (EI), involving the right arm, subsequent to surgical removal of a post-central arachnoid cyst. No averaging procedure was possible for focus localization as only small amplitude EEG signals were detectable mainly at 6 Hz.

The location of the equivalent source - depth 3.2 cm from the scalp - lies just anterior to the Rolandic fissure, as it is obtained from standard anatomical procedures (Gray, 1974), and in a reasonable portion of the motor homunculus. This well agrees with the clinical history of the patient. Fig.1d shows the iso-RC contour map on the basis of which localization was performed.

Frontal localizations

In both cases the best fitting sphere was to be chosen in the anterior portion of the head in order to appropriately represent the actual distribution of experimental sites over the scalp. From a theoretical point of view this is only partially satisfying, as the hypothesis of a medium of homogeneous conductivity is not likely to be fulfilled, depending on the particular skull architecture in this region. Fig.2 shows localizations obtained for patient RM both with the averaging and the RC procedure (large circle). It is evident that the evaluated locations of the focus agree fairly well. This patient presented a clinical history of frontal lobe epilepsy (absences and/or adversative seizures). The x-ray findings (see CT scan in Fig.2d) revealed a clearcut calcification inside the right frontal lobe.

Temporal localizations

Two out of the four studied cases will be described here, because of their significancy. They both were studied by means of ECoG before right temporal lobectomy. The investigation on the first patient (MF) was already described by Chapman and colleagues (1983). This patient presented a clinical history of right temporal lobe epilepsy with no pathological evidence provided by x-ray findings. Chapman and colleagues demonstrated a quite good agreement between localizations achieved by both the averaging and the RC methods. In order to localize the equivalent source they inserted the best fitting sphere in the posterior region of the head. Consequently, as the authors pointed out, the relatively flat temporal bone was unsatisfactorily represented by the sphere. The ECoG analysis successively performed on the exposed cortex, identified a "firing" area involving the first convolution and the tip of the right temporal lobe (see Fig.3). The successful surgical removal (recovery) of the involved area confirmed this finding, also revealing an underlying tumor. These results indicated a location of the active source displaced at least 3 cm anteriorly and probably shallower with respect to the site suggested by the neuromagnetic analysis. We ascribed this disagreement to the discrepancy between several experimental positions and their actual projection onto the fitting sphere. Fig.4(a-c) shows the modified localizations obtained using a "local" sphere for fitting the temporal region of the head. The small circles refer to the first two strong components evident in the averaged tracings shown in Fig.4d; the large circle refers to the localization achieved by means of the RC method. In all cases the location of the equivalent source agrees much better with the previously described anatomical findings.

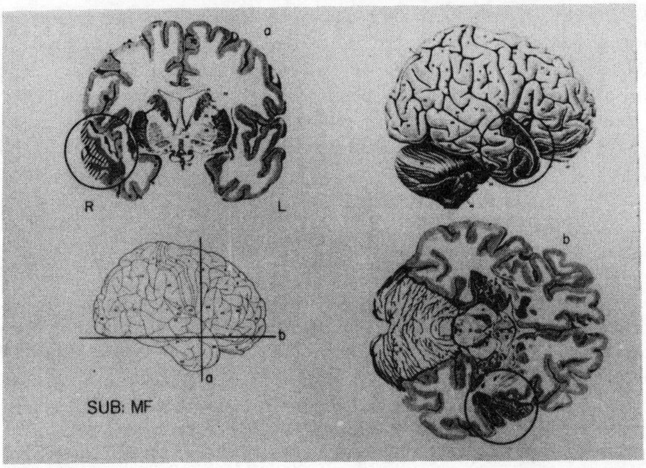

Fig.3. Portions of pathologically active cortex (shaded areas) as revealed by ECoG on subject MF.

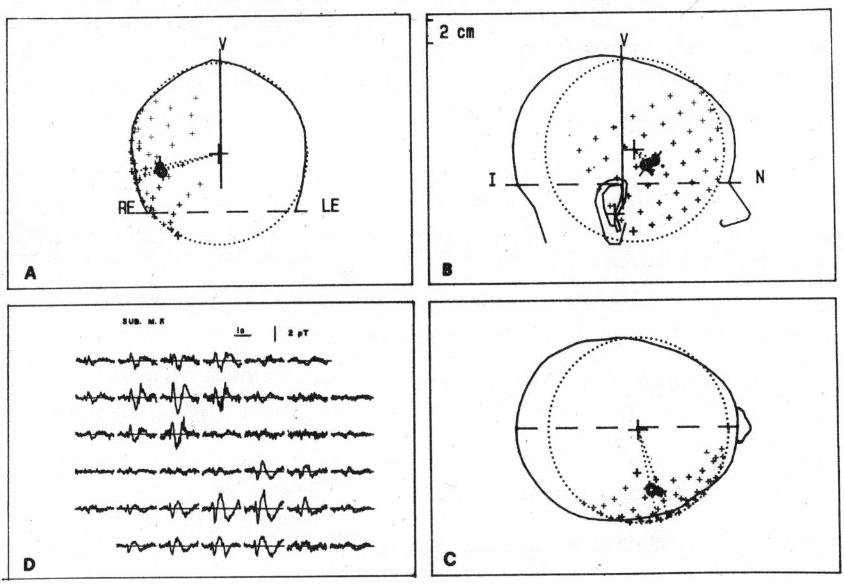

Fig.4. Localization for subject MF as obtained using the "local" sphere (see text). The large circle identifies the equivalent source localized by the RC method. Small circles represent source localization for the two first strong components of the averaged magnetic signals reported in D).

On the basis of this experience we decided to use to "local" sphere approximation also for another subject (VP) who showed a pathology relatively similar to that of MF, but with a right parieto-temporal tumoral calcification well identified by x-ray analyses. The localization was possible only by means of the RC method. The equivalent source was located in the lower portion of the anterior temporal lobe. The ECoG, performed before lobectomy, showed that the active cortical region covered an area of a few cm^2, involving the anterior third of the first and second convolutions and the tip of the right temporal lobe.

CONCLUSIONS

The possibility of 3-d localizing sources of interictal activity in focal epilepsy by the neuromagnetic method has been futher confirmed. The Relative Covariance method permits investigation of a number of clinical cases larger than that possible by means of the averaging procedure. Nevertheless, some of the studied cases point out that the simple dipole-in-a-sphere model, although satisfactory for central-parieto-occipital localizations, is questionable for frontal ones, and can provide incorrect source location in the temporal region. Particular care should be taken for localization of sources in this portion of the brain, which could be placed more posteriorly as they actually are. Finally, it seems confirmed that the neuromagnetic localization better identifies tangentially oriented sources, like those positioned inside cortical fissures, than radially oriented ones, like those in gyri.

REFERENCES

Barth D.S., W. Sutherling, J. Engel and J. Beatty (1982). Science, 218, 891-894.

Barth D.S., W. Sutherling, J. Engel and J. Beatty (1984). Science, 223, 293-296.

Carelli P., I. Modena, G.B.Ricci and G.L. Romani (1983). Biomagnetism: an interdisciplinary approach, S.J. Williamson, G.L. Romani, L. Kaufman and I. Modena Eds., Plenum Press, New York and London, 469-482.

Chapman R.M., G.L. Romani, S. Barbanera, R. Leoni, I. Modena, G.B. Ricci and F. Campitelli (1983). Lett. Nuovo Cimento, 38, 549-554.

Chapman R.M., R. Ilmoniemi, S. Barbanera and G.L. Romani (1984). Electroencephalogr. Clin. Neurophysiol., in press.

Gray H. (1974). Anatomy, T.P. Pick and R. Howden Eds., Running Press, Philadelphia.

Modena I., G.B. Ricci, S. Barbanera, R. Leoni, G.L. Romani and P. Carelli (1982). Electroencephalogr. Clin. Neurophysiol., 54, 622-628.

Ricci G.B., G.L. Romani, I. Modena, S. Barbanera, R. Leoni and P. Carelli (1981). Electroncephalogr. Clin. Neurophysiol., 52, 95.

Ricci (1983). Il Nuovo Cimento, 2D, 517-537.

Ricci G.B, I. Modena, S. Barbanera, F. Campitelli and G.L. Romani (1984). Acta Neurochirurgica, Suppl. 33, 85-89.

COMPARISON OF EEG, MEG, and ECoG IN EPILEPTIC PATIENTS

S. Sato*, P. Sheridan*, P. Smith‡, R. Bonner‡, H. Weinstock§, M. NissenoffΠ, D. Rose†, W. Theodore†, W. Friauf‡, and R. Porter†

*Epilepsy Br., and †Med. Neurol. Br., Nat. Inst. of Neurol. and Commun. Disorders and Stroke, and ‡Biomed. Eng. and Instr. Br., Nat. Inst. of Health, Bethesda, MD; §Ill. Inst. of Technol., Chicago, IL; ΠU.S. Naval Res. Lab., Washington, D.C.

ABSTRACT

Electroencephalography, magnetoencephalography, and electrocorticography were compared in two patients with epilepsy.

KEYWORDS

Electroencephalography; magnetoencephalography; electrocorticography; epilepsy; seizures.

INTRODUCTION

Recent reports suggest that the noninvasive method of magnetoencephalography (MEG) may provide as much three-dimensional information as electroencephalography (EEG) with depth electrode recording. With MEG, the mapping of magnetic fields generated by the brain can provide clues as to the location, orientation, polarity, and depth of a current dipole or dipoles (Barth et al., 1982, 1984). Modena et al. (1982) and Ricci (1983) have claimed that in all cases of Jacksonian motor epilepsy, the MEG was able to localize epileptic activity even in the absence of EEG abnormalities. They did not find the field extrema in focal epileptic activity, although they stated that the localized epileptic activity shows an area of maximal activity which fades as it gets farther away from the source (Ricci, 1983).

MATERIALS AND METHODS

The EEG, MEG, and electrocorticogram (ECoG) were compared in two patients with complex partial seizures. Our single-channel SQUID system (S.H.E. Corporation) for MEG recording has a second-order gradiometer with a coil diameter of 23.6 mm and baseline of 31.7 mm, and requires no shielded room. Magnetic fields were measured in a magnetically quiet room at the National Institutes of Health. A special wooden bed and SQUID support were constructed for this purpose. The MEG and EEG were always recorded simultaneously, using gold electrodes placed according to the international 10-20 system. Electrodes were affixed to the scalp with collodion. The EEG machine, FM tape recorder, and other monitoring equipment were placed about 6 m away from the SQUID system and the patient. The SQUID system was supported on a simple wooden sawhorse; the probe was positioned by moving the patient rather than the SQUID. The patient lay on her back or side during the recording. The EEG, MEG, and ECoG data were stored on magnetic tape.

RESULTS

The presurgical EEGs of a 34-year-old woman who underwent left temporal lobectomy for intractable complex partial seizures showed spike discharges in the left temporal region, but independently in the anterior (F7-T3) and posterior temporal (T5) areas, predominantly in the anterior area. The MEG also showed two foci similar to those on the EEG. The ECoG showed epileptiform discharges in the inferior and middle temporal gyri both anterior and posterior to the vein of Labbe. The discharging area posterior to the vein approximately corresponded to the scalp electrode position of T5. The temporal lobe just anterior to the vein of Labbe was excised. The postexcision ECoG showed persistent epileptiform discharges posterior to the vein (Fig. 1). The EEGs again showed a spike focus at T5. The MEG and EEG were in phase at the midpoint between C3 and T3 (Fig. 2) and out of phase at T5.

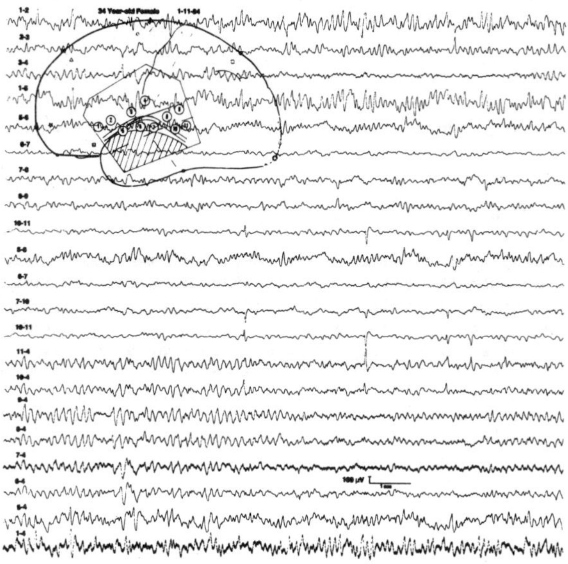

Fig. 1. Postexcision electrocorticogram of a 34-year-old woman who underwent left temporal lobectomy for intractable complex partial seizures. Insert diagram shows the exposed brain area; the shaded area indicates the excised area (5 cm from the tip of the temporal lobe) of the brain. The two parallel lines indicate the vein of Labbe. The number indicates electrode position. Note independent spikes at electrodes 10 and 11.

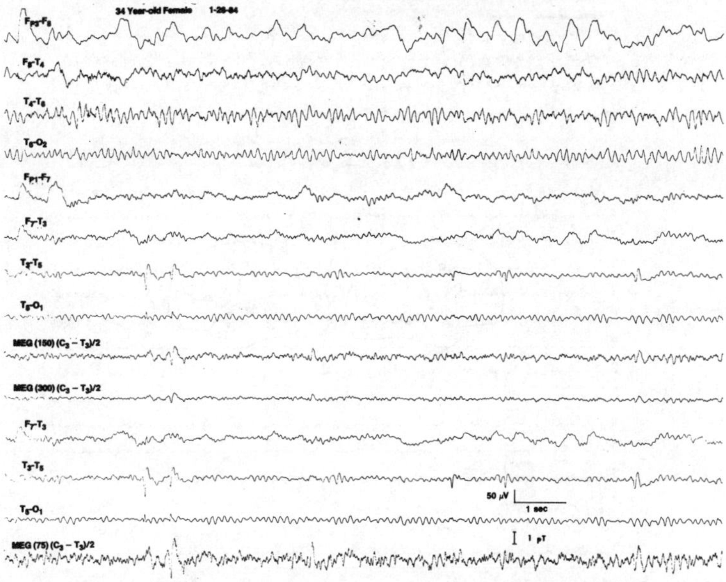

Fig. 2. Postoperative simultaneous EEG and MEG recording from the patient described in Fig. 1. The first eight channels are EEG, covering the right and left temporal areas. The next two channels are MEG at different sensitivity settings. The last four channels are a set-up for magnetic tape recording. Three of them are EEG, covering the left temporal area, and one is MEG. The EEG spikes at T3-T5 appear in phase with MEG spikes at the midpoint between C3 and T3. Both are morphologically similar.

Numerous EEGs in a 20-year-old woman with intractable complex partial seizures showed independent temporal spikes or sharp waves bilaterally. The discharges sometimes occurred predominantly on the right side and at other times predominantly on the left side. Because it was impossible to localize the discharges, it was decided to do subdural recording with flap electrodes. A set of three flap electrodes, each having four contact points, was placed subdurally through a burr hole in the temporal areas bilaterally. One electrode was placed on the frontal opercular area, one on the temporal tip area, and one on the superior and middle temporal area (Fig. 3). Epileptiform discharges occurred in all locations, but most frequently in the right temporal tip area, where 270 to 600 uV spikes or spikes and waves occurred spontaneously and frequently (0.5-2 per sec); episodic increases in the discharge rate were associated with clinical seizures (Fig. 3). The scalp-recorded EEG from the corresponding area showed blunted, saw-toothed, 75 to 120 uV slow waves occurring at about the same frequency as the spikes and spikes and waves in the subdural recording. The MEG from the anterior temporal area showed magnetic sharp waves similar to those on the EEG (Fig. 4).

The MEG and EEG were in phase at F8 but out of phase at point 13, which is about 5 cm toward the vertex from F8. Examination of the skull X-ray film taken at the time of subdural recording showed that the area of the right temporal lobe where the subdural electrode was placed corresponded to point 15 (the midpoint between F8 and T4) and 2 cm below this point. The MEG at these points showed blunted, saw-toothed waves shifting into phase and out of phase with the EEG waves.

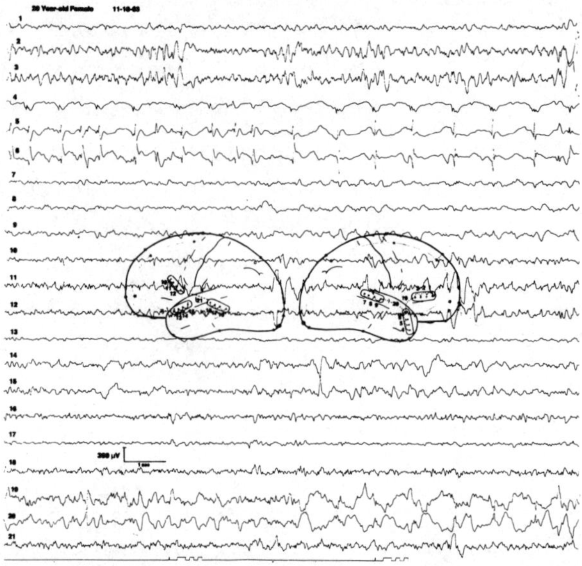

Fig. 3. Subdural recording with flap electrodes in a 20-year-old woman with intractable complex partial seizures. The position of the flap electrodes is shown in the brain diagram, and the adjacent electrode contacts are connected in sequence and given consecutive channel numbers. Frequent high-voltage spikes or spike-and-wave complexes located at the right temporal tip are seen at channels 4 through 6. The magnitude and sharpness of these spikes suggest that they originate superficially.

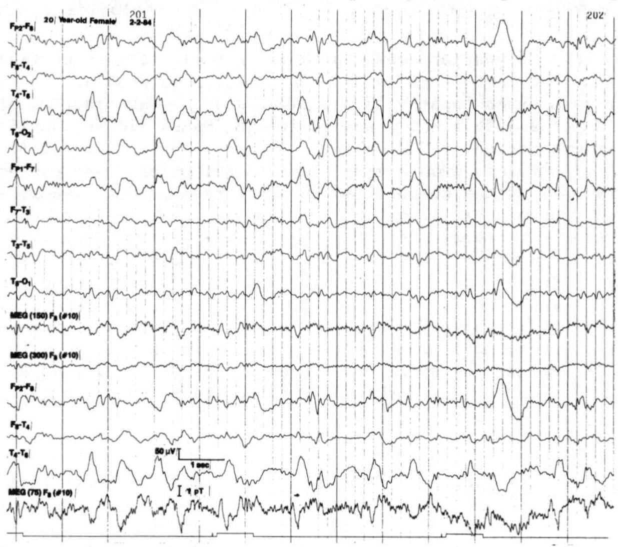

Fig. 4. Caption on next page.

Fig. 4. Simultaneous EEG and MEG recordings from the patient described in Fig. 3. The channel designations are the same as those in Fig. 2. In the right anterior to midtemporal area (Fp2-F8, F8-T4 and T4-T6), periodic high-voltage slow waves are seen. The Eeg (T4-T6) appears to be out of phase with the MEG at F8, although both re morphologically similar.

DISCUSSION

The phase relationship between EEG and MEG was not always constant; the in-phase relationship gradually shifted to the out-of-phase relationship or vice versa. When they were not in phase, there was a lag for the MEG. It has been reported that somatosensory magnetic fields are similar to cortically recorded somatosensory evoked potentials (Kaufman et al., 1981). In the first patient, the MEG spikes were morphologically similar to those on the EEG and ECoG. In the second patient, however, the MEG spikes were not as sharp as the ECoG spikes, but were similar to those on the EEG.

The magnitude and sharpness of spikes in the second patient strongly suggest that these discharges have a cortical origin. However, neither EEG nor MEG reflected such a cortical event. Since the spikes at the temporal tip were such a continuous event during the subdural recording, it is unlikely that they did not occur during the MEG monitoring. These spikes might have been missed by the MEG because the source dipole was perpendicular to the surface. This assumption, however, does not explain why the scalp-recorded EEG did not pick them up. It is regrettable that the scalp and subdural EEG were not recorded simultaneously in this patient. Modena et al. (1982) and Ricci (1983) have stated that MEG performs best with a superficial focus, but less well with medial and basal foci, although they did not map the magnetic fields. However, our findings in the second patient did not support this contention.

The morphological difference between the MEG and ECoG spikes may mean that these two modalities were recording from different sources, that MEG spikes might somehow be influenced by volume current, or that MEG spikes might somehow be modified through conductivity changes in the surrounding media, although the MEG is not supposed to be smeared or attenuated by intervening tissues (e.g., skull, scalp).

Although it is claimed that the EEG slow waves can correspond to the MEG spikes (Ricci, 1983), high-voltage slow sharp waves in our second patient were not associated with MEG spikes.

REFERENCES

Barth, D.S., W. Sutherling, J. Engel, Jr., and J. Beatty (1982). Neuromagnetic localization of epileptiform spike activity in the human brain. Science, 218, 891-894.

Barth, D.S., W. Sutherling, J. Engel, Jr., and J. Beatty (1984). Neuromagnetic evidence of spatially distributed sources underlying epileptiform spikes in the human brain. Science, 223, 293-296.

Kaufman, L., Y. Okada, D. Brenner, and S.J. Williamson (1981). On the relation between somatic evoked potentials and fields. Int. J. Neurosci., 15, 223-239.

Modena, I., G.B. Ricci, S. Barbanera, R. Leoni, G.L. Romani, and P. Carelli. (1982). Biomagnetic measurements of spontaneous brain activity in epileptic patients. Electroencephalogr. Clin. Neurophysiol., 54, 622-628.

Ricci, G.B. (1983). Clinical magnetoencephalography. Il Nuovo Cimento, 2, 515-537.

SEARCH FOR MEG SIGNALS DUE TO AUDITORY BRAINSTEM STIMULATION

David Cohen and B. Neil Cuffin

Francis Bitter National Magnet Laboratory
Massachusetts Institute of Technology
Cambridge, Massachusetts 02139 USA

ABSTRACT

A search is made for MEG signals over the human head due to the early auditory (brainstem) evoked response. No signals were seen above the MEG noise in the two subjects who were measured.

KEYWORDS

MEG, auditory evoked response, early auditory components, brainstem.

INTRODUCTION

EEG measurements of early (1-10 msec) components due to auditory response are used both clinically and in auditory research (Chiappa, 1983). Although the sources of these components are known to be located in the auditory nerve entering the brainstem and in the brainstem, source details are not known; therefore, MEG measurements of these early components could be useful in providing further source information. To this end, we searched for these MEG signals. Because the sources are deep, which would suppress their MEG (Cuffin and Cohen, 1979) and because the EEG signals are weak ($\sim 1\ \mu$volt), we expected the MEG signals to be very weak. We therefore searched for these signals at about the limit of our MEG sensitivity, which is close to the present day MEG sensitivity limit generally.

METHOD

The measurements were made in the MIT shielded room. A three channel MEG detector was used. One channel consisted of an asymmetric in-line first-order gradiometer. Essentially this was a monopolar coil (2.8 cm diameter) which measured the conventional field component normal to the head (B_z); the signal from this coil was sensed by an rf SQUID. The other two channels each consisted of a side-by-side first-order gradiometer. Each gradiometer, as previously described (Cohen, 1979) was a bipolar coil made up of two D's connected in opposition, and measured the transverse (x or y) gradient of B_z at the same location on the head as the monopolar coil; the signal from each of these coils was sensed by a dc SQUID. The monopolar sensitivity was about 10 $fT/\sqrt{Hz}$, limited mostly by Johnson noise in the rf shield. The bipolar sensitivity was about 5 $fT/cm/\sqrt{Hz}$, limited mostly by the dc SQUID noise. While the bipolar sensitivity falls off more rapidly than does the monopolar sensitivity with increasing depth of a source in the head, it was used here because the falloff loss is compensated by the fact that the bipolar coils see "straight ahead" instead of "to the side," and are less subject to rf shield noise hence benefit from the low-noise dc SQUID. Monopolar plus bipolar coils thus are a good combination for a search of this type.

Two subjects were measured; these were normal young men. An auditory click (about 75 db) was produced in the left ear via an air tube, fed by a loudspeaker. The click duration was about 0.5 msec, measured by a microphone at the tube exit. EEGs at various locations were measured simultaneously with the MEGs; these were essentially monitor channels, to see that the early components were indeed present at a normal level. The MEG and EEG outputs, in a bandwidth of 1 to 1000 Hz, were fed into a DEC MINC computer for signal averaging. Some MEG outputs in a bandwidth of 1 to 3000 Hz were also fed into an fm tape recorder for later playback at various reduced bandwidths to search for the MEG signals at optimal signal/noise. The stimulation rate was 8/sec, and as many as 10,000 sweeps were averaged; the computer sampling interval was about 0.2 msec.

MEG measurements were repeatedly made at locations about 3 and 6 cm above, forward, and behind the tragus of the left ear. In addition, MEGs were measured at some locations further removed from the left ear. The location grouping about the left ear was determined to be optimal, based on angle-of-view (lead-field) considerations of monopolar and bipolar coils, previously studied theoretically (Cuffin and Cohen, 1979) and verified experimentally (Cohen and Cuffin, 1983). In brief, because the sources are either in the left auditory nerve or in the brainstem, the grouping about the left ear using both monopolar and bipolar coils covers every possibility to see any non-radial sources at about the closest spacings to these

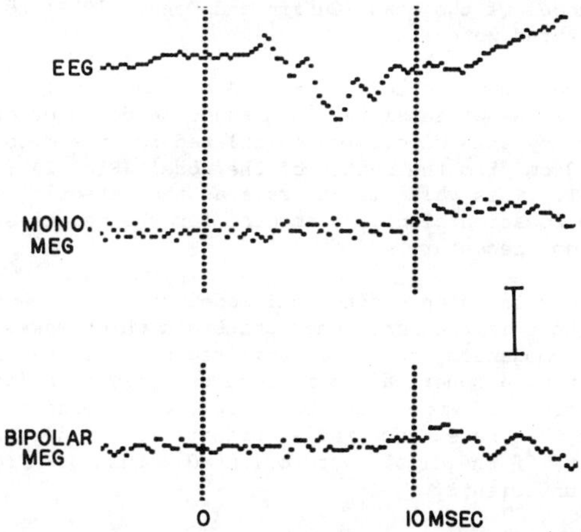

Fig. 1. Examples of EEG and MEG due to auditory evoked response in one of two subjects. The EEG and monopolar MEG (B_z) were recorded simultaneously; the EEG is due to the inion referenced to the vertex, while B_z was measured 6 cm posterior to the left ear tragus, below T_5. The bipolar MEG ($\Delta B_z/\Delta y$) is one of the two bipolar channels from a different run, measured at 1 cm below T_3. All traces are due to 10,000 sweeps. The scale bar is 1.0 μvolts for EEG, 100 fT for monopolar MEG, and 80 fT/cm for bipolar MEG.

sources. Stated otherwise, if the early auditory signals could not be seen at any of these locations with any of the coils, then the lead-field analysis showed they would not be seen at any other location around the head.

RESULTS

Although the normal brainstem signals were seen on the EEG (∿1 μvolt), no signal was ever seen above the noise in any of the MEG channels, at any of the locations measured. Some examples of the negative traces are shown in Fig. 1. Although there is a suggestion of a signal at about 2.5 msec in the particular monopolar MEG trace shown, it cannot be reproduced hence is only noise. The MEG signals seen here beyond 10 msec are believed to be largely myogenic.

Further, when the MEG data on the tape were played back into the computer for signal averaging at a variety of bandwidths, no signals were again seen above the noise of the traces. Examples of the bandwidths used were 500 to 1000 Hz and 400 to 800 Hz. From a consideration of all our data, we set an upper limit of 20 fT for the MEG signal from these early auditory components. We again note that these MEG measurements were made under very low noise conditions.

DISCUSSION

To understand the negative MEG result, we performed some simple theoretical modelling. We wished to see if the negative result was due to MEG suppression because of the deep location of a tangential source, or instead was due to the radial orientation of the source (at any location), or perhaps a combination of both. First, the recorded EEG values were used to determine the strength of a dipole in a four-layer spherical model of the head (Cuffin and Cohen, 1979) to produce potentials equal to the measured potentials.

The required strength was found to be ∿1 μa-cm; since the dipole was placed deep in the model, this strength is not sensitive to changes in depth or orientation. Then the B_z extrema produced by this dipole were calculated for the dipole oriented tangentially and located 2 cm from the center of the model (Fig. 2A); the extrema were found to be about 20 fT. This value is the same as the detection limit, hence in this simple model deep location alone can account for the negative MEG result; radial orientation is not necessary.

However, for deep sources the simple spherical model may not be accurate because of the lower brain case; this is the bone under the brain which makes the head more complex than a simple homogeneous sphere in this region. We therefore modelled this situation by locating a spherical brain eccentrically in a larger sphere of bone (Fig. 2B). For this head model, the B_z extrema were found to increase to about 40 fT. Thus, in this case, some radial orientation is necessary to account for the negative result; if the dipole were oriented within 30° from radial, this would reduce the MEG sufficiently.

In conclusion, using the detection system described here, we could find no MEG early auditory signal above the noise, at the locations measured. This negative result could be explained to first order by only the depth of the source. However, if model refinements such as the presence of the brain case are taken into account, it may be necessary to invoke some radial orientation of the source. Our results suggest that the MEG from a deep source is difficult or impossible to see, at least from the brainstem region.

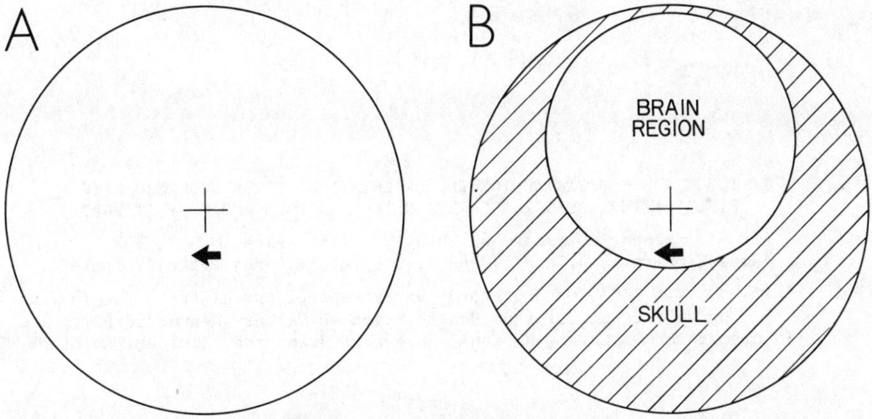

Fig. 2. (A) Homogeneous spherical head model, of 9 cm radius, used to first calculate the magnetic field B_z. The tangential dipole is placed 2 cm from the center. (B) Inhomogeneous spherical head model, which takes into account the lower brain case; the skull here has zero conductivity.

REFERENCES

Chiappa, K.H. (1983). Evoked Potentials in Clinical Medicine, Raven Press, New York, Chap. 5, pp 144-158.

Cohen, D. and B.N. Cuffin (1979). Magnetic Measurement and Display of Current Generators in the Brain, Part I. 12th International Conference on Medical and Biological Engineering, Israel, 15-16.

Cohen, D. and B.N. Cuffin (1983). Demonstration of Useful Differences between MEG and EEG. Electroenceph. clin. Neurophysiol., 56, 38-51.

Cuffin, B.N., and D. Cohen (1979). Comparison of the Magnetoencephalogram and Electroencephalogram. Electroenceph. clin. Neurophysiol., 47, 132-146.

TOPOGRAPHIC DISTRIBUTION AND ORIENTATION OF THE NEUROMAGNETIC FIELD PATTERN ASSOCIATED WITH CLICK AND TONE AUDITORY STIMULI

John T. Zimmerman, Ph.D.*, Martin Reite, M.D.*,
James Zimmerman, D.Sc.**, John A. Daspit, Jr.* and Peter D. Teale*

* EEG/MEG Laboratory, Department of Psychiatry
University of Colorado Health Sciences Center, Denver 80262
** Cryogenics Division, National Bureau of Standards, Boulder, Colorado

ABSTRACT

Simultaneously recorded magnetoencephalographic (MEG) auditory evoked fields (AEFs) and electroencephalographic (EEG) auditory evoked potentials (AEPs) were examined from 10 different positions over the lateral surface of the right hemisphere in six male adults and from the left hemisphere of four of these subjects. AEPs were identified over a widespread surface of the hemisphere with maximal amplitude near the vertex. AEFs had higher amplitudes in the general vicinity of the primary auditory cortex falling off rapidly in amplitude with increasing distance. This AEF pattern is consistent with a current dipole model oriented perpendicular to the Sylvian fissure with a possible additional contribution from a secondary source, oriented in a horizontal direction and located within the anterior temporal region.

KEYWORDS

Magnetoencephalography, auditory evoked field, topographic distribution, dipole orientation.

INTRODUCTION

Research from several laboratories has suggested that the neural sources responsible for the generation of the auditory evoked magnetic field (AEF) are located in the general vicinity of the auditory cortex (Reite et al., 1978; Elberling et al., 1982; Romani et al., 1982; Hari et al., 1980). Farrell et al. (1980) used a first derivative gradiometer with a coil diameter of 4.8 cm to systematically map the MEG AEF, in response to click stimuli. Their group inferred that the source of the auditory evoked field could be modeled via a vertically-oriented current dipole in the auditory cortex located approximately 3 cm beneath the surface of the scalp. Zimmerman et al. (1981) used a figure-8 gradiometer aimed over the auditory cortex region. In that study the magnetic field amplitude evoked by click stimuli was maximal when the gradiometer was oriented so as to detect a current dipole oriented perpendicular to the Sylvian fissure. In comparison, the EEG evoked potential to auditory stimuli tends to be somewhat more widespread in its distribution and of highest amplitude over the vertex. The present study examined the topographic distribution of AEFs evoked by click and tone stimuli over right and left hemispheres in normal subjects and compared these to AEPs recorded at the same time in response to the same stimuli.

METHODS

Simultaneously recorded EEG AEPs and MEG AEFs were compared with respect to waveform component latency and amplitude at 10 different positions over the lateral surface of the right hemisphere in six normal adult males. Subjects ranged in age between 23 and 68. Recordings from the left hemisphere were made in four of these same subjects. The target position, from which all other electrode locations were measured, was located at a point 15% of the tragus-to-tragus distance above the tragus. This represented a point 1/4th of the distance between electrode positions T4 and C4, above T4, over the right hemisphere or analogously over the left, above T3. This was estimated to approximately overlie the primary auditory cortex. A Cartesian coordinate system reference scheme was developed based upon published diagrams of Jasper (1958). The orientation of the X-axis was determined by a line on the lateral aspect of the head connecting Pz to the nasion. A second line parallel to this was drawn below it, intersecting the target position which served to orient the anterior/posterior line of electrodes, approximately along the Sylvian fissure. The Y-axis line, orthogonal to the first line was drawn through the target position. Electrode positions 6, 7, and 8 were position along this Y-axis line. Inter-electrode distances were 3 cm. Two additional electrode positions were studied later, at positions 9 and 10, as illustrated in Figure 1.

A gold cup EEG electrode was placed at each recording position, referred to a contralateral earlobe. At each location one recording was done with the axis of an off-diagonal SQUID figure-8 gradiometer oriented perpendicular to the anterior/posterior line of electrodes (designated the 0° orientation) and a second recording with the gradiometer axis oriented parallel to the anterior/posterior line of electrodes (labeled the 90° recording orientation). This gave 2 orthogonal components of the magnetic field gradient; one with respect to dx (along a more or less an anterior/posterior direction) and one with respect to dy (along a more or less a superior/inferior direction). All recordings were made in the numerical sequence shown in Figure 1. As the recordings progressed from one location to the next, the sequence of 0° and 90° measurements was alternated. Two EEG AEPs were obtained at each recording site, one while the gradiometer was in the 0° orientation, the second while it was in the 90° orientation. These two were averaged to get a single mean EEG AEP for each location.

Auditory stimuli were presented to the contralateral ear. Clicks and tones (92.5 dB sound pressure level) were presented alternately, about once every 700 msec. Tones were 88 msec duration and 981 Hz. Click stimuli were produced by a 1/10th millisecond positive square wave input to an audio amplifier. Sound was transmitted to the subject by a system of plastic tubing, terminating in an earpiece worn in the subject's ear. Sound transit time from the speaker to the earpiece was 23 msec.

The EEG and MEG were amplified, low pass filtered at 40 Hz, and monitored on a polygraph. Two dual-channel signal averagers were used to average 128 responses to the alternating clicks or tones. Both signal averagers were AC coupled and provided high pass filtering at 2 Hz. EEG and MEG signals were averaged for 500 msec after onset. The average waveform was plotted stimulus with an X/Y recorder.

The plotted waveforms hand digitized. The computer calculated and plotted an average waveform across subjects for both clicks and tones for each of the 10 electrode positions for both EEG and MEG. Separate averages were computed for both 0° and 90° orientations for the MEG data.

With respect to the MEG AEF waveforms, the computer performed a rectangular-to-polar conversion of the 0° and 90° samples in which the deviations of the 0° data points from the baseline were plotted along the Y-axis and the deviations of the 90° data points from its baseline were plotted along the X-axis. These diagrams approximate the current path during the 200 msec following sound onset

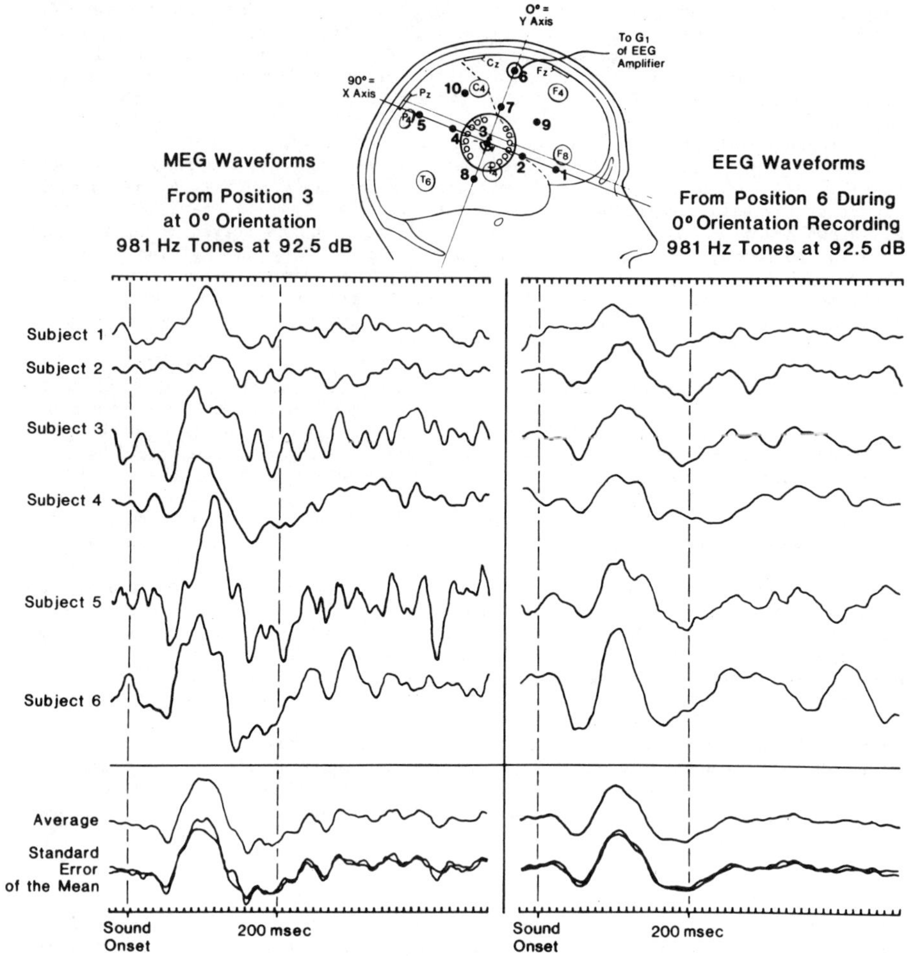

Figure 1. Top center: Right hemisphere diagram depicting the 10 electrode positions used in this experiment. The relative size of the SQUID double-D gradiometer is shown as it was aimed over recording position #3 where the largest MEG AEF was observed. An EEG electrode is shown at position 6 where the largest EEG AEP was recorded. Left: Intersubject variability of MEG AEF waveforms recorded from position 3 at 0° orientation with SQUID gradiometer positioned as shown. Right: Intersubject variability for EEG AEP waveforms recorded from position 6 with electrode positioned as illustrated. Average waveform and plus or minus standard errors of the mean shown at bottom.

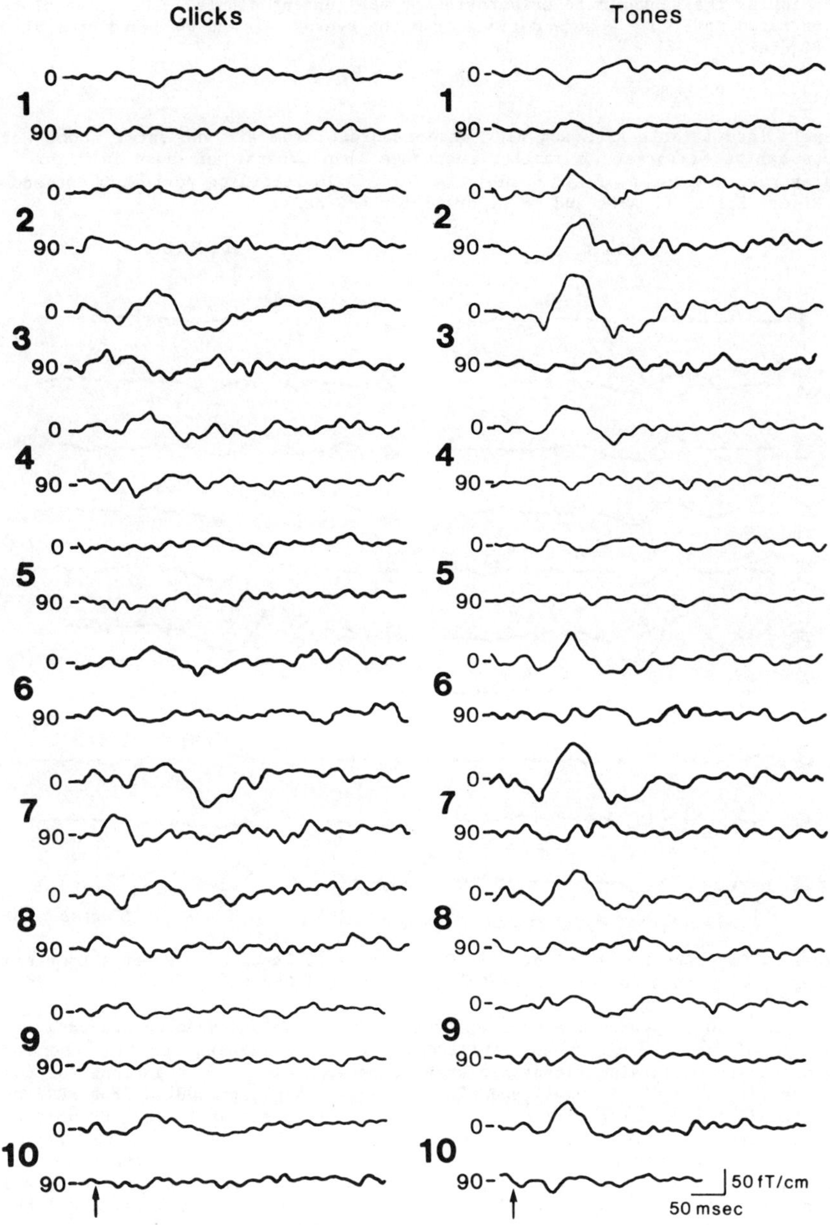

Figure 2. MEG AEFs averaged across all 6 subjects shown for each of the 10 recording positions comparing clicks (left) with tones (right).

and display the approximate orientation of the current dipole. The polar plots illustrated in Figure 4 were derived from the average 0° and 90° waveforms shown in Figure 2.

RESULTS

Clearly identifiable AEFs and AEPs were recorded from all subjects. Some subjects exhibited larger or smaller responses than others, as shown in Figure 1. The average waveforms of all 6 subjects for all 10 recording positions are shown in Figure 2 for MEG AEFs and in Figure 3 for EEG AEPs.

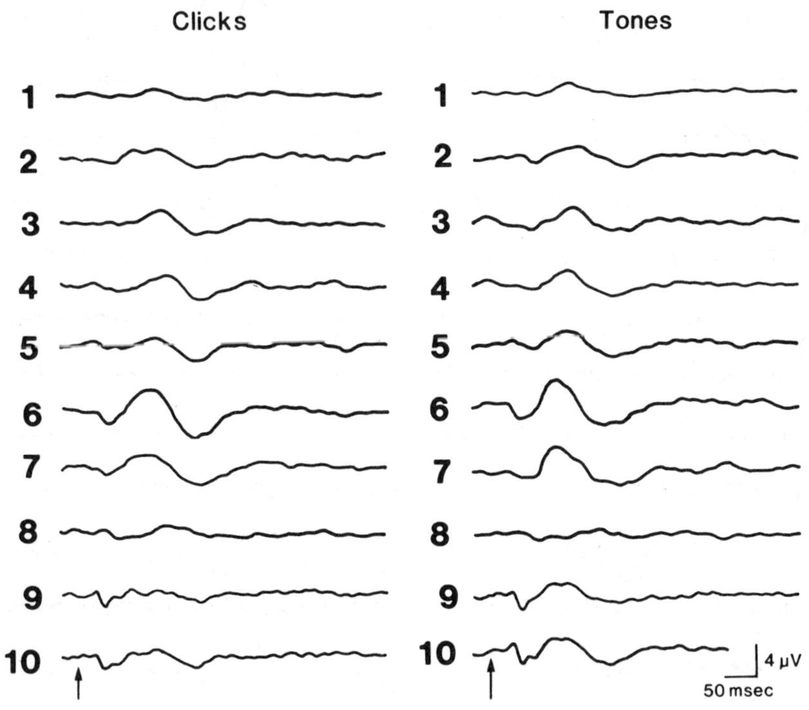

Figure 3. Average EEG AEPs for all 10 positions (0° and 90° waveforms combined) comparing waveforms for clicks (left) and tones (right).

A 2_w x 2_w x 10_w repeated measures analysis of variance was done on the data set. Analysis of the peak-to-trough amplitude differences between the N100 and P180 components revealed significant differences between electrode positions for both EEG (F=6.21, df=9/45, p<.0001) and MEG (f=6.65, df=9/45, p<.0001). For MEG, the largest AEF occured at position 3 for two subjects, position 4 for two subjects, and position 7 for two subjects. Averaged across all 6 subjects, the largest MEG AEF occured over position 3. This suggests a dipole source within an area between positions 3, 4, and 7. There were also significant differences between orientations for the MEG AEF. The 0o waveforms were almost always larger than the 90° waveforms (F=7.54, df=1/5, p<.05) at all locations (except for one subject at position 2). Tones evoked visually larger waveforms than clicks, but this effect was not statistically significant. However, tones did elicit a significantly shorter latency to the P50 component than clicks ($F=10.69$, $df=1/5$, $p<.03$). Left hemisphere responses at every location were smaller in amplitude than the homologous right hemisphere responses but the orientation of the polar plots formed a similar consistent pattern similar to that shown in Figure 4.

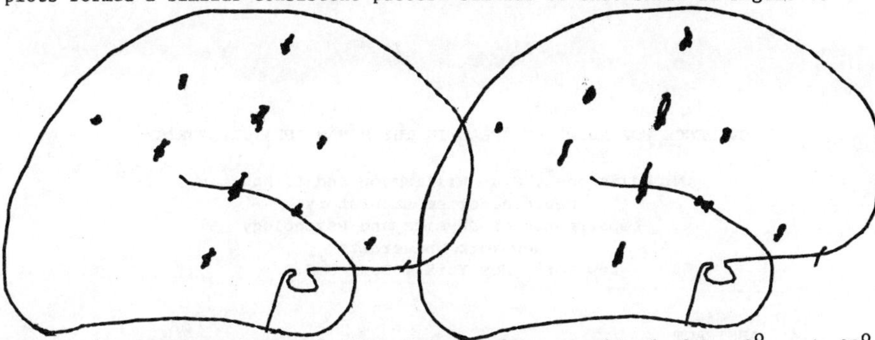

Figure 4. Polar plots of MEG AEF waveforms combined from 0° and 90° orientations comparing clicks (left) with tones (right).

DISCUSSION

The present study confirms previous reports suggesting that magnetic fields evoked by auditory stimuli are maximal in amplitude in the near vicinity of the primary auditory cortex whereas the simultaneously recorded EEG is of highest amplitude over the vertex. The rectangular-to-polar conversion plots suggest at least one source near position 3 oriented perpendicular to the Sylvian fissure and a secondary source oriented more horizontally near position 2. The presumption has generally been made that cortical sources contribute predominantly to extracranial magnetic fields. The possible contributions of underlying fiber tracks, whether they are thalamacortical or intracortical (e.g., uncinate fasciculus), remains to be elucidated.

REFERENCES

Elberling, C., Bak, C., Kofoed, B., Lebech, J. and Saermark, K. Auditory magnetic fields from the human cerebral cortex: Location and strength of an equivalent current dipole. Acta Neurol. Scandina., Vol. 65, 553-569, 1982.

Farrell, D.E., Tripp, J.H., Norgren, R., and Teyler, T.J. A study of the auditory evoked magnetic field of the human brain. Electroencephalography and Clinical Neurophysiol., Vol. 49, 31-37, 1980.

Hari, R., Kaila, K., Katila, T., Tuomisto, T. and Varpula, T. Interstimulus Interval Dependence of the Auditory Vertex Response and Its Magnetic Counterpart: Implications for Their Neural Generation. EEG and Clinical Neurophysiology, Vol. 54, 561-569, 1982.

Jasper, H.H. The ten twenty electrode system of the International Federation. Electroencephalography and Clinical Neurophysiology, Vol. 10, 371-375, 1958.

Reite, M., Edrich, J., Zimmerman, J.T. and Zimmerman, J.E. Human Magnetic Auditory Evoked Fields. Electroencephalography and Clinical Neurophysio., Vol. 45, 114-117, 1978.

Romani, G.L., Williamson, S.J. and Kaufman, L. Characterization of the human auditory cortex by the neuromagnetic methods. Exp. Brain Res., Vol. 47, 381-393, 1982.

Zimmerman, J.T., Reite, M. and Zimmerman, J.E. Magnetic Auditory Evoked Fields: Dipole Orientation. Electroencephalography and Clinical Neurophysio., Vol. 52, 151-156, 1981.

EVIDENCE FOR MULTIPLE AREAS IN THE HUMAN AUDITORY CORTEX*

M. Pelizzone[+], S.J. Williamson and L. Kaufman
Neuromagnetism Laboratory
Departments of Physics and Psychology
New York University
New York, New York 10003, U.S.A.

ABSTRACT

Neuromagnetic measurements of the transient response to tone burst stimuli of different frequencies and of the response evoked by amplitude modulated tones in the steady state paradigm are compared on the same subjects. The activity underlying the dominant 100 ms component of the transient response to a tone burst is insensitive to the frequency of the tone. The activity evoked in the steady state paradigm is tonotopically organized and its source is shifted laterally from that of the tone bursts. These results show that it is possible with the neuromagnetic measurements to differentiate spatially and functionally different regions across the human auditory cortex.

KEY WORDS

Neuromagnetism; evoked responses; magnetoencephalogram; SQUID

INTRODUCTION

The primate auditory cortex is divided into several distinct functional areas with tonotopic projections of the entire basilar membrane in some of them (Brugge,1982). Recently, neuromagnetic studies have shown that the location of cortical activity evoked by amplitude modulated tones in the steady state paradigm follows a tonotopic organization in which the activated region progresses along the cortex as a function of the stimulus frequency (Romani et al.,1982a,1982b). These results must be contrasted with other neuromagnetic studies measuring the transient response to tone bursts stimuli which did not reveal the same kind of systematic and precise functional organization when the carrier frequency was changed (Elberling et al.,1982; Arlinger et al.,1982; Tuomisto et al.,1983). This discrepency between the steady-state and transient data prompted us to directly compare responses to both kind of stimuli in the same subject.

METHODS

Our magnetic field sensor consisted of a second-order gradiometer with 2.4-cm diameter and 3.2-cm baseline between adjacent coils coupled to a SQUID. The output voltage from the SQUID electronics, which is simply proportional to the net magnetic flux sensed by the gradiometer, was passed through a comb filter to reduce the pickup of noise at the power line frequency and then through a bandpass filter with 48 db/octave rolloff to select the bandwith of interest.

Auditory stimuli were presented binaurally to the subjects by means of standard airline plastic earphones. Two different classes of stimuli were used. First, the transient responses were evoked by tone bursts of 200 or 400 ms duration, depending upon the subject, presented with a random interstimulus interval (ISI) of 3.5 ± 0.5 sec. Different carrier frequencies between 200 and 5000 Hz were used and also different rise and decay times for the envelope of the tone burst. The amplitudes were adjusted at the different frequencies so that they were subjectively equal in loudness. In general the stimuli remained in the range of 80-90 dB SPL. The measured power spectrum of the tone bursts at the output of the earphone was centered at the nominal frequency with typical width at half height corresponding to the inverse of the duration of the tone burst. The subject was asked to count the number of stimuli to ensure attentiveness. Responses were averaged for at least 100 trials in a filter bandwith of 0.5 to 45 Hz. In the second set of experiments, steady-state responses were evoked by continuous tones whose amplitudes were sinusoidally modulated at a frequency of 32 Hz. Responses were averaged for at least 2000 modulation cycles and the field patterns were deduced from the amplitude and phase lag of the response at 32 Hz (see Romani et al.1982b for more details).

We used both paradigms on three subjects and compared the evoked field patterns across their right hemisphere.

RESULTS

To detect in an efficient way small shifts in the location of the activity evoked by the different paradigms, we mapped the responses by steps of 1 cm at locations along a line crossing the two extrema and along two lines perpendicular to this direction at each extremum. This strategy gave at least 20 points for each map, each location being replicated at least twice.

The measured field patterns for subject PL and four different stimuli are presented in Fig.1. The isofield contours were calculated by computer using an interpolation fit of the data by the Laplacian method. The origin of the coordinates system is the ear canal, with the horizontal position measured toward the outer canthus of the eye and the vertical position measured perpendicular to this line. For this subject, we used three different 400 ms tone bursts of 200, 1000 and 2000 Hz with rise/decay time of 48, 12 and 6 ms respectively. The wave forms of the transient responses were very similar for the three different stimuli, with one dominant component occuring after about 100 ms (N1) and another one, with opposite polarity, after about 180 ms (P2). Nevertheless, our data indicate that due to the different rise times, the latencies of all the components of the response to a 200 Hz tone burst were longer by about 20 ms than for the 2000 Hz tone burst. The field patterns corresponding to the dominant component of the transient response occuring about 100 ms after stimulation were almost identical for the three different tone bursts (Fig.1,TB). On the other hand, the field pattern of the activity evoked by a continuous 1000 Hz tone amplitude modulated at 32 Hz had almost the same span between the extrema, but was clearly shifted laterally from that of the tone bursts (Fig.1,SSMT). Due to the weak signals in that particular case, we replicated the measurement at least six times at each location so that the signal to noise ratio was comparable to the transient measurements.

To compare in a systematic procedure the location of the cortical activity evoked

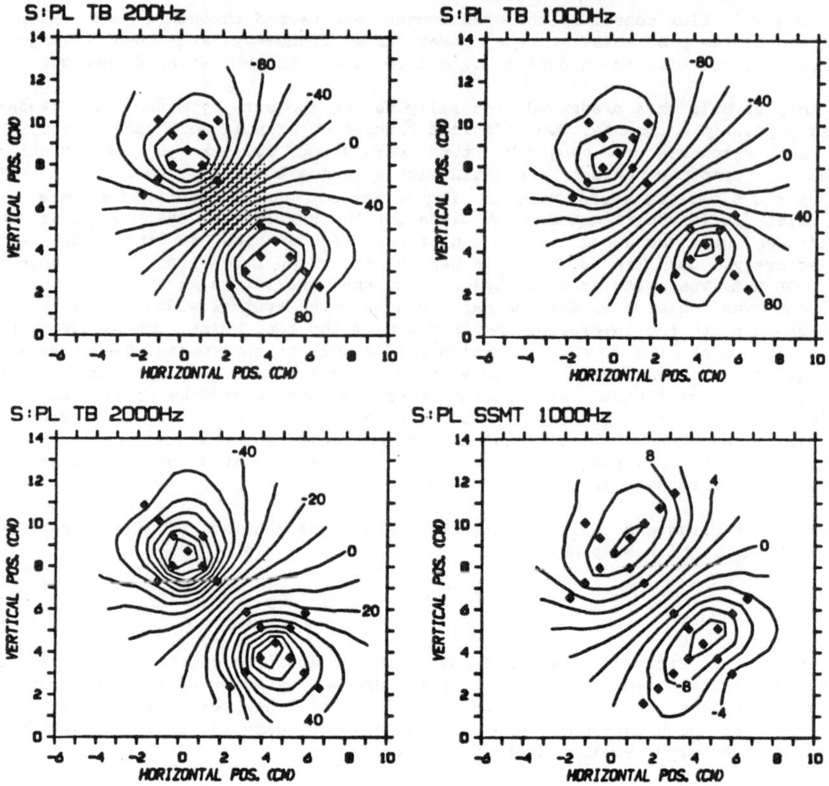

Fig. 1. (TB) Observed field patterns for the 100 ms component of the transient response to a tone burst of 200, 1000 and 2000 Hz respectively; (SSMT) observed field pattern for a steady-state amplitude-modulated tone of 1000 Hz. The origin of the coordinate system is the ear canal, with the horizontal position measured toward the outer canthus of the eye and the vertical position measured perpendicular to this line.

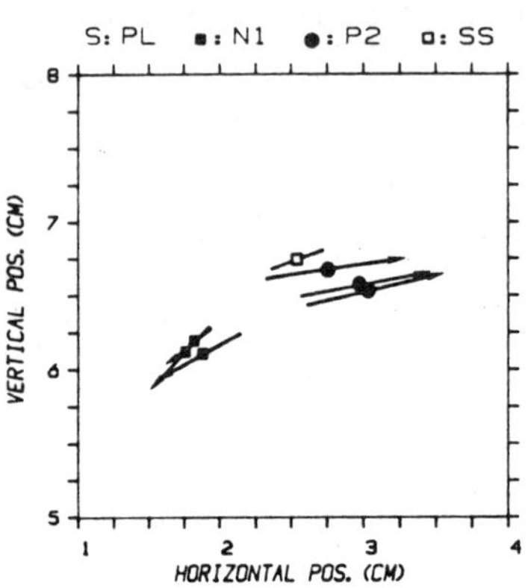

Fig. 2. Calculated best dipoles for the N1 and P2 components of the transient response to a tone burst and for the steady-state response (SS). This map is an enlargement of the central stippled region in one map of Fig. 1.

by these different stimuli a current dipole in a conducting sphere was used as a theoretical model in the following steps. The field due to a single current dipole in a 8 cm radius conducting sphere was calculated, as it is sensed by our second order gradiometer, and the sum of the squared deviations between the calculated and observed fields was computed. The parameters determining the dipole were then adjusted by computer until the sum of the squared deviations was minimal. Finally, the isofield contours of the best fit were plotted and compared with the data to make sure that no erroneous result was introduced by a local minimum in the sum of the squared deviations. For this subject with three different tone bursts, the parameters of the best fit gave the same surface location within ± 1 mm, the same depth within ± 1 mm and the same orientation within ± 8 deg for the source underlying the dominant 100 ms component of the transient response. As for the raw data, the location of the source evoked by a 1000 Hz amplitude modulated tone lies at the same depth but is shifted laterally by about 1 cm. The calculated dipoles are shown in Fig. 2, which is an enlargement of the central region of the maps. It is noteworthy that the same 1000 Hz carrier frequency is present in the amplitude modulated stimulus as in one of the transient stimuli.

Similar measurements on two other subjects confirmed the essential features of the findings. For subject MP, the activity evoked by a continuous 200 Hz amplitude modulated tone is also shifted anteriorly when compared to the activity underlying the 100 ms component of the transient response to a 200 Hz tone burst. For subject SW, the activity evoked by a continuous 1000 Hz amplitude modulated tone lies exactly on his tonotopic map (see Romani et al., 1982), but is shifted posteriorly when compared to the activity underlying the 100 ms component of the transient responses to clicks and to 1000 and 5000 Hz tone bursts.

Another dominant component of the transient response occured about 180 ms after stimulus presentation for subject PL. The variability of the data at this latency was somewhat greater than for the 100 ms component and the state of alertness of the subject did affect the measurements. This component also showed a larger intersubject variability. Nervertheless for subject PL, the field patterns at a latency of 180 ms showed two well defined extrema slightly shifted from that of the 100 ms component. The parameters of the best single current dipole fits gave the same surface location within ± 2 mm, the same depth within ± 2 mm and the same orientation within ± 5 deg for the three different tone bursts (see Fig.2). Thus, even if the intrinsic variability of the data was greater at this latency, the fits gave consistently a surface location also shifted anteriorly by about 1 cm when compared to the 100 ms component and no evident tonotopic organization for the 180 ms component of the transient response.

DISCUSSION

In contrast to the results reported by Elberling et al. (1982), but in agreement with the fine mapping on one subject in the work of Tuomisto et al. (1983), the present measurements reveal no significant shifts in the location of the source of the N1 component of the transient response when the carrier frequency of the tone burst is changed, whereas the sources evoked by steady-state amplitude-modulated tones lies in an area of the auditory cortex which is tonotopically organized. Thus the N1 source and the steady-state sources exhibit different functional properties. Even if the same carrier frequency is used in both paradigms, two spatially different areas are excited. The distinct character of these sources is also consistent with the different latencies of the responses: about 50 ms for the steady-state response and about 100 ms for the N1 component of the transient response.

Our measurements on the 180 ms component of the transient response are consistent with the results of Tuomisto et al.(1983) in the sense that they suggest different neural origins for N1 and P2. The form of the N1-P2 complex did show a considerable intersubject variability due to a poorly developped P2 in some of them. Measurements on a larger number of subjects are needed to confirm the findings on P2.

The excellent reproducibility of the field patterns, which has been observed by a careful mapping for different stimuli with sessions extending on a time scale of many weeks, allows to infer minor displacements in the location of multiples sources within the human auditory cortex. This prospect is particularly promising in view of the future application of large arrays of sensors.

CONCLUSION

These results indicate that: (1) it is possible with neuromagnetic measurements to differentiate spatially and functionally different regions across the human auditory cortex and (2) the 100 ms component of the transient response to a tone burst is insensitive to the pitch of the tone or to the exact shape of the envelope of the wave form. This component may well be associated primarily with the onset of a stimulus and be less sensitive on its details. This is consistent with the widely held notion that the pathways of the auditory system are divided among several functions and that some are tonotopically organized, while other are not.

ACKNOWLEDGEMENTS

We thank Dr. Yoshio Okada and Dr. Ken Schafer for their generous help. We also thank Prof. Anthony Movshon for kindly providing much of the computer software used in data analysis.

REFERENCES

*Supported in part by National Institute of Health Grant NS-19463-01 and Office of Naval Research Contract N00014-76-C-0568.

+Supported by the Swiss National Science Foundation.

Arlinger, S., C. Elberling, C. Bak, B. Kofoed, J. Lebech and K. Saermark (1982). Electroenceph. clin. Neurophysiol. 54:642-653.

Brugge, J.F. (1982). Auditory Cortical Areas in Primates, in Cortical Sensory Organization, Vol. 3, p. 59 (Ed. by C.N. Woolsey, Humana Press,Clifton, New Jersey).

Elberling, C., C. Bak, B. Kofoed, J. Lebech and K. Saermark (1982). Scand. Audiol. 11:61-65.

Romani, G.L., S.J. Williamson and L. Kaufman (1982). Science 216:1339-1340.

Romani, G.L., S.J. Williamson, L. Kaufman and D. Brenner (1982). Exp. Brain Res. 47:381-393.

Tuomisto, T.,R. Hari, T. Katila, T. Poutanen and T. Varpula (1983). Il Nuovo Cimento 2 D:471-483.

MAGNETIC RESPONSES TO PITCH CHANGES IN A SEQUENCE OF SHORT AUDITORY STIMULI

M. Sams*, R. Hari, E. Kaukoranta, K. Reinikainen, K. Alho*, M.S. Hämäläinen, R.J. Ilmoniemi, R. Näätänen*, and J. Salminen

Low Temperature Laboratory, Helsinki University of Technology, SF-02150 Espoo, Finland, and *Department of Psychology, University of Helsinki, SF-00170 Helsinki, Finland

ABSTRACT

Magnetic responses specific to rare frequency changes in a sequence of short auditory stimuli were recorded. The field pattern indicates activation of the primary auditory cortex.

KEYWORDS

Magnetic field; mismatch negativity, event-related brain potential; ERP; neuronal model.

INTRODUCTION

When a subject ignores an auditory stimulus sequence by engaging in some other task, an infrequent pitch change elicits an event-related brain potential (ERP) component called the mismatch negativity , MMN (for a review, see Näätänen and Gaillard, 1983). The peak latency of MMN is 150-200 ms depending on the size of the frequency change. It has been proposed (Näätänen and Gaillard, 1983; Näätänen, Sams and Alho, in press) that MMN reflects the mismatch process between the neuronal representation of the standard stimuli and the sensory inflow caused by the deviant stimulus. There is evidence that MMN might be generated in the auditory cortex (Simson, Vaughan and Ritter, 1977; Näätänen, Gaillard and Mäntysalo, 1978, 1980), but because of the ineffectiveness of electrical recordings in locating brain activity, the exact generator site of MMN has remained unknown. To determine the active brain areas during MMN more precisely, we recorded simultaneously magnetic and electrical responses to auditory frequency changes. The first report of this study has been presented elsewhere (Hari and co-workers, in press).

METHODS

The measurements were made in the Otaniemi magnetically shielded room with five experienced subjects. Sinusoidal tone bursts (70 dB SPL, 30 ms, rise time two cycles) were presented through a plastic tube and an earpiece to the subject's left ear once every 1300 ms. The number of standard tones between deviants varied randomly from 3 to 15 (even distribution) so that the probability of standards was 0.9 and that of deviants 0.1. During the stimulation the subject was reading a book.

Magnetic field was measured with a 4-channel first order gradiometer

(Ilmoniemi, Hari and Reinikainen, in press), the axis of which was perpendicular to the skull. Magnetic responses were measured at the ends of the Sylvian fissure to pick up activity from the primary auditory cortex. In addition, the field pattern over the right hemisphere was mapped from 60-75 points in two subjects. Simultaneously with the magnetic recordings, electrical responses were obtained from Fpz, Fz, Cz, and Pz, referred to the right ear. The signals were digitized and averaged on-line. The computer automatically deleted responses if the Fpz channel contained blinks, baseline shifts, or spikes. The analysis period of 1024 ms included a prestimulus interval of 128 ms.

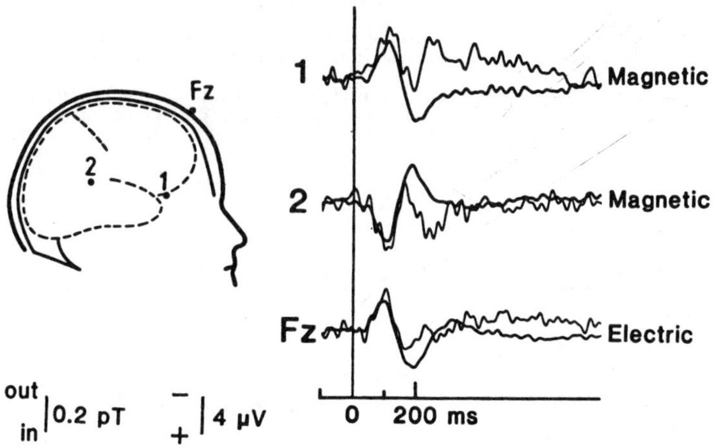

Fig. 1. Magnetic responses to the standard (thick lines) and deviant (thin lines) tones at the anterior and posterior ends of the Sylvian fissure, and the electrical response at Fz referred to the right ear. The recording locations are indicated in the schematic head. The passband was 0.05 - 40 Hz. The sensitivities of the magnetic channels were 22 fT/√Hz above 0.5 Hz.

RESULTS

Figure 1 shows simultaneously measured magnetic and electrical responses from one subject. In the electrical recordings, both the standards and deviants evoked a negative deflection peaking at about 100 ms (N100). After N100 the response to the deviants started to differ from that to the standards, and the difference peaked at about 220 ms. This difference, which was usually larger at Fz, is the MMN, elicited only by the deviants.

Magnetic responses showed a prominent deflection at the same latency as N100 (N100m). As in electrical recordings, responses to the deviants and standards started to differ thereafter, and the difference peaked at the same latency as MMN. This difference probably represents the magnetic counterpart of MMN, and will be called MMNm. In all five subjects, both N100m and MMNm were of opposite polarity between the ends of the Sylvian fissure. The mean peak latency of MMNm (± S.E.M.) for the five subjects was 218 ± 25 ms at the anterior and 196 ± 33 ms at the posterior end of the Sylvian fissure.

Figure 2 shows the field patterns of the MEG responses to the standards and deviants at the moment of stimulus presentation (0 ms), at the peak amplitude of N100m (110 ms), and at 240 ms (same subject as in Fig. 1). At 0 ms there was no organized field pattern. The more noisy pattern to the deviants is probably due to the smaller number of averages in the responses to the deviants than to the standards. At 110 ms the field pattern was dipolar both for the standards and deviants. The field came out of the skull at the anterior end of the right Sylvian fissure and re-entered at the posterior end.

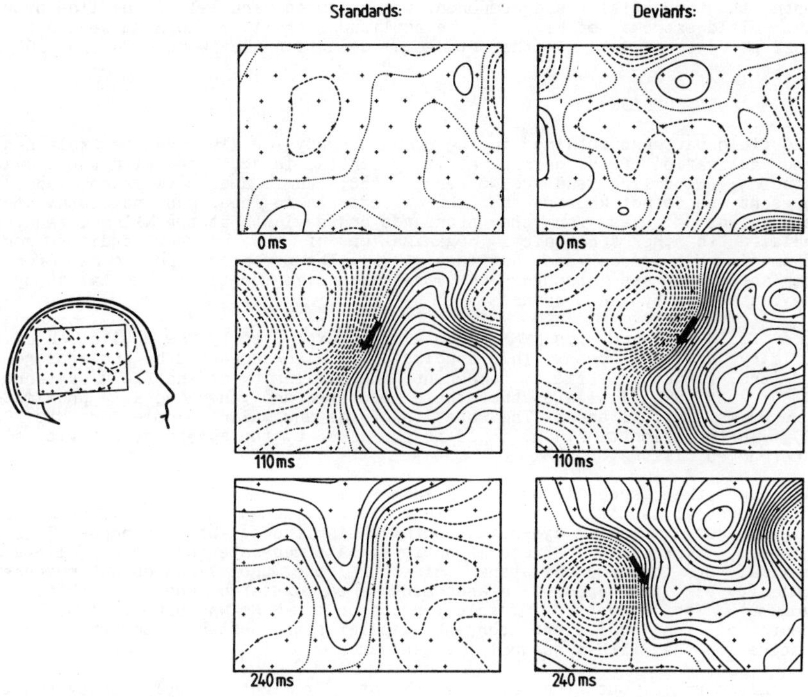

Fig. 2. Isocontour maps of responses of the standards and deviants in one subject at the moment of stimulus presentation (0 ms), at 110 ms and at 220 ms. The maps were calculated using a weighted least-squares approximation and bicubic spline interpolation. The isofield lines are separated by 35 fT. The continuous lines indicate flux out of the skull and the dotted lines flux into the skull. The interpoint distance in the measurement grid is 16 mm, and the crosses indicate the locations of pickup coils (three of them recorded simultaneously). Some positions of the probes are indicated by triangles in the schematic head. The arrows illustrate the locations of the equivalent current dipoles. The black dot indicates the location of T4.

The best-fit equivalent dipoles (see Hämäläinen and co-workers, this volume) for the standards and deviants were separated by 13 mm and the corresponding dipole moments were 13 nAm and 20 nAm. The depth of the dipole for the standards was 20 mm beneath the skull and for the deviants 25 mm. The local radius of curvature of the measurement sphere was estimated to be 13 cm.

For the standards, the dipolar field pattern started to vanish after N100m. However, the field pattern to the deviants continued to be dipolar and of the same polarity as that of N100m but now the equivalent current dipole was rotated about 50 degrees and was shifted 32 mm posteriorly.

When the tangential field component was measured parallel to the line between the field extrema of Fig. 2., the amplitudes of N100m and MMNm were 0.18 and 0.22 pT, respectively. In the orthogonal direction they were 0.04 and 0.01 pT.

DISCUSSION

The field patterns obtained in the present study for N100m can be explained by the activation of the primary auditory cortex, in agreement with the previous studies (Elberling and co-workers, 1980; Hari and co-workers, 1980). At present we do not know whether the differences in locations and depths of the equivalent dipoles for the standards and deviants at the N100m latency are related to the tonotopical organization of the primary auditory cortex (Romani, Williamson and Kaufman, 1982). The stronger dipole moment for the deviants suggests stronger or more extensive cortical excitation by the deviants than by the standards.

The field pattern for MMNm can also be explained by the activation of the primary auditory cortex. This conclusion is also supported by the measurement of the tangential field component. The similarity of the field patterns of N100m and MMNm indicates that these responses are generated at approximately the same cortical areas. The rotation of the equivalent dipole for MMNm (Fig. 2), in relation to N100m, could be explained by the spread of activity along the curved fissural cortex.

Although the cerebral origin of MMNm seems to be well established, possible extracerebral artefact sources must be considered. A rare deviant stimulus might cause a small eye-blink contaminating the cerebral response. This is, however, improbable because magnetic signals generated by the eye-blinks are maximal at the posterolateral sides of the orbits and do not reverse in polarity in the present measurement area (Antervo and co-workers, this volume). The effect of muscle tension on the MMNm was not studied, but the depth of the equivalent current dipole fits better with the cerebral activation than with a superficial generator.

We also ruled out the possibility that MMNm might be explained by the long inter-stimulus interval (ISI) between the deviant stimuli. We presented the deviants alone with ISIs evenly distributed between 4 and 16 s. There was no evidence of a response generated in the primary auditory cortex at the latency of MMNm. Furthermore, when the frequencies of the standard and deviant tones were interchanged, MMNm was elicited as before. Thus we conclude that MMNm is a specific response to a pitch change in a sequence of stimuli.

Both MMNm and MMN are responses to a stimulus change. Their similar time behaviour suggests that MMN, too, might get contribution from the activation of the primary auditory cortex. However, at present there are no detailed scalp distribution studies of MMN, so that firm conclusions can not be made. The direction of the equivalent dipole for MMNm in Fig. 2 would predict the maximal MMN amplitude at centro-parietal areas, whereas the measurements showed maxima at frontal areas. This indicates that the sources of MMN and MMNm cannot be identical. The electrical recordings evidently pick up activity also from other sources than the primary auditory cortex.

MMNm and MMN seem to be reflections of a mismatch process with the neuronal representation of the standard stimuli and the incoming sensory stimulation. Such representation could form the neurophysiological basis of the auditory echoic memory. Probing the features of this representation by measuring MMNm and MMN might thus provide an opportunity to study the features of this basic memory mechanism (Näätänen and co-workers, in press).

ACKNOWLEDGEMENTS

This study was supported by the Academy of Finland and the Emil Aaltonen Foundation (Helsinki, Finland).

REFERENCES

Antervo, A., R. Hari, T. Katila, T. Ryhänen, and M. Seppänen (1984). Magnetic field pattern produced by eye-blinking. This volume.

Elberling, C., C. Bak, B. Kofoed, J. Lebech, and K. Saermak (1980). Magnetic auditory responses from the human brain. Scand. Audiol., 9, 185-190.

Hari, R., A. Aittoniemi, M-L. Järvinen, T. Katila, and T. Varpula (1980). Auditory evoked transient and sustained magnetic fields of the human brain. Exp. Brain Res., 40, 237-240.

Hari, R., M. Hämäläinen., R. Ilmoniemi, E. Kaukoranta, K. Reinikainen, J. Salminen, K. Alho, R. Näätänen and M. Sams (1984). Responses of the primary auditory cortex to pitch changes in a sequence of tone pips: neuromagnetic recordings in man. Neurosci. Lett. , in press.

Hämäläinen, M., R. Ilmoniemi, J. Knuutila and K. Reinikainen (1984). Analyzing magnetoencephalographic data obtained with a four-channel magnetometer. This volume.

Ilmoniemi, R., R. Hari, and K. Reinikainen (1984). A four-channel SQUID magnetometer for brain research. Electroenceph. clin. Neurophysiol., in press.

Näätänen, R. and A. W. K. Gaillard (1983). The orienting reflex and the N2 deflection of the ERP. In A.W.K. Gaillard and W.Ritter (Eds.), Tutorials Event Related Potential Research: Endogenous Components. North-Holland, Amsterdam. pp. 119-141.

Näätänen, R., A.W.K. Gaillard, and S. Mäntysalo (1978). Early selective attention effect on evoked potential reinterpreted. Acta Psychol.,42, 313-329.

Näätänen, R., A.W.K. Gaillard and S. Mäntysalo (1980). Brain-potential correlates of voluntary and involuntary attention. In H.H. Kornhuber and L. Deecke (Eds.), Motivation, Motor and Sensory Processes of the brain: Electrical Potentials, Behaviour and Clinical Use, Progress in Brain Research (Vol. 54). Elsevier, Amsterdam. pp. 343-348.

Näätänen, R., M. Sams, and K. Alho (1984). The mismatch negativity. The ERP sign of cerebral mismatch process. Electroenceph. clin. Neurophysiol., Suppl., in press.

Romani, G., S. Williamson, and L. Kaufmann (1982). Tonotopic organization of the human auditory cortex. Science, 216, 1339-1340.

Simson, R., H.G. Vaughan, Jr., and W. Ritter (1977). The scalp topography of potentials in auditory and visual discrimination tasks. Electroenceph. clin. Neurophysiol., 42, 528-535.

NEUROMAGNETIC STUDY OF ENDOGENOUS FIELDS RELATED TO THE CONTINGENT NEGATIVE VARIATION*

R, Fiumara°, F. Campitelli°°, G.L. Romani, R. Leoni,
M. Caporali°, M. Zanasi°, A. Cappiello°, G. Fioriti
and I. Modena°°°

Istituto di Elettronica dello Stato Solido - C.N.R.
Via Cineto Romano 42 - 00156 Roma, ITALY
°Clinica Psichiatrica - Università "Tor Vergata", Roma
°°Elettronica S.p.A., Roma.
°°°Dipartimento di Fisica - Università "Tor Vergata", Roma

ABSTRACT

The neuromagnetic method has been used in the study of the Contingent Negative Variation (CNV). The investigation carried out on five subjects has shown a large inter- and intra-individual variability. A detailed mapping obtained in two of the five subjects confirms that signals associated to the later portion of the CNV (espectation or E-wave) are most likely produced by generators located in the frontal area of the brain. Data have been interpreted in terms of a single equivalent current dipole. However, the complexity of the measured field distribution requires further investigation, as it probably reflects a more complex source configuration.

KEYWORDS

Neuromagnetism; Contingent Negative Variation; E-wave; current dipole.

INTRODUCTION

CNV is an electrical slow biorhythm induced between two conditioned sense stimuli, connected by a semantic link. The largest amplitude CNV signals are detected over the scalp areas corresponding to the frontal cortices. As originally demonstrated by Walter(1964), CNV is relatively stable in time under circumstances of psychical calm. Its tight correlation with a number of exogenous and endogeneous variables - both physical and psychological - was demonstrated by many authors. Distraction, anxiety, stress and motivation have been shown to be the most important factors which affect the characteristics of this signal.

*Also supported by Progetto Finalizzato Tecnologie Biomediche e Sanitarie, CNR.

Previous attempts to use the neuromagnetic method to investigate fields associated with the CNV have been recently reported (Fiumara and colleagues, 1983; Weinberg and colleagues, 1983; Weinberg, 1984a; Weinberg and colleagues, 1984b). The investigation has been motivated by the impressive results achieved by the neuromagnetic approach in localizing sources of spontaneous and evoked cerebral activity (Chapman and colleagues, 1983; Chapman and colleagues; 1984, Okada, 1984; Romani, Williamson and Kaufman 1982). In agreement with EEG evidence, the studies of Weinberg and colleagues also confirmed that at least two generators influence the CNV in anterior positions of the scalp, one related to orientation and arousal (O-wave), and one to expectation (E-wave).

In this paper we report experimental measurements carried out on a few subjects with the aim of achieving a detailed mapping of magnetic signals generated in the CNV paradigm and possibly localizing the equivalent generator responsible for the measured signals.

METHODS

The typical experimental procedure consisted in measuring the magnetic field at different sites of the frontal, temporal and parietal regions of the scalp, according to a pre-determined grid. Both hemispheres of five normal subjects (3 males and 2 females) were examined. Magnetic signals were detected using a second-order gradiometer, 1.5 cm diameter pickup coil and 5 cm baseline, coupled to an improved performance rfSQUID. The experimental measurements were performed in the biomagnetic facility at the Istituto di Elettronica dello Stato Solido (wooden hut), where the measured field noise spectral density was typically $40 fT/\sqrt{Hz}$ above 0.1 Hz. Simultaneous electric recordings were collected using a lead configuration similar to that described by McCallum and Walter (1968): 5 electrodes were used, 2 placed along the midline on the vertex (+) and on the glabella; 2 reference electrodes on the mastoid processes and directly connected to the glabella lead; a further ground electrode was placed on the ear lobe contralateral to the studied hemisphere. Both electric and magnetic channels were filtered in the frequency band $3x10^{-2} \div 30$ Hz, using a 48dB/octave filter, with flat delay characteristics. No comb filter was used for the magnetic channel to get the proper response in the low frequency range. Signals were A/D converted (12 bits) at a sampling frequency of 120 Hz - 1024 points/channel, corresponding to a sample window of 8.5 s - and stored onto a disc. The computer used for data acquisition and elaboration was an HP1000F. For each sample, two stimuli were generated and binaurally presented to the subject by means of a non-magnetic airplane earphone: after 2.4 s of data acquisition a software generated trigger produced the first stimulus, while an analogically furtherly delayed trigger produced the second stimulus 1.8 s after the first. The first stimulus S1 consisted in a 1 KHz tone burst (rectangular shape), 50 ms duration at approximatively 70 dB SPL; the second S2 in either 1 KHz or 4 KHz tone burst, 80 ms long at approximatively 90 dB SPL. The 4 KHz stimulus was the imperative one. The attention of the subject was warned by S1. On receipt of the imperative 4 KHz stimulus S2, the subject was required to move the middle finger of the hand contralateral to the mapped hemisphere. The intertrial interval, i.e. the time interval between successive S1 stimuli, was 8.5 s. The 1 KHz and the imperative 4 KHz tone bursts were randomically provided to the subject but the latter were statistically more frequent (about 2 out of 3).

This choice of the experimental paradigm was motivated by several considerations: in particular a quite long pre-S1 period was chosen to have a good "dumb" segment for calculating the baseline with respect to which successive signals amplitudes were to be measured. A 4 s long post-S2 period was chosen to ensure the system to reset as large Post Imperative Variations (PINV) are generally observed. Finally, an S1-S2 interval of 1.8 s was chosen to reduce overlapping between O-wave and E-wave, without reducing the E-wave. Furthermore, a reduced amplitude and duration S1 stimulus should minimize the O-wave and increase the E-wave. Under these experimental conditions the E-wave is expected to be present starting 500 ms before S2 (Jervis and colleagues, 1984).

For each recording site 100 samples were collected. On-line average was monitored to follow the experiments. Off-line editing and averaging was successively performed. The editing criteria were: i) to discard all samples containing peak to peak variations greater than 4 pT; ii) to calculate the standard deviation σ in the dumb pre-S1 segment for each sample, to compare it with that (σ_M) of the corresponding segment of the result of the average of the residual N samples after i) and, finally to discard all samples for which $\sigma > \sigma_M x2x\sqrt{N}$. Typically from 30 to 70 samples were used up for final averaging.

DATA ANALYSIS AND RESULTS

Fig.1 shows a map of the averaged magnetic signals measured over the right hemisphere of subject FC. Row 1 is located about 1 cm above the nasion-inion line (dashed line in Fig.3b) and parallel to it. The sites in column C lie on the intersection line between the head and the plane identified by the ear canals and the vertex (solid line in Fig.3b). All other columns represent lines on planes parallel to the described one. All sites are equally spaced by 1.8 cm in both directions. Only 3.8 s portions of the 8.5 s samples are shown. For clearness the signals are additionally filtered by software in the bandwidth $3x10^{-2}\div10$ Hz. We remark that all the iso-field contour maps shown in the following are derived from the original signals in the bandwidth $3x10^{-2}\div30$ Hz. Vertical bars identify stimuli onsets. Horizontal bars represent the baseline calculated on all the pre-S1 segment. The simultaneously measured (D7) electric signal is shown in the inset. The lower part of the map shows clearcut auditory evoked signals approximately 120 ms after both S1 and S2. At least 800 ms before S2 a ramp pattern begins, which continues until the second auditory evoked signal and shows polarity reversal approximately in the regions F2 F3 and E8 F9. Fig.2 shows 10 isofield contour maps derived from the signals shown in Fig.1 and obtained in correspondance of successive time instants starting 120 ms after S1 up to 120 ms after S2. The first and the last map depict a clearcut dipolar-like pattern relative to the distribution of the auditory evoked field. The second and third maps do not show any significant distribution. Starting about 1.5 s after S1, a clearcut pattern shows up, becomes slightly stronger and keeps a substantially stable location. It must be remarked that this pattern looks more complex than the simple dipolar one. Fig. 3(A-C) illustrates the three-dimensional localization of the equivalent sources corresponding to the first, seventh and tenth map of Fig.2. For all

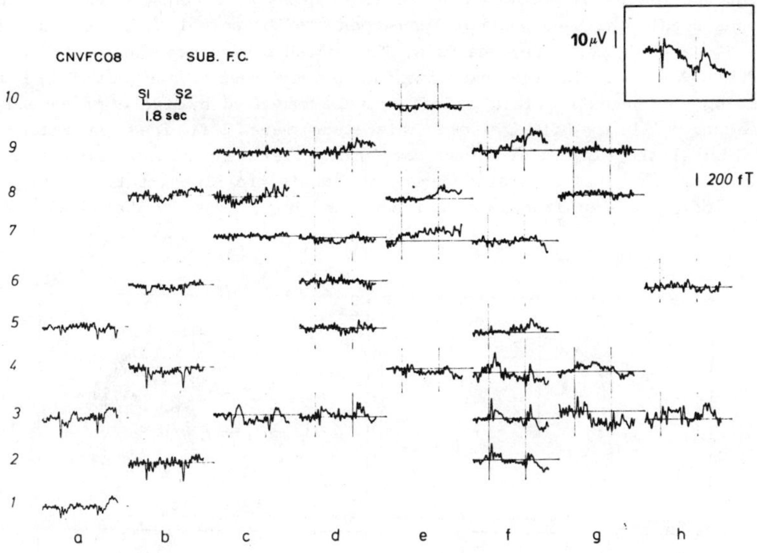

Fig.1. Averaged magnetic signals as recorded over the right hemisphere of subject FC. Grid spacing 1.8x1.8 cm. The ear canal is 3cm below C1 (for details see text). The inset shows the averaged electric signal simultaneous with D7. The displaying bandwidth is $3\times10^{-2}\div10$ Hz.

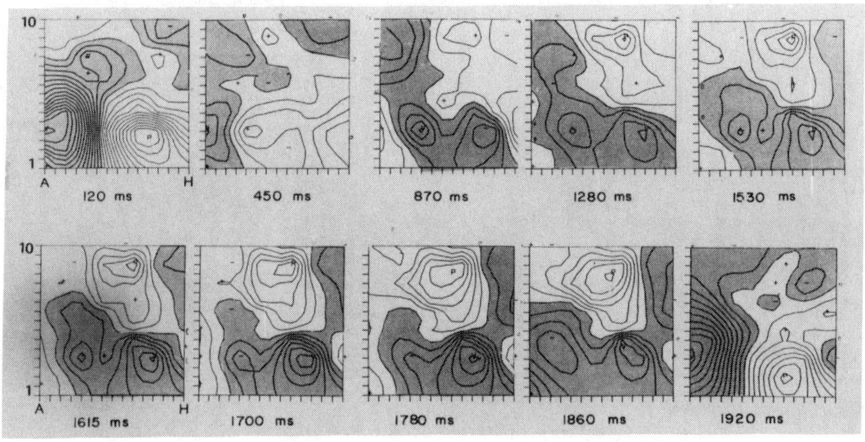

Fig.2. Isofield contour maps derived from fig.1. Time instants (after S1) in correspondance of which maps are obtained are reported below each map. Shaded areas represent negative fields (entering the scalp). The first and the last map correspond to the auditory evoked field pattern (A1, A2), the 3rd to the 9th to the E-wave.

localizations the simple model of a current dipole in a homogeneously conducting sphere was used. The head profiles correspond to the actual skull contours of the subject as derived from x-ray analysis. The dotted sphere represents the one which best fits the head in the experimental points area. The data were fitted calculating the magnetic field as effectively measured by the used second-order gradiometer. All the five dipole parameters were left free and iteratively adjusted until the minimum χ^2 was obtained. The standard deviation needed for evaluating the χ^2 was estimated during the whole pre-S1 segment. In Fig.3D the number ν of degrees of freedom relative to each fit, the evaluated χ^2 values,

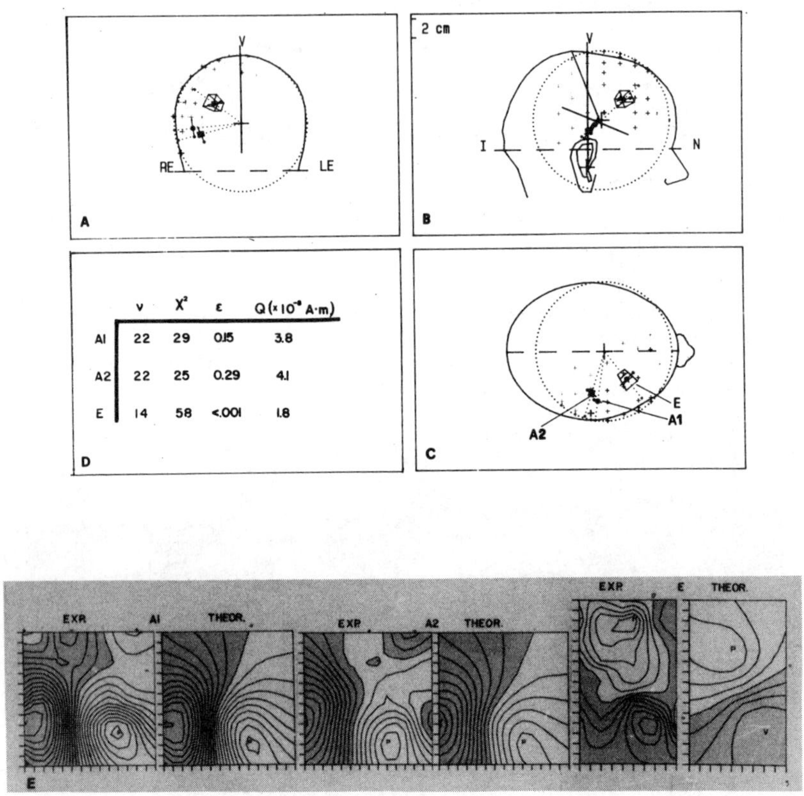

Fig. 3. A-C) 3-d localization and corresponding 95% confidence region for the equivalent sources corresponding to A1 (120 ms), A2 (1920 ms) and E (1700 ms) field distributions; D) statistical parameters relative to each fit, and equivalent dipole strength; E) comparison between experimental and theoretical distributions for A1, A2, E.

the consequent levels of statistical significance ϵ , and the dipole strengths Q are reported. Fig.3(A-C) also shows the 95% confidence regions for the position of each fitted dipole. Fig.3E shows the comparison between the experimentally derived maps and those reconstructed in the same frame on the basis of the estimated parameters. Mapping of the left hemisphere of a second subject (AC) gave similar results for the auditory evoked signals but somewhat more complex distribution for the E-wave, hardly interpretable in terms of a single equivalent dipolar source.

DISCUSSION AND CONCLUSIONS

The significance levels corresponding to the localizations of the equivalent sources of the auditory evoked fields (A1, A2) are particularly good and guarantee that the fitting model is adequate to the purpose. On the other hand the size of the confidence region in both cases supports the idea that an actual displacement of the equivalent source was measured. The estimated depth increase of about 0.6 cm is consistent with previous findings on the tonotopic organization of the auditory cortex (Romani, Williamson and Kaufman, 1982), despite the 4 KHz source was elicited only about 70% of the total number of trials. The anterior-posterior shift for increasing frequency does not agree with corresponding findings of the same reference; however, the second subject showed no lateral translation of the elicited source and a depth increase between A1 and A2 of 0.5 cm, thus suggesting that transverse shifts are likely to be due to inter-individual variations in the cortex shape. The observed variations in source location for increasing frequency are in contrast with recent findings on differentiation between cortical areas elicited by either stedy-state or transient stimulation which did not show any significant frequency dependent source shift for the transient modality (Pelizzone and colleagues, 1984). This could be due to the different duration of the pulse enveloping the pure tone - 400 ms vs 50-80 ms.

For what concerns the localization of the equivalent source of the field distribution depicted in the seventh map of Fig.2 we remark that the significance level is extremely low, thus supporting the idea of inadequacy of the model. Further support is provided by the quite large 95% confidence region. Equivalent source localization and poor statistical significance were confirmed by similar analyses performed on the other maps (4,5,6,8,9th) concerning the pre-S2 segment. Indeed, the shape of the field patterns measured during the expectation period is asymmetrical, particularly in the upper region close to the vertex. In all subjects a strong variability in the signals detected in this area of the scalp was observed. One possible explanation is that more than one source are simultaneously active during the expectation interval. On the other hand a quite good reproducibility was observed for locations in the temporal region. This suggests that the mentioned variability is not likely to be ascribed to instrumental causes. We must remark that too lengthy recording sessions, as those presently required to collect enough data for adequate mapping, can seriously affect the concentration of the subject.

In conclusion, mapping of magnetic signals related to the E-wave in the CNV paradigm, provides evidence that they are likely to be generated in the anterior region of the brain. Localization performed using the simple dipole-in-a-sphere model suggests an equivalent source in the frontal cortex. Nevertheless, further investigation is required to determine the actual source configuration.

REFERENCES

Chapman R.M., G.L. Romani, S. Barbanera, R. Leoni, I. Modena, G.B. Ricci and F. Campitelli (1983). Lettere al Nuovo Cimento, 38, 549-554.

Chapman R.M., R.J. Ilmoniemi, S. Barbanera and G.L. Romani (1984). Electroencephalogr. Clin. Neurophysiol. In press.

Fiumara R., I. Modena , G.L. Romani , F. Campitelli , M. Caporali, M. Zanasi, A. Cappiello, A. Pasini (1984). New Research Strategies in Biological Psychiatry, Kemali D., Morozov P.V., Toffano G. Eds., John Libbey Co. Ltd., London.

Jerwis B.W., E. Allen, T.E. Johnson, M.J. Nichols and N.R. Hudson (1984). IEEE Trans. Biomed. Eng. BME-31, 342-349.

McCallum W.C. and W.G. Walter (1968). Electroencephalogr. Clin. Neurophysiol. 35, 319-329.

Okada Y. (1984). Approaches to defining the intracranial generators of event related electrical and magnetic fields. Proceedings EPIC VII Meetings, Florence, Italy. In press.

Pelizzone M., S.J. Williamson, L. Kaufman and K.L. Schafer (1984). Annals of New York Academy of Science. In press.

Romani G.L., S.J. Williamson and L, Kaufman (1982). Science, 216, 1339-1340.

Walter W.G. (1964). Arch. Psychiat. Neverkr., 206, 309-322.

Weinberg H., P. Brickett, L. Deecke and J. Boschert (1983). Il Nuovo Cimento 2D, 495-504.

Weinberg H. (1984a). Aproaches to defining the intracranical generators of event related to electrical and magnetic fields. Proceedings EPIC VII Meetings, Florence, Italy. In press.

Weinberg H., P.A. Brickett, J. Vrba, A.A. Fife and M.B. Burbank (1984b). Annals of New YorK Academy of Sciences. In press.

FEW-TRIAL EVOKED FIELD STABILITY USING THE DC SQUID

G. Lewis, M. Blackburn, P. Naitoh* and M. Metcalfe

Navy Personnel Research and Development Center,
San Diego, CA 92152 USA
*Naval Health Research Center,
San Diego, CA 92138 USA

INTRODUCTION

A primary goal of our research is to improve personnel assessment. One procedure which has shown promise is to examine individual differences with emphasis on brain processes. Methods used to date have included evoked potential, and more recently, evoked field measurement (Lewis, 1983; Lewis & Blackburn, 1984). Application of evoked fields (EF) to the study of decision-making and workload requires within and between recording session EF reliability. Single epoch EF recording is desirable in these studies in order to capture short-term brain processes. Previously, averaged potentials and fields have been required due to low signal-to-noise ratio. The objective of the current research is to determine the reliability of visual evoked fields (VEFs) recorded under few trials. Due to improved signal-to-noise ratio, we can now report the recording of single epoch VEF for the first time.

METHOD

Three experienced laboratory personnel (two males, ages 35 and 44, and one female, aged 38) were placed in a prone position for VEF recordings repeated over several days. Recordings were obtained in an unshielded laboratory with a DC SQUID Biomagnetic Detection System (S.H.E. Corp. model 600B, second-derivative gradiometer) having an inherent white noise level of less than 20 fT per square root Hz. A custom counter-weighted system supported the biomagnetic dewar and probe. The pick-up coil was 2.4 cm diameter (3.2 cm between coils) and placed about 12 mm above the subject. The VEF signal (1000X gain on SQUID control unit) was band-passed (1-40Hz, Krohn-Hite, model 3343, 48 db/octave rolloff) and further amplified (100X, Grass P511J) prior to digital conversion. Data acquisition system consisted of a field-portable computer system (Data General NOVA 2/10) equipped with a three-drive floppy disk unit, video terminal, custom-made power supply with fluorescent tube for visual stimulus presentation, and oscilloscopes for monitoring ongoing and sampled EF activity. Each subject viewed (binocular, central fixation) a black and white checkerboard pattern subtending 5.6 degrees visual angle (69 cd per square meter mean luminance). Each check subtended 0.36 degrees visual angle. The checkerboard was back luminated by a 2 ms computer-generated aperiodic flash with a mean inter-flash interval of 750 (150 SD) ms. Recordings were obtained from two locations corresponding to 01 and 02 sites in the 10/20 System. Recording time base was 512 ms. Each waveform was normalized, therefore the waveform mean was the baseline described by dash lines in the figure. Separate VEF recordings were obtained for 40 and single flash presentations. For analyses, the first and second 20 flash records of the 40 flash presentations were averaged separately and combined. In each of the records, latency and amplitude measurements (between baseline and peak) were obtained for the four largest components with alternating polarity after 60 ms. Comparisons of VEF components were made within session and across days. In addition, comparisons were made over 40, 20, and 1 stimulus conditions.

RESULTS

Means (MN) and standard deviations (SD) of latency and amplitude data from each subject and recording site appear in Table 1. These data represent averages over days (N=2) for the 40 stimuli. The zero SD values represent identical latencies from day 1 to day 2. For the 20 stimuli, the data represent averages over days and records (first and second 20 stimuli presentation, N=4). For the 20 stimuli records, the average latencies were 105(15 SD), 176(18), 312(56), and 470(33) ms for the 01 site; and 115(15), 171(16), 289(22), and 429(44) ms for 02. Latency data in Table 1 show large individual differences. Differences between the means for 40 and 20 stimuli, however, were generally small, with the largest being 17 ms for subject 3, site 01, peak 3. Overall differences between the 40 and 20 stimuli records had a grand mean of 6 ms.

Table 1

Summary Statistics for Peak Latencies (ms) and Amplitudes (fT)

Subj No.	Site	Stim No.	Peak 1 MN	Peak 1 SD	Peak 2 MN	Peak 2 SD	Peak 3 MN	Peak 3 SD	Peak 4 MN	Peak 4 SD
					Latencies					
1	01	40	98	21	160	0	256	24	429	14
		20	93	16	158	6	254	32	436	35
	02	40	118	2	150	5	283	5	462	33
		20	116	3	152	4	286	5	467	26
2	01	40	126	0	180	9	323	5	482	33
		20	121	3	175	8	316	7	474	26
	02	40	110	24	166	38	269	71	366	47
		20	100	25	178	32	268	35	381	61
3	01	40	96	24	203	5	349	14	495	5
		20	100	14	195	17	366	17	500	15
	02	40	133	9	180	0	313	9	449	42
		20	130	4	181	3	313	11	437	46
					Amplitudes					
1	01	40	-82	20	92	23	-98	25	102	20
		20	-94	33	98	43	-106	24	113	37
	02	40	116	79	-193	82	170	42	-146	8
		20	122	79	-197	70	172	49	-156	33
2	01	40	-100	22	105	27	-91	7	88	28
		20	-102	51	118	35	-91	10	94	55
	02	40	74	8	-68	6	92	34	-76	17
		20	88	24	-93	18	103	46	-82	16
3	01	40	52	22	-106	31	182	37	-94	0
		20	68	32	-138	62	200	71	-123	30
	02	40	164	77	-244	15	117	11	-68	15
		20	195	77	-247	17	124	60	-83	15

Table 1 also presents amplitude data for each subject, site and the same four peaks described earlier. Amplitudes were measured from baseline to peak with positive and negative direction being maintained. Subjects 1 and 2 showed a waveform reversal form 01 to 02, but subject 3 did not. Amplitude differences between the 40 and 20 stimuli conditions were 7, 10, and 20 fT for subjects 1, 2, and 3 respectively. Overall mean of absolute amplitude differences was 12 fT. Coefficient of variation (CV) for latencies of peaks 1, 2, 3, and 4, when averaged over subjects and sites, were similar for the 40 and 20 stimuli records. The values were 13, 6, 8, and 7 percent and 11, 7, 7, and 8 percent for the 40 and 20 stimuli conditions respectively. Amplitude CV. values, also averaged over subjects and sites, were less for the 40 than the 20 stimuli condition. The values were 36, 23, 21, and 17 for peaks 1, 2, 3, and 4 under the 40 stimuli condition compared to 44, 30, 32, and 29 for the 20 stimuli condition.

Figure 1 shows averaged (40 and 20 stimuli) and single epoch records for the three subjects. Calibration was 50 fT and time base was 512 ms. Calibration scaling was adjusted to accommodate the different amplitudes between records in the same figure. The large amplitude of the single epoch records should be noted. Figure 1.A. shows the VEF records from subject 3, 02 site. The top overlay shows the similarity, and therefore stability in the 40 stimuli recorded two days apart. Stability is also observed in the 20 stimuli records. A well defined positive peak may be observed between 100-150 ms in each of the 40, 20, and single epoch records. In addition, a large negative component is seen between 150-200 ms in each record including the single epoch overlay records. The peak-to-peak amplitude for the positive-negative components between 100-200 ms was about 460, 540, and 915 fT for the 40, 20 stimuli and single stimulus records respectively.

Figure 1.B. shows VEF records for 40 and 20 stimuli, and single stimulus records for subject 1, 02 site. This figure gives an example of individual differences in waveforms when compared with Fig. 1.A. A large negative deflection may be seen in the 40 and 20 stimuli and single epoch records. The large positive deflection, present in the 40 and 20 stimuli records is still present in the 1 trial records, but in much smaller amplitude at about 130 ms. A later appearing (ca. 200-300 ms) positive waveform complex was smaller than those positive peaks around 100 ms in the 40 and 20 records. In the single epoch records, the relative amplitudes for these complexes reversed. For components in the 100-300 ms region, greatest peak-to-peak amplitudes were about 320, 350, and 1090 fT for the 40 and 20 stimuli and single epoch records respectively.

Subject 2 (Fig. 1.C.) showed similar VEF morphology to those of subjects 1 and 3 for the 40 stimuli condition, but had less amplitude. The lower amplitude may have been due to waveform instability over time suggested by large standard deviations in the Table 1 data. With the exception of the large positive component at about 50 ms, well defined components were not observed in the single epoch record for subject 2 compared with those from subjects 1 and 3. Such a record shows not only individual differences between records, but also the potential difficulties in measurement.

Figure 1.D. shows hardware noise recordings for the 40 stimuli and single epoch records.

DISCUSSION

The findings in this research were made possible by several factors. First, experienced personnel provided the necessary immobility, constant arousal level, and adequate selective and sustained attention. We have noticed in our earlier evoked potential and evoked field research that individual differences are large, even in experienced laboratory personnel subjects. Second, the DC SQUID provided the signal-to-noise improvement required for recording few-trial and single epoch VEFs; and third a substantial improvement in the support system allowed the subject to remain prone but be moved easily for different recording positions.

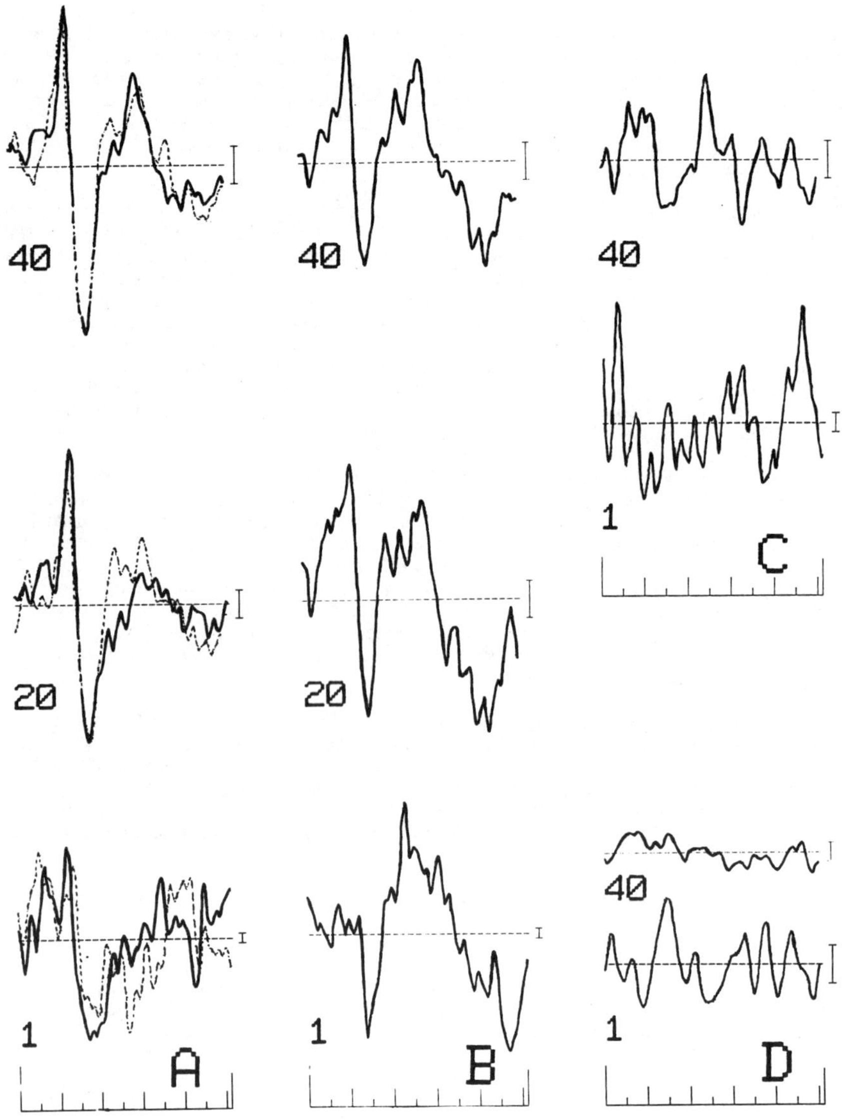

Fig. 1. Averaged and single epoch VEF recordings. A=subject 3, site 02; B=subject 1, 02; C=subject 2, 02; D=noise. Calibration 50 fT, time base 512 ms.

Quantifying the single epoch VEF is relatively straightforward for those similar to Figs. 1.A. and 1.B., however, techniques such as discriminant analysis (Squires & Donchin, 1976), preprocessing by Wiener filter (Naitoh & Sunderman, 1976), or frequency analysis (Lewis, 1973) may provide additional valuable information in the few-trial and single epoch VEF.

Our earlier bioelectric research applications to personnel assessment suggested best results were obtained when few stimuli were used per average (i.e., 10 stimuli per record). Because of large individual differences in the evoked potential records we also found an integrated amplitude measure (i.e., μVrms) to provide at least one number per averaged waveform where particular components may not be well defined (Lewis & Froning, 1981). Similar problems existed in the current research especially in the single epoch records for certain conditions. Variability over the three subjects was large. Several records showed high frequency activity in the single epoch that did not appear in the averaged record, even two stimuli records. Using an integrated measure (i.e., rms) may facilitate the amplitude measurement of few-trial records. Component detection and measurement is required however, to provide latency information, an important benchmark in brain processing.

Research will continue to further describe and quantify the few-trial and single epoch VEF. Such results raise the possibility of improved accurate monitoring and measuring of short-term brain processes. We are continuing to explore the implications and utility of VEF single epoch analyses for habituation, other learning processes, decision-making, and performance prediction.

REFERENCES

Lewis, G. W. (1973). Computer reduction and analysis of neurophysiological data with special emphasis on evoked potential frequency analysis. Computers in Biology and Medicine, 3, 427-436.

Lewis, G. W. (1983). Event related brain electrical and magnetic activity: Toward predicting on-job performance. International Journal of Neuroscience, 18, 159-182.

Lewis, G. W. and Froning, J. N. (1981). Sensory interaction, brain activity, and reading ability in young adults. International Journal of Neuroscience, 15, 128-140.

Lewis, G. W. and Blackburn, M. R. (1984). Biomagnetism: Possible new predictor of personnel performance. (NPRDC Tech. Rep. 84-43). San Diego, CA: Navy Personnel Research and Development Center.

Naitoh, P. and Sunderman, S. (1977). Before averaging: Preprocessing slow potential data with a Wiener filter. In: Otto, D. (Ed.), Multidisciplinary perspectives in event-related brain potential research. Proceedings of the Fourth International Congress on Event-Related Slow Potential of the Brain. EPA-600/9-77-043.

Squires, K. C. and Donchin, E. (1976). Beyond averaging: The use of discriminant functions to recognize event related potentials elicited by single auditory stimuli. Electroencephalography and Clinical Neurophysiology, 41, 449-459.

Partially funded by Defense Nuclear Agency (DNA). The views expressed in this paper are those of the authors and not necessarily those of the Department of the Navy or DNA.

VISUALLY EVOKED RESPONSES FROM NON-OCCIPITAL AREAS OF HUMAN CORTEX*

O.V. Lounasmaa+, S.J. Williamson, L. Kaufman, and R. Tanenbaum

Neuromagnetism Laboratory
Departments of Physics and Psychology
New York University, New York, New York, 10003, U.S.A.

ABSTRACT

Visually evoked neuromagnetic responses from the central area of the cerebral cortex in addition to the usual responses from the occipital areas of primary visual cortex were observed when the velocity of a moving grating pattern was modulated sinusoidally. The source of the central field has different functional properties than the source in primary sensory cortex. The position, depth, and orientation of the source are consistent with it lying in the Rolandic fissure near or in the eye representation area of motor cortex.

KEYWORDS

Neuromagnetism; magnetoencephalogram; visually evoked response; event-related responses

INTRODUCTION

It is now well established that many different areas of the brain respond to visual stimulation in various species of animals. These so-called extrastriate areas include the medial temporal (MT) region of the macaque monkey, and other areas as well, depending upon the species (Van Essen et al., 1981). Evidence exists that cells in some of these extrastriate areas are particularly responsive to moving patterns. We report visual experiments employing moving grating patterns that evoke magnetic fields whose distribution suggests the existence of similar extrastriate areas in humans.

METHODS

The stimuli were bar gratings created by a sinusoidal variation of the luminance across the screen of an oscilloscope. The speed of the grating was changed sinusoidally as it drifted, the modulation being expressed as a "velocity contrast" of 50%, giving the modulation amplitude as a percentage of the average velocity. Psychophysical thresholds for detecting a change in speed were obtained for this kind of stimulus by Kaufman and Williamson (1984). In the present neuromagnetic study the moving grating had a spatial frequency of 1.5 c/deg for one subject (SW) and 3 c/deg for another (PL). The average luminance was 49 cd/m^2, and the luminance contrast of the grating was nearly 100%. The gratings were caused to drift with an average speed of 3 or 5 deg/sec toward the left or the right of the subject. The frequency of modulation of the speed was 7 Hz. Thus when fixating at a black dot at the center of the screen, the subject saw a bar grating drifting in one direction and its speed changed from slow to relatively fast 7 times per second. This modulation frequency served as a reference signal for recovering responses through signal averaging with the steady-state paradigm.

The principles of neuromagnetic measurements are described in detail in a recent multi-authored text (Williamson et al., 1983). A SQUID sensor was used to measure the evoked field at various positions normal to the scalp. The detection coil was a second-order gradiometer of 2.4-cm diameter and 3.2-cm baseline separating adjacent coils. The output of the SQUID electronics was applied to a comb filter to reduce coherent noise at the power line frequency and then to a bandpass filter with 48 db/octave rolloff on either side of 7 Hz. The output of the filter was sent to a signal averager, and the phase lag and amplitude of each response, after one minute of averaging, were recorded.

Measurement positions are indicated by spherical coordinates θ,ρ of a reference system whose origin was judged to give the best constant-radius fit to the posterior half of the head. The corresponding Cartesian system has the +X-axis pointing posteriorly, +Y-axis toward the right shoulder, and +Z-axis upward. Plots shown in this paper present the axes in units of the radius R of the head, where R = 10 cm. The angle θ of a position is measured from the Z-axis, and the angle ρ from the +X-axis in a Z-plane. The orientation of a tangential vector, representing a current dipole model source, is indicated by the angle ψ by which its head is rotated counterclockwise from a plane of constant ρ, with ψ = 0 indicating that the dipole lies in the plane and points toward smaller values of θ.

RESULTS

The right hemispheres of two subjects (SW and PL) were studied in detail with two or more measurements at about 65 positions for each. Responses were observed over a wide area of the scalp from the occipital pole anteriorly to the central region. By studying how these responses varied with stimulus conditions it was possible to differentiate between activity associated with primary striate cortex in the visual area and extra-striate cortex in the more anterior areas. We shall describe the results on SW in detail; the essential features reported here are also seen in the data of the second subject PL.

Occipital Sources

Strong responses observed over the occipital area can be associated with sources in the primary sensory areas of the left and right hemispheres. Figure 1a is a view of the head seen from the back, with positions projected onto the Y-Z plane. Responses are denoted by phasers centered at the various measuring positions. Using data for responses to left and right visual field stimuli, the corresponding current dipoles in the right (RO) and left (LO) occipital regions were estimated by fitting their field patterns to the observed data. These patterns were calculated from the formulas given by Williamson and Kaufman (1981), including the corrections for finite diameter and baseline of the detection coil. Agreement is obtained with one dipole on each side of the midline, with approximately horizontal, opposing orientations. For simplicity we imposed the constraint that they have equal stength, depths and phase lags, and we adjusted the orientation of the LO source slightly so that the upper domain of its field pattern largely cancels the upper domain of the RO source, as is implied by the data in Fig. 1a for a full visual field presentation.

Figure 1b shows the phasers for the RO source, and Fig. 1c the phasers for the superimposed RO and LO souces. The dipole positions and other parameters are given in Table 1. Generally good agreement is achieved with the data in Fig. 1a. The one outstanding exception is the position Y/R=0.35, Z/R=-0.24, where the predicted field is more than twice the observed. This we believe is due to variability in the observed responses, since the data in Fig. 1a show the response at this position to be weaker than responses at positions to either side. Agreement is also less satisfactory in the area above this domain, where the upper domains of RO and LO (of opposite field direction) overlap. The anomalous phases observed here may well come from activity of another source in the right occipital area. Evidence of this was earlier found by Brenner et al. (1981) in the region between upper and lower domains of a source responding to contrast-reversal visual stimuli.

Central Source

The calculated field strength of the superimposed LO and RO field patterns is predicted to be barely detectable in the central area. In Fig. 2a are shown the measured responses in the parietal, temporal, and central areas, this time with positions projected onto the X-Z plane. In Fig. 2b are shown the observed responses with the predicted fields from LO and RO subtracted from the data. The most remarkable features of Fig. 2b are the two domains of relatively strong responses in the central area, one centered at about X/R=0.1, Z/R=0.5 and the other at about X/R=0.35, Z/R=0.9. The phasors in these two domains differ in phase by about 180 deg, and therefore it is reasonable to identify them as being associated with a source that may be modeled as a current dipole lying under the center of the pattern. Studies with left and right visual field stimulation show that response amplitudes are little changed at positions anterior to X/R = 0.4, which is direct evidence that the source of the central field has a different

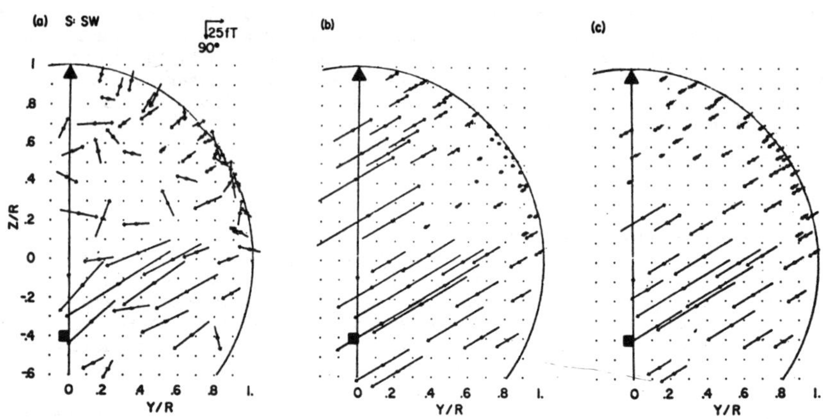

Fig. 1. (a) Observed responses over the occipital region (vertex is designated by a triangle and the inion by a square); (b) calculated contributions from RO source; (c) calculated contributions from both RO and LO sources.

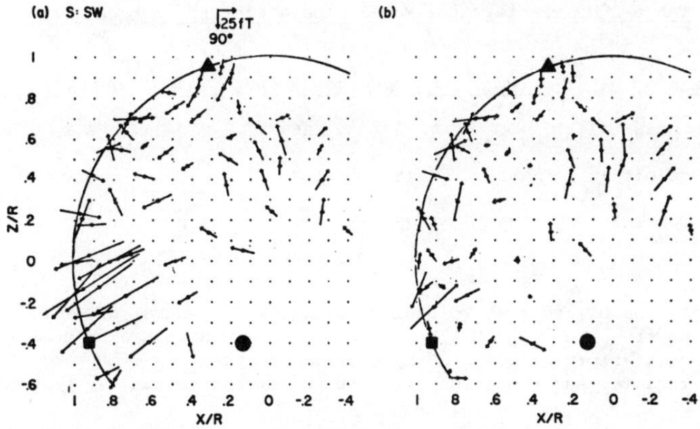

Fig. 2. (a) Observed responses over the right hemisphere (ear canal is designated by a circle); (b) observed responses with calculated contributions of RO and LO sources subtracted.

functional property than that exhibited by RO and LO. The medial temporal area of the macaque monkey also contains representations of ipsilateral and contralateral hemifields as does this one.

Figure 3a displays a dipolar field pattern that reproduces the essential features of the observations in Fig. 2b. When this pattern is subtracted from the phasers in Fig. 2b, what results is shown in Fig. 3b. The fit appears reasonable, aside from exceptional points at about X/R=-0.06, Z/R=0.66 and X/R=0.06, Z/R=0.7 which also display deviant phases in the original data of Fig. 2a. This right central (RC) source of the dipole pattern is characterized by the parameters in Table 1. We can relate the location of this source to the Rolandic fissure, which is well-established for this subject by sensorimotor studies (Okada et al., 1982).

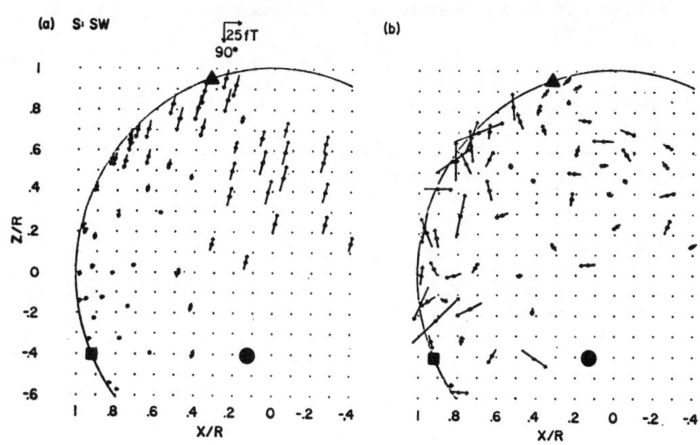

Fig. 3. (a) Calculated contributions of RC source; (b) observed responses with the contributions of RO, LO, and RC sources subtracted.

Table 1 Parameters for Model Dipole Sources (Subject: SW)

Source	θ (deg)	ρ (deg)	Ψ (deg)	d (cm)	Q (nA-m)	Phase Lag (deg)
Right Occipital	90	6	125	2.0	6.0	150
Left Occipital	85	-15	260	2.0	6.0	150
Right Central	45	75	135	3.2	2.2	285

Motor activity associated with voluntary ballastic flexure of the left index finger is observed at about X/R=0.4, Z/R=0.8, a position 3 mm posterior to the Rolandic fissure as prescribed by Gray (1977). The position of RC is 2 cm lower along this fissure, and the orientation of its dipole is exactly perpendicular to the fissure.

DISCUSSION

Observations over the central area clearly reveal a magnetic response to this visual stimulus. This is in addition to the response over the occipital area, whose sources can be modeled by two approximately opposing current dipoles, representing the activity of the primary visual cortex along opposite sides of the longitudinal fissure. The fields over the central area may be ascribed to a current dipole lying deep within the Rolandic fissure near the eye area. One hypothesis explaining this activity could be that it is associated with motor activity related to eye movements in tracking the velocity modulation of the grating. We argue that this is unlikely, since the eye is incapable of tracking oscillations as rapid as 7 Hz. More likely, this activity is in response to signals ultimately originating in the visual cortex that are provided to the eye motor area for comparison purposes (c.f. von Holst and Mittelstaadt, 1950). This conjecture has some support in the preliminary observations in PL that there is a consistent increase in apparent latency on going from the occipital to central field patterns, and this suggests a causal relationship. Thus it appears that there is sequential activity, beginning in the occipital area and shifting to the central.

ACKNOWLEDGEMENTS

We thank Prof. Y.C. Okada for helpful discussions and Prof. J. Anthony Movshon for generous help with software development.

REFERENCES

*Supported in part by Office of Naval Research Contract N00014-76-C-0568.
+Supported by the Academy of Finland. Permanent address: Low Temperature Laboratory, Helsinki University of Technology, SF-02150 Espoo 15, Finland.

Brenner, D., Y. Okada, E. Maclin, S.J. Williamson, and L. Kaufman (1981). In S.N. Erne, H.-D. Hahlbohm, and H. Lubbig (Eds.), Biomagnetism, Walter de Gruyter, Berlin. pp. 431-442.

Gray, H. (1977). Anatomy, Descriptive and Surgical. T.P. Pich and R. Howden, Eds., Bounty Books, New York.

Kaufman, L., and S.J. Williamson (1984). Vision Res., in press.

Okada, Y.C., S.J. Williamson, and L. Kaufman (1982). Magnetic field of the human sensorimotor cortex. Intern. J. Neuroscience, 17, 33-38.

Van Essen, D.C., J.A.R. Maunsell, and J.L. Bixby (1981). Organization of Extrastriate Visual Areas in the Macaque Monkey. In C.N. Woolsey, Ed., Cortical Sensory Organization, Vol. 2, Multiple Sensory Areas, Humana Press, Clifton, New Jersey. pp 157-168.

von Holst, E., and H. Mittelstaadt (1950). Das Reafferenz-princip. Die Naturwissenshaften, 20, 464-467.

Williamson, S.J., and L. Kaufman (1981). Evoked cortical magnetic fields. In S.N. Erne, H.-D. Hahlbohm, and H. Lubbig, Eds., Biomagnetism, Walter de Gruyter, Berlin, pp 353-402.

Williamson, S.J., G.L. Romani, L. Kaufman, and I. Modena, Eds. (1983). Biomagnetism: An Interdisciplinary Approach. Plenum, New York.

MAGNETIC FIELDS EVOKED BY DYNAMIC RANDOM-DOT STEREOGRAMS

H. Weinberg*, P. Brickett*, R.A. Neill**,
B. Fenelon**, and M. Baff*

*Brain Behaviour Lab., Dept. of Psyc., Simon Fraser Univ.,
Burnaby, British Columbia, Canada, V5A 1S6
**Dept. of Psyc., Univ. of Newcastle, New South Wales,
Australia, 2308

ABSTRACT

The visual evoked field was recorded for computer generated dynamic random-dot stereograms which when viewed binocularly gave the perception of a square area suddenly appearing in front of its background. Early and late components of the evoked field were identified which were coincident with the onset and offset of the simultaneously recorded evoked potential. Topographical analysis of the evoked field recorded from 41 positions for one subject shows a possible occipital localization of the earlier component and a right hemisphere predominance of activity for the less dipolar-like distribution of the later component.

KEYWORDS

Depth perception; random-dot stereograms; stereopsis; evoked field; neuromagnetism; magnetoencephalography; evoked potentials.

INTRODUCTION

The three-dimensional patterns seen in dynamic random-dot stereograms are unique visual stimuli in that the depth information is not contained in monocular input but exists only at the cortical level where information from the two eyes is combined. Evoked potentials in response to such stimuli have been reported by a number of workers in recent years, e.g. Mol and Caberg (1977), Lehmann and Julesz (1978). As also shown by more recent research, the evoked potential consistently features a single negative wave, peaking at 150 to 350 ms post stimulus, with a topographical distribution covering much of the posterior scalp (Herpers, Caberg, & Mol, 1981). This paper presents initial findings of the corresponding magnetic fields evoked by dynamic random-dot stereograms.

In the literature on depth perception, the terms "local" and "global" stereopsis are commonly used to refer to the stages of information processing which the visual system performs in converting binocular input into perceived depth. Local stereopsis is the point-by-point comparison of the images on the two retinae and the calculation of the associated disparity for every feature on the visual image. Global stereopsis is the integration of all the coded, local disparities into an overall perception of a three dimensional scene. Concerning the possible anatomical location of these mechanisms, Poggio and Fischer (1977) found that 84% of the neurons studied in area 17 of the visual cortex of the Rhesus monkey were sensitive to binocular disparity. The surrounding prestriate cortex is also a site of cells involved in processing local disparities. Cowey (1979) has reported that for the monkey, lesions in an area near the midline in the occipital lobe impairs stereoacuity, i.e. the ability to make fine discriminations, but does not affect global stereopsis. On the other hand the reverse conditions result from bilateral lesions of an area of the temporal lobe, the posterior inferotemporal cortex (Cowey and Porter, 1979). For humans, research with brain-damaged patients suggests further that the right hemisphere is dominant for global stereopsis (Benton & Hecaen, 1970; Carmon & Bechtoldt, 1969; Hamsher, 1978).

The idea that local stereopsis is a separate process from depth perception is also supported by studies of the responses to random-dot correlograms, visual stimuli for which the left and right patterns are alternately either identical or uncorrelated. Such stimulation is an effective cue for local but not global stereopsis and there is an electrical evoked response (Lehmann, Skrandies, and Lindenmaier, 1978). For the human infant, an evoked potential is detectable at 3 months of age for correlograms (Braddick & others, 1980), but not until 4.5 months for stereograms (Petrig & others, 1981). Furthermore, adults have been found to generate normal evoked potentials to correlograms presented at up to twice the limiting frequency for stereograms (Julesz, Kropfl, & Petrig, 1980). Assuming that two different processes are involved in the response to dynamic random-dot stereograms, it seems reasonable, at least in hindsight, to expect magnetic fields to be observed from at least two distinct sources.

METHOD

The equipment used to produce the dynamic random-dot stereograms has previously been described (Neill & Fenelon, 1981). A microcomputer-controlled generator displayed approximately 100 thousand pairs of points per second on two cathode ray tubes (CRTs). When the stereoscopic stimulus was off, all points were in the same position on both CRTs. When the stimulus was on, a square subsection of points in the central area was shifted to the right on the CRT viewed with the left eye and to the left on the CRT viewed with the right eye. This gave the appearance of a square area of dots located in front of the background plane of dots. The stimulus was presented for 200 ms once per second. In order to keep the magnetically noisy equipment at a sufficient distance (approximately 4 m), the subject viewed the CRTs, as reflected in a mirror, through specially made binoculars. The resulting approximate visual angle of the display area was 13 degrees, that of the square area was 4 degrees, and the interocular stimulus disparity was 0.6 degrees.

Electrical recordings were obtained from electrodes on the occipital scalp at 10/20 system locations 01 and 02, referenced to a location 10 cm above the nasion. As a monitor for eye movements, the electrooculogram (EOG) was also recorded from an electrode on the suborbital ridge. Magnetic fields were recorded from positions over the occipital, parietal, and temporal areas using a SQUID third order gradiometer manufactured by CTF Systems, Inc. (Vrba and others, 1982). This instrument has a sensing coil diameter of 38 mm and a distance of 55 mm between coils. Half second epochs of data were sampled with a PDP/12 computer and averages were calculated from 256 sweeps.

The location of each recording position was measured with respect to the vertical and horizontal distance to the inion. For the topographical mapping the horizontal axis was compressed by a factor of two so that the mapping could be carried out for a circular area approximating the distribution of recording positions. Although some positions fell outside of this circle, all positions were used for the interpolation of map contours. The procedure for computing contour maps was to first interpolate values for a 78 by 75 regularly-spaced grid, second to (spatial) lowpass filter those values, and third use linear interpolation to increase the map resolution to 156 by 225. The initial interpolation algorithm which was used computes a value on the grid by taking a weighted average of the data values from the surrounding positions within a given search radius which was set to one quarter of the diameter of the circular area mapped.

RESULTS AND DISCUSSION

Results for the one subject that we have studied extensively show a sizable magnetic evoked field coincident with the electrical response. Figure 1 presents evoked field waveforms which have been selected as examples of the two most obvious components observed in the total set of recordings. The first and third traces from the top at the right show an early component directed into the head at the midline and out of the head at the right temporal area, while the second trace shows the reduced amplitude at an intermediate location. The first two traces at the left show a later component which is most prominent at the right parietal area. Also shown in Fig. 1 is the O1 and O2 evoked potential, which is similar to others previously obtained with this stimulus (Neill & Fenelon, 1981).

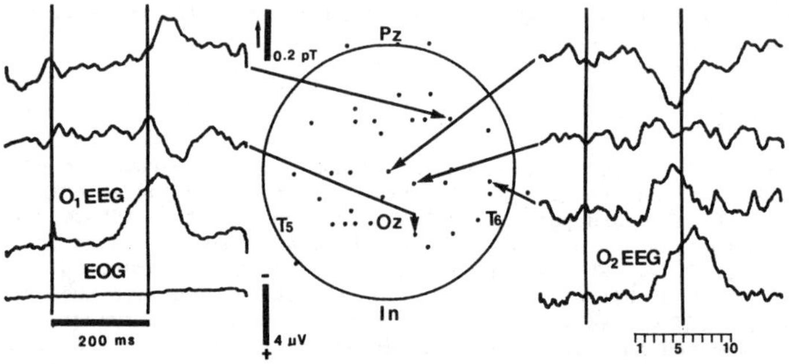

Fig. 1. Five evoked field waveforms recorded from the positions indicated on the diagram of the back of the head. Upward deflections correspond to fields exiting the head. Other traces are as labeled. Center of circle is 5.5 cm above inion, radius is 5 cm (vertical) and 10 cm (horizontal). Stimulus onset and offset marked by vertical lines. The 10 intervals indicated at lower right correspond to the 20 ms windows used for the topographical analysis presented in Fig. 2.

For the topographical analysis the mean magnetic response measured over 20 ms intervals was mapped for the 41 scalp positions from which we have recorded. Figure 2 presents 10 of these maps for time windows starting at 100 ms and ending at 280 ms after stimulus onset (1 to 10). The first map shows the lack of early activity; while number 4 at 160 to 180 ms shows the largest amplitudes for the early component with fields entering the head along the occipital midline and exiting from the right hemisphere. Maps of the later activity show a right parietal peak of fields exiting the head (maps 7-10). However, since the later field patterns do not appear dipolar, especially those shown in map 7, localization is more problematic. Although accurate localization is unclear, a right hemisphere preponderance is

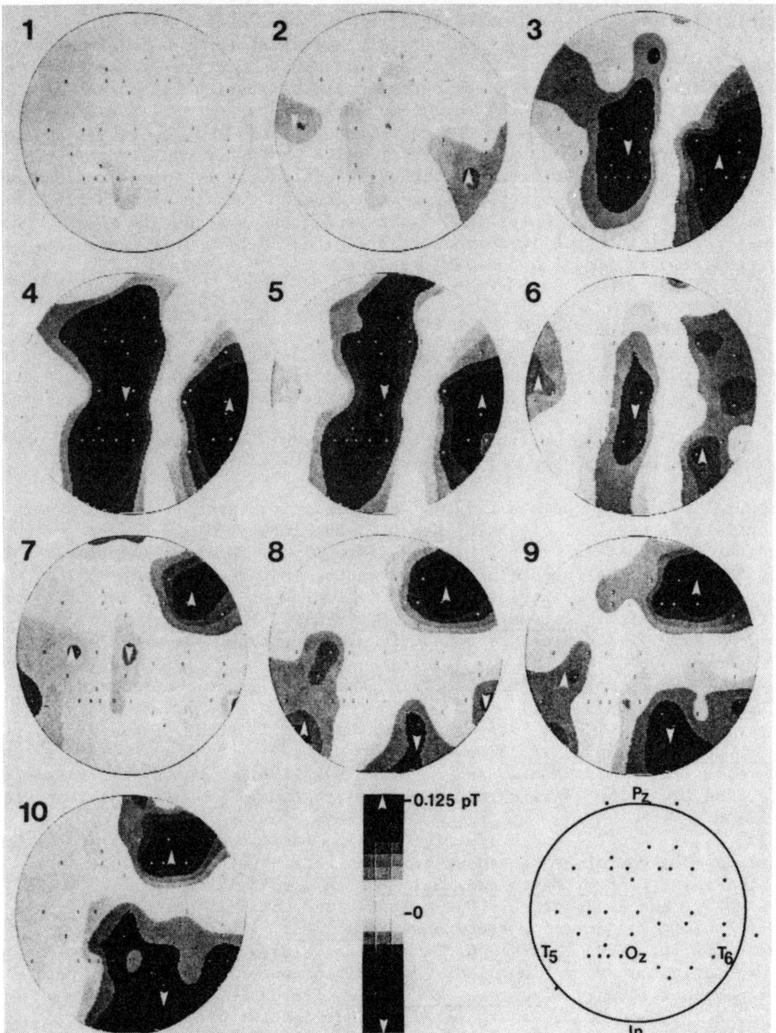

Fig. 2. Evoked field maps for 20 ms windows starting 100 ms after stimulus onset. Circle represents same area as in Fig. 1. Maximum amplitude on calibration bar corresponds to 0.125 picoTesla. Fields exiting the head indicated by upward arrows.

consistent with animal and human clinical studies which suggest that the right hemisphere is most important for global stereopsis.

The most interesting finding from this research is that, while the evoked potential generally features only a single negative wave, the evoked field shows two major components. Although we hesitate to apply a simple dipole model in order to allow calculation of a source location (which would be theoretically possible for the early component), the observed topographical distribution is consistent with a visual cortex location for the early component and a temporal lobe location for the latter component. It could be the case that the same two generators are producing both the recorded magnetic and electric fields, but only one electric component is observed on the scalp due to a summation effect caused by volume conduction. If this is true, then the magnetic recordings contain additional information about the stages of information processing. Future research is needed to test the idea that the early and late magnetic components are produced by neural mechanisms responsible for, respectively, local and global stereopsis.

ACKNOWLEDGEMENTS

Supported by B.C. Science Council, Ministry of Transport, N.S.E.R.C., N.R.C., and D.S.S., Canada; S.R.C., University of Newcastle, and N.H. & M.R.C., Australia.

REFERENCES

Benton, A. L., and H. Hecaen (1970). Stereoscopic vision in patients with unilateral cerebral disease. Neurol., 20, 1084-1088.

Braddick, O., J. Atkinson, B. Julesz, W. Kropfl, I. Bodis-Wollner, and E. Raab (1980). Cortical binocularity in infants. Nature, 288, 363-365.

Carmon, A. and H. P. Bechtoldt (1969). Dominance of the right cerebral hemisphere for stereopsis. Neuropsychologia, 7, 29-39.

Cowey, A. (1979). Cortical maps and visual perception: The Grindley Memorial Lecture. Q. J. Exp. Psyc., 31, 1-17.

Cowey, A., and J. Porter (1979). Brain damage and global stereopsis. Proc. R. Soc. Lond. B, 204, 399-407.

Hamsher, K. de S. (1978). Stereopsis and unilateral brain disease. Invest. Ophthal. Vis. Sci., 17, 336-343.

Herpers, M. J., H. B. Caberg, and J. M. F. Mol (1981). Human cerebral potentials evoked by moving dynamic random-dot stereograms. EEG Clin. Neurophysiol., 52, 50-56.

Julesz, B., W. Kropfl, and B. Petrig (1980). Large evoked potentials to dynamic random-dot correlograms and stereograms permit quick determination of stereopsis. Proc. Nat. Acad. Sci. USA, 77, 2348-2351.

Lehmann, D., and B. Julesz (1978). Lateralized cortical potentials evoked in humans by dynamic random-dot stereograms. Vis. Res., 18, 1265-1271.

Lehmann, D., W. Skrandies, and C. Lindenmaier (1978). Sustained cortical potentials evoked in humans by binocularly correlated, uncorrelated and disparate dynamic random-dot stimuli. Neurosc. Lett., 10, 129-134.

Mol, J. M. F., and H. B. Caberg (1977). Contingent stereoptic variation as a response to moving random-dot stereograms. EEG. Clin. Neurophysiol., 43, 513.

Neill, R. A., and B. Fenelon (1981). The objective evaluation of stereopsis. Aust. Phys. Eng. Sci. in Med., 4, 122-125.

Petrig, B., B. Julesz, W. Kropfl, G. Baumgartner, and M. Anliker (1981). Development of stereopsis and cortical binocularity in human infants: Electrophysiological evidence. Science, 213, 1402-1405.

Poggio, G. F. and B. Fischer (1977). Binocular interaction and depth sensitivity in striate and prestriate cortex of behaving rhesus monkey. J. Neurophysiol., 40, 1392-1405.

Vrba, J., A.A. Fife, M. B. Burbank, H. Weinberg, and P. A. Brickett (1982). Spatial discrimination in SQUID gradiometers and 3rd order gradiometer performance. Can. J. Phys., 60, 1060-1073.

NEUROMAGNETIC RESPONSES TO NOXIOUS STIMULATION

R. Hari[1], E. Kaukoranta[1], K. Reinikainen[1], and J. Mauno[2]
[1]Low Temperature Laboratory,
Helsinki University of Technology, 02150 Espoo, Finland
[2]Department of Physiology,
University of Helsinki, 00170 Helsinki, Finland

ABSTRACT

Neuromagnetic responses to electric stimulation of the tooth pulp were studied. The field pattern suggests activation of the frontal operculum.

KEYWORDS

Evoked responses; magnetic fields; tooth pulp; pain.

INTRODUCTION

Pain is a complex subjective experience which cannot be explained simply by the peripheral mechanisms. One possibility to get information on the central processes involved is to record cortical potentials evoked by well defined noxious stimuli. These potentials cannot be considered as specific correlates of pain but covary with the subjective estimates (Chapman, Chen, and Harkins, 1979). We have studied central mechanisms of pain by recording neuromagnetic responses to electric stimulation of the tooth pulp, which contains mainly nociceptive afferents. The first report of this work has been published (Hari and co-workers, 1983).

MATERIALS AND METHODS

Six healthy adult volunteers were studied in the Otaniemi magnetically shielded room. In all subjects, the component of the magnetic field perpendicular to the skull was measured with a 1-channel first order SQUID gradiometer. Later the measurements were continued in two subjects with a 4-channel gradiometer (Ilmoniemi and co-workers, 1984). The sensitivities of the gradiometers were 26 $fT/\sqrt{Hz}$ (1-Ch) and 22 $fT/\sqrt{Hz}$ (4-Ch). The pickup coils (areas 1.7 or 1.9 cm^2) situated 12 mm above the scalp. For two subjects, the right side of the head was extensively mapped. In the other subjects the main features of the field patterns and response waveforms were confirmed by less numerous measurements.

An intact tooth was stimulated electrically with non-invasive electrodes. The stimuli were monophasic constant current square-wave pulses of 10 ms duration. The cathode (diameter 4.5 mm) was hand-held against the crown of the tooth, and the anode was either on the skin beneath the nose or over the upper gingiva.

The measurement bandwidth was 0.05 - 50 Hz. All single responses contaminated by eye movement or muscular artefacts were omitted from the analysis. The analysis period of 1024 ms included a prestimulus baseline of at least 250 ms. In some experiments electrical responses were recorded simultaneously.

RESULTS

Dental stimulation elicited evoked responses in all subjects. The main deflection occured at latencies of 90-100 ms, reaching maximally 0.45 pT in amplitude. Stimulation of ipsi- and contralateral central incisors elicited similar responses, and clear responses were obtained also to stimulation of the ipsilateral canine tooth.

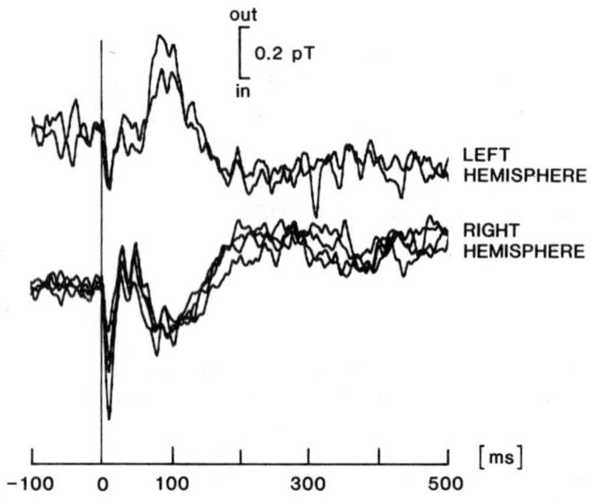

Fig. 1. Averaged magnetic responses (N = 50) of one subject at homologous measurement locations on the left and right anterior frontotemporal areas. The right central incisor was stimulated. Curves from consecutive measurements are superimposed. Stimulus artefacts can be seen after zero-time.

The polarities of the responses were opposite on the left and right hemispheres (Fig. 1). As shown in Fig. 2, the main deflection changed its polarity between the anterior and posterior frontal areas. On the right hemisphere, the flux came out of the skull at posterior and re-entered at the anterior recording locations. At the peak latency of the responses the estimated current dipole was directed upwards, and it was situated 30 mm beneath the skull (the radius of curvature of the measurement sphere was assumed to be 10 cm); the strength of the source was about 30 nAm. In two subjects recordings were made from both hemispheres: the sources were located more posteriorily on the left than on the right side. An increase of the interstimulus interval from 4 to 8 seconds increased the amplitude of the main deflection by about 50%. The main electric deflections (in the vertex-inion derivation) occured at latencies 140 ms (vertex negative) and 250 ms (vertex positive). The waveforms of the electric and magnetic responses were thus rather different.

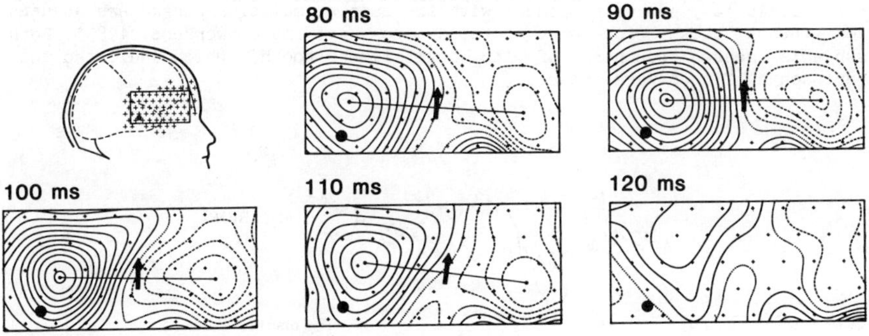

Fig. 2. Isofield contour maps of one subject at different latencies after stimulation of the contralateral central incisor. The maps were calculated using a weighted least squares approximation and bicubic spline interpolation (McLain, 1974). The solid lines indicate flux out of the skull and the dashed lines flux into the skull. The curves are separated by 40 fT. The interpoint distance of the measurement grid is 16 mm. The black dots show the location of T4 of the International 10 - 20 system. The arrows denote the equivalent current dipoles placed to the midpoints of the lines connecting the field extrema.

Control recordings

I. **Extracerebral current sources**

Vertical eye movements and eye blinking cannot be responsible for the recorded responses because the signals produced by them are maximal at the posterolateral side of the orbit and do not reverse polarity in our measurement area (Katila and co-workers, 1981; Antervo and co-workers, this volume). Muscular artefacts can be excluded because increased tension of the temporalis and masseter muscles did not increase the amplitude of the response. In measurements over the anterior extremum of the field, when the dewar was separated by 15 mm from the skull (the pickup coils were about 30 mm from the outer surface of the skull), the signal was about 70% of its original value measured close to the skull. This kind of slow decrease of the field as a function of distance fits better with a deep (cerebral) than a superficial (muscular) source. Neither does the form of the field pattern - with the steepest slopes near the polarity reversal line - suggest activation of a broad superficial dipole layer (like the temporal muscle).

II. **Activation of extrapulpal afferents**

Several reasons suggest that extrapulpal afferents were not activated. First, the subjective sensation caused by the stimulation was sharp pain, localized to the tooth. Second, electric stimulation of the trigeminal skin areas (branches II and III) elicited reflectory blinks and blink artefacts peaking at 50 ms; the polarity of these artefacts was opposite to that of magnetic responses in the same area. The possible contribution from activation of periodontal fibers was excluded by recording evoked responses to tapping of the tooth, which elicited a clear sensation of movement. The evoked responses were minimal as compared to responses to electric stimulation of (Fig. 3). Further, stimulation of a devitalized (pulpotomized) tooth in two subjects did not elicit any evoked responses with the same stimulation parameters as used for intact teeth (for more details see Hari and co-workers, 1983). Both subjects sensed touching of the devitalized tooth, which indicated that periodontal fibers were intact.

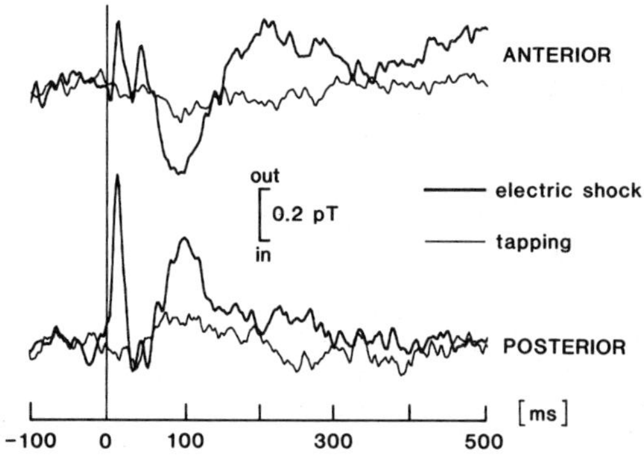

Fig. 3. Responses of one subject to the electric stimulation or tapping of a healthy central incisor. The measurements are from the anterior and posterior field extrema on the right hemisphere.

DISCUSSION

The extensive artefact controls lead us to suggest that the responses elicited by painful dental stimulation were not due to activation of cutaneous, gingival or periodontal fibers but rather transmitted by tooth pulp afferents. The pattern of the magnetic responses suggests bilateral activation of brain areas at the rostral end of the Sylvian fissure, indicating that tooth pulp has cortical representation in man. Anatomically, the activated area corresponds to frontal operculum; in one subject this approximate location was confirmed by NMR-imaging. We have earlier suggested (Hari and co-workers, 1983) that dental stimulation activates the rostral end of the secondary somatosensory cortex SII. Later mappings in one subject showed, however, that the cortical representation area of tooth pulp is clearly (about 3 cm) anterior to the SII activated by peroneal nerve stimulation. Very little is known about the function of the frontal opercular cortex in man. In monkey, the anterior opercular-insular cortex receives taste nerve afferents (Benjamin and Burton, 1968; Sanides, 1968). In cat, oral-facial afferents project to the presylvian and anterior orbital gyri where the cells have large and often bilateral receptive fields (Lund and Sessle, 1974). The spatial relations of the cortical representation areas of the tooth pulp and other oral-facial sites remain to be resolved by future studies.

No responses were found around the primary somatosensory cortex which in monkey is activated by dental stimulation (Melzack and Haugen, 1957). This negative finding might be due to stimulus artefacts which prevented us from studying the magnetic counterparts of the early cortical responses. The considerable waveform differences between the simultaneously measured electric and magnetic responses are probably due to the fact that radial current sources contribute only to the electrical signals.

Neuromagnetic responses might be useful in studies of acute experimental pain. Further work is, however, needed to evaluate the relationship between the observed neural activity and the perceived pain.

ACKNOWLEDGEMENTS

This work has been supported by the Academy of Finland. T. Huopaniemi participated in the early parts of the study. The magnetometers were constructed by R. Ilmoniemi and the dental stimulator by P. Vikberg. The computer programs were written by M. Hämäläinen and J. Salminen. J. Sipponen took the NMR-pictures.

REFERENCES

Antervo, A., R. Hari, T. Katila, T. Ryhänen, and M. Seppänen (1984). Magnetic field pattern produced by eye blinking. This volume.

Benjamin R.M., and H. Burton (1968). Projection of taste nerve afferents to anterior opercular-insular cortex in squirrel monkey (Saimiri sciureus). Brain Res., 7, 221-231.

Chapman, C.R., A.C.N. Chen, and S.W. Harkins, (1979). Brain evoked potentials as correlates of laboratory pain: a review and perspective. In J.J. Bonica, J.C. Liebeskind and D. Albe-Fessard (Eds.), Advances in Pain Research and Therapy, Vol. 3, Raven Press, New York, pp. 791-803.

Hari, R., E. Kaukoranta, K. Reinikainen, T. Huopaniemi, and J. Mauno (1983) Neuromagnetic localization of cortical activity evoked by painful dental stimulation in man. Neurosci. Lett., 42, 77-82.

Ilmoniemi, R., Hari, R., and Reinikainen, K. (1984). A four-channel SQUID magnetometer for brain research. Electroenceph. clin. Neurophysiol., in press.

Katila, T., R. Maniewski, T. Poutanen, T. Varpula, and P. Karp (1981). Magnetic fields produced by the human eye. J. Appl. Phys., 52, 2565-2571.

Lund, J.P., and B.J. Sessle (1974). Oral-facial and jaw muscle afferent projections to neurons in cat frontal cortex. Exp. Neurol., 45, 314-331.

McLain, D.H. (1974). Drawing contours from arbitrary data points. Computer J., 17, 318-324.

Melzack, R. and F.P. Haugen (1957). Responses evoked at the cortex by tooth stimulation. Am. J. Physiol., 190, 570-574.

Sanides, F. (1968). The architecture of the cortical taste nerve areas in squirrel monkey (Saimiri sciureus) and their relationships to insular, sensorimotor and prefrontal regions. Brain Res., 8, 97-124.

MAGNETIC FIELDS EVOKED BY PERONEAL NERVE STIMULATION

E. Kaukoranta, R. Hari, and K. Reinikainen

Low Temperature Laboratory, Helsinki University of Technology, SF-02150 Espoo, Finland

ABSTRACT

Magnetic fields evoked by peroneal nerve stimulation were measured. The field patterns indicate activation of contralateral primary and ipsi- and contralateral secondary somatosensory cortices. At the secondary somatosensory cortices the temporal behaviour of ipsi- and contralateral responses was different.

KEYWORDS

Primary somatosensory cortex; secondary somatosensory cortex; magnetic fields; evoked responses; peroneal nerve.

INTRODUCTION

Electric stimulation of the median nerve elicites evoked magnetic responses around the contralateral primary somatosensory hand area (Kaufman and co-workers, 1981) and around the ipsi- and contralateral secondary somatosensory cortices (SII) at the upper banks of the Sylvian fissures (Hari and co-workers, 1984). Correspondingly, the field pattern elicited by stimulation of peroneal nerves is compatible with the activation of the primary somatosensory foot areas on the mesial surfaces of the postcentral gyri and with the activation of SII.

In the present work, special emphasis was put on the comparison of field patterns elicited by left- and right-sided peroneal nerve stimulation.

METHODS

Six healthy volunteers (ages 24-35 years) from the laboratory personnel served as subjects. Rectangular constant-voltage pulses of 0.3 ms duration were delivered alternately to the left and right peroneal nerves at the ankles through surface electrodes. The interstimulus interval for each nerve was 2.3 s and the stimulus intensity was just above the motor threshold. The magnetic field component perpendicular to the skull was measured either with a

1-channel or with a 4-channel first-order gradiometer (see Hari and co-workers, 1984 and Ilmoniemi, Hari and Reinikainen, in press). In order to get responses from the primary somatosensory cortices the measurements were made along the midline in all six subjects. In addition, responses were mapped from 60 points on the top of the head in one subject. Further, to study the activity of the second somatosensory cortices the field patterns on the lateral sides of the head were mapped from about 60 locations in three subjects. The signals were digitized (sampling frequency 500 Hz), and artefact-free responses were averaged on-line. The analysis period of 1024 ms included a prestimulus interval of 300 ms. Isofield contour maps were calculated using a weighted least-squares approximation and bicubic spline interpolation.

RESULTS

Figure 1 shows averaged magnetic responses to stimulation of the left and right peroneal nerves in one subject. At this recording location the polarities of the responses were opposite. In midline measurements (Fig. 2), the responses reversed in polarity between the anterior and posterior locations. In 5 out of the 6 subjects the polarity reversal point was posterior to the vertex. For left-sided stimulation the flux came out from frontal parts and re-entered the skull in posterior parts during the main deflection of the responses at latencies around 100 ms. For right-sided stimulation the direction of the field was the opposite. In 5 out of the 6 subjects the polarity reversal for the right peroneal nerve stimulation occured 1-2 cm posterior to that for the left sided stimulation. In some subjects the responses were preceded by slow shifts which also reversed in polarity between the anterior and posterior measurement locations. The field pattern revealed by mapping around the primary somatosensory cortex was in agreement with measurements along the midline. The pattern was, however, more complicated than that for the median nerve stimulation measured around the Rolandic fissure (Hari and co-workers, 1984).

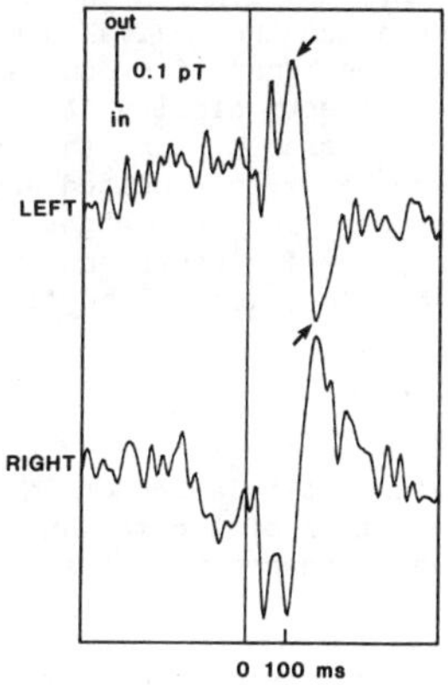

Fig. 1. Averaged magnetic responses (N = 90) to stimulation of the left and right peroneal nerves. The recording is from the vertex (Cz). The passband is 0.05-45 Hz. Arrows indicate peak amplitudes. An upward deflection indicates flux out of the skull.

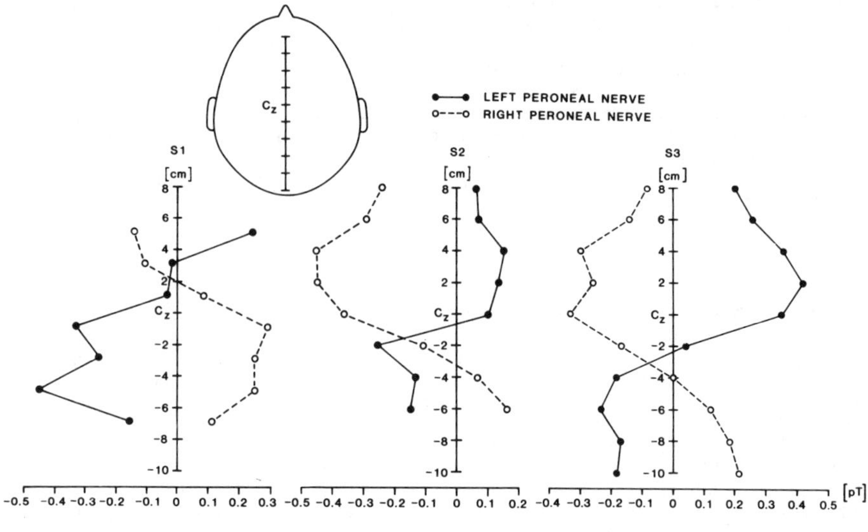

Fig. 2. Amplitude distribution along the midline for responses to stimulation of left (solid line) and right (dashed line) peroneal nerves in three subjects. The peak-to-peak amplitudes (see Fig. 1) were used in constructing this figure.

Mappings on the lateral sides of the head revealed dipolar field patterns with maximal amplitudes at 100-150 ms, depending on the hemisphere and the subject (Fig. 3). For ipsilateral stimulation the maximal amplitudes were smaller and they were reached 10-40 ms later than for contralateral stimulation. The locations of the equivalent current dipoles for contra- and ipsilateral stimulation do not differ considerably. In the right hemisphere of this subject the current dipoles were, however, located about 2 cm more anteriorly compared to the left hemisphere. In the right hemisphere the field was directed out of the skull in posterior parts and into the skull in anterior parts. In the left hemisphere the direction of the field was the opposite.

DISCUSSION

Although the field pattern for stimulation of the peroneal nerves is more complicated than that for stimulation of the median nerves it is in agreement with activation of the primary somatosensory foot areas at the mesial surfaces of the hemispheres.

Vera, Perot, and Fountain (1983), recording evoked potentials to posterior tibial nerve stimulation, suggested that the post-central gyrus could be located more posteriorly in the left than in the right hemisphere. This agrees with our finding that the polarity reversal point for the right peroneal nerve stimulation was usually posterior to that for the left-sided stimulation.

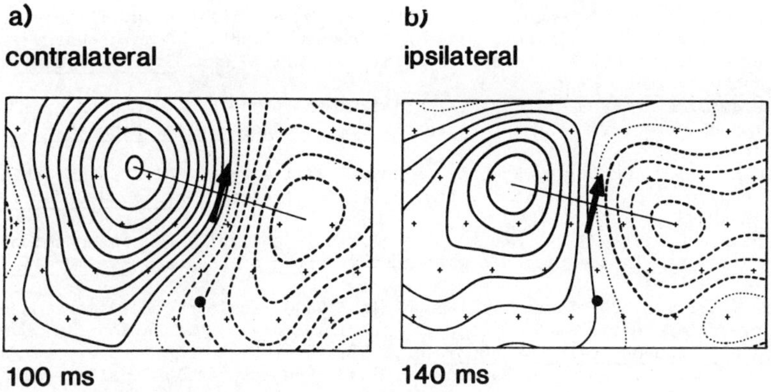

Fig. 3. Isofield maps to contralateral and ipsilateral peroneal nerve stimulations over the left hemisphere during the maximal field amplitudes, which were reached later for ipsi- than contralateral stimulation. The distance between measurement points is 16 mm. The dot indicates the location of T3 (international 10-20 system) on the same horizontal line as the eyes. Solid lines indicate flux out of the skull. Curves are separated by 40 fT. The arrows indicate the equivalent current dipoles in the middle between the field extrema.

The dipolar field patterns over the sides of the head were at some latencies asymmetric indicating that all of their features cannot be explained by a simple current dipole model. However, the patterns support the activation of SII on the upper bank of the Sylvian fissure. In contrast to the primary somatosensory cortex, which is activated mainly contralaterally, SII can be activated by both contra- and ipsilateral stimuli (Woolsey and Fairman, 1946; Robinson, 1973). Our results support this view. The evoked potentials measured from pig's SII to ipsilateral forelimb stimulation are smaller and have longer latency than those to contralateral stimulation (Woolsey and Fairman, 1946). Single cell studies from cat's SII (Robinson, 1973) indicate that the activity elicited by ipsilateral stimulation is generally weaker than that produced by contralateral stimulation. The ipsilateral responses are suggested to be mediated by the contralateral somatosensory cortex via corpus callosum (Robinson, 1973). Our results indicate that, also in humans, the input to the contralateral SII is stronger and faster than that to the ipsilateral SII.

ACKNOWLEDGEMENTS

The magnetometers were constructed by R. Ilmoniemi, and the computer programs were written by M. Hämäläinen and J. Salminen. This study was supported by the Academy of Finland and by the Emil Aaltonen Foundation.

REFERENCES:

Hari, R., K. Reinikainen, E. Kaukoranta, M. Hämäläinen, R. Ilmoniemi, A. Penttinen, J. Salminen, and D. Teszner (1984). Somatosensory evoked cerebral magnetic fields from SI and SII in man. Electroenceph. clin. Neurophysiol., 57, 254-263.

Ilmoniemi, R., R. Hari, and K. Reinikainen (1984). A four-channel SQUID magnetometer for brain research. Electroenceph. clin. Neurophysiol., in press.

Kaufman, L., Y. Okada, D. Brenner, and S.J. Williamson (1981). On the relation between somatic evoked potentials and fields. Int. J. Neurosci., 15, 223-239.

Robinson, D. (1973). Electrophysiological analysis of interhemispheric relations in the second somatosensory cortex of the cat. Exp. Brain Res., 18, 131-144.

Vera, C.L., P.L. Perot, and E.L. Fountain (1983). Scalp recorded somatosensory evoked potentials to posterior tibial nerve stimulation in humans. Electroenceph. clin. Neurophysiol., 56, 159-168.

Woolsey, C. N., and D. Fairman (1946). Contralateral, ipsilateral, and bilateral representation of cutaneous receptors in somatic areas I and II of the cerebral cortex of pig, sheep, and other mammals. Surgery, 19, 684-702.

MAGNETOENCEPHALOGRAPHIC EVIDENCE FOR POSSIBLE SUPPLEMENTARY MOTOR AREA PARTICIPATION IN HUMAN VOLUNTARY MOVEMENT

L. Deecke*, J. Boschert*, P. Brickett**, and H. Weinberg**

*Abteilung Neurologie, Universitat Ulm, D-7900 Ulm, West Germany

**Brain Behaviour Laboratory, Simon Fraser University, Burnaby, B.C., V5A 1S6, Canada

ABSTRACT

The magnetic activity of the brain was recorded during the repetitive performance of a complex series of finger movements. A slow magnetic shift was found to occur starting up to 1 s prior to the initiation of movement. The topography of this activity was related to the location of structures known to be involved in the organization of the movements, i.e., the hand area of the rolandic motor cortex, and the supplementary motor area in the mesial frontal cortex.

KEYWORDS

Event-related field; Bereitschaftspotential; supplementary motor area; movement programming.

INTRODUCTION

A slow magnetic field shift of 50-100 fT on the average has been detected in the human brain occurring in the foreperiod of a human voluntary movement. This magnetic field accompanies a slow negative electrical cerebral potential of 5-15 µV on the average which occurs in the same foreperiod, the Bereitschaftspotential (BP) or readiness potential (Kornhuber and Deecke, 1965). Consequently, the analogous magnetic field has been termed Bereitschaftsmagnetfeld (BM) or readiness magnetic field preceding simple voluntary finger (Deecke, Weinberg, and Brickett, 1982) and foot or toe movements (Deecke and others, 1983; Hari and others, 1983). A similar BM has been detected prior to human speech production (Weinberg and others, 1983).

Evidence accumulated hitherto regarding supplementary motor area (SMA) activation accompanying human voluntary movement stems from electroencephalographic (EEG) and regional cerebral blood flow (rCBF) studies: In parkinsonism, the BP prior to voluntary finger movements is greatly reduced over the lateral precentral regions but normal in amplitude over the midline (Cz), suggesting an active generator in the depth of the sagittal fissures between the hemispheres, the site of the SMA (Deecke and Kornhuber, 1978). Independently, during (perpetuous) contralateral hand movements in normal subjects, rCBF studies reveal - besides the activity of the hand area of the primary motor cortex (MI) - an additional site of increased blood flow which corresponds to the SMA (Roland and others, 1980).

This communication reports on magnetoencephalographic recordings of slow field shifts preceding a more complex finger tapping task, which is assumed to require stronger SMA activation than does simple finger movements (cf. Eccles, 1982). The field topography found is compared to the one for simple finger movements and that for foot movements, and the results will be discussed with reference to the possible role of the SMA.

METHODS

Movement-related magnetic field changes in relation to a complex finger tapping task were recorded in three healthy right-handed subjects from different positions over the head. The magnetoencephalogram (MEG) was recorded with a SQUID third-order spatial gradiometer, manufactured by CTF Systems, Inc., (Vrba and others, 1982). The sensing coil of this gradiometer is 3.8 cm in diameter and the coil separation is 5.5 cm. At each recording location the sensing coil was positioned normal to the scalp. At least one channel of EEG was always recorded simultaneously as a monitor for the consistency of the BP. The EEG was recorded from vertex except when that was the MEG location. The rectified surface electromyogram (EMG), bipolarly recorded from the right superficial flexor digitorum muscle, served as the trigger. EEG and electro-oculogram (EOG) were referred to a linked ears reference.

The complex finger movements for each trial consisted of a series of seven brief touches of the thumb and the opposing four fingers of the right hand. The first finger to be touched progressed in the same order as the sequence of touches, so that the series for the first two trials would be 2,3,4,5,4,3,2; 3,4,5,4,3,2,3. The subject was instructed to perform the movements as fast and precisely as possible, at irregular intervals of approximately 10 s, and to hold their breath before, during, and after each trial. In order to check for movement artifacts, head displacement was simultaneously measured by means of a strain gauge activated by tension in a rayon cord attached to the subject's forehead. For each trial a total of 256 points were sampled over an epoch of 3 s, starting 2.25 s before EMG onset. Artifact-contaminated trials were rejected on line by the experimenters. At least one block of 80 trials was collected for each MEG recording location and averages were computed of the first and second 40 trials of each block. Further details have been described by Weinberg and others (1983).

RESULTS AND DISCUSSION

Figure 1 shows the averaged data preceding and during the complex finger tapping task for the one subject for which we have the most data. The vertical lines indicate the start of muscle activity (at time 0). The (numbered) MEG averages from the 11 positions as indicated on the head diagram show to varying degrees the Bereitschaftsmagnetfeld (BM), or readiness magnetic field, preceding movement. Upward deflections indicate fields exiting the head. The grand-average rectified EMG in the lower left shows how accurately the finger tapping was performed, as EMG peaks are evident which correspond to the first four taps. The superimposed strain gauge recordings (SG) of Fig. 1, lower right, do not show significant head movements prior to the onset of tapping. The EOG, as well, indicates stable eye position. Vertex (Cz) EEG at lower middle represents the grand average of all the vertex EEG averages and shows the negative-going Bereitschaftspotential (BP).

Considering the topography of the BM, most of the MEG recordings show a field directed into the head, most prominently seen at all the midline positions

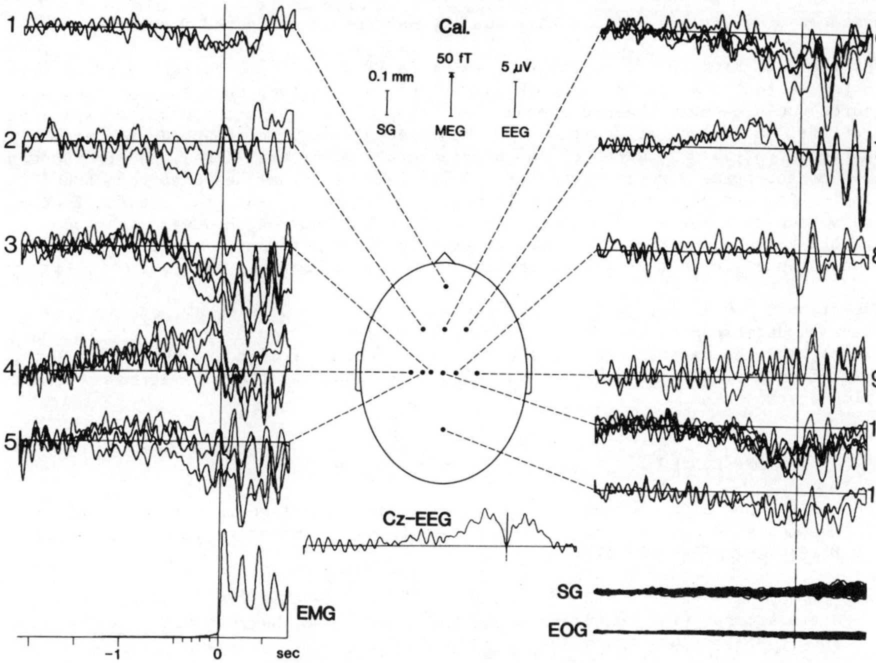

Fig. 1. Bereitschaftsmagnetfeld (BM) or readiness magnetic field preceding a complex right finger tapping task in a right-handed S. Each trace represents an average of 40 artifact-free trials. Recording positions (in percentages of 10-20 system, A = anterior, P = posterior, R = right, L = left) from Cz (vertex) as follows: (1)20%A = Fz; (3)6%L; (4)16%L = roughly C3; (5)10%L; (6)10%A = FCz; (7)10%A, 10%R; (8)6%R; (9)16%R = roughly C4; (10)Cz; (11)11%P. SG = strain gauge recordings of head displacement. Bandwidth 0.03 - 15 Hz for all, except EMG 5 - 70 Hz. Upward deflections correspond to fields exiting the head (MEG) or negative potentials (EEG).

including the vertex (position 10), and just anterior of vertex (6); and somewhat less for the far anterior (1) and posterior (11) positions. Similar shifts are also evident to the left of vertex (3) as well as left anterior (7). The right hemisphere, however, shows no shift at the position to the right of vertex (8) and a reversal of polarity for an early component at the right anterior position (7). For these data, the only other evidence of field directed out of the head is at the far left of vertex near C3 (4), while no similar shift is seen at the corresponding position on the right hemisphere (9). Although the outward-going BM at position 4 was quite variable for the subject whose data are shown in Fig. 1, it was more clearly defined and consistent for the one other subject for whom we have data at this location. Although we did no recording from the right hemisphere, for the other four midline positions which we investigated for this second subject, the inward-going BM was only clearly evident at position 6. However, it was also the case that this second subject did not perform the finger tapping task as accurately, as reflected in the averaged EMG. This was even more the case for out third subject, as the EMG did not show consistent peaks for four averages of 40 trials, and so this subject was not investigated further.

For the data from the first subject, shown in Fig. 1, if two current dipoles were to be located so as to account for as much of the observed data as possible, one source would be near the midline just anterior of Cz, while the other would be at the left, just right of position 4. These sources would correspond, respectively, to the approximate location of the supplementary motor area (SMA), and to the hand area of the primary motor cortex (MI). The later reversal has also been found for simple finger movement (Deecke, Weinberg, and Brickett, 1982) but not for foot or toe movements (Deecke and others, 1983). The SMA reversal, however, is also present between position 7 and the midline for foot movements, and the presumed dipole source is in the same orientation (Deecke and others, 1983).

The present MEG investigation adds another piece of suggestive evidence to previous data such as the BP investigation in parkinsonism of Deecke and Kornhuber (1978), and the regional cerebral blood flow study of Roland and others (1980), which together indicate an important role of the SMA for the initiation of human voluntary movement.

REFERENCES

Deecke, L., and H. H. Kornhuber (1978). An electrical sign of participation of the mesial "supplementary" motor cortex in human voluntary finger movements. Brain Res., 159, 473-476.

Deecke, L., H. Weinberg, and P. Brickett (1982). Magnetic fields of the human brain accompanying voluntary movement: Bereitschaftsmagnetfeld. Exp. Brain Res., 48, 144-148.

Deecke, L., J. Boschert, H. Weinberg, and P. Brickett (1983). Magnetic fields of the human brain (Bereitschaftsmagnetfeld) preceding voluntary foot and toe movements. Exp. Brain Res., 52, 81-86.

Eccles, J. C. (1982). The initiation of voluntary movements by the supplementary motor area. Arch. Psychiat. Nervenkr., 231, 423-441.

Hari, R., A. Antervo, T. Katila, T. Poutanen, M. Seppanen, T. Tuomisto, and T. Varpula (1983). Cerebral magnetic fields associated with voluntary limb movements. Il Nuovo Cimento, 2D, 484-494.

Kornhuber, H. H., and L. Deecke (1965). Hirnpotential anderungen bei Willkurbewegungen und passiven Bewegungen des Menschen: Bereitschaftspotential und reafferente Potentiale. Pflugers Arch., 284, 1-17.

Roland, P. E., B. Larsen, N. A. Lassen, and E. Skinhoj (1980). Supplementary motor area and other cortical areas in organization of voluntary movements in man. J. Neurophysiol., 43, 118-136.

Vrba, J., A. A. Fife, M. B. Burbank, H. Weinberg, and P. A. Brickett (1982). Spatial discrimination in SQUID gradiometers and third order gradiometer performance. Can. J. Phys., 60, 1060-1073.

Weinberg, H., P. Brickett, L. Deecke, and J. Boschert (1983). Slow magnetic fields of the brain preceding movements and speech. Il Nuovo Cimento, 2D, 495-504.

MAGNETIC FIELD PATTERN PRODUCED BY EYE BLINKING

A. Antervo, R. Hari[1], T. Katila, T. Ryhänen and M. Seppänen

Dept. of Technical Physics and [1]Low Temperature Laboratory,
Helsinki University of Technology, SF-02150 Espoo, Finland

ABSTRACT

We have studied the magnetic signals produced by eye blinking. The signals were maximal in amplitude at the postero-lateral sides of the orbits. The signal reversed its polarity over both eyes and in the midline. The amplitude changes of the signals produced by eye movements and blinks were similar during the dark and light adaptation. The results support the idea that the primary current source of the blink signals is the transretinal current density.

KEYWORDS

Eye blinks; magnetic fields; MEG; simulation.

INTRODUCTION

Eye blinking produces large electric signals which can contaminate evoked potential recordings. It has been proposed (Barry and Melville Jones, 1965) that these signals are due to the changes in the relative position of the upper lid and eyeball. Deflections assosiated with eye blinks have also been observed in the real-time magnetoencephalogram: these deflections have opposite polarities in the right and left hemisphere (Hughes and co-workes, 1977). We have studied field patterns of averaged signals produced by voluntary blinking. Simulations were used to clarify bioelectric phenomena underlying magnetic and electric blink signals. A detailed description of the work will be published elsewhere (Antervo and co-workers, 1984).

EXPERIMENTAL METHODS

Two healthy members of the laboratory personnel served as subjects. They were instructed to fixate to a mark on the wall while blinking in a natural manner once every 2 to 5 seconds. The head was supported by a vacuum cast. The measurements were carried out in the Otaniemi magnetically shielded room with a first-order SQUID-gradiometer (pickup coil ϕ=2 cm, intercoil distance 14 cm). The pickup coil was situated approximately 2 cm above the skull and the magnetic field perpendicular to the skull was measured. Extensive mappings were made on the right hemisphere, and at some locations on the left hemisphere. The electro-oculogram was recorded simultaneously with magnetic blink signals. The effect of light and dark adaptation on the magnetic and electric signals produced by eye movements and blinks was also studied.

The magnetic and electric signals were stored on a FM-tape recorder and analysed off-line with an Eclipse S/140 -minicomputer. The peak of the electric signal was used for timing the averaging. The sampling frequency was 300 Hz, and the analysis period was 1 s. Approximately 40 artefact free epochs were averaged for each recording location.

EXPERIMENTAL RESULTS

Figure 1a shows magnetic blink signals on the right hemisphere of subject 1. The simultaneously measured electric and magnetic signals were monophasic and their waveforms were alike. The duration of blink signals ranged from 200 to 400 ms. No systematic differences were found in the onset times or peak latencies of the electric and magnetic signals. In both subjects, the maximal magnetic signals, approximately 3-4 pT in amplitude, were obtained on the postero-lateral sides of the orbits. The amplitude of the signal decreased sharply in all directions away from the maximal point. The flux was directed out of the skull at all measured locations on the right temporal area; the polarity of the signal was opposite on the left temporal area. The polarity of the magnetic blink signal reversed also over both eyes and in the midline.

In normal illumination the amplitude of the magnetic blink signal was on the average 3.9 pT, and the mean peak-to-peak amplitude of the magneto-oculographic signals, corresponding to vertical gaze deviations of 60 degrees, was 10.6 pT. During the dark adaptation the amplitudes decreased to 2.6 pT and 6.2 pT, respectively. After switching on the lights the amplitudes of the signals reached the preadaptation levels again.

SIMULATION RESULTS

The nature of the current distributions related to the eye blinking was studied theoretically by using electric and magnetic multipole expansions. The field patterns arising from the dipole terms, i.e. the lowest order terms of the expansions, were calculated and compared. The head was approximated with a spherical, homogeneous volume conductor. In the model the transretinal current flow was assumed to be the primary current density. The blink signals were described with the aid of secondary current densities. Two equivalent sources, representing the eyes, were placed at the equator plane of the sphere. The angle between them was 45°.

Figure 1b represents the electric and magnetic field patterns created by the dipole terms of the current multipole expansion (two current dipoles). The measured electric field pattern (Zao, G. Rémond and A. Rémond, 1952) was the basis in this simulation. First, the orientation of the dipoles (see Fig. 1b) was chosen to explain the asymmetric electric field pattern. Then the corresponding pattern for the magnetic field component perpendicular to the skull was calculated. The measured distribution of the magnetic field matched with the model within our resolution. The direction of the difference dipole suggests that eye blinking causes an upward current flow. This could be due to the increased conductivity in the upper part of the orbit. The field patterns associated with blinking could also be simulated by modelling the changes of secondary current densities on the surface boundaries between different conductivities.

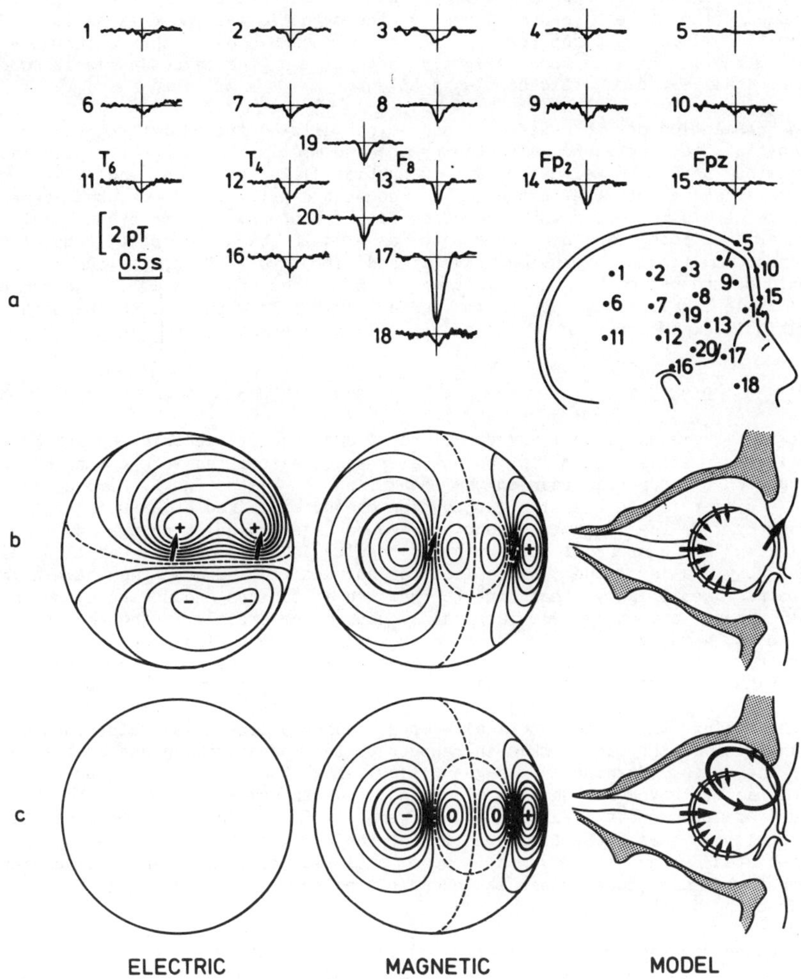

Figure 1. (a) Averaged magnetic blink signals in subject 1. The passband is 0.03-45 Hz. (b) The calculated field patterns of the electric and magnetic blink signals for the current dipole model. The current dipoles in the "orbits" of the homogeneous, spherical head are indicated by the arrows and the polarity reversals by the dashed lines. For further explanation, see text. (c) The calculated field patterns of the electric and magnetic blink signals for the magnetic dipole model. For further explanation, see text.

During blinking the increase of the current density in the upper part of the orbit could give arise to a current loop which can be modelled with a magnetic dipole. Figure 1c represents the magnetic field pattern created by magnetic dipoles (the lowest order terms of the magnetic multipole expansion). The magnetic field patterns of Fig. 1b and 1c resemble each other although e.g. the magnetic maxima were slightly sharper in the magnetic dipole model. Electrically, the magnetic dipole is silent.

The resolution of our measurements does not allow differentiation between the magnetic field patterns generated by the two models. The current dipole model explains both the electric and magnetic field patterns and gives more information about the nature of the current densities, whereas the magnetic dipole explains only the magnetic field pattern. On the other hand, the magnetic dipole can be represented as a sum of two quadrupole terms of the current multipole expansion (Katila and Karp, 1983). Thus, the current loop could be understood as a contribution of the higher order terms to the current multipole model. Yet we prefer the current multipole expansion to the magnetic multipole expansion when modelling physiological events.

DISCUSSION

Eye lid movements as well as eyeball rotations produce electric changes on the scalp and facial skin. The potential changes associated with eye blinks are larger on the forehead than on the cheek (Zao, G. Rémond and A. Rémond, 1952). The asymmetric potential distribution can be explained by assuming that the geometry of the volume conductor around the eyeball changes during eye blinking: the closure of the upper lid results in an upward current flow. Also the bony walls of the orbit affect the current flow because they have higher electric resistivity than other extracerebral tissues around the eyes. Thus, the currents evidently circulate mainly within the orbits. In the present work a model with a slightly upwards pointing current dipole in each eye gave electric and magnetic field patterns similar to the measured field distributions produced by blinking.

When moving the gaze vertically the measured potential distribution is symmetric on the facial skin in respect to the transversal plane at the level of the eyes (Overton and Shagass, 1969). Magnetic field pattern produced by vertical eye movements shows a clear amplitude extremum at the postero-lateral side of the orbit (Karp and co-workers, 1976; Katila and co-workers, 1981) resembling, thus, the field pattern of the blink signals. The field patterns associated with eye movements were satisfactorily explained by a simple current dipole model where the current dipoles were placed along the optical axis of both eyes.

Variation in the retinal illumination is known to affect to the corneoretinal potential. In the present work the amplitudes of both electric and magnetic eye movement and blink signals changed in similar manner during dark and light adaptations. This finding suggests that the transretinal current density is the primary source of all the four signals. However, the generation mechanisms underlying the signals seem to differ: while moving the eyes the direction of the primary current density changes in respect to the volume conductor whereas blinking changes the secondary current density.

The results indicate that eye blinking may easily contaminate neuromagnetic recordings. The danger of contamination is pronounced in the posterior parts of the brain because blink signals cannot be seen in the real time recordings. Simultaneous EOG recordings are thus nessessary to rule out blink artefacts on these areas.

ACKNOWLEDGEMENTS

This study was financially supported by the Academy of Finland. We thank Mr. T. Poutanen for his aid at the early stages of the measurements.

REFERENCES

Antervo A., R. Hari, T. Katila, T. Ryhänen and M. Seppänen (1984). Magnetic fields produced by eye blinking. To be published.

Barry W. and G.M. Melvill Jones (1965). Influence of eye lid movement upon electro-oculographic recordings of vertical eye movements. Aerosp. Med., 36, 855-858.

Hughes J.R., J. Cohen, C.I. Mayman, M.L. Scholl and D.E. Hendrix (1977). Relationship of the magnetoencephalogram to abnormal activity in the electroencephalogram. J. Neurol., 217, 79-93.

Karp P.J., T.E. Katila, P. Mäkipää and P. Saar (1976). Magneto-oculography detection of the DC magnetic field of the eye. Proceedings of the International Conference on Medical and Biological Engineering, Ottawa, pp. 504-507.

Katila T. and P.J. Karp (1983). Magnetocardiography: morphology and multipole presentations. In S.J. Williamson, G.-L. Romani, L. Kaufman and I. Modena (Eds.), Biomagnetism. An Interdisciplinary Approach. Plenum Press, New York. Chapt. 9, pp.237-263.

Katila T., R. Maniewsky, T. Poutanen, T. Varpula and P.J. Karp (1976). Magnetic fields produced by the human eye. J. Appl. Phys., 52, 2565-2571.

Overton D.A. and S. Shagass (1969). Distribution of eye movement and eye blink potential over the scalp. Electroenceph. clin. Neurophysiol., 27, 546.

Zao Z.Z., G. Rémond and A. Rémond (1952). Le champ electrique de l'oeil. La Semaine de Hopitaux, 28, 1506-1513.

BIOMAGNETISM IN THE STUDY OF LUNG FUNCTION

J.D. Brain,* P. Gehr,** P.A. Valberg,* S.B. Bloom*, and I. Nemoto***

*Department of Environmental Science and Physiology,
Harvard University School of Public Health, Boston, MA, USA 02115

**Current Address: Anatomisches Institut, Universitat Bern,
Postfach 139, CH-3000 Bern 9, Switzerland

***Current Address: Tokyo Denki University, College of Natural Sciences
and Engineering, Dept. of Mathematical Sciences,
Hatoyama-Mura, Hiki-Gun, Saitama-Ken, Japan 350-03

ABSTRACT

Magnetic forms of iron oxide are easily generated. When inhaled by experimental animals or humans, they are relatively non-toxic and can be easily detected non-invasively by magnetometry. This technique can be used to describe the distribution of retained particles in the respiratory tract. Repeated measurements describe clearance of particles. Since iron oxide is ingested by pulmonary macrophages, magnetometry can also be used to follow the progression of phagocytosis in vivo, to describe organelle motion in macrophages, and to estimate cytoplasmic rheology.

KEYWORDS

Magnetometry; pulmonary clearance; macrophages; phagocytosis; cell motility; cytoplasmic rheology; hematite.

INTRODUCTION

Before describing our work using magnetic forms of iron oxide as a probe of the lung, let us briefly consider its biological fate in the respiratory tract. Iron oxide can be used as a non-toxic test aerosol for studies of particle deposition and clearance in the lungs of animals and humans. It can be readily generated in various forms by combustion of iron pentacarbonyl (Brain and colleagues, 1974; Valberg and Brain, 1979). It is easily seen in the light microscope when stained by the Prussian blue reaction and in the electron microscope by virtue of its electron density (Figure 1).

We have used both of these microscopic techniques to study particle clearance in the lungs of mice after they were exposed to an aerosol of submicrometric non-magnetic hematite (Fe_2O_3) (MMAD = 0.31 um, σ g=1.25) (Sorokin and Brain, 1975). The animals were serially sacrificed up to 14 months after inhalation exposure. Acutely, particles which deposited on airways were swept toward the pharynx by mucociliary transport. Examination of the lung sections from animals serially sacrificed clearly showed three phases of clearance. Particles depositing on alveolar surfaces were rapidly ingested by alveolar macrophages. This phase was virtually complete in 24 hrs. The second phase was characterized by the gradual migration of particle-laden alveolar macrophages toward ciliated bronchiolar surfaces. These cells were also swept toward the pharynx by ciliary action, but appeared to move much more slowly than uningested particles cleared initially. Clearance by this pathway may continue for several months. Finally, the third, or chronic phase began when particles appeared within macrophages in the connective tissue of the lungs; these particles persisted for long periods of time but were slowly solubilized. We suspect that the rate of solubilization depends critically on the particle size; smaller particles with a correspondingly greater surface area-to-volume ratio dissolve faster. Light microscopy was the most useful in delineating these phases, while electron microscopy permitted the description of particles within phagosomes and phagolysosomes. Recently, Sorokin (Sorokin, 1983a, 1983b) has further characterized the dynamics of lysosomal formation and fusion following exposure to inhaled iron oxide aerosol. This same hematite aerosol was also used in experiments which described (a) altered biochemical activity of lysosomal enzymes in pulmonary macrophages from rabbits breathing iron oxide (Grant and colleagues, 1979), (b) changes in the phagocytic behavior of pulmonary macrophages from hamsters breathing iron oxide (Kavet and colleagues, 1978), and (c) the uptake of iron oxide aerosols by the airway epithelium of normal (Watson and Brain, 1979) and sulfur-dioxide injured lungs (Watson and Brain, 1980).

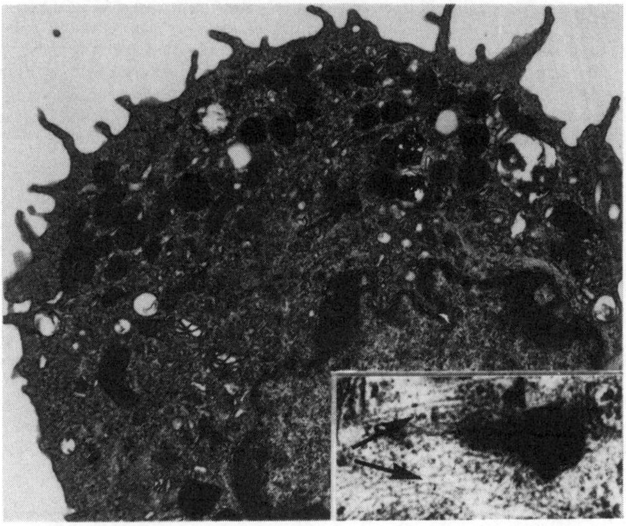

Fig. 1. An alveolar macrophage lavaged from a Syrian hamster lung. The arrow indicates Fe_3O_4 particles; magnification x12,500. Inset shows Fe_3O_4 particles wrapped in a membrane in close association with microfilaments (arrows) (x50,000). From Gehr and colleagues, 1983c.

We have also used iron oxide particles as a non-toxic negative control in an *in vivo* hamster bioassay designed to assess the pulmonary toxicity of inhaled particulates (Beck and colleagues, 1982). The assay utilizes bronchoalveolar lavage performed at various times after particle exposure; cellular and biochemical constituents relating to lung injury are analyzed and interpreted in order to characterize the nature and extent of lung damage (Brain and colleagues, 1983). We have used our bioassay to compare the toxicity of iron oxide with that of other dusts such as alpha-quartz, aluminum oxide, volcanic ash from Mt. St. Helens (Beck and colleagues, 1981) and emissions from coal and wood stoves. Although high doses of iron oxide caused some nonspecific inflammatory changes one day following exposure (elevated neutrophil counts and albumin levels), the response was always less than that observed for toxic particulates. Occupational exposures to pure iron oxide also appear to be innocuous. Furthermore, most of the cellular and biochemical parameters returned to control values in seven days or less after iron oxide exposure. These studies confirm the results of morphologic and epidemiologic studies which also suggest that iron oxide has low toxicity. Alpha-quartz, in contrast, caused many of the indicators of pathology to be elevated for up to twenty-eight days after exposure.

METHODS AND RESULTS

When magnetic forms of iron oxide are instilled or inhaled into the lungs, the retained particles can be used as an *in vivo* tracer for phagocytosis, cytoplasmic motility, cytoplasmic viscosity, and particle clearance. These particles can be magnetized and aligned by an external magnetic field; the remanent magnetism coming from them can then be measured at the surface of a human's or animal's chest. Immediately following magnetization, the field from the lungs begins to decay (relaxation). Relaxation is caused by the random reorientation of particles away from their initially aligned state. Some physical and biological aspects of this phenomena have been described by Cohen and colleagues (1984). Interestingly, the characteristics of this decay change with time after exposure (Gehr and colleagues, 1983c, Brain and colleagues, 1984a). As shown in Figure 2A, relaxation curves from the same hamster show changes in shape and in initial field strength with time. The latter is due to clearance. In Figure 2B, these curves are normalized with respect to their B_o's to emphasize the changes in their rates of decay. Also shown in Figure 2B is the relaxation half-time (the time for the field to decrease from B_o to 1/2 B_o). Similar data obtained from rabbits exposed to a gamma-Fe_2O_3 (maghemite) aerosol are shown in Figure 3. We believe that the early increase in relaxation rate (decrease in relaxation half-time) seen in our data represents the transition of free particles to phagocytized particles. Our studies suggest that ingested particles relax faster than free particles; therefore, non-invasive magnetometry might be used to follow the progression of *in situ* phagocytosis. These results correlate well with our morphologic findings showing that phagocytosis proceeds rapidly toward completion during the first twenty-four hours after aerosol deposition. In addition to the morphologic evidence for this, we have quantified the progress of phagocytosis using our lambda technique (Brain and Corkery, 1977). One hour following instillation of colloidal gold into rabbits, 27% of it had been phagocytized, while after 16 hours, 91% had been ingested (Brain et al., 1984a) Relaxation parameters, such as the half-time for the decay, continued to change for weeks after initial deposition of the particles. This may reflect changes in the proportions of the total particle burden located in various pulmonary compartments (e.g., macrophages, connective tissue) (Brain and colleagues, 1984a; Gehr and colleagues, 1983c).

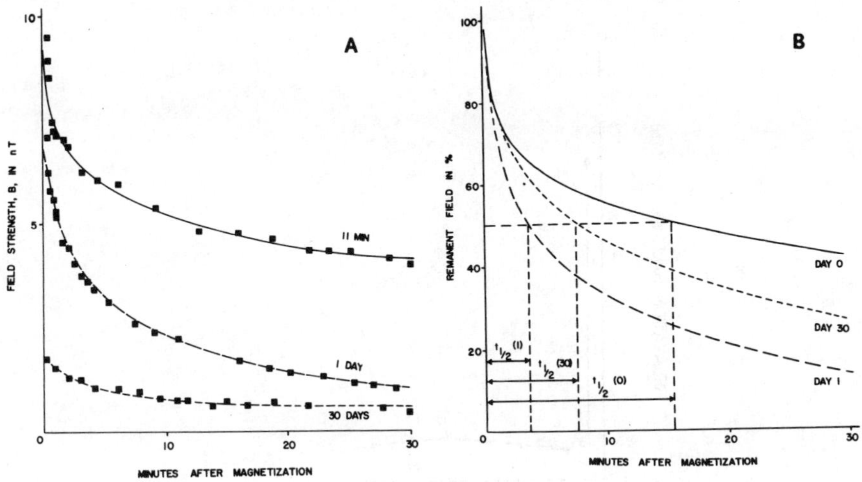

Fig. 2A. Relaxation curves. Data from hamster No. 1 at 3 different times after intratracheal instillation of Fe_3O_4 particles.

Fig. 2B. Relaxation curves. Same data with initial field strength (B_o) of each curve normalized to 100%. $t_{1/2}$ (0), $t_{1/2}$ (1), and $t_{1/2}$ (30) are relaxation half times 11 min, 1 day, and 30 days after instillation, respectively. From Gehr and colleagues, 1983c.

Relaxation curves can be interpreted by considering the following factors. First, the phenomenon reflects actual physical movement (rotation) of the magnetic particles. If the particles are embedded in a frozen lung or a lucite block, no relaxation occurs. Thus, a decreasing magnetic field represents the physical reorientation of the dust, as the particles themselves cannot spontaneously demagnetize. Secondly, we must consider the forces available to move particles within the lungs. These forces include mucociliary action (mucus transport as well as the shearing and tumbling action of the cilia), macrophage mobility, intracellular movement within macrophages, fluid movements in the alveoli, shape and area changes associated with ventilation, cardiac motion (and arterial pressure pulses), and Brownian movement (Table 1). Thirdly, we must consider the milieu in which the particles are found. The rheologic properties of the medium surrounding the particles at any given site will influence the effectiveness of the applied forces. The medium may either resist the applied forces or help to transmit them to the particle-containing phagosomes. Particles not only initially deposit at a number of different sites within the lungs, but change their location with time. For example, extracellular particles initially may float freely in alveolar or airway lining fluid or may be adherent to epithelial cells. With increasing time, more and more will be ingested by airway and/or alveolar macrophages; a few may be found within epithelial cells or in macrophages within connective tissue and lymph nodes.

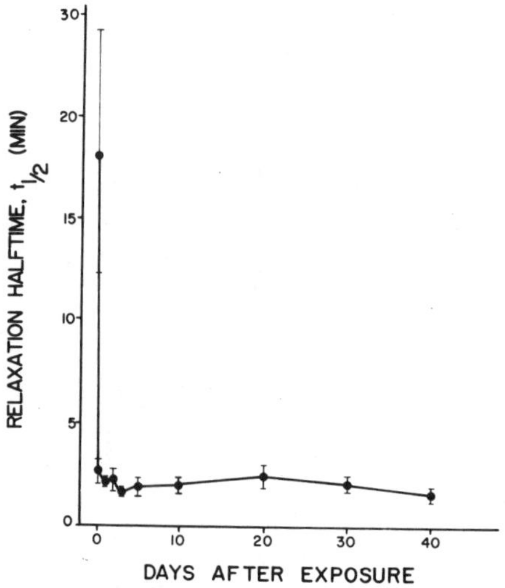

Fig. 3. Relaxation half-time ($t_{1/2}$) as a function of days after exposure. Half-time decreases as time after exposure increases from 0 to 30 days. All values represent mean ± standard error based on five rabbits. From Brain and colleagues, 1984a.

TABLE 1 Possible mechanisms of relaxation.

1. Mucociliary transport; local mixing produced by cilia and mouthward movement of airway lining fluid
2. Macrophage locomotion
3. Intracellular movement of macrophage organelles, primarily phagosomes and phagolysosomes
4. Elastic recoil of organelles in the cytoplasmic matrix
5. Fluid movements in alveoli, such as alveolar bronchiolar transport of the alveolar lining layer
6. Changes in shape and area associated with breathing
7. Cardiac-induced motion (either heart motion or arterial pressure pulse)
8. Brownian movement

Our _in vivo_ experiments suggest that the rotational forces applied to the particles (after phagocytosis is complete) come largely from intracellular movements of phagosomes and phagolysosomes that contain the particles (Brain and colleagues, 1984a,b; Gehr and colleagues 1983b, 1984c). We believe that these movements arise from contractions of the cytoskeleton, which orchestrates cell motion required for functions such as phagocytosis, secretion, or ameboid movement. Even in the motionless lung in a dead animal, relaxation takes place as long as the _in situ_ alveolar macrophages are kept warm and oxygenated (Brain and colleagues, 1984b). The hypothesis that organelle motion is a dominant mechanism for relaxation has now been confirmed in studies of hamster pulmonary macrophages observed in vitro (Gehr and colleagues, 1983a; Valberg, 1983, 1984; Nemoto and colleagues, 1984). Macrophages which had previously ingested magnetic particles exhibited relaxation which was quantitatively similar to that seen in vivo, demonstrating that cardiac and respiratory movements are not essential for particle misalignment. Nemoto and colleagues (1984) demonstrated temperature dependence of relaxation over a wide range.

The interventions used in these studies shed additional light on the mechanisms of relaxation. Cytochalasin B and D, known to disintegrate the microfilament system (actin), slowed relaxation as did cold and cyanide (uncoupler of the respiratory chain in mitochondria), whereas colchicine and nocodazole (disintegrator of the microtubules) had no effect. Formalin and glutaraldehyde fixation resulted in a complete halt of relaxation. These interventions, most of which compromise the contractile capabilities of the cytoskeleton, demonstrate that organelle motions are driven by an energy dependent, actin-based mechanism (Gehr and colleagues, 1983a, 1984, this volume; Valberg 1983, 1984).

The rheologic properties of cytoplasm can also be measured using magnetic particles and magnetometry (Crick and Hughes, 1950; Valberg, 1984; Valberg, this volume). After magnetizing the particles, a magnetic field is applied perpendicular to the axis along which the particles were aligned by the initial magnetization. The ease with which the particles realign toward this "twisting" field is related to the viscosity of the medium surrounding the particles. Valberg (this volume) has shown that cooling, but not cytochalasin D, increases the viscosity of the cytoplasm of hamster pulmonary macrophages.

Clearance of particles can also be described by magnetometry. The remanent field immediately following magnetization, before relaxation occurs, is proportional to the amount of magnetic material remaining in the lungs. This measurement, repeated over weeks or months, describes a clearance curve. It requires several seconds to move an animal or subject from the magnetizing field to the probe of the measuring device, and, of course, some relaxation has ensued when the initial reading is taken. In order to estimate the field before relaxation, we fit the data from the first one or two minutes of relaxation to the equation $B = B_o e^{-\lambda_o t}$. B is the field strength at time t after magnetization, and λ_o is the decay constant. B_o is the field strength at the time the magnet was removed. We have described clearance in rabbits and hamsters (Brain and colleagues, 1984a; Gehr and colleagues, 1983). Figure 4 shows clearance of magnetic dust from rabbits. Halpern and colleagues (1981) obtained similar data in donkeys.

Experimental studies of clearance of magnetic dust have also been performed on human subjects. The magnetic properties of such minerals as magnetite (Fe_3O_4) and gamma hematite (maghemite) are suitable. Amounts as small as 0.05 mg can be detected and repeated measurements describe the clearance of magnetic dust from the lungs (Cohen and colleagues, 1979). In that study, the clearance kinetics of magnetic dust in smokers was compared to that of non-smokers; the former were shown to clear their magnetic dust burden significantly more slowly over a one year period. In long term studies such as these, magnetic iron oxide is less harmful than radioactively tagged particles.

Some materials in the workplace also contain magnetic contaminants making noninvasive measurements of body burden possible. Exposed workers frequently exhibit significant retained magnetic dust which can be detected with a fluxgate or a superconducting magnetometer. The industries involved include foundry workers, iron miners, refiners, welders and workers with various iron products.

Also, asbestos miners have magnetic material in their lungs, as Fe_3O_4 particles are usually attached to asbestos fibers (Cohen and colleagues, 1981; Stroink and colleagues, 1981a). Freedman and colleagues (1980) have shown that magnetic material also may be associated with coal dust. Kalliomaki and colleagues (1980) have described magnetic contamination in steel welders. Such measurements could be superior to estimates of retained dust based on air concentrations in the workplace. Magnetometric measurements reflect individual differences in collection efficiency and clearance.

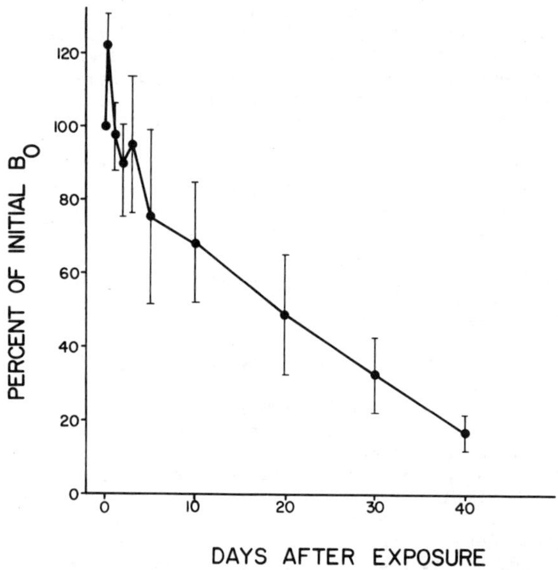

Fig. 4. Disappearance of gamma-iron oxide from lungs of five rabbits as indicated by remanent magnetic field normal to chest wall immediately after magnetization. The reading, shown as a percentage of the field strength measured immediately after exposure, is plotted versus time after aerosol exposure. At each time indicated, the rabbit was magnetized by an external magnetic field. The decrease in the magnetic field with time is related to physical clearance of iron oxide from the lungs and/or conversion to nonmagnetic forms. All values represent mean ± standard error. From Brain and colleagues, 1984).

Noninvasive measurements of body burden are possible. There are, however, crucial problems. It is important to know whether the magnetic component is a constant fraction of the parent material. For coal this varies from mine to mine and from seam to seam. The magnetic content of asbestos may be less variable (Stroink and colleagues, 1981b). Another crucial question is whether the magnetic and nonmagnetic compounds have the same biologic fate and, therefore, the same half-life in the body. For example, are magnetite and asbestos fibers handled in the same way? Another aspect of this question is whether the magnetic constituents are converted into nonmagnetic forms. Then iron may still be present in the body but no longer detectable. For example, iron oxide may be solubilized and resynthesized by the body into ferritin and hemosiderin (Watson and Brain, 1979); then the iron atoms are still present in the lungs but are no longer ferromagnetic. Finally, there is the problem of confounding exposures to magnetic particles from different sources. Many iron, coal and asbestos miners are exposed to welding fumes. Many miners moonlight or may have had previous jobs in these industries. Cohen and colleagues (1981) have measured the amount of magnetic material in the lungs of asbestos miners and millers in Quebec. They found that measurements at an industrial site were feasible and that the amounts of magnetic contamination in the lungs were clearly detectable. However, they had difficulty with miners who had significant welding exposures. For the nonwelders, however, they found a modest correlation between the amount of magnetic material in the lungs and aerometric estimates of the dust exposure.

CONCLUSIONS

Magnetometry can be used to monitor magnetic particle clearance from the respiratory tract of animals and humans. In addition to describing the anatomic fate of inhaled particles, we have shown that magnetometry can also be used to assess _in situ_ particle phagocytosis and events reflecting cytoplasmic viscosity and motility. Current studies of deposited magnetic iron oxide are correlating biochemical, histological, and morphometric parameters with the magnetic phenomena. We also hope to be able to assess non-invasively the long-term effects of various environmental and pathological conditions on phagocytosis, cytoplasmic motility, cytoplasmic rheology, and clearance in living animals.

ACKNOWLEDGMENTS

This work was supported by Grant HL-31029, HL-29175, and ES-00002 from the National Institutes of Health and Swiss National Foundation Grant 83.734.9.79. We thank Dr. David Cohen, Senior Scientist at the Francis Bitter National Magnet Laboratory, Massachusetts Institute of Technology, for introducing us to biomagnetometry and for his advice and help with regard to magnetometric techniques.

REFERENCES

Beck, B.D., J.D. Brain, and D.E. Bohannon (1982). An _in vivo_ hamster bioassay to assess the toxicity of particulates for the lungs. _Toxicol. Appl. Pharmacol._, _66_, 9-29.

Beck, B.D., J.D. Brain, and D.E. Bohannon (1981). The pulmonary toxicity of an ash sample from the Mt. St. Helens volcano. _Exp. Lung Res._, _2_, 289-301.

Brain, J.D., and B.D. Beck. (1983). Assessment of particle toxicity. In W.L. Wagner, W.N. Rom, and J.A. Merchant (Ed.), _Health Issues Related to Metal and Non Metallic Mining_. Butterworth Publishers, Boston. pp. 51-61.

Brain, J.D., S.B. Bloom, P.A. Valberg, and P. Gehr. (1984a). Behavior of magnetic dusts in the lungs of rabbits correlates with phagocytosis. _Exp. Lung Res._, _6_, 115-131.

Brain, J.D., and G.C. Corkery. (1977). The effect of increased particles on the endocytosis of radiocolloids by pulmonary macrophages in vivo: competitive and toxic effects. In W.H. Walton (Ed.), Inhaled Particles and Vapours, IV, Pergamon Press, Oxford, pp. 551-564.

Brain, J.D., T. Hu, and S.B. Bloom. (1984b). Effects of gas composition, temperature and ventilatory movements on macrophage organelle motion. Fed. Proc., 43, 1001.

Brain, J.D., P.A. Valberg, S.P. Sorokin, and W.C. Hinds. (1974). An iron oxide aerosol suitable for animal exposures. Environ Res., 7, 13-26.

Cohen, D. (1973). Ferromagnetic contamination in the lungs and other organs of the human body. Science, 180, 745-748.

Cohen, D. (1978). Report of the low-field group: The magnetic field of the lung. Publication No. MIT/FBNML-78/1. Springfield, VA, National Technical Information Service.

Cohen, D., S. Arai, and J.D. Brain. (1979). Smoking impairs long-term dust clearance from the lungs. Science, 204, 514-517.

Cohen, D., T.S. Crowther, G.W. Gibbs, and M.R. Becklake. (1981). Magnetic lung measurements in relation to occupational exposure in asbestos miners and millers of Quebec. Environ. Res. 26, 535-550.

Cohen, D., I. Nemoto, L. Kaufman, and S. Arai. (1984). Ferrimagnetic particles in the lung. Part II: The relaxation process. IEEE Trans. Biomed. Engineering, BME-31, 274-285.

Crick, F.H.C., and A.F.W. Hughes. (1950). The physical properties of cytoplasm: A study by means of the magnetic particle method. Exp. Cell Res., 1, 336-338.

Freedman, A.P., S.E. Robinson, and R.J. Johnston. (1980). Non-invasive magnetopneumographic estimation of lung dust loads and distribution in bituminous coal workers. JOM, 26, 613-618.

Gehr, P., J.D. Brain, and S.B. Bloom (1983). Is lysosome movement an actin-based mechanism? A cytomagnetometric study. Experientia, 40, 624 (Abstr.).

Gehr, P., J.D. Brain, and S.B. Bloom. (1983a). Magnetometry: a tool to study intracellular movement. J. Cell Biol., 97, 194a (Abstr.).

Gehr, P., J.D. Brain, S.B. Bloom, and P.A. Valberg. (1983b). Magnetic particles in the liver: a probe for cell function. Nature, 302, 336-338.

Gehr, P., J.D. Brain, I. Nemoto, and S.B. Bloom. (1983c). Behavior of magnetic particles in hamster lungs: estimates of clearance and cytoplasmic motility. J. Appl. Physiol., 55, 1196-1202.

Grant, M.M., S.P. Sorokin, and J.D. Brain. (1979). Lysosomal enzyme activities in pulmonary macrophages from rabbits breathing iron oxide. Am. Rev. Respir. Dis., 120, 1003-1012.

Halpern, M., S.J. Williamson, D.M. Spektor, R.B. Schlesinger, and M. Lippmann. (1981). Remanent magnetic fields for measuring particle retention and distribution in the lungs. Exp. Lung Res., 2, 27-35.

Kalliomaki, K., P.L. Kalliomaki, V. Kelha, and V. Vaaranen. (1980). Instrumentation for measuring the magnetic lung contamination of steel workers. Ann. Occup. Hyg., 23, 175-184.

Kavet, R.I., J.D. Brain, and D.L. Levens. (1978). Characteristics of pulmonary macrophages lavaged from hamsters exposed to iron oxide aerosols. Lab. Invest., 38, 312-319.

Nemoto, I., H. Toyotama, J.D. Brain, and P. Gehr. Behavior of ferromagnetic particles in hamster pulmonary macrophages in vitro: magnetometric estimation of cell motility. J. Cell Physiol. (Submitted)

Sorokin, S.P. (1983). Dynamics of lysosomal elements in pulmonary alveolar macrophages. I. The postactivation lysosomal cycle. Anat. Rec., 206, 117-143.

Sorokin, S.P. (1983). Dynamics of lysosomal elements in pulmonary alveolar macrophages. II. Transitory intracytoplasmic events in activated cells. Anat. Rec., 206, 145-170.
Sorokin, S.P., and J.D. Brain. (1975). Pathways of clearance in mouse lungs exposed to iron oxide aerosols. Anat. Rec., 181, 581-625.
Stroink, G., D. Dahn, and J. Holland (1981a). Magnetopneumographic estimation of lung dust loads and distributions in asbestos miners and millers. Am. Rev. Respir. Dis., 123, 144 (Abstr.).
Stroink, G., R.A. Dunlap, and D. Hutt (1981b). Room-temperature magnetization measurements of some Canadian chrysotile and UICC asbestos samples. Can. Mineral., 19, 519-524.
Valberg, P.A. (1983). Endosome motion and rheology in lung macrophages after maghemite phagocytosis. J. Cell Biol., 97, 426a (Abstr.).
Valberg, P.A., and J.D. Brain. (1979). Generation and use of three types of iron oxide aerosol. Am. Rev. Respir. Dis., 120, 1013-1024.
Valberg, P.A. (1984). Magnetometry of ingested particles in pulmonary macrophages. Science, 224, 513-516.
Watson, A.Y., and J.D. Brain (1980). The effect of SO_2 on the uptake of particles by mouse bronchial epithelium. Exp. Lung Res., 1, 67-87.
Watson, A.Y., and J.D. Brain (1979). Uptake of iron oxide aerosols by mouse airway epithelium. Lab. Invest., 40, 450-459.

MAGNETIC PARTICLES USED AS ACTIVE AND PASSIVE PROBES OF INTRACELLULAR PROPERTIES OF LIVING CELLS

Peter A. Valberg

Department of Environmental Science and Physiology
Harvard School of Public Health
Boston, Massachusetts 02115

SUMMARY

A non-optical means of sensing cytoplasmic movement and of measuring physical properties of cytoplasm is presented. Motions of aligned magnetic particles can be detected from the fields they produce, so that when such particles are within cells, motions of organelles can be followed magnetometrically. The particles can also be manipulated by magnetic fields so that structural characteristics of the cytoplasm can be probed. These measurements were made on isolated lung macrophages after the cells had ingested magnetic particles. Videomicroscopy was used to confirm that magnetization aligned ingested particles, and that cytoplasmic activity randomized the alignment. Magnetic fields were used to reorient intracellular particles. The low-shear viscosity of the cytoplasm was determined by applying a twisting field perpendicular to the direction of alignment and then magnetometrically detecting the rate of particle rotation. The data demonstrated that magnetic particles in isolated phagocytic cells can be used to sense cell and phagosome motion, to probe cell rheology, and to assess the integrity of cytoplasmic elements

KEYWORDS

Lung macrophages; magnetometry; inhalation toxicology; cell motility; cytoplasmic movement; cell viscosity; cell rheology; relaxation; videomicroscopy.

INTRODUCTION

Considerable interest is currently focused on determining the mechanical properties of the cytoplasm within living cells. Those intracellular functions involving movement that have received recent attention are vesicle translocations (Herman and Albertini, 1984), cell viscosity (Sato and co-workers, 1984), actin-myosin force generation (Sheetz and Spudich, 1983), phagocytosis (Steinman and co-workers, 1983), cytoskeletal properties (Beckerle and Porter, 1983), and motility (Allen, 1981). A novel means of assessing these functions is to measure the movements of magnetic particles that have been phagocytized by cells. When a magnetic field is applied briefly, the magnetic directions of these particles are aligned, and they produce a detectable field, proportional to the total mass of particles. This initial remanent magnetic field decreases if the particle orientation moves away from the initial direction of alignment. The motion of magnetic particles within a living organism was studied optically by Crick and Hughes (1950), and later observed magnetometrically for the first time by Cohen (1973). The magnetometric approach for detecting particle motion has recently been applied both to intact animals (Brain and co-workers, 1984; Gehr and co-workers, 1983a and 1983b), to humans (Cohen and co-workers,

1984), and to isolated cells (Valberg, 1984). Magnetic particles can be used to monitor spontaneous cytoplasmic motion without the need for optical observation. Furthermore, aligned particles can be rotated by external fields, and their rate of rotation is a measure of intracellular viscosity, thus providing additional information about the cytoplasm. This report describes the results of such measurements on isolated lung macrophages.

METHODS

Pulmonary macrophages were loaded with γ-Fe_2O_3 (maghemite) particles by having hamsters breathe an aerosol (particle aerodynamic diameter = 0.7 μm) generated by a technique previously described (Valberg and Brain, 1979). These lung cells are known to ingest material deposited on alveolar surfaces from inhaled air (Sorokin and Brain, 1975). One day after inhalation exposure, pulmonary macrophages were harvested using thirteen 3-ml saline washes of the lungs. The first wash was discarded because of a high content of mucus, surfactant, and cell debris; the remaining washes were pooled and the macrophages were pelleted by centrifugation. Cells were resuspended in Hanks medium and allowed to adhere to the bottom of glass vials. Motion of the intracellular particles was then studied in the attached macrophages.

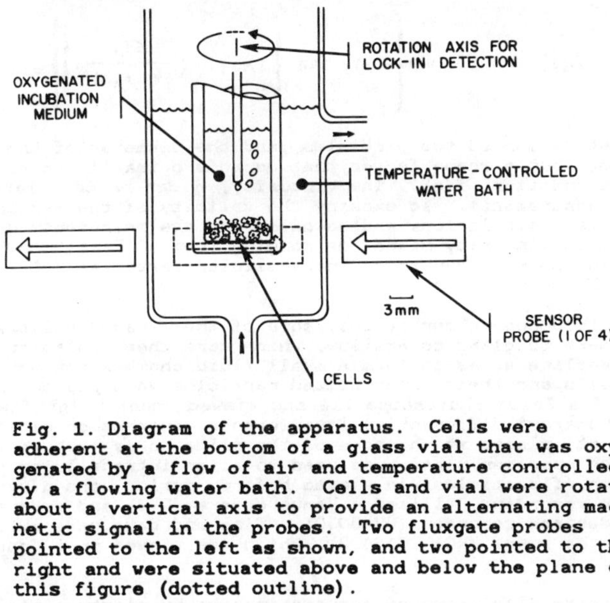

Fig. 1. Diagram of the apparatus. Cells were adherent at the bottom of a glass vial that was oxygenated by a flow of air and temperature controlled by a flowing water bath. Cells and vial were rotated about a vertical axis to provide an alternating magnetic signal in the probes. Two fluxgate probes pointed to the left as shown, and two pointed to the right and were situated above and below the plane of this figure (dotted outline).

Alignment of the ingested-particle magnetic moments was produced by a brief pulse (10 μsec) of a 0.1 T (Tesla) magnetizing field. Calculations show that even in a medium with viscosity as low as that of water, particles will not rotate significantly in such a short time. A pulse is preferable to using a permanent magnet, as done in previous studies, which physically rotates the particles, may introduce distortions in the cytoskeleton, and enhances magnetic particle aggregation. To detect the weak remanent field produced by the particles (~ 4 nT), the use of noise-reduction techniques was necessary to prevent interference from much larger environmental magnetic fields (~10^4 nT). A Förster 1.107 fluxgate magnetometer was used with two sets of special field and gradient probes connected in a second-order gradiometer configuration as depicted in Fig. 1.

The probes were oriented parallel to field lines from the local magnetic dipole to optomize their sensitivity to the field from the cells (i.e., internalized particles). The pulse magnetization was along a direction indicated by the arrows on the outer probes. The adherent cells and the vial containing them were rotated about a vertical axis at approximately 10-13 rotations per second. A phase-sensitive detector combined the magnetometer signal with a reference signal synchronized to cell vial rotation, thereby amplifying only those signals in synchrony with sample rotation. The use of these techniques made the apparatus maximally sensitive to the dipole magnetic field from the cells, and magnetic noise from outside sources was effectively cancelled.

Intracellular viscosity was probed by aligning the ingested magnetic particles, and then using a current-carrying coil to apply a magnetic field (H_i) along the axis of cell vial rotation, i.e. perpendicular to the direction in which the particles were aligned by magnetization. As the particles realign toward this "twisting" field, the vector component of the remanent magnetization sensed by the fluxgate probes, B, decreases, and the time course of particle realignment can be monitored. The sensed field (B) depends on time, t, in a way that is related to both the strength of the twisting field, H_i, and the viscosity, η, of the medium which retards particle motion:

$$B(t) = B_0 \sin\left\{2 \arc\tan\left[\exp -\left(\frac{MH_i t}{6\eta\ f(e)}\right)\right]\right\} \tag{1}$$

M is the magnetization of the particles, and the diameter of the particles is not a factor, but a shape factor that depends primarily on eccentricity, e, enters as a function, f(e). The viscosity, η can be calculated from the experimental measurements. To examine the validity of the assumptions involved in these calculations and to calibrate the measurements, particles were suspended in viscosity standards ranging from 60 Pa-s to 600 Pa-s, and were twisted in the same manner as intracellular particles (viscosity of water is ~0.001 Pa-s).

For the videomicroscopy observations, some of the lavaged cells were allowed to adhere to glass coverslips, which were then sealed to microscope slides with vaseline so as to form a small fluid chamber between cover slip and slide. Cells and their internalized particles were placed on the warm (37°C) stage of a Zeiss Photoscope III and viewed under brightfield, phase-contrast, or Nomarski differential-interference contrast conditions. A permanent magnet, placed at the edge of the slide so as to have field lines parallel to the specimen plane, was used to align intracellular particles. The disposition of particles was viewed before, during, and after magnetization. Video recording utilized a Venus camera (DV2) and a Panasonic time-lapse videotape recorder (NV-8030). Time and date were automatically combined with the image (RCA TC1440B), which was viewed on a Conrac (SN/9) monitor.

Lavaged cells were also prepared for examination by electron microscopy. Both non-magnetized and magnetized cells were pelleted, fixed, dehydrated in graded ethanols and propylene oxide, and embedded in Epon. The pellets were sectioned and poststained (uranyl acetate and lead citrate). Sections were examined in a Philips 300 transmission electron microscope.

RESULTS

Macrophage content of particles was about 24 μg γ-Fe_2O_3 per 10^6 cells. Electron micrographs (Fig. 2A) showed the electron-dense γ-Fe_2O_3 to be intracellular and within phagosomes or secondary lysosomes. Transmission electron micrographs were made of cells either with the magnetic particles in random alignment or after alignment with a permanent magnet. Figures 2

A and B show this comparison and illustrate how the permanent magnet technique for aligning particles can result in chains of particles.

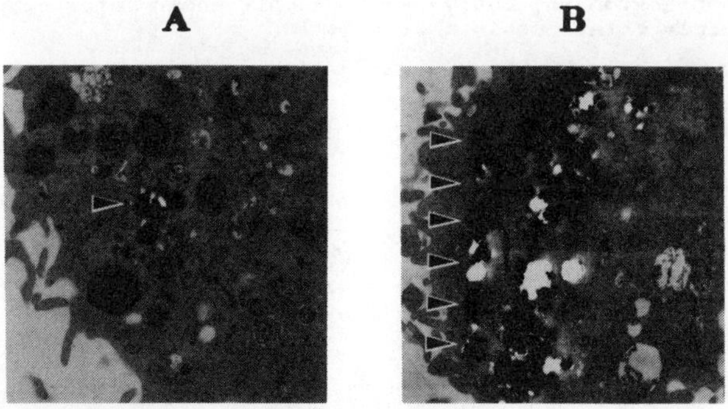

Fig. 2A and 2B. Electron micrographs show the electron-dense, crystalline maghemite to be intracellular and within membrane-bound vesicles. The cell in A had not been magnetized, but the cell in B (which also contains a larger amount of maghemite) had been magnetized for 2 min with the field lines going vertically. Note aggregation of maghemite vesicles into a linear array (arrows).

The magnetic iron oxide was resolved in living cells under bright-field and Nomarski conditions. Application of a magnetic field could be seen to cause lineup of particles within cells. Use of a Zeiss Planapo ×63 objective with 1.4 aperture allowed optical sectioning of the cells so that the intracellular location of maghemite particles could be appreciated. Time-lapse video sequences of cells under the microscope at 37°C showed alignment of the particles as the magnet was applied and then gradual loss of alignment as the cytoplasm was rearranged through active motions. Compression of time by a factor of 9 showed pseudopod formation and retraction, veil movement from periphery to center, membrane ruffling, and the inclusion of the maghemite-containing endosomes in this general motion.

Figures 3A,B, and C show recordings of the magnetic field from the cells. In each graph, the upper curve is the time course of the remanent field with no external force applied, and the lower curve was produced with a twisting field of 2.5 mT applied perpendicular to the original direction of magnetization. If cells were formalin-fixed, the remanent field was constant (not shown), but for oxygenated macrophages at 37°C, the field decreased rapidly (Fig. 3A), to about one-half of its initial value in 5 min. The spontaneous decrease in remanent field seen for living cells is due to random rotation of particles away from their original direction of magnetization as driven by active cytoplasmic rearrangements that move and rotate the particles. The externally-applied twisting force causes additional rotation as evidenced by the increased decay of the remanent field. Figures 3B and 3C show that active movements within the cytoplasm are inhibited by either cooling the cells (10°C), or by incubation with cytochalasin D (cytD=20μM). Moreover, a dramatic difference can be seen in the effect of a twisting force on each preparation; cooled cytoplasm resists shear, but cytD-treated cytoplasm does not. Elasticity could also be observed, as shown in Figure 3D. The twisting field was applied for one minute during the interval between by the two arrows. The yield and corresponding elastic recoil of the cytoplasm are evident. By comparing these results to the effect of twisting particles suspended in a series of viscosity standards, we calculated that the apparent cytoplasmic viscosity of oxygenated cells at 37°C ranges from 1.2 kPa-s at shear rates of 0.003 s^{-1} to 2.7 kPa-s at shear rates of 0.001 s^{-1}. The cytoplasm is undoubtedly non-Newtonian, but we applied only shear rates comparable to observed particle rotations in active cells.

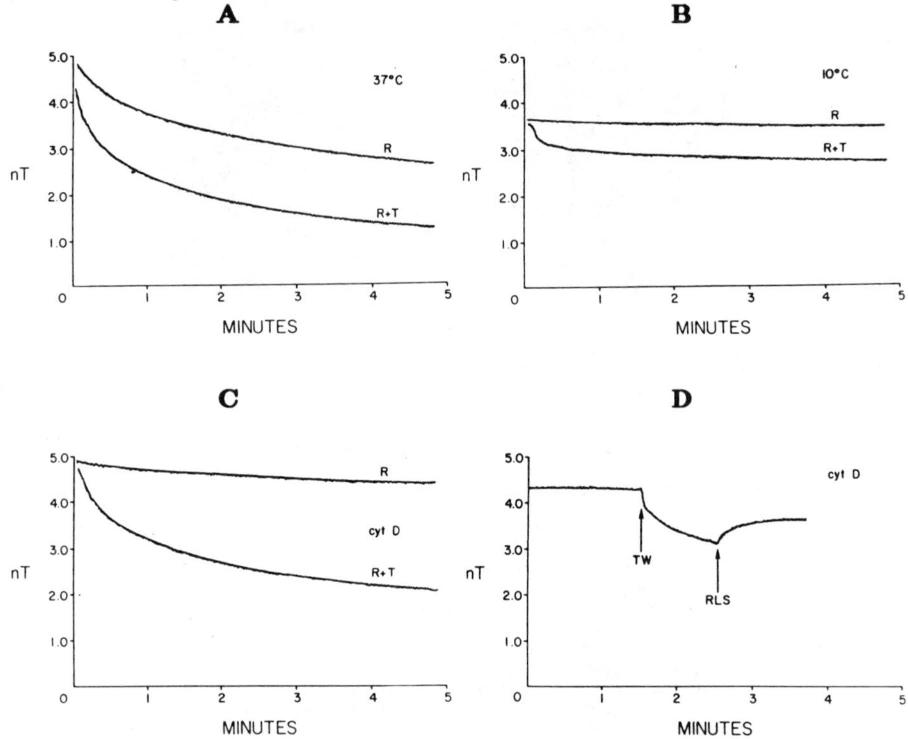

Fig. 3A-D. A: Remanent-field for oxygenated cells at 37°C in the case of unassisted relaxation (R) and relaxation in the presence of a twisting field (R+T). B: Same as A for cells at 10°C. C: Same as A for cells in 20 µM cytD. D: Application (TW) and release (RLS) of the twisting field showing the elastic recoil of cytoplasm.

DISCUSSION

These results are important in the interpretation of measurements made on magnetic particles within intact organisms (Gehr and co-workers, 1983a and 1983b; Brain and co-workers, 1984; Cohen and co-workers, 1984). Even though various conditions influence remanent-field decay ("relaxation"), effects on the driving force could not be distinguished from effects on the viscous properties of the medium. Also, the contribution of Brownian movement in the observed relaxation could not be ascertained without analyzing magnetization curves. However, the application of a known external torque can help answer these questions. The contribution of Brownian movement to decay can be calculated from the apparent viscosity to be of the order of

9.2×10^{-4} min^{-1}, whereas relaxation is observed to proceed at about 7.8×10^{-2} min^{-1} over the first fifteen minutes. Thus, intracellular particles appear not to be driven primarily by thermal energy. A similar conclusion was reached by Nemoto (1982) through the analysis of magnetization curves for particles contained in the lungs of humans. The twisting results of Fig. 3A,B, and C suggest that whereas cooling the cytoplasm may both inhibit force generation and increase cytoplasmic viscosity, cytD interferes primarily with force generation. Whole-animal studies have also shown that the remanent-field decay from macrophages magnetized *in situ* is changed when toxic particles are instilled into the lung (Brain, Hu, and Bloom, 1984). Further work will be needed to determine whether the changes are due to a direct effect of the toxic particles on force generation by macrophages, or whether toxic particles cause translocation of the magnetite and/or macrophages into a new environment where viscosity and freedom of movement are altered.

In comparison to earlier work on cytoplasmic viscosity, these techniques use much smaller particles and are also non-optical and easier to quantify. Furthermore, rather than using reconstituted actin gels (Pollard, 1981) or non-mammalian cells (Sato and co-workers, 1984), we are using intact mammalian macrophages examined at physiological conditions. In conclusion, magnetometry of isolated living cells can yield additional information about the cytoplasm because intracellular particles can be manipulated by external magnetic fields.

REFERENCES

Allen, R. D. (1981). Motility. *J. Cell Biol., 91,* 148s-155s.

Beckerle, M. C., and K. R. Porter (1983). Analysis of the role of microtubules and actin in erytrhrophore intracellular motility. *J. Cell Biol., 96,* 354-362.

Brain, J. D., T. Hu, and S. B. Bloom (1984). Effects of SiO_2 and Fe_2O_3 on the clearance of magnetic particles from hamster lungs. *Am. Rev. Respir. Dis., 129,* A144.

Brain, J. D., S. B. Bloom, P. A. Valberg, and P. Gehr (1984). Correlation between the behavior of magnetic iron oxide particles in the lungs of rabbits and phagocytosis. *Exp. Lung Res., 6,* 115-131.

Cohen, D., I. Nemoto, L. Kaufman, and S. Arai (1984). Ferrimagnetic particles in the lung Part II: The relaxation process. *IEEE Trans. Biomed. Engineering., BME-31,* 274-285.

Cohen, D. (1973). Ferromagnetic contaminants in the lungs and other organs of the body. *Science, 180,* 745-748.

Crick, F. H. C., and A. F. W. Hughes (1950). The physical properties of cytoplasm: A study by means of the magnetic particle method. *Expt. Cell Res., 1,* 37-80.

Gehr, P., J. D. Brain, S. B. Bloom, and P. A. Valberg (1983). Magnetic particles in the liver: a probe for intracellular movement. *Nature, 302,* 336-338.

Gehr, P., J. D. Brain, I. Nemoto, and S. B. Bloom (1983). Behavior of magnetic particles in hamster lungs: estimates of clearance and cytoplasmic motility. *J. Appl. Physiol.: Respirat. Environ. Exercise Physiol., 55,* 1196-1202.

Herman, B., and D. F. Albertini (1984). A time-lapse video image intensification analysis of cytoplasmic organelle movements during endosome translocation. *J. Cell Biol., 98,* 565-576.

Nemoto, I. (1982). A model of magnetization and relaxation of ferrimagnetic particles in the lung. *IEEE Trans. Biomed. Eng., BME-29,* 745-752.

Pollard, T. D. (1981). Cytoplasmic contractile proteins. *J. Cell Biol., 91,* 156s-165s.

Sato, M., T. Z. Wong, D. T. Brown, and R. D. Allen (1984). Rheological properties of living cytoplasm: A preliminary investigation of squid axoplasm (Loligo pealei). *Cell Motil., 4,* 7-23.

Sheetz, M. P., and J. A. Spudich (1983). Movement of myosin-coated fluores-

cent beads on actin cables in vitro. *Nature, 303,* 31-35.
Sorokin, S. P., and J. D. Brain (1975). Pathways of clearance in mouse lungs exposed to iron oxide aerosols. *Anat. Rec., 181,* 581-626.
Steinman, R. M., I. S. Mellman, W. A. Muller, and Z. A. Cohn (1983). Endocytosis and the recycling of plasma membrane. *J. Cell Biol., 96,* 1-27.
Valberg, P. A. (1984). Magnetometry of ingested particles in pulmonary macrophages. *Science, 224,* 513-516.
Valberg, P. A., and J. D. Brain (1979). Generation and use of three types of iron-oxide aerosol. *Am. Rev. Respir. Dis., 120,* 1013-1024.

ORGANELLE MOVEMENTS OF ALVEOLAR MACROPHAGES STUDIED BY CYTOMAGNETOMETRY

P. Gehr*, J.D. Brain**, I. Nemoto*** and S.B. Bloom**
*Department of Anatomy, University of Berne, Berne, Switzerland
**Department of Environmental Science and Physiology, Harvard School of Public Health, Boston, USA
***Department of Mathematical Science, Tokyo Denki University, Saitama-Ken, Japan

ABSTRACT

Monolayer cultures of alveolar macrophages which contain Fe_3O_4 particles in their phagosomes or secondary lysosomes were studied by magnetometric means. Evidence was found that motion of these organelles is an actin-based mechanism.

KEYWORDS

Magnetic particles; cell cultures; cytoskeleton; actin; lysosomes; cytocholasin B; colchicine; p-trifluoromethoxy carbonyl cyanide phenylhydrazone.

INTRODUCTION

Previous studies have shown that magnetic fields from microgram quantities of ferrimagnetic particles (Fe_3O_4, magnetite) instilled _in vivo_ into lungs can be detected at the chest surface (Cohen, 1973; Valberg and Brain, 1979; Halpern and co-workers, 1981; Gehr and co-workers, 1983b; Högstedt, Kadefors and Näslund, 1983; Brain and co-workers, 1984). A magnetic field external to the body magnetizes and aligns these particles. Following removal of the magnetizing field the particles produce a remanent magnetic field that is measurable above the chest with a fluxgate magnetometer. Its field strength is an estimate of the amount of iron oxide retained in the lungs. Sequential measurements over long periods of time therefore enable us to study non-invasively the clearance of the particles.

Interestingly, the remanent field is not stable but decays with time. This is due to random progressive rotation of the particles away from their initial aligned state. This process is called relaxation and was first observed by Cohen (1973).

After their instillation into the lungs, magnetic particles are avidly taken up by alveolar macrophages where they ultimately reside in phagosomes and secondary lysosomes (Fig. 1, left). Cellular movements were therefore proposed to be responsible for the relaxation seen in _in vivo_ studies (Brain and co-workers, 1984;

Nemoto and co-workers, 1984; Gehr and co-workers, 1983a; Gehr and co-workers, 1983b; Cohen, 1978). Phagosomes and secondary lysosomes have been described as being an integral part of the intracellular filament system (Fig. 1, right; Wolosewick and Porter, 1979). Moore and co-workers (1976) even showed that they are connected to actin filaments. We therefore speculated that the particle rotations might be the result of forces imparted to these organelles by contractile ele-

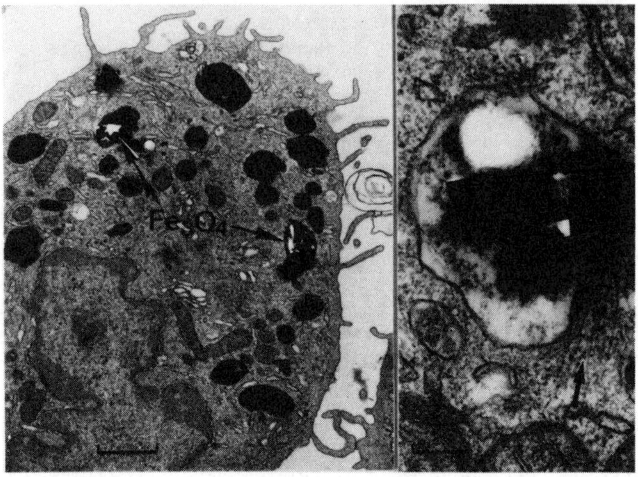

Fig. 1 Left: Alveolar macrophage containing Fe_3O_4 particles in secondary lysosomes (bar = 1 μm). Right: Fe_3O_4 particles in a secondary lysosome in close association with filaments (arrow) (bar = 0.02 μm).

ments of the filament system; the phagosomes and secondary lysosomes containing magnetite particles would be rotated by contracting filaments (sliding filaments, i.e. interaction of actin and myosin), much like puppets manipulated by their strings. Such movements, described as saltatory, or dancing, movements (Rebhun, 1972) would rotate the ingested particles in a myriad of directions different from their original orientation received during magnetization, resulting in a progressively decreasing net field at the chest surface (Gehr and co-workers, 1983a, b).

If relaxation is primarily caused by cell motility, rather than lung movement, it should be possible to demonstrate the phenomena in cultured cells. We report here *in vitro* measurements of the magnetic field of alveolar macrophages that had previously phagozytized magnetite particles*.

---

* Abstract 23rd Ann. Meeting Amer. Soc. Cell Biol.: Gehr, P., J.D. Brain, and S.B. Bloom (1983). Magnetometry: A tool to study intracellular movement. *J.Cell Biol.*, *97*, 194a.

MATERIALS AND METHODS

Submicrometric Fe_3O_4 particles (magnetite, type MO-7029 Pfizer; 0.2 mg suspended in 0.15 ml saline per 100 g body weight) were instilled into the lungs of Syrian golden hamsters. After 24-36 hours, macrophages were collected by repeated pulmonary lavage (Brain and Frank, 1969), and the lavages were pooled and centrifuged. The cells were then resuspended in culture medium (medium 199, 10% calf serum, gentamycin, mycostatin) and incubated in Petri dishes (10^6 cells per 1 ml culture medium, 2.5 ml cell suspension per dish) at 37 °C for 1 hour. Non-adherent cells were rinsed off and the remaining monolayer culture (approximately 2000-3000 cells per mm^2) reincubated. Four hours later the dish was placed in a temperature controlled well. After applying a magnetic field of 0.15 T for 5 sec to the monolayer culture the well was moved into a cylindrical moly-permalloy shield to reduce external magnetic interference, and the culture repeatedly presented to a fluxgate magnetometer probe (Magnetoscop F 1.067, Foerster Instruments). The remanent field was measured this way above the cells for 20 min. Since it took a few seconds to move the culture dish containing the well to the shielded magnetometer probe and since the remanent field of the cells was decaying during this time (relaxation), we estimated the initial remanent field strength, B_0, by fitting the data from the first minute to an exponential function and extrapolating back to time zero (Nemoto, 1982). The rate of decay (relaxation rate) was characterized by the relaxation half-time, $t_{1/2}$, the time required for the remanent field to decay to one-half of its initial field strength (one-half of B_0), the initial relaxation rate, λ_0, which represents the fractional change of B_0 during the first minute of decay (Nemoto, 1982) and B_{20}/B_0, which is the fraction of B_0 after 20 min of decay (period of measurement).

Some of the cultures were also treated with drugs for 10 min. at 37 °C, such as 20 µM cytochalasin B (cyt. B), which is a disintegrator of the actin filaments, 60 µM colchicine (colch.), which is a disintegrator of the microtubules, and 100 µM p-trifluoromethoxy carbonyl cyanide phenylhydrazone (FCCP), which is an uncoupler of the respiratory chain; other cultures were fixed with 2.5% potassium phosphate buffered glutaraldehyde (GA) for 10 min at 37 °C or cooled to 20 °C for 10 min prior to magnetizing and measuring the monolayer culture.

RESULTS

The remanent magnetic field measured above the monolayer culture of alveolar macrophages decayed in a manner very similar to that of lungs in live animals as can be seen in Fig. 2. B_0, however, was smaller (between 6 and 7 nT on the average as compared to about 20 nT in live animals) because only a few hundred thousand cells could be measured in a monolayer culture.

Figure 3 summarizes the findings of the cultures treated with drugs, glutaraldehyde and low temperature. The relaxation was found to be significantly slower in cultures treated with cyt. B, FCCP and low temperature, whereas it was not significantly influenced by colch. No relaxation could be observed after GA-treatment.

DISCUSSION

The findings confirm our hypothesis that the relaxation phenomenon is a vital process which is taking place at a subcellular level. The following conclusions can be drawn. (1) Relaxation is caused by intracellular forces. (2) The drug and temperature experiments provide evidence that these forces which randomize the orientation of lysosomes containing the particles (and hence cause the remanent field to decay) originate from contractile proteins. Cyt. B known to disintegrate the actin filament system, decreased the relaxation rate significantly, whereas colch., known to disintegrate the microtubular system, hardly had any effect. (3) The uncoupler FCCP and low temperature both cause the relaxation rate to decrease, i.e. lysosomal motion to slow down, suggesting that the process is energy dependent. Similar results were recently reported by Valberg (1984).

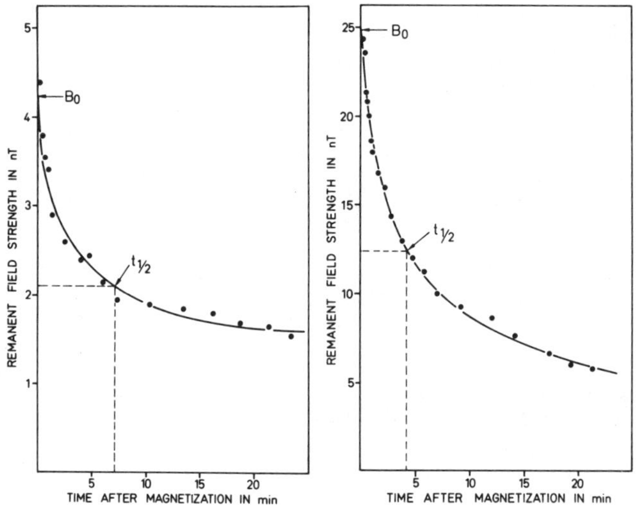

Fig. 2 Comparison of *in vitro* (left graph) and *in vivo* (right graph) relaxation curves. Closed circles represent individual readings of the magnetometer; B_0 is the initial field strength; $t_{1/2}$ is the relaxation half time (see text, for explanation); nT = nanotesla.

Since it has been shown that lysosomes are connected to actin filaments (Moore and co-workers, 1976), our hypothesis of lysosomes being moved by the acto-myosin system (sliding filament mechanism) like puppets which are manipulated by their strings, is taking shape. The cytomagnetometric measurements presented here provide evidence that motion of phagosomes and secondary lysosomes is an actin-based mechanism.

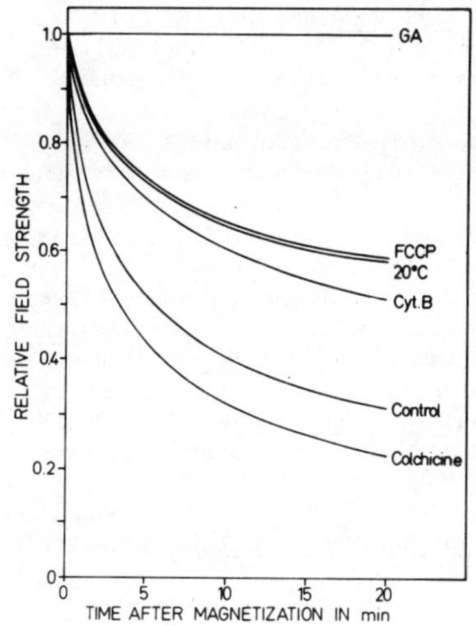

Fig. 3 Normalized relaxation curves of monolayer cultures treated with different drugs, fixative and low temperature (see text for explanation).

ACKNOWLEDGEMENTS

We thank Dr. D.A. Begg for his advice regarding preservation of the cytoskeleton and Ms Ch. Furter, Mrs. K. Babl, and Ch. Lehmann for technical assistance. We are also thankful to Dr. K.E. Conley for reading the manuscript and to Ms R. Fankhauser for typing the manuscript. This study was supported by SNSF grant 3.130-0.81 and NIH grant HL-31029.

REFERENCES

Brain, J. D., and N. R. Frank (1969). Recovery of free cells from rat lungs by repeated washings. J. Appl. Physiol., 25, 63-69.

Brain, J. D., S. B. Bloom, P. A. Valberg, and P. Gehr (1984). Correlation between the behavior of magnetic iron oxide particles in the lungs of rabbits and phagocytosis. Exp. Lung Res., 6, 115-131.

Cohen, D. (1973). Ferromagnetic contamination in lungs and other organs of the human body. Science, 180, 745-748.

Cohen, D. (1978). Report of the low-field group: the magnetic field of the lung. National Technical Information Service, Springfield, Va. MIT/FBNML-78/1.

Gehr, P., J. D. Brain, S. B. Bloom, and P. A. Valberg (1983a). Magnetic particles in the liver: A probe for intracellular movement. Nature, 302, 336-338.

Gehr, P., J. D. Brain, I. Nemoto, and S. B. Bloom (1983b). Behaviour of magnetic particles in hamster lungs: Estimates of clearance and cytoplasmic motility. J. Appl. Physiol., 55, 1196-1202.

Halpern, M., S. J. Williamson, D. M. Spektor, R. B. Schlesinger, and M. Lippman (1981). Remanent magnetic fields for measuring particle retention and distribution in the lungs. Exp. Lung Res., 2, 27-35.

Högstedt, P., R. Kadefors, and P. E. Näslund (1983). Magnetopneumography in welders: Observations on relaxation and clearance. Il Nuovo Cimento, 2, 608-616.

Moore, P. L., H. L. Bank, N. T. Brissie, and S. S. Spicer (1976). Association of microfilament bundles with lysosomes in polymorphonuclear leukocytes. J. Cell Biol., 71, 659-666.

Nemoto, I. (1982). A model of magnetization and relaxation of ferrimagnetic particles in the lung. IEEE Trans. Biomed. Eng., BME-29, 745-752.

Nemoto, I., H. Toyotama, J. D. Brain, and P. Gehr (1984). Behavior of ferrimagnetic particles in hamster pulmonary macrophages in vitro; magnetometric estimation of cell motility. J. Cell. Physiol., (submitted).

Rebhun, L. I. (1972). Polarized intracellular transport: Saltatory movements and cytoplasmic streaming. Int. Rev. Cytol., 32, 93-137.

Valberg, P. A. (1984). Magnetometry of ingested particles in pulmonary macrophages. Science, 224, 513-516.

Valberg, P. A., and J. D. Brain (1979). Generation and use of three types of iron-oxide aerosol. Am. Rev. Respir. Dis., 120, 1013-1024.

Wolosewick, J. J., and K. R. Porter (1979). Microtrabecular lattice of the cytoplasmic ground substance: artifact or reality? J. Cell Biol., 82, 114-139.

MEASUREMENTS OF THE BEHAVIOR OF MAGNETIC FINE PARTICLES IN A MACROPHAGE-LIKE CELL (J774.1)

H. Toyotama* and I. Nemoto**

*Research Development Corp. of Japan, Ultra-Fine Particle Project, Shingidan Office, Kyobashi 1-11-2, Chuo-ku, Tokyo 104, Japan

**Tokyo Denki Univ., School of Science and Engineering, Hatoyama, Saitama 350-03, Japan

ABSTRACT

Motions of submicrometric particles in cultured macrophage-like cells J774 were studied. Particles were introduced into the cells by phagocytosis. Rotational movements of the particles were measured by a magnetometric method. The particles were magnetized and aligned by applying an external magnetic field. After the external field was removed, the particles produced a weak remanent magnetic field which decreased due to random misalignment of the particles (relaxation) in the cells. Microscopic observation showed that initial remanent field was proportional to the number of the particles which were attached to the cell surface and/or phagocytized by the cells. When the temperature was lowered, the relaxation became slower. The effect of low temperature was reversible. When metabolic inhibitors, KCN and MIA (monoiodoacetic acid), were given to the cells simultaneously, the relaxation became slow immediately. Intracellular ATP concentration was also measured after the treatment of the cells by the metabolic inhibitors. The results showed close correlation between intracellular ATP concentration and the rate of relaxation. When cytochalasin B (which is known to depolymerize actin filament and prevent actin-myosin interactons) was applied, the decay rate of the remanent field became faster than that of the control cells. From these results, we conclude that (1) magnetometric measurements can be used to quantify the rate of phagocytosis, (2) relaxation is due to random rotational movements of the phagosomes which are related to active ATP-dependent intracellular mechanisms such as those in the actin-myosin system.

KEYWORDS

Macrophage-like cell (J774.1); phagocytosis; magnetic fine particles; magnetometry; relaxation; metabolic inhibitors; intracellular ATP.

INTRODUCTION

A magnetic field produced by ferrimagnetic particles in the lungs of man was first observed by D. Cohen (1973). It was demonstrated by Nemoto and colleagues (submitted for publication) that relaxation occurred in lavaged pulmonary macrophage *in vitro*, confirming the hypothesis that relaxation of the magnetic field observed in the lungs is primarily related to processes in pulmonary macrophages (Gehr and colleagues

1983). In the present paper, we present the results of relaxation measurements of cultured macrophage-like cells J774.1. With these cells always at hand, it is easier to provide various experimental situations than it is with pulmonary macrophages which have to be harvested and are limited in number for each animal.

MATERIALS AND METHODS

Cell Culture

Macrophage-like cells J774.1 are derived from mouse peritoneal macrophages and are endowed with remarkable phagocytic activity similar to native macrophages. The cells were maintained in Ham's F12 culture medium supplemented with 10% new born calf serum, penicillin (100 units/ml) and streptomycin (100 μg/ml). They were incubated in water-saturated air with 5% CO_2. Cells cultured to a subconfluent state in dishes 90 mm in diameter were used throughout the present experiments. Approximately 10^6 cells in 1.5 ml of culture medium were placed for 1 hour at 37°C on the bottom of glass sample vials (15 mm in diameter, 40 mm in height). After the monolayer cultures were established, the cells were washed twice with culture medium before adding the particles. Magnetic fine particles were dispersed in F12 medium (without serum) by sonication (Branson, Sonifier 200) immediately before use. For the assay, 0.03 mg of particles suspended in 0.1 ml of culture medium were applied to each sample vial. At various times after addition of the particles, the sample vial was measured to assay the rate of pagocytosis (time course of phagocytosis). To study the effect of inhibitors on the rate of relaxation, the cells were incubated for 24 hours after addition of the particles before measurement. The cells were washed twice to remove remaining particles which had not been phagocytized.

Particles

In the present experiment, we used five kinds of magnetic fine particles as summarized in Table 1. Magnetite particles (Fe_3O_4) were purchased from Pfizer, U.S.A. Other particles (Fe-UFP, Ni-UFP and γ-Fe_2O_3) were prepared in our laboratory by the Gas Evaporation Method (GEM). These particles are significantly free of impurities. Furthermore, the distribution of their sizes is narrower than that of Fe_3O_4 particles which are chemically produced.

Magnetometry

Magnetization and orientation of the particles were achieved by placing the cells midway between the electric coils. The external field (1000 gauss) was applied for 5 seconds. Within 3 seconds after cessation of the applied field, the vial was moved to a position immediately above the probe of a fluxgate magnetometer (Förster) located in a magnetic shield. The shield consisted of double Permaloy pipes. The center of the bottom of the sample vial was positioned above the axis of the upward probe. The reading was compared to a baseline obtained when the vial was placed far from the probe. Temperature of the vial was controlled by circulating water around the vial. When 0.03 mg of magnetite particles (Fe_3O_4) were phagocytized, the initial remanent field ranged from 70 to 100 μG. We obtained an initial relaxation rate, λ (expressed as the fraction which decays per second), by fitting a straight line to the log of the magnetic field during the first 30 seconds.

Chemicals

All chemicals used were of reagent grade and were obtained from Sigma, St. Louis, U.S.A. KCN, MIA and colchicine were dissolved in water for use. Cytochalasin B were dissolved in DMSO and added to the culture medium, the final concentration of DMSO being kept below 0.1%.

APT Assay

10^6 cells in 3 ml culture medium F12 was boiled for 5 minutes without homogenization to extract ATP from the cells. The amount of ATP in the extract was determined by measuring the intensity of the light emitted when luciferin-luciferase was added to the extract and comparing the reading with that for an ATP solution of known concentration. The light emission was measured by the photomultiplier of a bioluminescence reader (Biolumicounter 100, NITION, Japan). The intracellular concentration of ATP was calculated by assuming a spherical cell shape with diameter 10 μm.

RESULTS

Time Course of Phagocytosis by Magnetometry

Figure 1(a) shows relaxation curves measured at different times after the addition of particles (Octahedral-Fe_3O_4, 0.03 mg/10^6 cells). The extrapolated initial value of remanent field increased with time within 150 min. (b). From the light microcopic observation, the initial value was roughly proportional to the number of the particles which attached to the surface of the cells and/or phagocytized by the cells. Measurements were done under the presence of remaining particles (not phagocytized) on the bottom of the sample vial. The particles which deposited on the bottom were aligned by the earth's field and formed chains. Therefore, the direction of the magnetic dipoles of these particles were perpendicular to the axis of the probe of the magnetometer. Because of the geometry, the contribution of the remaining particles to the remanent magnetic field was small. Within 5 hours, all particles were phagocytized by the cells under control conditions. The rate of phagocytosis measured by magnetometry showed strong temperature dependence (data not shown here). Metabolic inhibitors affected phagocytosis remarkably. When the cells were treated with both 1 mM KCN and 1 mM MIA, the initial value did not increase even 150 minutes after the particles were applied. After all the particles were phagocytized, the rate of relaxation contined to become somewhat faster during the first day and then became constant.

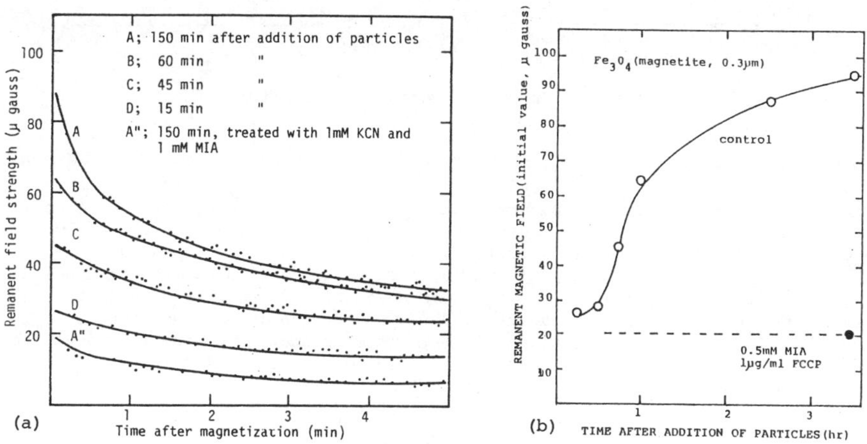

Fig. 1. Time course phagocytosis.

Effects of Drugs (Inhibitors of ATP Synthesis and Cytoskeletal Motion) on Relaxation

Figure 2(a) shows the effects of metabolic inhibitors on the rate of relaxation. When either kind of inhibitor (1 mM KCN; inhibitor of oxidative phosphorylation or 1 mM MIA; inhibitor of glycolysis) was given individually, the effect was slight. When both inhibitors were applied simultaneously, the relaxation became much slower immediately. Figure 2(b) shows the results of measurement of intracellular concentration of ATP after treatment of the same drugs. Intracellular ATP level decreased immediately after simultaneous treatment with both 1 mM KCN and 1 mM MIA, but did not change so much by addition of the individual inhibitor. This result shows close correlation between intracellular ATP level and the rate of relaxation. Cytochalasin B (2 μg/ml) increased the rate of relaxation significantly. This is a rather curious result, as cytochalasin B is an inhibitor of cytoplasmic motion. Colchicine (5 mM) did not affect relaxation. Furthermore, relaxation showed remarkable temperature dependence (data not shown).

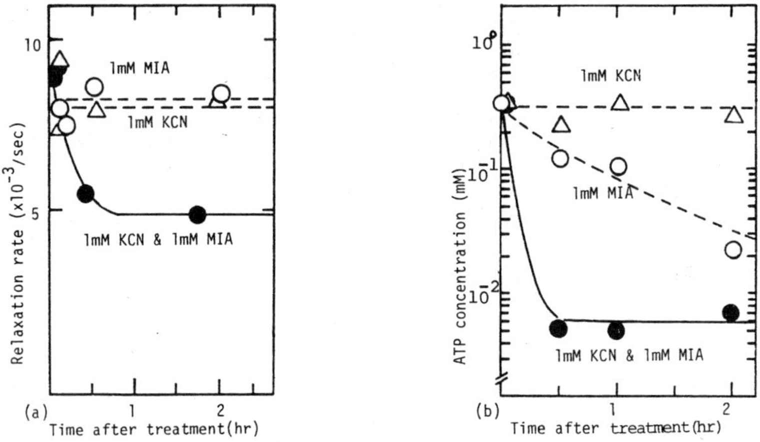

Fig. 2. Effect of metabolic inhibitors.

DISCUSSION

Table 1 shows the summary of magnetic fine particles used in the present experiment. Among them, iron oxide particles, i.e., γ-Fe_2O_3 and Fe_3O_4 were easily phagocytized, but metal ultra fine particles such as Fe and Ni were phagocytized to a much smaller degree. Initial relaxation rate did not depend so much on the size and shape of the particles. If rotational movements of the particles in the cells were simply caused by thermal agitation, λ would have a much larger scatter according to the variety of the particles. This is another piece of evidence that relaxation is not the result of Brownian motion due to thermal agitation. The relaxation showed a tendency to increase with time within 24 hours after exposure. We observed microscopically that many phagocytized particles existed mainly in the peripheral regions of the cells during the first few hours after addition of the particles. Then they became dispersed uniformly in the cells. The increase of λ seems to reflect the change of the locations of phagosomes in the cells. The effects of inhibitors of oxidative phosphorylation and glycolysis were very interesting. When KCN and MIA were applied individually, both relaxation rate and intracellular ATP level seemed not to be affected at least for 2 hours. But when both inhibitors were applied simultaneously, the relaxation rate became slower and intracellular ATP level decreased immediately. This suggests enhancement of one of the two pathways of ATP synthesis, that is, oxidative phosphorylation and glycolysis, to keep the intracellular concentration constant when the other pathway is blocked. The effects of cytochalasin B suggest the contribution of actin fibres to the rotational movements of particles. Gehr (1983) found that phagocytized particles were surrounded tightly by the membrane of phagosomes coated with filamentous structure in pulmonary macrophages. Stossel (1975) discovered that in peritoneal macrophages these filamentous structures contained actin, actin-binding protein and the regulatory protein gelsolin. Low concentration of intracellular ATP is known to cause rigor-state in muscle tissues and to stiffen muscle fibres. The rigor-state of muscles is due to actin-myosin interaction under low concentrations of ATP. We believe that the slow relaxation due to low ATP concentrations reflects the rigor-state of the cyto-skeleton.

From these results, we conclude that this technique has considerable potential for the study of cell motility.

TABLE 1 Magnetic Particles

Type	Shape	Size	Phagocytosis	$\lambda\,(10^{-3}/s)$
Fe_3O_4 (magnetite)	octahedral	dia. 0.3 μm	+	-6.09 ± 0.43
Fe_3O_4 (magnetite)	acicular	0.1 μm × 0.8 μm	+	-7.19 ± 0.54
γ-Fe_2O_3 (ferrite)	sphere	dia. 0.04 μm	+	-8.00 ± 1.28
Fe-UFP (metal)	chain	0.03 μm × 1 μm	-	-
Ni-UFP (metal)	chain	0.03 μm × 1 μm	-	-

ACKNOWLEDGMENTS

The authors are indebted to Dr. Y. Akamatsu and Dr. T. Kitagawa of the National Institute of Health of Japan for providing the cells and teaching us the method of cell culture.

REFERENCES

Cohen, D. (1973). Ferromagnetic contamination in lungs and other organs of the human body. Science, 180, 745-748.

Gehr, P., J. D. Brain, I. Nemoto, and S. Bloom (1983). Behavior of magnetic particles in hamster lungs: estimation of clearance and cytoplasmic motility. J. Appl. Physiol., 55, 1196-1201.

Nemoto, I., H. Toyotama, J. D. Brain, and P. Gehr. Behavior of ferrimagnetic particles in hamster pulmonary macrophages in vitro: magnetometric estimation of cell motility (submitted for publication).

Stossel, T. P., and J. H. Hartwig (1975). Interaction of actin, myosin and a new actin-binding protein of rabbit alveolar macrophages. J. Biol. Chem., 250, 5706-5712.

AN INVERSE SOLUTION IN MAGNETOPNEUMOGRAPHY

G. Stroink, F. Brauer, C. Purcell and P. Krieger

Physics Department, Dalhousie University
Halifax, Nova Scotia, Canada B3H 3J5

ABSTRACT

Based on a multipole analysis of the measured field profile of a uniformly magnetized lung we explore several methods to determine the dipole moment of the lung. We find that the method that uses the dipole and quadrupole coefficients to find the origin of the multipole expansion is the most accurate. This is confirmed by using this method to determine the dipole moment of field profiles produced by objects that simulate the shape of the lung.

KEYWORDS

Magnetopneumography, multipole analysis, dipole moment, gradiometer, lung model, phantom lung.

INTRODUCTION

Several research groups have obtained an estimate of the total magnetic dust load in lungs by comparing the measured magnetic field profile with that of a uniformly magnetized geometric model of the lung. We have shown previously (Stroink and co-workers 1982) that even in the ideal situation where the lung has a uniform magnetization, large errors in such a forward solution are introduced by the uncertainties in the distance between lung and probe. If the lung, as can be expected, does not have a uniform magnetization, then the estimated total magnetic moment based on the geometric models can differ from the true value by as much as a factor of three (Kalliomäki and co-workers 1983).
An inverse solution, based on a multipole analysis of the measured field profiles has been discussed by Attoniemi and co-workers (1981). In this study we assess the effectiveness of such a multipole analysis to obtain an inverse solution for the dipole moment of the lung. Given the remanent magnetization of the dust in the lung, this dipole moment can be directly related to the total magnetic dust load (Stroink and co-workers 1981).

THEORY

Outside a sphere containing the magnetic sources the multipole expansion of the magnetic field can be written as

$$\underline{H}(\underline{r}) = \frac{1}{4\pi} \sum_{n=0}^{\infty} \sum_{m=0}^{n} [A_{nm} \underline{H}^{e}_{mn}(\underline{r}) + B_{nm} \underline{H}^{o}_{mn}(\underline{r})] \quad . \qquad \text{Eq. 1}$$

The even $\underline{H}^{e}_{mn}(\underline{r})$ and odd $\underline{H}^{o}_{mn}(\underline{r})$ unit fields contain the radial dependence characteristic for the various multipoles, and the coefficients A_{nm} and B_{nm} determine the strength of a particular multipole component.

In magnetopneumography (MPG) the field $B_z = \mu H_z$ is measured. Using Eq. 1, this z-component of the field at a particular location can be written as the vector product $B_z = \underline{T}\cdot\underline{m}$, where $\underline{T}$ is the vector of unit fields at the measuring location and $\underline{m}$ is the multipole vector containing the coefficients A and B. If a gradiometer system is used, $\underline{T}$ has to be replaced by a linear combination of vectors calculated at the positions of the gradiometer coils. For M measuring points we can write:

$$\underline{B}_z = \hat{T}\cdot\underline{m} \; , \qquad \text{Eq. 2}$$

where $\underline{B}_z$ is the vector of measured field values, $\hat{T}$ a N x M matrix, where N is the dimension of the multipole vector (e.g. 3 for a dipole, 8 when the quadrupole terms are included). For M>N, Eq. 2 is overdetermined and can be solved for $\underline{m}$, the coefficients of the multipoles, including those of the dipole. A more complete treatment of the multipole analysis is given by Karp and co-workers (1980), and Brauer and Stroink (1984).

METHOD

The general outline of the program used to calculate the dipole moment is given in Fig. 1. Variations of the procedure include the calculation of the multipole coefficients up to dipole, quadrupole, and octupole orders to find the origin that produces a minimal residual. They will be referred to as dipole, dipole and quadrupole, and dipole, quadrupole and octupole origin search, respectively. The iterative search used to find this optimal origin applies a finite differences Levenberg-Marquardt algorithm. In addition, the origin calculated by this iterative process can be shifted to minimize the quadrupole contribution by using the dipole and quadrupole coefficients of the original origin. Such an origin shift has been successfully applied in a multipole analysis in electrocardiography (Savard and co-workers 1982). Finally, by using different $\hat{T}$ matrices appropriate for the different gradiometer systems, one can determine which measuring system gives the best results in determining the dipole moment of the object measured. The gradiometer systems considered here are a second order gradiometer, and a two gradiometer system consisting of one first order gradiometer above and another underneath the subject.

Because of all the possible variations described above, several methods to search for the origin of the multipole expansion were tested:

(A) Search for a dipole origin, for a second order gradiometer;
(B) Search for a dipole origin, for a two gradiometer system;
(C) Search for a dipole origin, for a second order gradiometer, followed by an origin shift;
(D) Search for the origin using dipole and quadrupole terms (dipole and quadrupole origin), for a second order gradiometer;
(E) Search for a dipole and quadrupole origin, for a two gradiometer system;
(F) Search for a dipole and quadrupole origin, for a second order gradiometer followed by an origin shift as in (C).

As the measured field data we used the field of a uniformly magnetized lung model (Stroink and co-workers, 1982), consisting of 4654 dipoles pointing in the z-direction. The field of these dipoles is calculated on a grid of 11 x 5 points in a x-y plane above the object. The spacing of the grid is 4 cm in the x-direction and 5 cm in the y-direction. For the second order gradiometer, we used, as a baseline, b=4 cm. For the two gradiometer system we chose b=10 cm, with a distance of 50 cm between the nearest sensing coils of the two gradiometers.

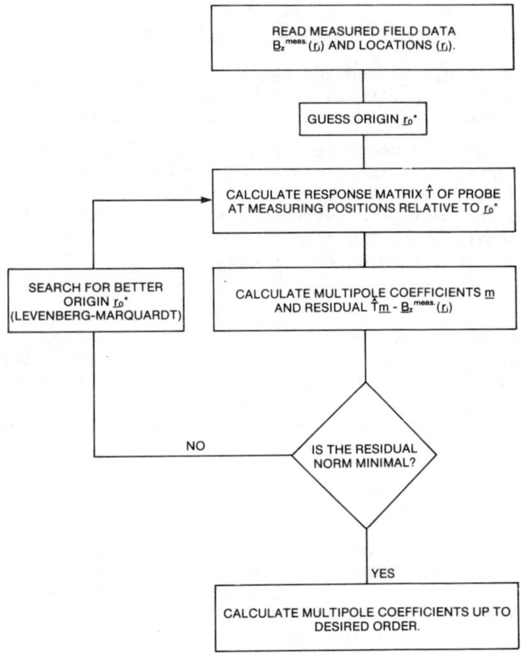

Fig. 1: General outline of the program used to calculate the multipole moments.

RESULTS

Theoretical: Fig. 2 shows the results of the various methods when no random errors are added to the measurements. The results are expressed as a ratio of the calculated value p_z and the theoretical value $p_{theor.}$, known from adding all the dipoles of the lung model. For the second order gradiometer and for the two gradiometer system, the best results are achieved with a dipole and quadrupole origin search. Obviously, in the near field of the lung, because of its extension, quadrupole terms are important. For the second order gradiometer (D in Fig. 2), at z=25 cm, the solution is already within 3% of the true value, and at z=40 cm, within 1.5%. For the two gradiometer system (E in Fig. 2), the error is about 0.2% with the lung in the centre of symmetry of the two probes (z=25 cm). It remains below 1% in the entire region between the two probes. Fig. 3 shows the same methods but now with random errors with a maximum of 2 mm added to the x-y positions used to calculate the test field data. No errors are assumed in the field data. Again methods D and E show the best results. For large z the results for the 2nd order gradiometer deviate from the true value. This is largely due to the fact that the field values decrease rapidly with z (Stroink and co-workers, 1982), whereas the position errors remain constant. This enhances the overall uncertainty in determining the dipole moment. A detailed discussion of these results will be published elsewhere (F. Brauer and G. Stroink, 1984).

Experimental: In addition to these model calculations, the accuracy of the computer program was tested using magnetic field data obtained from circular coils of dimensions (diameter 10.7 cm, length 8.9 cm) that are of the same order of magnitude as the lung. Data sets B_z^{meas} were obtained with a 2nd order gradiometer (baseline 4 cm), by making 5 scans in a plane perpendicular to the axis of the coils. During each scan 11 points were sampled. For a single coil, eight such data sets were obtained in a plane at distances between 18 and 40 cm from the centre of the coil. The results are shown in Table 1. The results for two such coils with their centres separated by 15.2 cm are also shown in the table.

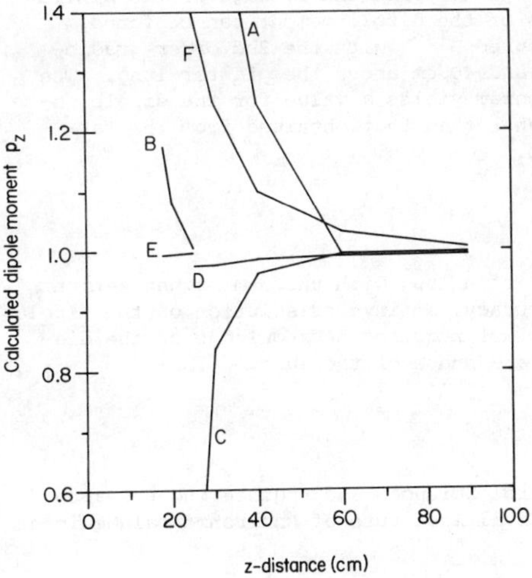

Fig. 2: Calculated magnetic dipole moment $p_z = p_{z\,calc}/p_{z\,theor.}$ for the ellipsoid lung model using methods A-F.

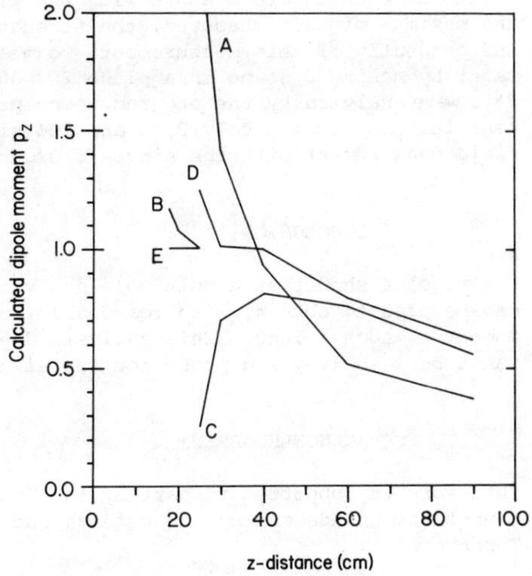

Fig. 3: calculated magnetic dipole moment for a set of random errors of less than 2 mm in the x-y measuring positions.

	$p_{z\,calc}/p_{z\,coil}$	z (±.5 cm)	$z_{cal.}$ (cm)
single coil	.69	18	18
	.75	20	22
	.79	22	23
	.83	24	25
	.92	26	28
	.85	30	33
	.82	34	37
	.77		40
double coil	1.06	18	20
	1.03	26	28
	.73	38	40

Table 1: Calculated moment and distance between the coil and probe with the centre of the coil placed at a distance z from the probe.

As can be seen, the calculated moment agrees reasonably well with the value calculated from the dimensions and windings of the coil. The location of the origin of this moment is also found with reasonable accuracy.

Finally, we measured with the second order gradiometer the field of a plaster phantom lung of realistic shape. Magnetite (Fe_3O_4) was mixed uniformly in the plaster as a magnetic contaminant. The dipole moment of the magnetized plaster lung was measured with a fluxgate sensor 1 m away from the rotating lung, the fluxgate pointed towards the centre of the lung, perpendicular to the axis of rotation. At such a distance the moment is mainly dipolar in character, whereas the signal is still large enough to measure with the fluxgate. From the average of the minimum and maximum signals measured, the strength of the dipole moment can be found. Independently of this measurement, we measured B_z^{meas} with the 2nd order gradiometer by making 5 scans in a plane 20, 30 and 40 cm above the plaster lung. The data were analyzed by the program. The program yields a value for the dipole moment that is respectively 2, 5 and 22% higher than that obtained from the far-field measurement with the single fluxgate.

CONCLUSION

The results show that a multipole analysis of a lung with uniform magnetization can be used to obtain, with reasonable accuracy, an inverse solution of the dipole moment of such a lung. This analysis does not require the knowledge of the distance between lung and probe nor the size and shape of the lung.

ACKNOWLEDGEMENT

This work is supported in part by the Natural Sciences and Engineering Research Council of Canada and by the Defence and Civil Institute of Environmental Medicine, Toronto.

REFERENCES

Aittoniemi, K., K. Kalliomäki, T. Katila, T. Varpula (1981). "Practical magnetopneumography using fluxgate magnetometers", in S.N. Ernè, H.-D. Hahlbohm, H. Lübbig (eds.) "Biomagnetism", Berlin: DeGruyter, pp. 475-484.

Brauer, F. and G. Stroink (1984). Multipole analysis in magnetopneumography: A model study, IEEE Trans. on Biomed. Eng. (to be published).

Kalliomäki, K., P.-L. Kalliomäki, K. Aittoniemi and M. Moilanen (1983). "Magnetic technique for measuring lung contamination", in S.J. Williamson, G.L. Romani, L. Kaufman, and I. Modena (eds.), "Biomagnetism," New York: Plenum Press, pp. 545-557.

Karp, P.J., T.E. Katila, M. Saarinen, P. Siltanen, and T.V. Varpula (1980). "The normal human magnetocardiogram. II. A multipole analysis", Circulation Res. 47: 117-130.

Savard, P., G.E. Mailloux, F.A. Roberge, R.M. Galrajami, and R. Guards (1982). "A simulation study of the single moving dipole representation of cardiac electrical activity", IEEE Trans. Biomed. Eng. BME-29: 700-707.

Stroink, G., R.A. Dunlap and D. Hutt (1981). "Room-temperature magnetization measurements of some Canadian Chrysotile and UICC asbestos samples", Can. Mineral. 19: 519-624.

Stroink, G., D. Dahn and F. Brauer (1982). "A lung model to study the magnetic field of lungs of miners", IEEE Trans. Mag. MAG-18: 1791-1793.

A MOBILE MAGNETOPNEUMOGRAPH WITH DUST QUALITY SENSING

M-L Junttila*, K. Kalliomaki*, P-L Kalliomaki** and
K. Aittoniemi***

*Oulu University, SF-90570 Oulu, Finland
**Institute of Occupational Health, Haartmaninkatu 1,
SF-00290 Helsinki, Finland
***Outokumpu Company, SF-00100 Helsinki, Finland

ABSTRACT

This paper presents a new mobile magnetopneumograph. It applies the coercive force of magnetized particles to dust quality sensing and has approximately constant sensitivity on the chest area.

KEYWORDS

Magnetopneumography; coersive force; sensitivity of magnetopnemograph.

INTRODUCTION

The number of persons exposed to metal aerosols is high and increasing along with the development of the metallurgical industry. Many of the aerosols contain Cr and Ni which may cause hazardous effects due to their harmful compounds. Assuming that the magnetic properties of the metal aerosols are known, the level of exposure to them can be estimated using magnetopneumography (MPG).

Several methods of detection of MPG (Cohen, 1973; Robinson, 1979; Kalliomaki,1980; Stern, 1982) have been applied to many and partly unknown metal dusts retained in the lungs. None of the methods detects the quality of dust, a result which is necessary if the amount of dust retained is to be solved. The coercive force of particles seems to give this information.

Some of the methods use complicated lung models, as the sensitivity of the methods depends strongly on the distance from the detector to the indefinite source. To prevent the need for lung models the sensitivity of the measuring system has to be constant in the chest area. In this paper the sensitivity of the opposing magnetizing coil and sensor system applied in the mobile MPG is studied.

ON THE COERCIVE FORCE OF FERRIMAGNETIC PARTICLES

Particles containing iron usually contain the ferrimagnetic iron oxide magnetite (Fe_3O_4), which has a permanent magnetic moment. The magnetizing curve of ferri-magnetic material is presented in Fig. 1. When the external field stength H(kA/m) is removed, there remains a fraction of the internal flux density (B) in the material called remanenet flux density (B_r), which is used to estimate the amount of material. To remove the remanence ($B \rightarrow 0$) an external magnetizing field strength, coercive force (H_c), is needed. The coercive force varies from 5 to 50 kA/m depending on the process generating the particles. When the coercive force of dust is measured the origin of the particles can be determined.

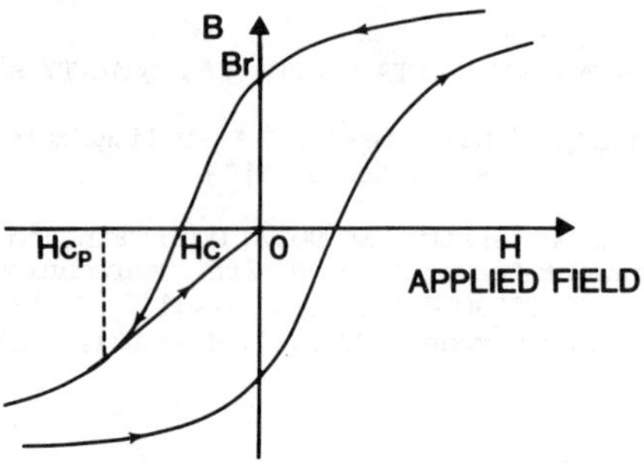

Fig. 1 Hysteresis Loop.

Measurement of the coercive force of dust samples is relatively easy with a well known vibrating sample magnetometer (VSM). Measurement of the coercive force of dust retained in the lungs is more complicated: the object to be measured is large, the dust concentration is quite low, and the orientation of particles (relaxation) due to static demagnetizing field causes errors.

The application of the static demagnetizing field can be avoided if a pulse coercive force (H_{cp}, Fig. 1) is measured instead of the coercive force. The pulse coercive force is the amplitude of a short demagnetizing pulse, which totally demagnetizes a previously magnetized sample. Measurements of dust samples show that the pulse coercive force depends on the process generating the dust, in the same way as the coersive force. As the duration of the demagnetizing pulse can be short (less than 1 s) the orientation error is minimized.

MEASUREMENT OF THE PULSE COERCIVE FORCE IN VIVO

The pulse coercive force of the dust in the lungs is the first parameter to be measured during the measurement procedure. First the chest area of a subject is magnetized with short pulse (50 mT, 0.5s), and the remanent flux density of the lungs (B_{r1}) is measured several times. One minute after the magnetization a fixed demagnetization pulse (15mT, 0.5s) is applied, and the remaining flux density (B_{r2}) is measured as earlier. The ratio B_{r1}/B_{r2} is the measure of the coercitivity or magnetic hardness of the dust. This ratio varies from 0.00 to 0.40 depending on the quality of the dust (Kalliomäki, 1981). Stainless steel (SS) grinding dusts are magnetically very hard (B_{r1}/B_{r2} 0.36 - 0.46), whereas the ratio of manual metal arc (MMA) fumes is less than 0.10. Tungsten inert gas (TIG) welding of stainless steel produces quite hard fumes (B_{r1}/B_{r2} 0.21 - 0.26).

CONSTRUCTION DETAILS OF THE NEW MAGNETOPNEUMOGRAPH

To magnetize with a flux density of 50 mT, the electric power level has to be about 20 kW depending on the size of the coils. This level exceeds the capacity of a common electric wiring and therefore a large capacitor (2.4Farads, 60V) was selected to store the needed magnetizing energy. Critical damping was applied to prevent oscillation and to minimize the duration of magnetizing/demagnetizing pulse. The capacitive time constant (RC) was 0.48 s, the inductive time constant (L/R) 0.12s. Figure 2 shows the principle of the magnetizing/demagnetizing circuit.

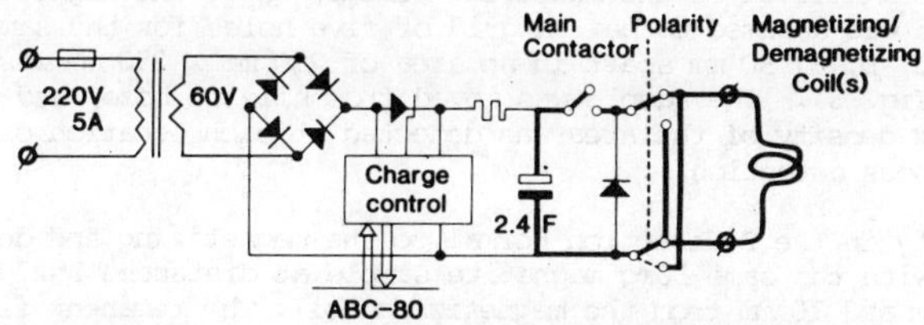

Fig. 2 Principle of the magnetizing/demagnetizing circuit.

Furthermore the instrument consists of a fluxgate gradiometer (Dr. Förster), a wooden bed driven by an electric motor and a microcomputer(ABC 80), which controls the measuring procedure. All the instruments are installed in a caravan to form a mobile healthcare unit.

SENSITIVITY DISTRIBUTION OF THE OPPOSING MAGNETIZING COIL AND SENSOR SYSTEM

To study the sensitivity of the magnetopneumograph system, the magnetizing and detection methods have to be studied simultaneously. Sensitivity increases with an increase in the magnetizing field and detector sensitivity, assuming that a linear relationship holds between the remanent flux density caused by the contaminants and the magetizing field. As the goal is to obtain uniform sensitivity, the product of the magnetizing field and the detector system has to be constant in the chest area.

Theoretical calculations

The magnetizing coil was estimated mathematically by calculating the field created by a cylinder symmetric solenoid. The first order gardiometer of the instrument was modelled with two loops, the size of which (200 mm X 300 mm) corresponds to the actual area of detection. On the X-Y plane the sensitivity was calculated at points stepping 30 mm in X- direction and 50 mm in Y-direction. The average sensitivity of each X-Y plane was determined.

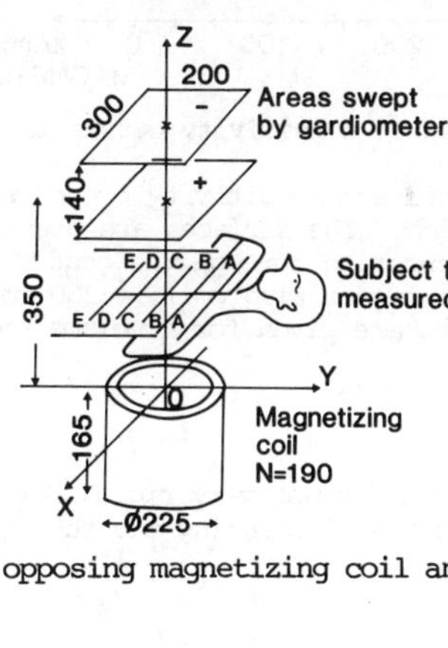

Fig. 3 Dimensions of the opposing magnetizing coil and sensor magnetopneumograph.

Empirical measurements

The sensitivity of the opposing magnetizing coil and sensor magnetopneumography was tested with a 50 mg sample of magnetite powder. The spatial sensitivity was studied at the mid-level of the measuring volume, where the highest sensitivity of the instrument was assumed to be. A grid of five holes for the samples in five lines (A, B, C, D, E) 50 mm apart in an area of 250 mm X 250 mm was used for the measurement (Fig. 3). The sample was moved from hole to hole, and the average remanenet flux density of the area was detected at each location of the sample to assure homogenous detection.

The sensitivity in the Z-direction normal to the magnetizing and detecting coils was measured with the same 50mg magnetite sample at distances Z=270 mm, 230 mm, 170 mm, 120 mm and 70 mm from the magnetizing coil. The remanent flux density in the measuring area was obtained at each X-Y plane according to the ordinary measuring procedure as before.

RESULTS

According to the theoretical calculations the spatial sensitivity decreased to less than 30% of the maximum in an area the radius of which was 100 mm. In the actual measurement the 70 % sensitivity contour was elliptic. In the direction of the detection lines the sensitivity decreased beneath the 70% level at 50 mm from the mid-point, whereas in the direction normal to the lines the decrease occured at 100 mm from the mid-point. The highest sensitivity was found to be at the mid-point of the measured plane. Theoretical and measured sensitivity curves as a function of the distance from the magnetizing coil are presented in Fig. 4. The results are presented as the precentage of the maximum sensitivity level.

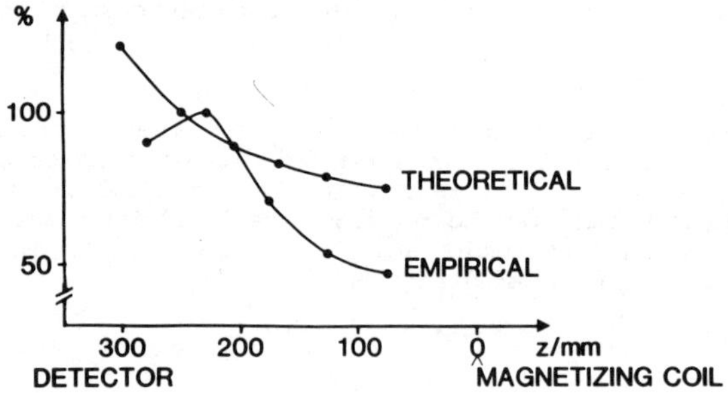

Fig. 4 Theoretical and measured sensitivity curves in Z- direction

In Fig. 5a the calculated constant sensitivity contours of 90% and 80% of the maximum are in the Z-direction. The surfaces are presented with the sensitivity volume cut along the lines entering through the center of the volume. The distance between the detector and the magnetizing coil is 300 mm. In Fig. 5b corresponding constant sensitivity contours are given for a volume the height of which is 250 mm.

DISCUSSION

The construction of the caravan is not very rigid. During the measurements the moving bed causes some vibration, increasing the the noise level (160 pT_{rms} of one sweep) of the instrument.

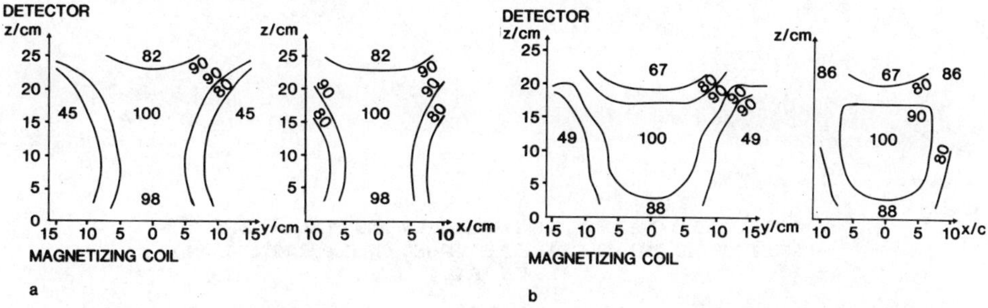

Fig. 5 Constant Sensitivity Contours of the Measuring Volume in Z-X and Z-Y planes. Fig.5a The Height of the Volume is 300mm. Fig.5b The Height of the Volume is 250mm.

The spatial sensitivity at the center level of the calculated volume was quite homogenous over a circular area (radius 100 mm) that would cover the chest of an adult in theoretical calculations.

The measured constant sensitivity contours proved to be more elliptic than circular ones. One reson for this findigs may be that the detector has a longer sweep in the X-direction than in the Y-direction. Although the center of the plane was found to have the highest value of the remanent flux density (100% sensitivity), the origin of the grid may have been a little to one side, thus distorting the measured sensitivity pattern.

In the Z-direction the calculated sensitivity curve differed from the one measured near the detector and the magnetizing coil. The theoretical calculations do not take into account saturation of the magnetite near the magnetizing coil, where the remanent flux density of the sample does not increase with the increase in the field created by the coil. Beneath the detector the detection loop distorts the calculated sensitivity pattern, giving higher values at the circular of the loop than at the center (Fig. 4). However the detected remanent flux density of a small sample decreases as the distance from the detector becomes too short.

The distance separating the detector and the magnetizing coil affects the sensitivity pattern (Fig.5). The sensitivity is more homogenous over the measured area when the distance between the detectos and the magnetizing coil is shorter.

REFERENCES

Cohen,D. (1973). Ferromagnetic contamination in the lungs and other organs of the human body, Science, 180, 745-748

Kalliomaki, K., K. Aittoniemi, P-L Kalliomäki, and M. Moilanen (1981). Measurement of lung-retained contaminants in vivo among workers exposed to metal aerosols. Am. Ind. Hyg. Assoc. J., 42, 234-238

Kalliomaki,K., P-L Kalliomaki, and M. Moilanen (1983). A new magnetic method for studying the quality of lung-retained dust in-vivo. A.Polytecn.Scand., 138, 72-75

Robinsson, S.-E., and A.P. Freedman (1979). Direct magnetopneumografic measurement of particle concentrations and clearance in the lung. Proc. IEEE Conf. Frontries of Eng. in Health Care, 183-188,

Stern R., and K. Drenk (1982). Susceptpbplity Bridge Magnetopneumography of Inviduals with and without Occupational Dust Exposure. SVC Report 82.05

MAGNETOPNEUMOGRAPHY - LUNG RETENTION AND CLEARANCE OF MANUAL METAL ARC WELDING FUMES BASED ON EXPERIMENTAL AND HUMAN DATA

P-L. Kalliomäki*, K. Kalliomäki**, E. Rahkonen* and M-L. Junttila**

*Institute of Occupational Health, Haartmaninkatu 1, SF-00290 Helsinki, Finland
**University of Oulu, Dept. of Electrical Engineering, SF-90570 Oulu, Finland

ABSTRACT

Using magnetopneumography, retention and clearance of different welding fumes have been studied both experimentally in rats and *in vivo* among welders. The clearance of stainless steel fume proved to be slower than that of mild steel.

KEYWORDS

Magnetopneumography; manual metal arc welding fumes; lung retention; lung clearance.

INTRODUCTION

Traditionally respiratory clearance has been studied mainly with isotope techniques. Clearance rates of different materials have been investigated mostly after short-time exposures. The fast components, utilizing the tracheobronchial pathway, have been extensively studied. On the other hand, there are only few investigations concerning the slow mechanisms important in long-term exposure. There have been no safe and easy methods for monitoring the respiratory clearance rate after long-term exposure.

Magnetopneumography provides a non-invasive method of studying lung retention and clearance patterns in subjects continuously exposed to industrial metal aerosols with a magnetic component (Kalliomäki and others, 1978; Kalliomäki and others, 1983d, 1983e).

According to the theory presented by Stöber and others (1965) in continuous exposure, the lung retention will not exceed a certain level unless the clearance mechanism breaks down. At a constant alveolar deposition rate (dD_A/dt), the lung retention (R_L) is

(1) $R_L = dD_A/dt\ (1-e^{-Kt}/K)$

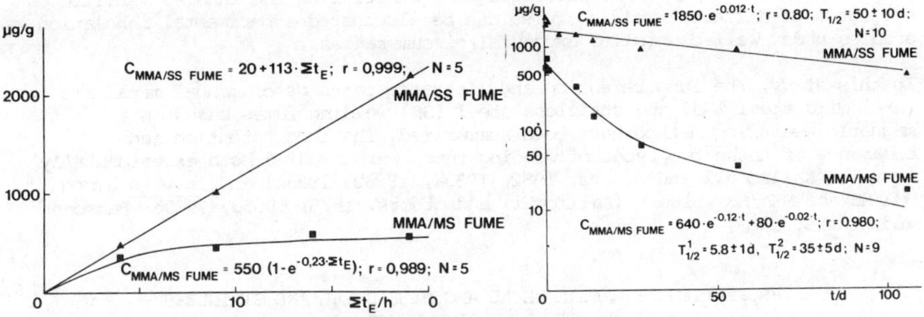

Fig. 1. Retention and clearance of MMA/MS and MMA/SS welding fumes in rat lungs.

LUNG RETENTION AND CLEARANCE OF MANUAL METAL ARC MILD STEEL AND STAINLESS STEEL WELDING FUME IN WELDERS

Manual Metal Arc Mild Steel Welders

Lung retention and clearance of MMA/MS welding fumes and the lung function of 47 shipyard arc welders was followed during the period 1976 - 1982. A group of retired welders was also included. The tests were conducted from three to six times during the follow-up period. The amount of lung-retained dust was measured with magnetopneumography. The results of this study have been reported in detail earlier (Kalliomäki and others, 1978, 1983d,e). Among the young MMA/MS welders (length of exposure t_E < 5a) the amount of lung contamination increased linearly as a function of the length of exposure during the first 5 years. The average lung retention rate of this group was about 70 mg/year. A constant level of lung contamination was achieved in about 5 - 10 years of continuous exposure, which indicates that a balance between retention and clearance will be obtained during long-term exposure. The clearance rate after chronic exposure to MMA/MS welding fumes was typically 0.2 a^{-1} which corresponds to a half-time of 3.5 years.

Following equation (2) can be fitted to the observations concerning shipyard welders

$$(2) \quad B = B_{max}(1 - e^{-0.2t_E})$$

where B_{max} is the saturation value (typically 6 nT) and t_E is the length of exposure in years.

The lung retention rate of the younger welders (about 70/year) seemed to be lower than the expected retention rate of the older active welders (190 mg/year). This observation indicates that the exposure levels of welders have been decreasing.

where the clearance rate factor K_t is material-specific constant.

The chemical and physical structure of industrial metal aerosols such as welding fume is complicated. The biological kinetics of the inhaled metallic compounds of different welding fumes can be elucidated experimental inhalation studies under well-controlled exposural circumstances.

In this study, the lung retention and clearance patterns of manual metal arc (MMA) mild steel (MS) and stainless steel (SS) welding fumes based on a magnetic measuring method have been summarized. The lung retention and clearance of these two types of welding fumes were studied both experimentally in rats (Kalliomäki and others, 1982, 1983a, 1983b, 1983c) and *in vivo* among MMA/MS and MMA/SS welders (Kalliomäki and others, 1978, 1983d, 1983e; Rahkonen and others, 1983).

PROPERTIES OF MANUAL METAL ARC MILD STEEL AND STAINLESS STEEL WELDING FUMES

The properties of welding fumes depend on the material welded and on the welding method. Manual metal arc welding is still one of the most important methods of welding mild steel and alloyed steel.

The MMA welding fumes consist of large particles ($d_{AD} > 1 \mu m$), agglomerates, and chains. There are chains consisting of very small particles ($d_{AD} < 50$ nm) only and chains consisting mainly of large particles ($d_{AD} < 400$ nm). The length of the chains may be several micrometers (Tuomisaari and others, 1983; Loberg and Harrysson, 1980). A summary of MMA/MS and MMA/SS fume analyses found in literature is presented in Table 1 (Tuomisaari and others, 1983; Minni and others, 1984). The table also includes the main iron (Fe) compounds in the MMA/MS and MMA/SS welding fumes as suggested by X-ray diffraction studies (Loberg and Harrysson, 1980; Tuomisaari and others, 1983) and X-ray photoelectron spectroscopy (Minni and others, 1984). Magnetic (Fe_3O_4)

TABLE 1 The elemental concentrations of the alloyed elements in the manual metal arc (MMA) mild steel (MS) and stainless steel (SS) welding fumes. The iron (Fe) compounds suggested to be present in these welding fumes and the remanent magnetic moments $(M_R(Am^2/kg))$ are also given (Tuomisaari and others, 1983; Minni and others, 1984). (AWS E 316L-16 rutile coated electrode was used in MMA/SS welding, and AWSE 7018 basic coated electrode in MMA/MS welding)

Alloyed elements	Mild steel	Stainless steel
Fe	20	4.0
Mn	2.8	2.2
Cr	0.012	3.0
Ni	0.05	0.4
Remanent magnetic moment (Am^2/kg	0.4	0.05
Suggested iron (Fe) compounds	Fe_3O_4 Fe_2O_3 $MnFe_2O_4$	$K_2(Fe,Mn,Cr)O_4$ $K(Fe,Mn,Cr)F_3$

serves as a "tracer" in MMA/MS welding fumes whereas the magnetic properties of the compounds identified in MMA/SS fume are not known. The remanent magnetic moments of the welding fumes measured in bulk samples are given in Table 1 (Moilanen and others, 1983).

COMPARISON OF RETENTION AND CLEARANCE PATTERNS OF MANUAL METAL ARC MILD STEEL AND STAINLESS STEEL WELDING FUMES IN RAT LUNGS

The lung clearance parameters for various inorganic elements such as iron (Fe), chromium (Cr), nickel (Ni) and manganese (Mn) are usually determined with the compound in the form of a monodispersive synthetic aerosol. There are few studies concerning the possible synergistic effects of simultaneously inhaled metal compounds on lung retention and clearance; the multi-element exposure is a situation which actually exists among welders.

Oberdoerster and others (1980) showed that continuous inhalation of nickel oxide (NiO) causes significant impairment of alveolar clearance for the inert test aerosol, hematite (Fe_2O_3). Stainless steel welding fumes consist chromium (Cr) and nickel (Ni) compounds (Table 1).

Fifty-two male Wistar rats (weight 300 ± 15 g) were exposed to either MMA/MS or MMA/SS welding fume. The duration of exposure was 1 h per workday for 1, 2, 3 or 4 weeks. The exposural situation simulated actual MMA welding as closely as possible. The system and procedure of exposure has been described previously (Kalliomäki and others, 1982, 1983a, 1983b, 1983c).

The lung retention and clearance of total iron (Fe), chromium (Cr), nickel (Ni), and manganese (Mn) in rats based on neutron activation analyses have been reported earlier (Kalliomäki and others, 1983a, 1983b, 1983c). The magnetic, i.e., exogenous iron, of the lungs was measured with a SQUID magnetometer in a magnetically shielded room at the Helsinki University of Technology. By comparing the remanent magnetic field of the lungs to the magnetic field of a calibration coil, the results could be immediately expressed as magnetic moments. The calculated magnetic moment of the coil was 110 nAm^2. The measuring distance was approximately 5 cm which greatly exceeds the size of the rat lungs. Therefore the shape of the lungs was not critical.

The total lung retention and clearance patterns of the welding fumes, based on the measurement of inhaled magnetic Fe compounds are presented in Fig. 1. The initial particle retention rates for the two welding fumes were nearly identical, but, owing to a fast clearance mechanism, the MMA/MS fume particles soon achieved a saturation level. The clearance patterns of the two welding fumes were different (Fig. 1). The fast clearance component with a half-time of 6 d, found for the MMA/MS welding fume, was lacking in the clearance pattern of the MMA/SS fume. However, 49 days after the exposure the clearance of the two of welding fumes were almost identical.

For the MMA/SS fume the length of exposure was too short to produce a saturation of the retention, because its clearance rate was low. These experimental welding fume inhalation studies show that the magnetic measuring method can be applied to estimation of the amount of lung retained welding fume contamination. The magnetic iron (Fe) present in inhaled complex metal aerosols seems to serve adequately as a "tracer" in measurements of the total lung retention. The sensitivity of the magnetic method using the SQUID magnetometer depends on the measuring distance, being approximately 0.03 μg of magnetite in studies of rat lungs.

Manual Metal Arc Stainless Steel Welders

A cross sectional study examined 37 manual metal arc stainless steel welders (length of exposure ranged from 1 to 22 years) employed at an engineering plant specializing in stainless steel metal products. Exposure to welding fumes was estimated with magnetopneumography. The revealed respiratory symptoms and results of lung function tests among these welders have been reported earlier (Rahkonen and others, 1983).

In the case of the MMA/SS welders, the following equation (3) can be fitted to the magnetopneumographic results.

$$(3) \quad B = B_{max}(1 - e^{-0.08t_E})$$

In this case the half-time of lung clearance is lower ($T_{1/2} = 8.7$ a) than that observed among MMA/MS welders Eq 2). B_{max} was typically 2.2 nT.

The lung retention rate (220 mg/a) in the MMA/SS welders seemed to be a little higher than the retention rate (190 mg/a) determined in MMA/MS welders. The difference in lung retention and clearance patterns observed in MMA/MS and MMA/SS welders is in accordance with the results of experimental studies. The maximum alveolar retention in rat lungs after a 19 h intermittent exposure was about four times higher after exposure to MMA/SS welding fume than after exposure to MMA/SS fume.

Neither in MMA/SS nor in MMA/MS welders did the magnetopneumographic results correlate with lung function (Kalliomäki and others, 1983; Rahkonen and others, 1983). There are several explanations for the poor correlations.

1. Owing to clearance of dust from the lungs, two subjects having the same amount of retained dust could have a very different exposural histories.
2. The observed changes in lung function were relatively small.
3. Several other factors, i.a., smoking affect lung function.

REFERENCES

Kalliomäki, P-L., O. Korhonen, V. Vaaranen, K. Kalliomäki, and M. Koponen (1978). Int. Arch. Occup. Environm. Hlth, 42, 83-90.

Kalliomäki, P-L., E-L. Lakomaa, K. Kalliomäki, M. Kiilunen, and R. Kivelä (1982). J. Toxicol. Environm. Hlth, 10, 223-232.

Kalliomäki, P-L., E-L. Lakomaa, K. Kalliomäki, R. Kivelä, M. Kiilunen, and V. Vaaranen (1983a). Brit. J. Ind. Med., 40, 229-234.

Kalliomäki, P-L., M. Tuomisaari, E-L. Lakomaa, K. Kalliomäki, and R. Kivelä (1983b). Am. Ind. Hyg. Ass. J., 44, 649-654.

Kalliomäki, P-L., M-L. Junttila, K. Kalliomäki, E-L. Lakomaa, and R. Kivelä (1983c). Am. Ind. Hyg. Ass. J., 44, 733-738.

Kalliomäki, P-L., K. Kalliomäki, E. Rahkonen, and K. Aittoniemi (1983d). Ann. Occup. Hyg., 27, 449-452.

Kalliomäki, P-L., M-L. Junttila, K. Kalliomäki, O. Korhonen, and E. Rahkonen (1983e). International Institute of Welding, Annual Assembly, IIW/IIS Doc VIII-2036-83, 32 p.

Loberg, B., and R. Harrysson (1980). University of Luleå, ASF-79-268.

Minni, E., T. Gustafsson, M. Koponen, and P-L. Kalliomäki (1984). J. Aerosol Sci., 15, 57-68.

Moilanen, M., K. Kalliomäki, P-L. Kalliomäki, and R. Kivelä (1983). J. of Magnetism and Magnetic Materials, 38, 165-171.

Oberdoerster, G., and G. Hochrainer (1980). Academic Press, London, p. 125.

Rahkonen, E., P-L. Kalliomäki, M-L. Junttila, and O. Korhonen (1983). VI International Pneumoconiosis Conference, Bochum, West-Germany.

Tuomisaari, M., E. Minni, K. Kalliomäki, and P-L. Kalliomäki (1983). International Institute of Welding, Annual Assembly, IIW/IIS Doc VIII-2037-83, 37 p.

USING MAGNETOPNEUMOGRAPHY FOR STUDYING THE INFLUENCE OF CIGARETTE SMOKING AND LUNG DISEASE ON ALVEOLAR CLEARANCE IN MAN

A. P. Freedman, S. E. Robinson, and M. R. Street

Division of Pulmonary Diseases
Hahnemann University, Philadelphia, Pennsylvania

ABSTRACT

Magnetopneumography of an inhaled magnetite tracer was used for measuring lung clearance in volunteer human subjects. Clearance was significantly impaired in smokers, as well as those volunteers with existing pulmonary disease.

KEYWORDS

Magnetopneumography; biomagnetism; alveolar lung clearance; magnetite particles.

INTRODUCTION

Magnetopneumography (MPG) is a noninvasive method for measuring the amount and distribution of inhaled ferrimagnetic particles in the chest (Cohen, 1973; Freedman, Robinson, and Green, 1982; Robinson and Freedman, 1979; Williamson and Kaufman, 1973). This method has been used for assessing the level of occupationally acquired lung dust in workers, and has extended the observation time of long-term lung clearance beyond that which is practical using radio-labelled particles. The measurement of lung clearance function is important in evaluating the safe limits of exposure to inhaled substances. A weakly toxic substance may have detrimental biological effects if it is retained in the lungs for a long period of time. Highly toxic fumes may be tolerated if they are removed rapidly from the lungs. Because of this, pre-existing lung diseases which impair lung clearance may have a profound influence on the safe limits for environmental and occupational pollutants and irritants. In the study presented here, we use the MPG as a tool to evaluate the relative effects of cigarette smoking, cystic fibrosis, and emphysema on long-term lung clearance, in human subjects.

MPG was used by Kalliomaki and co-workers (1978) to measure the alveolar clearance half time of welding fume, and we have studied clearance of coal mine dust in miners. Animal studies using a magnetite tracer in the donkey, the rabbit, and the rat, have shown that magnetic measurements and concommitant radiotracer measurements of clearance to be similar (Barbee, Robinson, and Lippmann, 1982; Halperin and co-workers, 1981; Oberdoerster and Freedman, 1982; Valberg and Brain, 1979). Cohen and Brain (1979) reported measurement of long-term alveolar clearance of inhaled magnetite ($FeO \cdot Fe_2O_3$) in humans, and concluded that smoking impaired long-term clearance of particles. Previously, studies of alveolar clearance of particles had been difficult to perform in human subjects because of the need to use a radiotracer with a high initial level of activity and a slow decay, permitting measurements over several months. Radiotracer studies utilizing whole body counters have found reduced alveolar clearance in patients with obstructive lung disease and in smokers, with great overlap with control subjects (Bailey, Fry, and James, 1982; Bohning, Atkins, and Cohn, 1982).

METHODS

Lung clearance function in 14 smokers and 15 nonsmokers free of pulmonary disease, 6 patients of similar age with cystic fibrosis, and 4 older patients with emphysema, was measured after a one-time inhalation exposure to magnetite dust. Clearance was determined by measuring the lungs' retention of the magnetite by localized-field MPG, at intervals of time following the exposure.

A Dautrebande generator was used to produce a magnetite aerosol (1.1 μm MMAD; $\sigma_g = 1.41$) at a concentration of 60 mg/m^3. Subjects wore nose clips and were instructed to inhale for thirty minutes at a respiratory rate of 7-8 breaths per minute with an inspiratory/expiratory time ratio of 1.0 to enhance alveolar deposition.

MPG measurements were performed with a localized-field technique using a 9 point grid over the posterior chest with 7 cm between adjacent points. The spinous process of the 7th cervical vertebra was used for referencing the scan. The remanent magnetic field (RMF) was measured using a 4.4 cm diameter second order SQUID gradiometer of 8 cm baseline. The RMF was measured before magnetization (baseline), after a 5 second magnetization with 135 mT DC at the skin, after shallow erasure with 30 mT RMS AC field, and after deeper erasure with a 50 mT RMS AC field. Partial erasures were employed to alter the distribution of sensitivity throughout the lung and thereby permit distinction between superficial and deeper concentrations of magnetite. The relatively weak damped AC magnetic field used demagnetizes material close to the surface of the chest without much disturbance of remanance of the deeper particles (Fig. 1) (Robinson, 1981). After this sequence was carried out at each point, relaxation was measured over a 10 minute period. The total measurement time per subject was 30 minutes.

We averaged the RMF values for the nine locations to derive mean values for the contributions of each of the lung compartments: total lung and outer, middle, and deep lung. All RMF values were corrected for pre-inhalation values (some background level of magnetic dust is always found in the lung). A single exponential analysis of clearance for each of these compartments was performed from the peak value after day 2 out to the 5, 8, and 12 month value, according to the available data. The Student T test and Spearman's rank correlation were used for statistical comparisons.

RESULTS AND DISCUSSION

The inhalation of magnetite resulted in RMF's ranging from 97-994 pT in contrast to pre-inhalation RMF's averaging 25 pT for both smokers and nonsmokers. The range of RMF's after inhalation reflects variations in inhalation time and deposition efficiency, as well as chest wall thickness and lung geometry. The rate at which the RMF changed with subject to probe distance, dB/dZ, was not evaluated, but could have been used to correct for the variations in chest wall thickness. Calibration factor was 713 pT/mg/liter based on phantom lung models with homogeneous distribution of magnetite and a 2 cm chest wall thickness.

There was no rapid, tracheobronchial clearance phase evident over the first two days in any group. This may have been due to the size of our aerosol and its consequent deposition primarily in the alveoli. Over the first week, we found that the RMF's after full magnetization tended to rise. This artifact has been seen in other studies, and has been primarily assumed to be a combination of dust redistribution and the spatial sensitivity of the probe to magnetic fields (Robinson, 1981). In some of our cases, a slow increase in RMF continues for up to 1 to 2 weeks. Preliminary compartmental analysis does not suggest that migration to pleural lymphatics is significantly affecting our measurements. Another possible explanation is that the RMF per unit mass of magnetite powder depends on the dispersion of the powder. Changes in particle agglomeration resulting from phagocytosis may affect the overall RMF. Clusters of magnetite particles tend to undergo rapid and spontaneous magnetic relaxation.

After 1 to 2 weeks, a slow decline in RMF is seen that correlates well with slow alveolar clearance kinetics. Clearance half-times out to 5 months based on total lung signal had a mean value of 65 ± 10 days (mean ± s.d.) for 13 nonsmokers and 108 ± 28 days for 12 smokers (Fig. 2). These values are in the range reported for this interval by others and have a similar overlap (Cohen, Arai, and Brain, 1979; Bohning, Atkins, and Cohn, 1982).

Despite patient dropouts, differences between smokers and nonsmokers remained significant at 8 months ($p < 0.005$) and at 12 months ($p < 0.05$). Clearance half-times out to 8 months based on total lung signal had a mean value of 81 ± 18 days for 12 nonsmokers and 126 ± 29 days for 8 smokers. Clearance half-times out to 12 months based on total lung signal had a mean value of 113 ± 41 days for 6 nonsmokers and 172 ± 60 days for 7 smokers. This slowing of clearance with time is well known and felt to be due to different mechanisms (alveolar macrophage transport, lymphatic transport, dissolution, etc.) coming into play. See Table 1.

For the 6 patients with cystic fibrosis, clearance half-times out to 5 months based on total lung signal had a mean value of 86 ± 15 days, which is significantly slower than nonsmokers and faster than smokers. For the 4 patients with emphysema, clearance half-times out to 5 months based on total lung signal had a mean value of 79 ± 33 days.

Compartmental analysis for the healthy smokers and nonsmokers was then employed. Within each compartment (outer, midlung, deep), clearance was slower in smokers ($p < 0.02$) with the largest differences in the deepest compartment ($p < 0.002$). Within the group of smokers, clearance was found to be significantly slower in deep lung structures ($p < 0.02$). Data are shown in Table 2.

A correlation between signal remaining after deep erasure and time after inhalation ($p < 0.02$) was found for nonsmokers, compatable with central migration of dust with time. This was not seen in the smokers, suggesting slower movement toward central structures in smokers. This represents the first *in vivo* observation in humans of dust translocation during long-term clearance.

The early half time of relaxation after magnetization (first 5 minutes) was not different between smokers and nonsmokers. However, among the smokers, it became significantly slower ($p < 0.05$) as time after inhalation increased perhaps suggesting an increase in viscosity of the media (macrophage and interstitium) or clumping of particles making them less amenable to relaxation. Data show wide ($p < 0.01$) in nonsmokers, but no correlation of this phase of relaxation was seen with time after inhalation in either group.

CONCLUSIONS

1. Alveolar clearance is significantly slower in smokers.
2. Translocation to deeper structures during clearance can be demonstrated with magnetopneumography.
3. Among smokers, clearance is slowest in deeper lung regions.
4. Some aspects of magnetic relaxation, loss of magnetic alignment of particles with time, are different between smokers and nonsmokers.
5. Magnetopneumography is suitable for the noninvasive study of alveolar clearance and dust translocation in humans..

ACKNOWLEDGEMENTS

Supported by grant #1321 from the Council for Tobacco Research and grant #1 RO1 HL 26678-01 AFY from the National Institute of Health, NHLBI Division.

REFERENCES

Bailey, M. R., F. A. Fry, and A. C. James (1982). The long-term clearance kinetics of insoluble particles from the human lung. In A. Critchlow (ed), Inhaled Particles V. Pergamon Press, Oxford. pp. 273-290.

Barbee, W. K., S. E. Robinson, and M. Lippmann (1982). Comparison of lung retention of magnetite dust by magnetic and radiotracer methods. Fourth International Workshop on Biomagnetism, Workshop Digest. Rome. pp. 66-67.

Bohning, D. E., H. L. Atkins, and S. H. Cohn (1982). Long-term particle clearance in man: normal and impaired. In A. Critchlow (ed), Inhaled Particles V. Pergamon Press, Oxford. pp. 259-271.

Cohen, D., S. F. Arai, and J. D. Brain (1979). Smoking impairs long-term dust clearance from the lung. Science, 204, 514-517.

Cohen, D. (1973). Ferromagnetic contamination in the lungs and in other organs of the human body. Science, 180, 745-748.

Freedman, A. P., S. E. Robinson, and F. H. Y. Green (1982). Magnetopneumography as a tool for the study of dust retention in the lungs. In A. Critchlow (ed), Inhaled Particles V. Pergamon Press, Oxford. pp. 319-335.

Halperin, M., S. J. Williamson, D. M. Spektor, R. B. Schlesinger, and M. Lippmann (1981). Remanent magnetic fields for measuring particle retention and distribution in the lungs. Exptl. Lung Res., 2, 27-35.

Kalliomaki, P-L., O. Korhonen, V. Vaaranen, K. Kalliomaki, and M. Koponen (1978). Lung retention and clearance of shipyard arc welders. Int. Arch. Occup. Environ. Health, 42, 83-90.

Oberdoerster, G., and A. P. Freedman (1982). Lung clearance measurements in rats using magnetopneumography. J. Aerosol Sci., 13, 160-165.

Robinson, S. E., and A. P. Freedman (1979). Direct magnetopneumographic measurement of particle concentration and clearance in the lung. Frontiers of Engineering in Health Care. IEEE, New York. pp. 183-188.

Robinson, S. E. (1981). Non-invasive imaging of magnetic particulate in the lung and other organs. IEEE Trans. Nucl. Sci., NS-28, 272-274.

Valberg, P. A., and J. D. Brain (1979). Generation and use of three types of iron-oxide aerosol. Am. Rev. Respir. Dis., 120, 1013-1024.

Williamson, S. J., and L. Kaufman (1981). Biomagnetism. J. Magnetism Magnetic Materials, 22, 129-202.

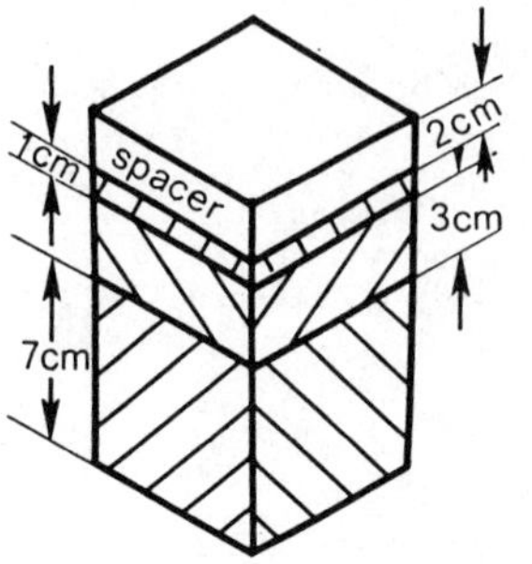

FORWARD EQUATIONS FOR PHANTOM LUNG

$$B_{FM} = 712.9\ \mathrm{pT}/\mu\mathrm{g/cm^3}\ (.547\rho_1 + .416\rho_2 + .037\rho_3)$$

$$B_{SE} = 245.6\ \mathrm{pT}/\mu\mathrm{g/cm^3}\ (.411\rho_1 + .530\rho_2 + .059\rho_3)$$

$$B_{DE} = 94.4\ \mathrm{pT}/\mu\mathrm{g/cm^3}\ (.318\rho_1 + .593\rho_2 + .089\rho_3)$$

INVERSE EQUATIONS FOR PHANTOM LUNG

$$\rho_1 = 6.73\times10^{-2}\ \mu\mathrm{g/cm^3/pT}\ (\ .127B_{FM} - .465B_{SE} + .408B_{DE})$$

$$\rho_2 = 1.82\times10^{-1}\ \mu\mathrm{g/cm^3/pT}\ (-.069B_{FM} + .420B_{SE} - .511B_{DE})$$

$$\rho_3 = 7.32\times10^{-1}\ \mu\mathrm{g/cm^3/pT}\ (\ .073B_{FM} - .054B_{SE} + .873B_{DE})$$

Fig. 1. Phantom lung model, and forward and inverse equations derived from measurements. (B) is remanent field in pT after magnetization (FM), shallow demagnetization (SE), and deep demagnetization (DE). (ρ) is the concentration of magnetite in $\mu g/cm^3$ for compartments 1 (superficial lung - 1 cm thick), 2 (intermediate lung - 3 cm thick), and 3 (deep lung - 7 cm thick).

MONOEXPONENTIAL CLEARANCE T-1/2

	SMOKERS	NONSMOKERS
To 5 Months:	65 ± 10 Days (n=13)	108 ± 28 Days (n=12)
To 8 Months:	81 ± 18 Days (n=12)	126 ± 29 Days (n=8)
To 12 Months:	113 ± 41 Days (n=6)	172 ± 60 Days (n=7)

Table 1. Clearance half-times of smokers and nonsmokers remain significantly different at 8 months and 12 months. Slowing of clearance with time is evident in both groups.

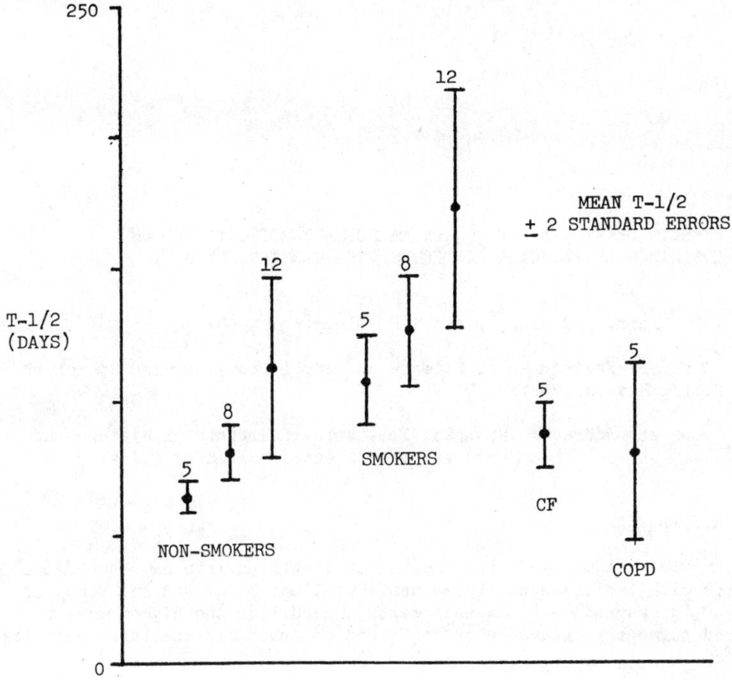

Fig. 2. Alveolar clearance half-times calculated from MPG retention data for smokers, non-smokers, cystic fibrosis (CF) patients, and emphysema (COPD) patients. Error bars are 95% confidence limits (plus and minus two standard errors); number above bar is the number of months of observation.

	SMOKERS	NONSMOKERS	
TOTAL LUNG	108 ± 28 Days	65 ± 10 Days	($p < 0.001$)
SUPERFICIAL LUNG	114 ± 41 Days	59 ± 10 Days	($p < 0.005$)
MIDDLE LUNG	96 ± 30 Days	62 ± 7 Days	($p < 0.02$)
DEEP LUNG	156 ± 61 Days	70 ± 7 Days	($p < 0.002$)

Table 2. Comparison of clearance curves out to five months for smokers and nonsmokers, by lung compartment. Clearance half-times are given as mean ± s.d.

EFFECTS OF SiO_2 AND Fe_2O_3 ON MACROPHAGE MOTILITY AND ON CLEARANCE OF MAGNETIC PARTICLES FROM HAMSTER LUNGS.

J.D. Brain,*, T. Hu,**, and S.B. Bloom*

*Dept. of Environmental Science and Physiology, Harvard School of Public Health, Boston, MA.

**Current Address: Shanghai Institute of Industrial Hygiene and Occupational Disease, Shanghai, China

ABSTRACT

Magnetic iron oxide (Fe_3O_4) was intratracheally instilled into hamsters alone or in combination with two doses of alpha-quartz (silica; SiO_2) and two doses of hematite (Fe_2O_3). Magnetometric measurements showed that the higher dose of SiO_2 inhibited pulmonary clearance of Fe_3O_4 and enhanced intracellular motility of macrophages.

KEYWORDS

Magnetometry; pulmonary clearance; macrophages; cell motility; silica; hematite.

INTRODUCTION

When magnetic iron oxide (Fe_3O_4) is inhaled, particles retained in the lungs can be magnetized and aligned by an external magnetic field. The resulting remanent magnetism of the particles can serve as a noninvasive probe of particle clearance (Cohen, 1978; Cohen and colleagues, 1979), phagocytosis and cell motility (Gehr and colleagues, 1983a,b; Brain and colleagues, 1984). The field produced is measurable at the body surface and decreases rapidly after magnetization (relaxation). The remanent field immediately following magnetization is proportional to the amount of magnetic material present; repeated measurements describe clearance of the dust from the lungs.

Relaxation is caused primarily by reorientation of the magnetized particles by intracellular rotational forces. Our experiments suggest that particles phagocytized by macrophages are rotated by actions of the cytoskeleton on phagosomes (Gehr and colleagues 1983a,b) and secondary lysosomes. Changes in relaxation with time correlate with the progression of phagocytosis indicated by other *in situ* studies (Brain and Corkery, 1977); as a greater fraction of the particles are ingested, the relaxation rate increases, implying that intracellular particles relax more rapidly than extracellular ones (Brain and colleagues, 1984;

Gehr and colleagues, 1983b). Relaxation may also yield information on the pathology of cellular pulmonary defenses during fibrogenesis.

In the experiments reported here, we examined to what extent a fibrogenic toxic dust (SiO_2, threshold limit value [TLV] = 0.1 mg/m^3) and a nontoxic dust (non-magnetic Fe_2O_3, TLV = 5 mg/m^3) (ACGIH, 1983) modified Fe_3O_4 clearance and macrophage organelle motion.

METHOD

Male Syrian golden hamsters (110 ± 8g, $\overline{x}$ ± S.D.) were intratracheally instilled with 0.2 mg Fe_3O_4 alone or in combination with one of the following: 0.15 mg SiO_2, 3.75 mg SiO_2, 0.15 mg Fe_2O_3, or 3.75 mg Fe_2O_3. The particle size of both the SiO_2 and Fe_2O_3 ranged from 0.5-3 um (projected area diameter). Each dust mixture was suspended by sonication in 0.2 cc saline. From 1 hr to 84 days post-instillation, each animal was periodically magnetized with a cobalt-samarium magnet and the remanent field measured at its chest with a fluxgate magnetometer.

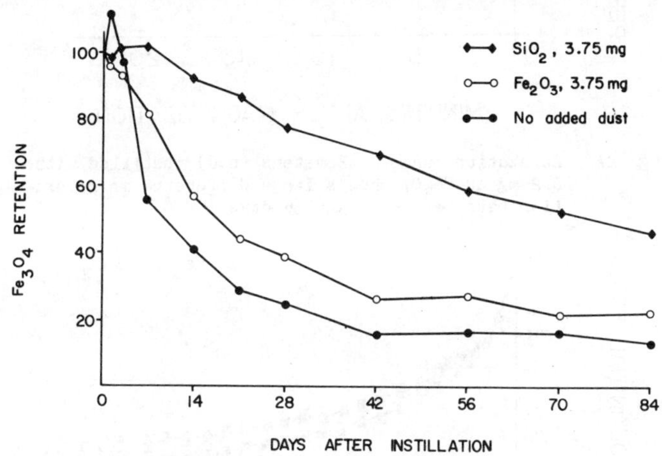

Fig. 1. Clearance of Fe_3O_4 from hamster lungs (n=6) using magnetometry. Each point is the mean of the field strengths immediately after magnetization on the day shown. SiO_2 significantly slows the clearance of Fe_3O_4.

RESULTS AND DISCUSSION

Fe_3O_4 alone had a clearance half-time ($T_{1/2}$) of 8.8 ± 1.0 days ($\overline{x}$ ± S.E.). However, in combination with 0.15 mg doses of Fe_2O_3 and SiO_2, $T_{1/2}$ rose to 12.2 ± 1.6 and 16.4 ± 0.9 days, respectively. Fig. 1 shows the mean clearance curves (n=6) for Fe_3O_4 alone and in combination with the 3.75 mg doses of the other two dusts. The 3.75 mg dose of Fe_2O_3 caused the $T_{1/2}$ to rise to 18.4 ± 2.4 days. The 3.75 mg dose of SiO_2 slowed clearance markedly; $T_{1/2}$ increased to 74.1 ± 9.0 days.

Following each magnetization, the remanent field detected at the hamster's chest decayed due to relaxation. Figure 2A shows relaxation curves measured at 7d and 56d after instillation of Fe_3O_4 alone; Figs. 2B and 2C show curves measured of the same two times for Fe_3O_4 combined with 3.75 mg doses of SiO_2 and Fe_2O_3, respectively. In each case, relaxation at 56d is slower than at 7d. This is

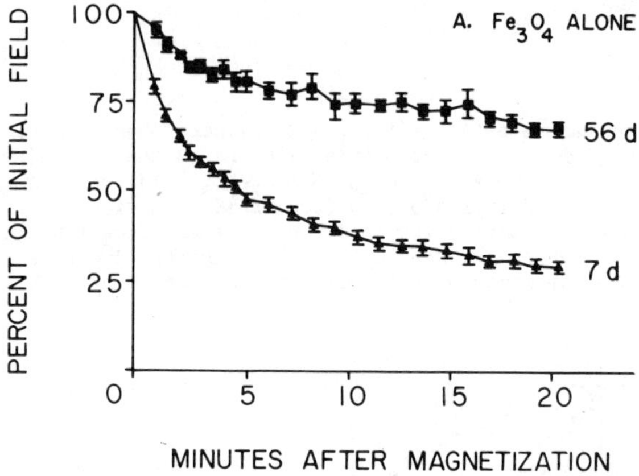

Fig. 2A. Relaxation curves. Hamsters (n=6) instilled with 0.2 mg of Fe_3O_4 show a large difference in relaxation rate between 7 and 56 days.

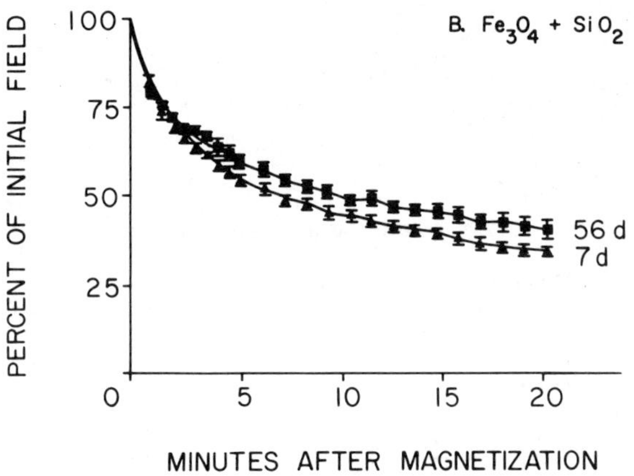

Fig. 2B. Relaxation curves. Hamsters (n=6) instilled with 3.75 mg SiO_2 in addition to the Fe_3O_4 show much much less change with time. This may be due to inhibited clearance of particle-containing macrophages, to activation of the remaining macrophages, or to the presence of Fe_3O_4 locations different from those of the controls.

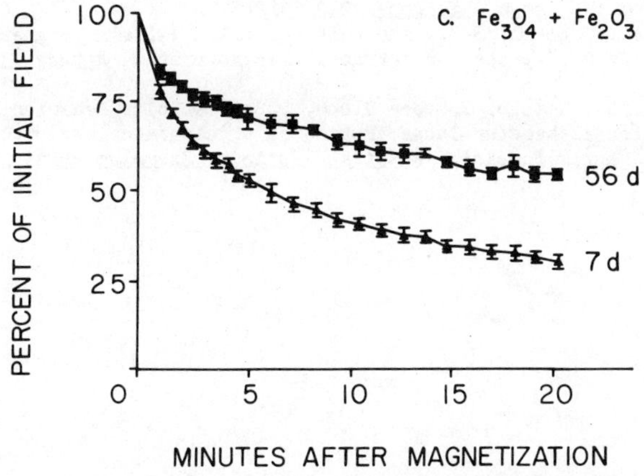

Fig. 2C. Instillation of 3.75 mg Fe_2O_3 in addition to the Fe_3O_4 (n=6) results in a 56 day curve that decays more rapidly than the controls, but much less than the silica-instilled animals. This suggests that some but not all of the changes caused by silica are related to increased particle burden.

due to clearance of particle-containing macrophages and the probable sequestration of particles in pulmonary connective tissue. Note that relaxation changes least for SiO_2, suggesting either inhibition of clearance of macrophages, a high level of activation of the particle-containing cells that remain, or the presence of Fe_3O_4 in compartments different from that seen in the hamsters given Fe_3O_4 alone and Fe_3O_4 + Fe_2O_3.

We conclude that Fe_3O_4 and magnetometry can be used to evaluate the effects of mineral fibers and particulates on pulmonary clearance and macrophage function. (Supported by NIH Grant #HL-31029)

REFERENCES

American Conference of Government Industrial Hygienists (1983). Threshold Limit Values for Chemical Substances and Physical Agents in the Work Environment with Intended Changes for 1983-84. ACGIH, Cincinnati.

Brain, J.D., Bloom, S.B., Valberg, P.A., and Gehr, P. (1984). Correlation between the behavior of magnetic iron oxide particles in the lungs of rabbits and phagocytosis. Exper. Lung Res., 6, 115-131.

Brain, J.D., and Corkery, G.C. (1977). The effect of increased particles on the endocytosis of radiocolloids by pulmonary macrophages in vivo: Competitive and toxic effects. In W.H. Walton (Ed.), Inhaled Particles and Vapors IV, Unwin, London, pp. 551-564.

Cohen, D. (1978). Report of the low-field group: The magnetic field of the lung. Publication no. MIT/FBNML-78/1. National Technical Information Service, Springfield, VA.

Cohen, D., Arai, S.F., and Brain, J.D. (1979). Smoking impairs long-term dust clearance from the lungs. Science, 204, 514-517.

Gehr, P., Brain, J.D., Bloom, S.B., and Valberg, P.A. (1983a). Magnetic particles in the liver: A probe for intracellular movement. Nature, 302, 336-338.

Gehr, P., Brain, J.D., Nemoto, I., and Bloom, S.B. (1983b). Behavior of magnetic particles in hamster lungs: Estimates of clearance and cytoplasmic motility. J. Appl. Physiol.: Respirat. Environ. Exercise Physiol., 55, 1196-1202.

IN VIVO AND IN VITRO MEASUREMENTS OF MAGNETIC RELAXATION IN HAMSTER PULMONARY MACROPHAGES

I. Nemoto*, H. Toyotama**, P. Gehr*** and J. D. Brain****

*Tokyo Denki Univ., School of Sci. and Engng, Hatoyama, Saitama, Japan
**Research and Development Corp., Satte, Saitama, Japan
***Univ. Bern, Anatomisches Inst., Bern, Switzerland
****Harvard School of Public Health, Boston, MA, U.S.A.

ABSTRACT

Relaxation of the magnetic field produced by the particles within pulmonary macrophages was measured both in alive hamsters and in harvested macrophages to study phagocitic activities and intracellular movements of the macrophages.

KEYWORDS

Pulmonary macrophages, magnetic relaxation, motility, ATP synthesis.

INTRODUCTION

Relaxation of magnetic field produced by magnetic particles has been observed in the lungs of man (Cohen, 1975; Cohen and colleagues, 1984; Cohen and Nemoto, 1984) and of animals such as hamster (Gehr and colleagues, 1983). The relaxation is the decay of the magnetic field produced by magnetic particles which have been strongly magnetized by an external field. It is caused by random movements of the particles which have been recently associated with the movements of subcellular organelles (Gehr and colleagues, 1983; Nemoto, 1982). The purpose of the present report is to show some new data measured of alive hamsters with procedures somewhat different from those used in the previous work and also the data measured of pulmonary macrophages lavaged from the lungs. The animals were used to study the change of relaxation rate during the first 24 hours after their exposure to the magnetic particles. The change of relaxation rate is believed to reflect the process of phagocytocis. Lavaged macrophages showed relaxation curves similar to those obtained from alive hamsters. With these macrophages in vitro, we studied the effects of incubation temperature and of some chemicals known to interfere with the ATP synthesis.

METHODS

Syrian golden hamsters were given Fe_3O_4 particles (Pfizer #7029, 0.3 μm diameter) suspended in saline by intratracheal instillation described by Brain and colleagues (1976). The magnetization and measurement of the animal was done with the apparatus shown in Fig. 1. The external field of up to 0.1 T was produced by two coils driven by a DC current. After cessation of the external field, the animal attached to a board was pulled into the Permaloy shield to be measured. The magnetic field at the middle of the chest was measured by pulling the animal to the probe of the fluxgate magnetometer and then out of the range of detection of the probe so that the difference between the readings at the two positions was regarded as the lung field. The animals were divided into 3 groups: A, B and C. Groups A and C consisted of 4 animals each, and B of 36 animals. Each of group A was magnetized and measured at 0, 1.5, 2, 3, 6, 8 and 24 hours after instillation of the particles, therefore being anesthetized 5 to 7 times (depending on the depth of anesthesia) during the 24 hours. Each of group B was measured only once during the 24 hours at one of the following 9 time points: 0, 1, 1.5, 2, 2.5, 4, 6, 17 and 24 hours after instillation. Four animals were used for each of the nine times. For measurement of macrophages, hamsters were given the particles by the method mentioned above. Lung lavage was performed 1 to 2 days after the instillation by the way described by Brain and colleagues (1976). The lavage fluid was centrifuged (200-300 g) twice The yield of macrophages was between $2\text{-}10\times10^6$. The suspension of macrophages was then transferred to bottles 1.8-2 cm in diameter and 2.5 cm in height. Each bottle contained 1-1.5 ml of sample comprising about $1\text{-}2\times10^6$ pulmonary macrophages. The bottles were incubated at 37°C and exposed to air. After 1.5-2 hours a monocellular layer of macrophages was formed on the bottom of the bottles. Figure 2 shows the apparatus used for magnetization and measurement. Except for the differences in the direction (vertical magnetization and measurement for *in vitro* macrophages as opposed to horizontal for the animals), the procedure was identical to that for animal measurement.

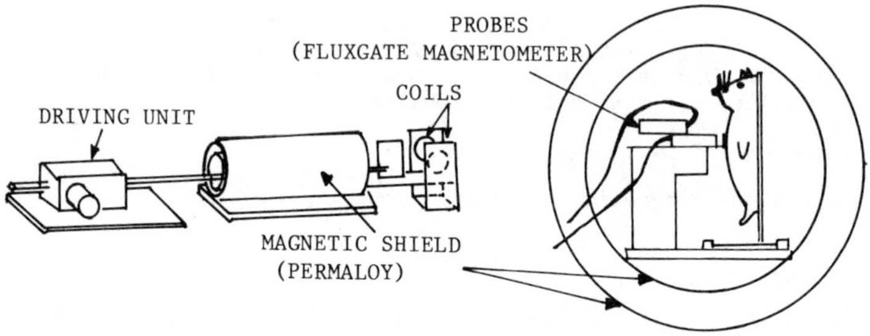

Fig. 1. The apparatus for magnetization and measurement of alive hamsters.

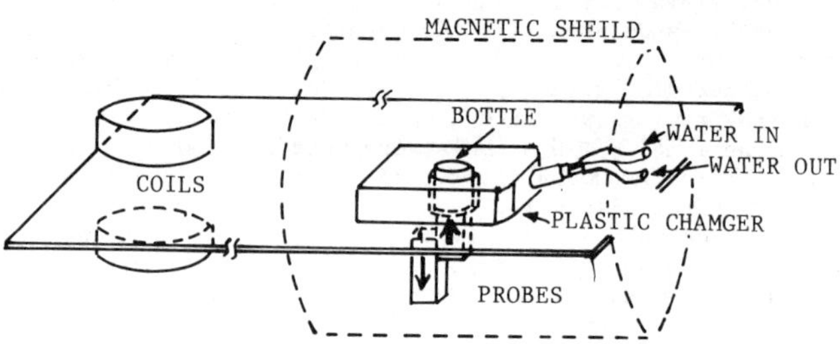

Fig. 2. Magnetization and measurement of lavaged macrophages.

RESULTS

Figure 3 is a summary of the time course of the relaxation characteristic $T_{0.7}$, the time required for the remanent field from the lungs to decrease to 70% of its

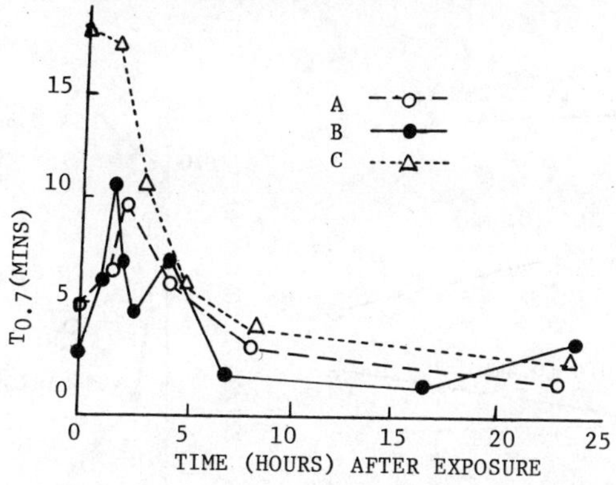

Fig. 3. Change of relaxation characteristic $T_{0.7}$. Three curves are shown for three groups of animals measured by different methods.

initial value. The three curves stand for the 3 groups of animals, A, B and C. Each point in each curve is the mean of the data obtained from the animals in the corresponding group. Curves A and B share similar characteristics: The relaxation becomes slowest at around 2 hours after exposure and then becomes faster. The curve C has a different feature during the first five hours after exposure. It shows the slowest relaxation rate immediately after exposure and gradual increase throughout the 24 hours.

Figure 4 shows the relationship between the temperature and the initial relaxation rate λ_o for the in vitro macrophages. λ_o was obtained by approximating the relaxation curve during the first one minute by an exponential function $\exp(-\lambda_o t)$. The rate λ_o is plotted against the reciprocal of the absolute temperature T°K. The equation for the regression line is $\ln\lambda_o = -3220/T + 12.2$. From the slope of this line, the activation energy was calculated to be 6.3 kCal/mol by the conventional way.

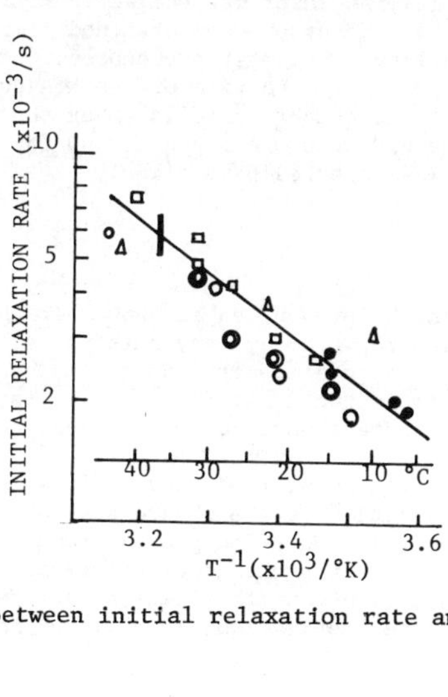

Fig. 4. Relationship between initial relaxation rate and incubation temperature.

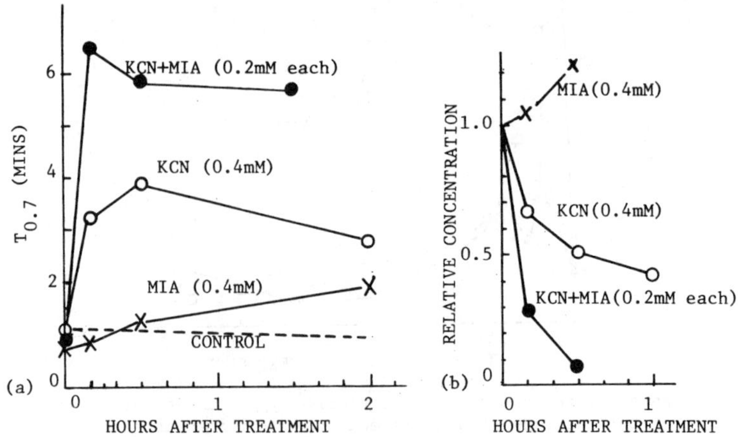

Fig. 5. (a): Change of $T_{0.7}$ after treatment with one or both of KCN and MIA. $T_{0.7}$ is plotted against time after treatment. (b): Change of ATP concentration after treatment with the same chemicals.

Figure 5 shows the effects of the inhibitors of ATP synthesis on the relaxation rate of the in vitro macrophages and on the ATP concentration in them. In (a) $T_{0.7}$ is plotted against the time after treatment by the inhibitor. It shows the clear effects of the inhibitors when they were given to the cells simultaneously. The effects were less pronounced when either of them was given without the other, although the concentration of each inhibitor was twice as high as that when both of them were given. In (b), the change of ATP concentration in the cells after treatment with the same inhibitors is shown. Pulmonary macrophages from one animal were divided into 9 bottles, two of which were used as controls; that is 0 min after treatment. Each point in (b) was obtained for each of the other seven bottles. For example, the black circle at 10 mins represents the sample which was given KCN and MIA simultaneously and 10 mins later was boiled to kill the cells and extract ATP. The other data points in the graph were obtained similarly. When both inhibitors were given, the influence was most pronounced. KCN had a smaller effect and MIA, none at least for 30 mins. The rise of the MIA curve is not significant, considering the variation of the number of cells among the bottles and errors in the measurement of ATP. The bar at 0 min indicates SD for the measurement of one of the control bottles (it was measured five times).

DISCUSSION

All the results mentioned in the present paper show that the relaxation of the lung magnetic field which has been reported by many authors is a phenomenon in pulmonary macrophages, although relaxation can result from some other processes not directly related to macrophages in some situations; for example, the slow relaxation observed immediately after exposure cannot be due to macrophages.

Previous results (Gehr and colleagues, 1982) have left some room for possible effects of successive anesthetization on the change of the rate of relaxation during the first 24 hours after instillation. The present results (Fig. 3, curve A and B) show that the overall features of the change of the relaxation rate did not change significantly when the frequency of anesthetization was reduced, indicating the small effects of successive anesthetization, if any, and thus the validity of the previous experiments.

The short-time magnetization was tried to exclude possible contribution of elasticity to relaxation: When an external field is applied to the particles, they align themselves so that their easy axis of magnetization coincide with the direction of the applied field. The rotation of the particles twist the medium in which they reside. If the medium is visco-elastic, it would recoil the particles and some part of relaxation would be due to the elasticity. On the other hand, if the external field lasts for a short time, the particles would not be able to align themselves. Curve C of Fig. 3 was measured with a 25 msec pulse magnetization. It does not have a peak whereas the other two curves have clear peaks. The result indicates that the relatively fast relaxation seen immediately after the instillation (0-1 hour) may well be due to visco-elastic properties of the medium in which the particles reside before being phagocytized by alveolar macrophages.

The effects of the inhibitors of ATP synthesis indicate, as well as the temperature experiments, that the relaxation is related to a process which is strongly energy-dependent. KCN is an inhibitor to the oxidative phosphorylation and MIA, the glycolysis. The results indicate that when both of the processes are inhibited, the intracellular motility decreases due to the decrease in ATP concentration. When one of the two processes is inhibited, the effects were less pronounced.

ACKNOWLEDGEMENTS

This work has been supported by Tokyo Denki University Center for Research Grant (#Q56) and by NIH grant HL-31029. The authors are grateful to Dr. David Cohen, Senior Scientist at MIT Magnet Laboratory for his continuous encouragement and helpful discussions. We also thank Mr. Kazuhiro Ogura at Tokyo Denki University for his help with the experiments.

REFERENCES

Brain, J. D., D. E. Knudson, S. P. Sorokin and M. A. Davis (1976). Pulmonary distribution of particles given by intratracheal instillation. Environ. Res., 11, 13-33.

Cohen, D. (1975). Measurements of the magnetic fields produced by the human heart, brain, and lungs. IEEE Trans. Magn., 11, 694-700.

Cohen, D., S. F. Arai and J. D. Brain (1978). Smoking impairs long-term clearance from the lung. Science, 204, 514-516.

Cohen, D. and I. Nemoto (1984). Ferrimagnetic particles in the lung. Part II: The relaxation process, IEEE Trans. Bio-Med. Engng., 31, 274-284.

Gehr, P. D., J. D. Brain, I. Nemoto and S. Bloom (1983). Behavior of magnetic particles in hamster lungs: Estimates of clearance and cytoplasmic motility. J. Appl. Physiol., 55, 1196-1201.

Nemoto, I. (1982). A model of magnetization and relaxation of ferrimagnetic particles in the lung. IEEE Trans. Bio-Med. Engng., 29, 745-752.

CLEARANCE OF MAGNETITE PARTICLES FROM THE LUNGS OF RATS DURING EXPOSURE TO COAL DUST AND DIESEL EXHAUST

M.C.R. Street*, A.P. Freedman*, G. Oberdoerster**, F.H.Y. Green***, T. Lewis***, N.L. Marcello**, S.R. Drago**

*Pulmonary Division, Hahnemann Univ., Philadelphia, PA; **Dept. Radiation Biology and Biophysics, Univ. Rochester Med. Ctr., NY; ***Nat. Institute for Occup. Safety and Health, Morgantown, WV and Cincinnati, OH.

ABSTRACT

The effect of a simulated work environment containing chronic amounts of diesel exhaust and coal dust on the pulmonary clearance of an inhaled inert tracer was studied in rats. The rats exposed to diesel exhaust had an increase in clearance rate of the tracer, while the coal dust group showed no change from the control group.

KEYWORDS

Clearance; diesel exhaust; coal dust; magnetite; alveolar; magnetopneumography.

INTRODUCTION

Diesel powered equipment in underground coal mines is used extensively in several countries, but little is known about possible health effects of a combined exposure to coal mine dust and diesel exhaust. Some effects of large exposures to diesel exhaust or coal dust have been studied. Le Bouffant (1971) has found that in rats the exposure to a high concentration of 300 mg/m^3 of coal mine dust from 60 hours to 3 months significantly depressed the long term clearance of $^{59}Fe_2O_3$ particles from the lung. In contrast, mucociliary clearance was found to be stimulated by inhalation of coal particles in humans (Camner and others, 1973). Experiments designed to study the effect of inhaled diesel exhaust on lung clearance of either inert tracer particles or diesel particles themselves indicate that a prolonged exposure (3 to 16 months) at a particulate exposure level of 4 mg/m^3 depressed long term clearance significantly (Heinrich and others, 1981; Griffis and others, 1983). These results suggest that both coal mine dust and diesel exhaust inhalation may lead to an impairment in alveolar clearance when inhaled at high concentrations. Whether exposure to lower concentrations of these pollutants and to a combination of both will result in similar effects is not known. In the USA, the Mines Safety and Health Administration (MSHA) enforces a dust standard of 2 mg respirable dust per m^3 for the protection of miners. Therefore, the purpose of this study was to investigate lung clearance of an inert tracer particle, $^{59}Fe_3O_4$, in rats exposed to a chronic inhalation of coal dust, diesel exhaust, and a combination of both at concentrations of 2 mg/m^3 of respirable particles representing the present MSHA standard.

METHODS

Three groups of 80 male Fischer 344 rats, mean body weight 110 g at the beginning of the study, were exposed for 7 hours per day, 5 days per week for a period of 6 months to coal dust, diesel exhaust, coal dust plus diesel exhaust. Two control groups of 67 and 40 rats were exposed to filtered air. Exposure chambers of the design of Hinners and colleagues (1968) were used. The airflow through the chambers was maintained at 0.71 l/min and the chamber loading was minimal.

The coal dust exposure had a total particulate level of 4.98 ± 0.82 mg/m^3 with 2.00 ± 0.41 mg/m^3 in the respirable range (diameter < 7 μm). The count median diameter (CMD) of the coal dust was 0.63 μm with σ_g=2.62. The coal dust was generated from a micronized Pittsburgh seam sample. The diesel exhaust exposure at a particulate concentration of 1.95 ± 0.25 mg/m^3 was generated by a 4 cylinder Caterpillar diesel engine (a type used in some underground coal mines) operated in a simulated duty cycle with varying engine load and speed. All particles of the diesel emission were in the respirable range (0.09 μm CMD with σ_g=2.00) For the combined exposure, a concentration of 1 mg/m^3 coal dust (respirable) plus 1 mg/m^3 diesel exhaust (2.02 ± 0.30 mg/m^3 total) was achieved by passing one half of the split diesel exhaust through a filter to remove particles. The particle free diesel exhaust gases were then recombined with the particle carrying fraction so that the animals in both groups, diesel exhaust alone and diesel exhaust plus coal dust, were exposed to the same concentrations of gases.(Green and others, 1983)

Two months after commencing the study, all animals were exposed in "nose-only" exposure units to magnetite ($^{59}Fe_3O_4$) dust of particle size 1.5 μm with σ_g=1.8 for 2 hours at a mean concentration of 15.4 ± 4.5 mg/m^3, except for the second control group (cold control) which was exposed to nonradioactive magnetite. Preparation and generation of the aerosol is described by Oberdoester and Freedman (1981). Following exposure, thoracic ^{59}Fe activity in the live animals was measured. The animals were not exposed in the magnetometric facility and so initial in vivo magnetopneumographic (MPG) measurements could not be performed. The animals of each group were divided into 8 subgroups of 10 animals each based on ^{59}Fe activity. Each subgroup had the same mean thoracic activity which was assumed to represent the same lung burden of magnetite. These subgroups of rats were sacrificed on days 0,1,2,7,15,30,60, and 120 by exsanguination under pentobarbital anesthesia. The lungs were excised and air dried in an inflated state (20 cm H_2O).

For the MPG measurements, a uniform dc magnetic field of 155 milliTesla (coil diameter of 7 cm) was applied for 5 seconds. The remanent magnetic field was measured at a distance of one centimeter with a second order SQUID gradiometer of 4.4 cm diameter and 8 cm baseline. After the lateral surface of the lung was measured, the signal was erased in an ac magnetic field and the procedure was repeated for the medial side. The mean of the lateral and medial measurements was calculated to give the lung burden of magnetite on the day of sacrifice. Lung volume was determined by water displacement after the lung was sealed in polyurethane. The total lung volumes of the rats ranged from 5.5 ± 0.8 ml on the day of exposure to the magnetite to 7.3 ± 1.3 ml for the last group sacrificed (day 120).

The magnetic measurements were calibrated using phantom lung models of plastic foam ranging from 0.5 to 6.0 ml in size and contained 53.8 μg/ml of magnetite. Over this range of volumes the calibration factors varied from 22.1 to 32.5 pT/ηg at one centimeter from the base of the SQUID. The magnetite retention data were normalized for all the groups to their respective initial lung deposition of ^{59}Fe. It had previously been determined that lungs from rats similarly exposed to diesel exhaust or coal dust but without exposure to magnetite had undetectable remanent fields.(Slatkin and others, 1981)

RESULTS

The alveolar clearance was modeled with a single exponential curve beginning on day one. Regression lines of clearance patterns were determined with the use of the least square method. The long term clearance half-times (T 1/2) of magnetite particles were 40 days (coal), 37 days ($^{59}Fe_3O_4$ control), 30 days (diesel/coal), and 26 days (diesel).(Fig. 1) The slopes of the magnetite lung clearance curves were compared by the F-test.(Sokal and Rohlf, 1969) The rate of magnetite clearance from the diesel exposed animals was significantly different from the other groups ($p < 0.05$), indicating a faster alveolar clearance of the deposited magnetite particles during exposure to diesel exhaust. The group of animals exposed to both 1 mg/m^3 diesel exhaust and 1 mg/m^3 coal dust had a mean half-time that was greater than that of the diesel group ($p < 0.05$), but significantly less than the coal dust group ($p < 0.001$). The coal dust group was not statistically different from the ^{59}Fe control group.(Table 1) The expected short term clearance of the magnetite test particles was not apparent in any of the groups.

The long term clearance half-times of the two control groups, ^{59}Fe and the cold control, were 37 days and 33 days respectively. The values were not significantly different ($p > 0.3$).

DISCUSSION

This study investigated the effect of a chronic inhalation exposure to coal dust, diesel exhaust, and a combination of both coal dust and diesel exhaust on lung clearance of an inert tracer particle, $^{59}Fe_3O_4$, in rats. The exposures were at concentrations of 2 mg/m^3 of respirable particles which is the present MSHA standard. The results showed an increase in the alveolar clearance rate of magnetite in the diesel exhaust and coal dust plus diesel exhaust groups.

This contradicts a previous study on the effects of diesel exhaust exposure on lung clearance of an inert tracer particle. Heinrich and colleagues (1981) showed that 16 months of exposure to 4 mg/m^3 of diesel exhaust slowed the alveolar clearance of $^{59}Fe_2O_3$ by a factor of two. The present study exposed one group of rats to 2 mg/m^3 of diesel exhaust particles and one group to 1 mg/m^3 diesel plus 1 mg/m^3 coal dust for 6 months. The alveolar clearance of magnetite (1.5 µm MMAD) in these groups was increased by 42 percent and 23 percent respectively. Heinrich's study used a greater concentration of diesel exhaust for a longer period of time than the present study and the tracer particles were much smaller (0.5 - 0.8 µm). These differences in design between the two studies might account for the disparate results.

Oberdoerster and colleagues (1984) suggested that under lower exposure concentrations diesel exhaust particles act like inert particles in the lung. Thus the diesel exhaust may stimulate the alveolar macrophages as do inert particles.(Lehnert, 1983) A prolonged exposure to diesel exhaust at a moderate concentration might also stimulate the clearance of magnetite particles from the lung.(Oberdoerster and others, 1984) On the other hand, prolonged exposure to high concentrations of diesel exhaust was reported to lead to an accumulation of particle filled macrophages in peribronchial and subpleural regions in the lungs (White and Garg 1981). Chan and colleagues (1984) suggested this sequestration of alveolar macrophages as being responsible for the observed depression of long term clearance of diesel particles. Possibly, different clearance mechanisms exist for removing magnetite particles and diesel particles from the lung.

In this study exposure to coal dust particles alone showed no significant effect ($p > 0.4$) on the alveolar clearance of the magnetite tracer. Differences in particle size between coal dust and diesel exhaust may have resulted in lower alveolar burdens in the coal dust group.

Finally, comparison of the Fe_3O_4 and the $^{59}Fe_3O_4$ control groups with regard to MPG measurements showed no significant difference between the two clearance rates. This similarity implies that this level of gamma emissions does not affect the rate of alveolar clearance of the magnetite.

In conclusion, this study showed: 1. A chronic exposure of rats to diesel exhaust alone or combined with coal dust at a total concentration of 2 mg/m^3 leads to a stimulation of alveolar clearance mechanisms. This may be due to an adaptive reponse of the lung to a low particle load. 2. Coal dust alone at a concentration of 2 mg/m^3 does not affect the clearance of the inert tracer magnetite. 3. Using MPG as the standard, the amount of radioactivity used in this type of study does not in itself affect alveolar clearance.

ACKNOWLEDGEMENTS

Research performed partly under U.S. Department of Energy Contract No. DE-AC02-81ER60029-A002 and partly under NIOSH Contract No. 212820015.

REFERENCES

Camner, P., P.A. Helstrom, and K. Philipson (1973). Carbon dust and mucociliary transport. Arch Environ Health, 26, 294-296.

Chan, T.L., P.S. Lee, and W.E. Hering (1984). Pulmonary retention of inhaled diesel particles after prolonged exposure to diesel exhaust. Fundam Appl Toxicol, 4, 624-631.

Green, F.H.,R.L. Boyd, J. Danner-Rabovsky, M.J. Fischer, W.J. Moorman, T-M. Ong, J. Tucker, V. Vallyathan, W-Z. Whong, J. Zoldak, T. Lewis (1983). Inhalation studies of diesel exhaust and coal dust in rats. Scand J Work Environ Health, 9, 181-188.

Griffis, L.C., R.K. Wolff, R.F. Henderson, W.C. Griffith, B.V. Moker, and R.O. McClellan (1983). Clearance of diesel soot particles from rat lung after a subchronic diesel exhaust exposure. Fundam Appl Toxicol, 3, 99-103.

Heinrich, U., L. Peters, W. Funcke, F. Pott, W. Mohr, and W. Stoeber (1981). Investigations of toxic and carcinogenic effects of diesel exhaust in long-term inhalation exposures of rodents. EPA Diesel Emissions Symposium, Raleigh, N.C.

Hinners, R.G., J.K. Burkart, and C.L. Punte (1968). Animal inhalation exposure chambers. Arch Environ Health, 16, 194-206.

Le Bouffant, L. (1971). In WH Walton (Ed.), Inhaled Particles III. Gresham Press, Old Woking, Surrey, England. pp. 227-237.

Lehnert, B. (1983). Ph.D. Thesis, Rochester, NY.

Oberdoerster, G., and A.P. Freedman (1981). Lung clearance measurements in rats using magnetopneumography. J Aerosol Sci, 9, 19.

Oberdoerster, G., F.H.Y. Green, and A.P. Freedman (1984). Clearance of $^{59}Fe_3O_4$ particles from the lungs of rats during exposure to coal mine dust and diesel exhaust. J Aerosol Sci, 15, 235-237.

Slatkin, D., F.H.Y. Green, and A.P. Freedman (1981). Unpublished data.

Sokal R.R., F.J. Rohlf (1969). Biometry, W.H. Freeman and Company, San Francisco. pp. 448-458.

White, H.J., and B.D. Garg (1981). Early pulmonary response of the rat lung to inhalation of high concentration of diesel particles. J Appl Toxicol, 1, 104-110.

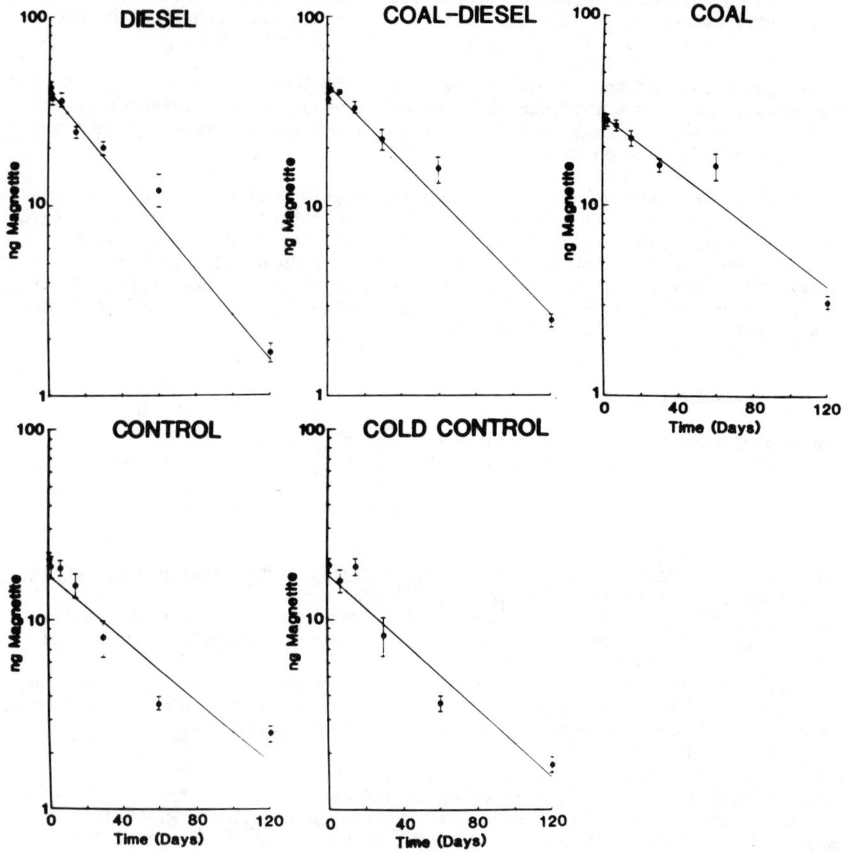

Fig. 1. Lung clearance of $^{59}Fe_3O_4$ particles inhaled during exposure to diesel exhaust, coal dust, or diesel exhaust plus coal dust as measured by MPG. The cold control group inhaled Fe_3O_4. The regression lines are from day one through day 120 and the error bars are one standard error from the mean.

TABLE 1 MPG Clearance Half Times Compared to Control

Exposure Group	T 1/2*	Significance	n
Diesel	26 days	p .001	69
Coal/Diesel	30	p .05	67
Coal	40	p .4	67
Control	37		59

* Monoexponential T 1/2 from day 1 through day 120

BIOLOGICAL EFFECTS OF EXTREMELY LOW FREQUENCY ELECTROMAGNETIC FIELDS

J. M. R. Delgado

Dpto. Investigación, Centro Ramón y Cajal, Madrid 34, Spain

ABSTRACT

Our experiments demonstrate that the application of extremely low frequency (10-100 Hz), low intensity (μT-mT) electromagnetic fields, produces powerful biological effects, including: (a) inhibition of bacterial growth; (b) mutagenic lethal effects in the X chromosome of the Drosophila; (c) modification of the firing rate in the neuron of the crab stretch receptor; (d) morphological and histological disturbances in the chick embryo; and (e) changes in the excitability of specific brain areas in monkeys.

Magnetic contamination of the environment is increasing due to the growing use of electrical energy, and preventive medicine will be necessary to avoid possible risks. Important medical benefits should be expected in the near future derived from electromagnetic research.

KEYWORDS

Electromagnetic fields; biological effects; bacterial growth; genetic effects; Drosophila; neuronal unitary activity; crayfish; embryological development; teratogenic effects; chick embryos; brain excitability; monkeys; environmental magnetic contamination.

INTRODUCTION

There is increasing interest in studying the possible biological effects of extremely low frequency (10-100 Hz) electromagnetic fields (EMF) because of their (a) intrinsic biophysical importance, (b) postulated effects on membranes and on transmembrane mechanisms, (c) beneficial treatment of pseudo-arthrosis, (d) expanding therapeutic applications to wound healing, nerve growth and other problems, (e) use in long range transmission (e.g., the earth antenna for communication with submerged submarines), and (f) potential EMF hazard of high power lines and industrial equipment producing electromagnetic contamination.

Published information about biological effects of magnetic fields is growing very rapidly but results are often contradictory due in part to differences in experimental conditions, use of different parameters of stimulation, lack of precision in reporting data, and the need for suitable experimental models. During the last ten years exposure to magnetic fields has been reported to affect growth rate of bacteria (Blackman and others, 1975), eucaryote (Adey, 1981; Grundler and others, 1977), tumoral cell cultures (Kim, 1976; Rockwell, 1977), mutagenic factors (Diebolt, 1978), metabolic activities (Weissbluth, 1979), cerebral calcium ion binding (Bawin and Adey, 1976), osteogenesis (Bassett and others, 1974a,b), and a variety of other biological systems.

The present article summarizes recent results of some of the investigations performed by our group in Madrid concerning the biological effects of EMF. These studies were characterized by the use of extremely low frequencies with intensities in the range of micro- to milli-Teslas (μT-mT) and by the application of similar EMF parameters to different biological models from bacteria to monkeys.

BACTERIAL GROWTH

This study was an evaluation of possible inhibitory effects of EMF on bacterial growth and an attempt to develop a model which could be used for parametric studies including dose-response effects (Monteagudo and others, 1984). A total of 225 Lactobacillus acidophilus cultures were grown in a Memmert incubator maintained at 37°C (±0.1°C) with continuous temperature monitoring of air and culture vessels by means of a scanning YSI telethermometer, model 47. This temperature was selected for maximum growth rate which was determined experimentally.

Bacterial growth in the nutrient medium was detected continuously without manipulation by measuring AC conductivity with the two-wires technique (Hause and others, 1981). An 8-channel device was used for this purpose with automatic scanning and value printing at a rate of 10 min/channel. Changes in the medium impedance produced by the metabolic activity of the bacteria can be sensed when the concentration of micro-organisms exceeds a threshold level of 10^6-10^7 cells/ml (Firstenberg-Eden, 1983). The time threshold reached was designated as "Mean Detection Time" (MDT) that depends on the initial concentration and the generation time. MDT plotted against the logarithm of the initial concentration is represented by a straight line, and its slope reflects the organism's generation time. Growth measurements were checked by parallel experiments using standard dilution and plating techniques. After preliminary testing, an inoculum concentration of 24. 10^3/ml of organisms was selected as appropriate for investigation of possible EMF effects.

Each experiment consisted of 16 culture vessels measuring 15 x 45 mm with 7 ml of liquid selective growth media. From this batch, 8 vessels were innoculated and exposed to EMF while 4 vessels were innoculated and not exposed. The other 4 vessels were not innoculated and not exposed. Magnetic fields were similar to those used for bone healing therapy (Magnos stimulator) and consisted of rectangular pulses, 26 μsec pulse duration, 4.4 KHz frequency, 5 msec trains, train repetition of 26 Hz, and flux density 0-40 Gauss peak. The magnetic field was monitored by an inductive probe.

Experimental results showed that at our concentration of 24,000 bacillus/ml, the maximum growth rate was obtained at about 37°C with an MDT of 21 hours. Culture vessels exposed to EMF intensities of 0, 5, and 12 Gauss had an MDT of 21 hours, indicating a lack of effect and reinforcing the value of controls. Cultures exposed to intentsities of 18, 25, 32, and 37 Gauss, however, had a progressive increase of MDT proportional to the strength of the magnetic field, as shown in Fig. 1, demonstrating a dose-response effect. The shape of the curve suggested that it would be possible to obtain a total inhibition of Lactobacillus acidophilus growth with higher intensity magnetic fluxes. Other microorganisms including harmful strains, may also be sensitive to EMF.

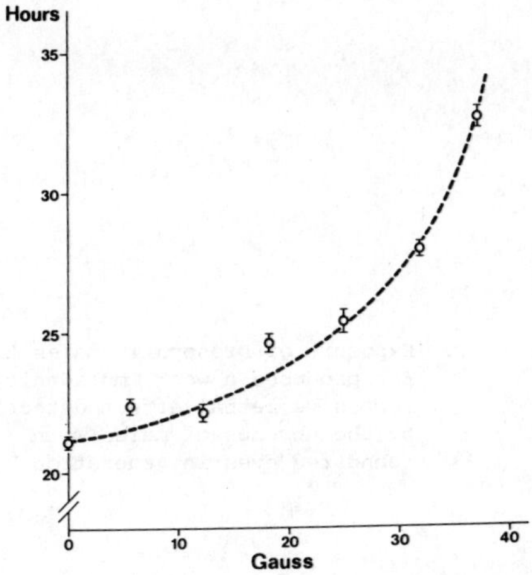

Fig. 1. Dose-response effect on bacterial growth. Inhibition is significant above 18 Gauss.

GENETIC EFFECTS IN DROSOPHILA

Experiments have been performed to test possible electromagnetic influences on the production of sex-linked recessive lethal mutations in mature sperm of *Drosophila melanogaster*, wild-type Canton-S strain (Ramírez and others, 1984; see also Ramírez and others, 1983). The marker chromosome used in evaluating the EMF treated X-chromosome was the Müller-5, also called Basc. This chromosome, according to Müller (1927) and Spencer and Stern (1948), is suitable for the study of mutations for two reasons: (1) the recessive (white-apricot eye colour) and semi-dominant (bar shaped eye) gene markers allow phenotypic identification of both males and females carrying X-chromosomes derived from those of EMF treated parent (P_1) males; and (2) mutations induced in the treated X-chromosomes are fixed by a double inversion which inhibits crossing over between paired X-chromosomes. This is achieved with a high level of efficiency and withouth significant loss of fertility or viability.

The sex-linked recessive lethal assay (Basc test) began with exposure of males (round red eyes) for 70 hours to pulsated EMF similar to those used clinically to heal non-union fractures. Parameters were rectangular pulses with pulse duration of 26 msec, 4.4 KHz frequency, 5 msec trains, train repetition rate of 16 Hz, and flux density 3.5 mT peak. Controls were treated in a similar way in the absence of EMF. The males were then mated to untreated, virgin, Basc females characterized by bar shaped, white-apricot eyes. Females of the resulting progeny (Fig. 2) carried the exposed X-chromosomes, some of which could have recessive lethal mutations produced by male exposure to EMF. Sibling matings of F_2 flies were then established and each set of progeny in the F_2 generation was examined. Induction of a recessive lethal mutation in the treated X-chromosome of the P_1 male was detected by the total absence of wild-type males (round red eyes) in the F_2 progeny. All vials which did not have wild-type males were rechecked by collecting the heterozygous F_2 females (kidney shaped red eyes) from the vial and mating them to their F_2 brothers (bar shaped white-apricot eyes). If no wild-type males appeared from the retested females, the vial was considered to contain an X-chromosome with a sex-linked recessive lethal mutation, since a single mating represents a single X-chromosome testing.

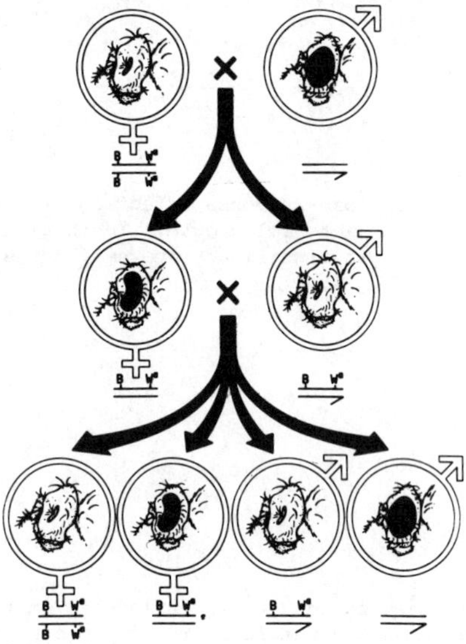

Fig. 2. Exposure of Drosophila males to EMF produced a weak, mutagenic, recessive lethal effect detected by the absence of males with round red eyes in generation F_2.

A total of almost 7,000 X-chromosomes were examined in this study. Each vial of the F_2 generation had over 60 flies with mean values as follows: 21% bar shaped, white-apricot eye females; 30% kidney shaped red eye females; 20% bar shaped, white-apricot eye males; and 29% round red eye males (wild-type). The control group (3994 chromosomes) had 0.26% recessive lethals, whereas the EMF treated group (2960 X-chromosomes) had 0.60% recessive lethals. The difference indicates a mutagenic effect which, although weak, was significant at the $P < 0.05$ level. Statistical significance of the data was determined with the method of Kastenbaum and Bowman (1970).

Mutation is an irreversible process which may persist for many generations, transmitted by the females which carry but do not suffer the consequences of the lethal agent. Mutagenic results in Drosophila should be interpreted with caution but they indicate the need for further studies to evaluate EMF parameters, minimum doses, and factors which could have mutagenic effects in animals and in humans.

UNITARY ACTIVITY OF THE CRAYFISH

The tonic stretch receptor of the crayfish (Procambarus Clarkii G) is characterized by a constant rhythmic activity with frequency related to the stretch of the receptor muscle. It is a classical preparation for investigation of neuronal physiology and coding of signals (Buño and others, 1984; Chaplain and others, 1971;

Eyzaguirre and Duffler, 1955). The firing rate is rather stable and could be a good model for investigation of possible effects of EMF on spontaneous activity of neurons. Our experiments (Bustamante and others, in preparation) were performed by isolating parts of the cephalothorax and first abdominal segments and placing them in a Van Harreveld saline bath thermostatically controlled at 12-14°C and oxygenated. The tail was pinned to a paraffin block. Under a dissecting microscope, the stretch receptor organ and dorsal nerve were exposed, dissected, and disconnected from the central nervous system by cutting the dorsal nerve. The anterior end of the muscle was held with the tip of fine forceps and could be mechanically stretched at precise levels with the aid of a galvanometer. Action potentials were recorded from the peripheral stump of the nerve with an "oil electrode" (Wilkens and Wolfe, 1974).

Magnetic fields were applied by means of a coil 5 mm in diameter and a ferrite mounted on a glass tube which could be placed close to the receptor with the aid of a micromanipulator. The characteristics of the coil were: 10 turns, R = 43.3 mΩ, L = 2 μH, Q = 0.3. The intensity of the field on the receptor was from 0 to 100 μT. Frequencies of magnetic pulses were from 0 to 30 Hz. Neuronal activity was recorded on tape (Hewlett Packard) and analyzed off-line by a PDP 11/40 computer.

Taking the peak of stimulus as reference and quantifying the fruequency of neuronal discharges before and after the peak (peri-stimulus or cross-correlation histogram), we may detect when the neuronal firing synchronized with the frequency of the magnetic pulses, indicating a phase correlation, i.e., locking phase. One example of the raw data photographed on the oscilloscope is shown in Fig. 3A. Recording 1 is the discharge in the dorsal nerve and 2 is the magnetic stimulus. The cross-correlation histogram is shown in Fig. 3B, demonstrating the synchronization between firing rate and magnetic frequency.

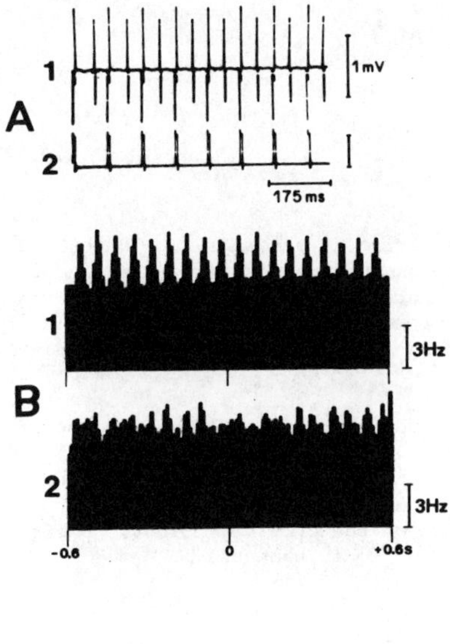

Fig. 3. Magnetic pulses beamed at the crayfish stretch receptor cause synchronization of neuronal firing.

The reliability of this data was tested by the non-parametric test of Siegel and Tukey ($P < 0.2\%$), comparing real (Fig. 3B-1) and shuffling (Fig. 3B-2) histograms. Results indicate the existence of magnetic-neuronal synchrony for EMF frequencies of 4, 8, and 12 Hz, and no detectable effects for frequencies of 16, 20, 24, and 28 Hz. These findings should not be generalized to other experimental conditions or to other species, but they do indicate functional correlations between frequencies of magnetic pulses and frequencies of neuronal discharges.

In another series of experiments (Fig. 4), we used a mechanical over-stretching of the muscle because in this situation the cell is very depolarized with most K^+ channels open (they are Ca^{++} dependent) and magnetic fields could modify the number of activated K^+ channels influencing the receptor reactivity in a detactable manner. In this study, the crab receptor was mechanically stimulated from its initial length of stretching with ramps of overstretching lasting 15-20 seconds and returning to the initial stretch for several minutes.

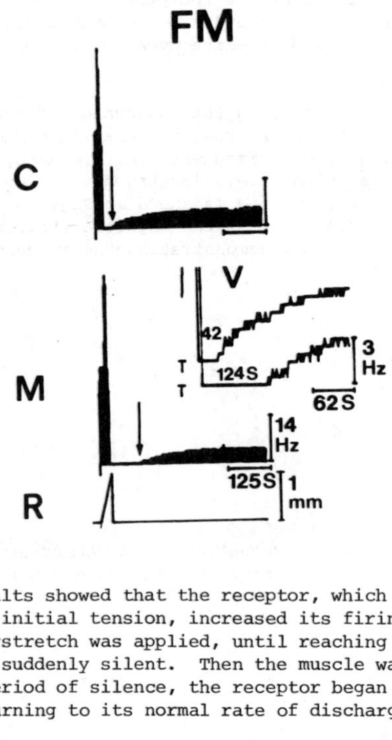

Fig. 4. The silence in neuronal firing produced by overstretch of the stretch receptor is considerably increased by exposure to MF, indicating a possible magnetic effect on the opening of ionic channels of the membrane.

Results showed that the receptor, which was firing at a normal rate depending on the initial tension, increased its firing frequency progressively when the ramp of overstretch was applied, until reaching a point of overstretch when the receptor was suddenly silent. Then the muscle was returned to its inital length and after a period of silence, the receptor began firing again, slowly at first, and then returning to its normal rate of discharge.

Application of magnetic fields during this experiment prolonged the receptor's recovery time considerably, as shown in Fig. 4. While in controls the silent interval was 42 seconds, when EMF was applied the silent interval increased to about 124 seconds. This model provides clear and consistent results and indicates that magnetic fields have an effect on neuronal physiology, possibly influencing the opening and closing of membrane ionic channels.

CHICK EMBRYOS

Fertilized eggs from white Leghorn hens were placed in Memmert incubtaors and exposed to magnetic fields with frequencies of 10-1000 Hz and intensities of 0.12-12 μT (Delgado and others, 1982). Control eggs were incubated in the absence of EMF. After 48 hours of incubation the embryos were analyzed morphologically and histologically, examining the (a) cephalic nervous system, (b) truncal nervous system, (c) heart, (d) vascularization, and (e) somites. Normality required morphological development to stages 10-12 of the Hamburger and Hamilton scale and adequate hisotological differentiation and growth. Results were as follows.

Experiments with 10 Hz (6 control, 12 exposed embryos)

The lowest intensity of 0.12 μT did not produce significant effects. Intensity of 1.2 μT caused some abnormality in the heart of one embryo. With the highest intensity of 12 μT, most embryos still appeared normal but two presented alterations of the nervous system. The somites were normal in number and development in all exposed embryos. Thus exposure to 10 Hz caused little or no disturbance.

Experiments with 100 Hz (10 control, 19 exposed embryos)

The most powerful and uniform effects were obtained with 100 Hz, 1.2 μT which produced widespread malformations (Fig. 5). General development was considerably delayed. The cephalic nervous system was reduced to a thickening of the asymmetrical ectodermal tissue and the truncal nervous system was represented by only a thin plate.

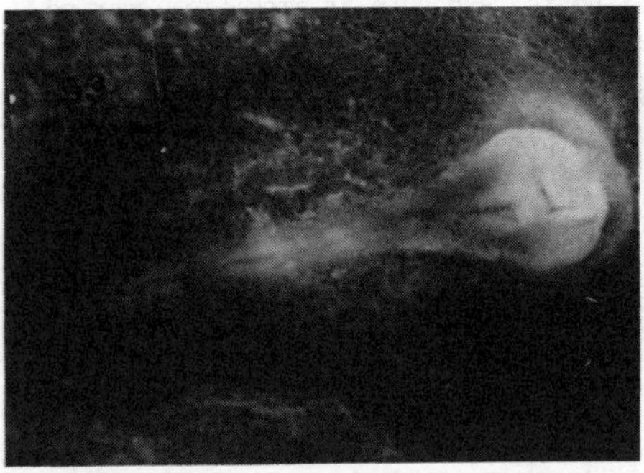

Fig. 5. Teratogenic effects may be produced by exposing developing chick embryos to EMF.

Alcian blue staining of the preparations revealed the absence of glycosaminoglycans which are essential for cell development. The neural ectoderm showed cells with nucleoli of high density, indicating a low metabolic cellular activity. Some areas had local necrosis indicating lethal effects of EMF. At the lower intensity of 0.12 μT, the teratogenic effect was milder, lacking uniformity, and in several embryos some organs seemed normal. At the highest intensity of 12 μT, embryological disturbances were evident but less pronounced than those caused by 1.2 μT.

Exposure to 1000 Hz (10 control, 11 exposed embryos)

Embryos exposed to 12 μT presented widespread delays in development, especially evident in organs of mesodermal origin. In most cases, the neural tube was still open along its entire length. The notocord was rather primitive and the pharynx had poorly organized walls. Mesenchymal cells were scarce with few mitoses and dense nucleoli. The greatest histological disturbance of vessel development was obtained with 1000 Hz and 12 μT which completely blocked formation of blood and vessels. With intensities of 1.2 μT the number of abnormal embryos was even greater than with 12 μT, whereas effects were weaker when 0.12 μT were used.

Importance of Pulse Shape

Pulse shape may be a decisive parameter for EMF effects (Ubeda and others, 1983). With 100 Hz, pulse rise time of 100 μsec produced teratogenic efects when intensities of 1.0 and 13.9 μT were used, but not with lower or higher intensities, demonstrating a "window" effect. Pulse rise time of 2.0 and 42 μsec, with intensities of 0.4 and 1.0 μT, proved to have a greater teratogenic effect than 100 μsec, causing morphological and histological alterations involving all developing systems. The most powerful effects were obtained with 1.0 μT and 42 μsec rise time.

Critical Time of Development

In another set of experiments, chick embryos were exposed to EMF of 100 Hz and 1 μT for only 5 hours at stages 7 and 9, corresponding to 25 and 30 hours of incubation time (Trillo and others, 1983). Some embryos were examined after exposure while others were left to develop for 5 more hours without EMF. Most embryos exhibited anomalies, showing that 5 hours' exposure at a critical period may be more detrimental than longer exposure for 48 hours, after which some embryos appeared normal. Abnormalities included increased fagility of tissues, arrested neural crest migration, changes in acid mucopolysaccharides, and an increased neuroepithelium mitotic index which was expressed as the number of dividing cells per 1000 counted cells.

Immediately after EMF exposure, the mitotic index increased in the prosencephalon, rombencephalon, and trunk. After the additional 5 hours of development without EMF, all embryos showed considerable increase of mitotic index in the prosencephalon. However, the mitotic index in the rombencephalon and trunk in half of the embryos was increased, contrasting with the other half in which it was similar or inferior to unexposed controls.

The fact that rather weak EMF intensities can produce powerful teratogenic effects is of considerable importance, but it should be emphasized that these results were obtained with pulsated fields and should not be generalized to sine waves which consitute the majority of instrumental, industrial, and power line exposures.

MAGNETIC ACTIVATION OF THE CENTRAL NERVOUS SYSTEM IN MONKEYS

The excitability of specific neuronal structures in nonmedicated, awake primates can be tested by application of electrical stimulation using permanently implanted electrodes (Delgado, 1981). In some of our studies we have had monkeys with electrodes implanted in the brain for over 10 years with reliability of spontaneous electrical activity, thersholds of excitability, and evoked behavioural responses. This preparation is an excellent model for investigation of possible modifications of brain physiology induced by external application of EMF.

Monkeys were equipped with electrodes impanted in the motor system, limbic structures, thalamus, and other areas of the brain. After establishing reliability of thresholds and responses during months of experimentation, studies of EMF began. Animals were restrained in a chair and a plastic helmet was placed over the head to hold two coils aimed at both cerebellar hemispheres. The coils were flat spirals 13 mm in external diameter, activated by a square wave generator driven simultaneously with 2 msec square pulses 50·Hz and 1.0A, producing fields of 2.5-1.1 Gauss at a depth of 5-8 mm where the cerebellar surface of the monkey is located. At this depth, the induced electrical field ranged from -110 to +50 mV/m and the calculated circulating eddy current density varied between -20 mA/m^2 and +9 mA/m^2, assuming a brain tissue resistance of 5.5 ohm/m.

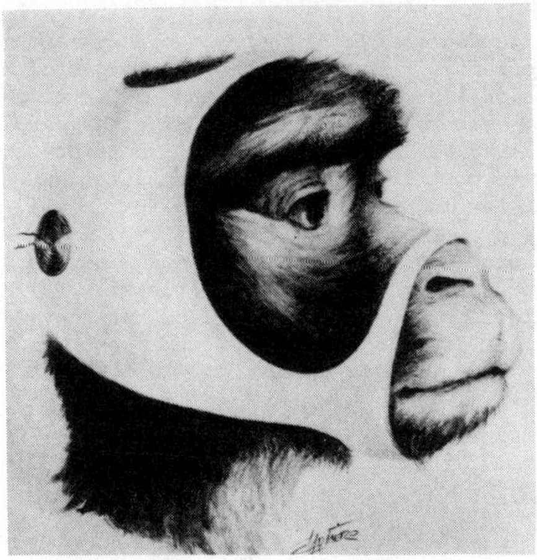

Fig. 6. Plastic helmets may be used in monkeys to apply EMF beamed at the cerebellum. This non-invasive, non-surgical technology may influence the excitability of the brain.

After establishing daily baselines of motor thresholds, EMF was applied for 9 hours and motor thresholds were measured again. Experiments were repeated for weeks in three monkeys. Statistical significance of results was evaluated by the correlated "t" test and by the Mann-Whitney test. Results showed that frequencies below 50 Hz influenced spontaneous behaviour of the monkey which appeared more relaxed, less reactive to external sensory stimulations, and with increased periods of sleepiness with eyes closed. In contrast, when 100 Hz were used, restlessness and tachypnea appeared and the animals were more alert and reactive. No changes, however, were observed in any of the experiments with respect to spontaneous activity of the brain or to food intake, weight, or health.

The most obejctive changes induced by EMF were demonstrated in some of the thresholds of motor responses. For example, in one of the monkeys, electrical stimulation of the left hippocampus with 117 μA ± 22 (cathodal, square waves 100 μsec, 100 Hz) produced a response of closing both eyes. This effect was reliable for over 1 year of repeated testing. Then a series of experiments was performed (n = 12) applying EMF for 9 hours and retesting thesholds. Results showed that there was an increase in values up to 299 μA, demonstrating a statistically significant ($P < .005$) increase in thresholds which represented a decrease in local excitability. In the same animal, two other cerebral points located in the internal capsule showed a statistically significant decrease in motor thresholds after EMF exposure.

Results in the three monkeys showed that cerebellar magnetic activation may induce an increase in motor thresholds of the left hippocampus and caudate nucleus, a decrease in motor cortex and internal capsule, and no effect in other structures. Local excitability usually returned to normal after night rest but in some cases cumulative effects and rebound were produced.

In summary, our studies demonstrate the technical possibility of using magnetic fields as a non-invasive methodology to produce activation of the central nervous system with different effects in different brain structures. This methodology may have important applications in research and in therapy.

MAGNETIC CONTAMINATION OF THE ENVIRONMENT

With the increasing awareness of biological effects of EMF, there is concern about the amount of existing fields and several scientists have measured the levels of EMF present in high power plants, transmission lines, industries, and the environment (Bridges and Preache, 1979; Marino and Becker, 1977; Presman, 1970; Sheppard and Eisenbud, 1977; Wertheimer and Leeper, 1978). This knowledge is important for the evaluation of occupational hazards and to contribute to this information our group has measured some of the existing EMF levels in Spain, as shown in Table 1.

For this purpose it was necessary to solve problems of instrumentation: laboratory Gaussmeters are bulky, heavy, and must be powered from electrical outlets. The few available portable units can measure electric and magnetic fields, but they require an inductive sensor with a high number of turns and they measure not magnetic flux density but its derivative with respect to time.

TABLE 1 Environmental Magnetic Levels

Instrument	Gauss
Oscilloscope, Tektronix 511	2.2
Power supply, Hewlett Packard 6200B	5.2
Transformer, 220/12 + 12, 2A	8.8
Drill, Black and Decker D720	7.3
Hair dryer, Moulinex	9.7
Disposal Moulinex - start	20.0
- stationary	3.0
Egg beater, Braun Miniprimer	25.0
Personal computer, Rockwell AIM 65	0.3
Computer room, PDP/10, PDP/11	Undetectable
Oil pump inside, IMB 1403	28.0
X-rays, Siabilipan - Siemens	0.2
Video monitor	0.3
Power substation, 400 Kv (height 20m)	0.1
(at floor level)	

In this case, high frequency harmonics and artefacts are multiplied by the value of their frequency, masking real magnetic values. With the use of 50 Hz filters part of this problem may be solved, but active filters with acceptable "Q" are rather complex. Another problem is the high number of turns required in the sensor which increase capacity coupling and therefore simultaneous sensitivity to electrical fields.

In order to solve some of the above mentioned problems, we have designed and constructed a portable instrument for the detection and measurement of magnetic flux density based on the Hall effect. Its characteristics are as follows: dimensions 15 x 11 x 5 cm, weight 215 g, power 2 batteries of 9 V, type 6F22, sensitivity 3 μT up to 40 mT in four scales with a maximum resolution of 0.5 μT; no special calibration is needed (see Fig. 7).

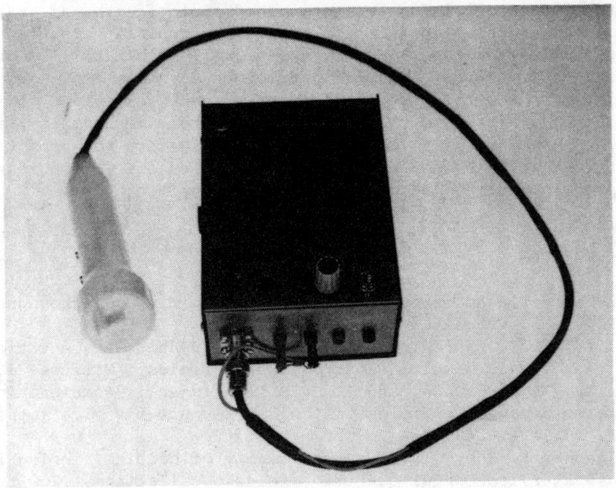

Fig. 7. Instrument to measure magnetic flux density based on the Hall effect.

The Hall sensor produces a signal proportional to the magnetic flux density with a lineal response between 0 and 100 KHz. This signal is amplified and its RMS value is visualized in a crystal liquid display of three digits. Some of the representative values obtained in our study are shown in Table 1. Motors with plastic covers gave higher values than powerful generators with good metallic shielding such as power plant turbines. Partly for this reason, values were low in the proximity of X-ray equipment, video monitors, and computer rooms, contrasting with the high levels of some home appliances.

We lack adequate knowledge of parametric characteristics which may produce beneficial or detrimental biological effects and, considering that sinusoidal waves seem less effective than pulsated fields, it is difficult to draw conclusions for preventive medicine in relation to existing EMF levels. More studies are certainly needed to measure these levels in homes and in industries, estimating the effects of both acute and chronic exposures. Special attention must be paid to industrial equipment and packing because female employees may stand for many hours daily with their abdomens a few centimeters from poorly shielded motors. Pregnancy and embryological development could be affected by weak EMF and this subject requires immediate attention.

CONCLUDING REMARKS

Life has evolved under the influence of two omnipresent forces: gravity and electromagnetic interactions, and it should be expected that both may play important roles in the functional activities of organisms. The biological importance of EMF has been well substantiated in numerous publications and meetings, but our understanding about mechanisms of action is poor and evaluation of some of the experimental data is also controversial. Adding to the confusion, a variety of machines has recently appeared on the market and different models are being offered to the medical profession and to the public claiming beneficial therapies for many illnesses from skin diseases to asthma and vascular problems. Knowledge of mechanisms of action is essential for possible beneficial effects, adequate preventive measures, and for reasonable legislation to clarify contradictions and to separate fact from fiction. Better reporting of results will be helpful, specifying details of instrumentation, procedures, and EMF parameters to allow duplication by different investigators. For this purpose it is also very convenient to arrange scientific exchange visits permitting study of records, methodology, and findings. Our Madrid group has already invited several investigators to spend time with us and in this article we take the opportunity to express our willingness to participate in reciprocal planning and performance of experimentation.

REFERENCES

Adey, W. R. (1981). Tissue interactions with nonionizing electromagnetic fields. Physiol. Rev., 61, 435-514.

Bassett, C. A. L., R. J. Pawluk, and A. A. Pilla (1974a). Augmentation of bone repair by inductively coupled electromagnetic fields. Science, 184, 575-577.

Bassett, C. A. L., R. J. Pawluk, and A. A. Pilla (1974b). Acceleration of fracture repair by electromagnetic fields. A surgically non-invasive method. Ann. N.Y. Acad. Sci., 238, 242-249.

Bawin, S. M., and W. R. Adey (1976). Sensitivity of calcium binding in cerebral tissue to weak environmental electric fields oscillating at low frequency. Proc. Natl. Acad. Sci. (USA), 73, 199-203.

Blackman, C. F., S. G. Bennane, C. M. Weil, and J. S. Ali (1975). Effects of nonionizing electromagnetic radiation on single-cell biologic systems. Ann. N.Y. Acad. Scie., 247, 352-366.

Bridges, J. E., and M. Preache (1981). Biological influences of power frequency electric fields. A tutorial review from a physical and experimental viewpoint. Proc. IEEE, 69, 1092-1120.

Bunõ, Jr., W., J. Bustamante, and J. Fuentes (1984). White noise analysis of pace-maker-response interactions and non-linearities in slowly adapting crayfish stretch receptor. J. Physiol., 350, 55-80.

Chaplain, R. A., B. Michaelis, and R. Coenen (1971). Systems analysis of biological receptors. I. Quantitative description of the input-output characteristics of the slowly-adapting stretch receptor of the crayfish. Kybernetik, 9, 85-95.

Delgado, J. M. R. (1981). Depth stimulation of the brain. In M. M. Patterson and R. P. Kesner (Eds.), Electrical Stimulation Research Techniques. Academic Press, New York. pp. 105-140.

Delgado, J. M. R., J. Leal, J. L. Monteagudo, and M. García Gracia (1982). Embryological changes induced by weak, extremely low frequency electromagnetic fields. J. Anat., 134, 533-551.

Diebolt, J. R. (1978). The influence of electrostatic and magnetic fields on mutation in Drosophila melanogaster spermatozoa. Mutation Res., 57, 169-174.

Eyzaguirre, C., and S. W. Kuffler (1955). Processes of excitation in the dendrites and in the soma of single isolated sensory nerve cells of lobster and crayfish. J. gen. Physiol., 39, 87-119.

Firstenberg-Eden, R. (1983). Rapid estimation of the number of micro-organisms in raw meat by impedance measurement. Food Technol., Jan., 64-70.

Grundler, W., F. Keilmann, and H. Fröhlich (1977). Resonant growth rate response of yeast cell irradiated by weak microwaves. Physics Lett., 62A, 463-466.

Hause, L. L., R. A. Komorowski, and F. Gayon (1981). Electrode and electrolyte impedance in the detection of bacterial growth. IEEE Trans. bio-med. Engng, BME-28, 403-410.

Kastenbaum, M. A., and K. O. Bowman (1970). Tables for determining the statistical significance of mutation frequencies. Mutation Res., 9, 527-549.

Kim, Y. S. (1976). Some possible effects of static magnetic fields on cancer. Tower int. Technomed. Inst. J. Life Sci., 6, 11-28.

Marino, A. A., and R. O. Becker (1977). Biological effects of extremely low frequency electric and magnetic fields: a review. Physiol. Chem. Physics, 9, 131-147.

Monteagudo, J. L., E. Ramírez, and J. M. R. Delgado (1984). Magnetic inhibition of bacterial growth. In Proc. XXI Gen. Assembly U.R.S.I. Florence, Aug.

Muller, H. J. (1927). Artificial transmutation of the gene. Science, 66, 84-87.

Presman, A. S. (1970). Electromagnetic Fields and Life. Plenum Press, New York and London.

Ramirez, E., J. L. Monteagudo, M. Garcia Gracia, and J. M. R. Delgado (1983). Oviposition and development of Drosophila modified by magnetic fields. Bioelectromagnetics, 4, 315-326.

Ramirez, E., J. L. Monteagudo, J. C. Medrano, and J. M. R. Delgado (1984). Drosophila mutation induced by a pulsed magnetic field. In Proc. XXI Gen. Assembly U.R.S.I. Florence, Aug.

Rockwell, S. (1977). Influence of a 14,000 Gauss magnetic field on the radiosensitivity and recovery of EMT6 cells in vitro. Int. J. Radiat. Biol., 31, 153-160.

Sheppard, A. R., and M. Eisenburg (1977). Biological Effects of Electric and Magnetic Fields of Extremely Low Frequency. New York Univ. Press, New York.

Spencer, W. P., and C. Stern (1948). Experiments to test the validity of the linear r-dose/mutation frequency relation in Drosophila at low dosage. Genetics, 33, 43.

Trillo, M. A., M. A. Jiminez, J. Leal, A. Ubeda, and J. M. R. Delgado (1983). Alterations and fractional recovery of chick embryos exposed to electromagnetic fields. Trans. 3rd Ann. BRAGS Meeting, III, 49.

Ubeda, A., J. Leal, M. A. Trillo, M. A. Jiminez, and J. M. R. Delgado (1983). Pulse shape of magnetic fields influences chick embryogenesis. J. Anat., 137, 513-536.

Weissbluth, M. (1979). Enzyme-substrate reaction in high magnetic fields. In T. S. Tenforde (Ed.), Magnetic Field Effects on Biological Systems. Plenum Press, New York. pp. 48-49.

Wertheimer, N., and E. Leeper (1979). Equipement du reseau electrique et cancer infantile. Am. J. Epidemiol., 109, 273-284.

Wilkens, L. A., and G. E. Wolfe (1974). A new electrode design for en passant recording, stimulation and intracellular dye infusion. Comp. Biochem. Physiol., 48, 217-220.

A 2-D MULTIPOLE ANALYSIS OF CURRENT SOURCES-APPLICATION TO MAGNETIC FIELD MEASUREMENTS OF THE HUMAN LEG

D.I.F. Grimes, T.B. Smith and S.J. Swithenby

Physics Department, The Open University, Milton Keynes, Buckinghamshire, England

ABSTRACT

We describe a method of relating measured quasi-dc magnetic fields back to their source currents: it is suitable for use when the source currents are axially uniform. Measurements on contrived current distributions that are not axially uniform indicate that the method still works provided the local variation of currents along the 'uniform' axis is not too fast. We apply the method to biomagnetic signals from the human leg.

KEYWORDS

Multipole; biomagnetism; current dipole; SQUID; human legs.

INTRODUCTION

We have previously reported observations of quasi-dc magnetic fields produced by the human leg (Grimes et al., 1983). The data is obtained with the subject lying on a padded bed with his leg parallel to the horizontal y-axis. As in all the experiments in this paper, the bed is moved horizontally in the transverse or x-direction under a SQUID gradiometer ($\partial^2 B_z/\partial z^2$), where the z-axis points vertically up. Figure 1 shows the SQUID output during a set of such scans across the anterior surface of the left leg of a typical normal subject who had just started to relax in a prone-supine position. The lines of scan are equally spaced along the y-axis, each y- position being defined by the number at the start of the trace.

The shapes of the individual traces vary only slowly with y-position, which suggests that, for each scan, modelling assuming an axially uniform current distribution of the form $J_y(x,z)\hat{\underset{\sim}{j}}$ will be more successful than the more usual current-dipole model used in other biomagnetic studies. This model has been developed in detail elsewhere (Grimes et al., 1984) and is only briefly summarised in the next section. Our main purpose here is to report the results of tests of the applicability of the model to current distributions that only approximate to the assumptions of the model. We present also an application to leg data.

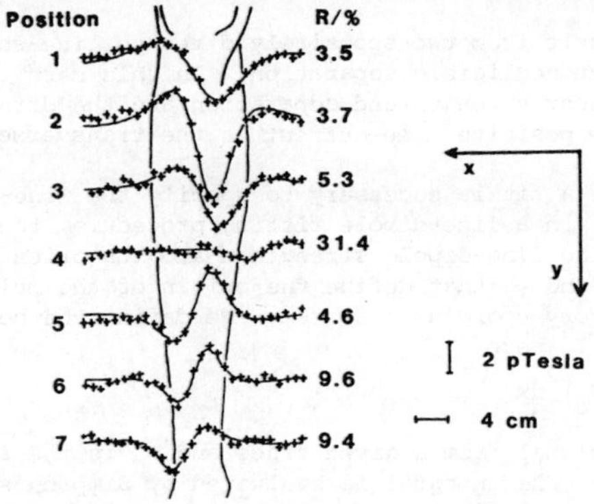

Figure 1 SQUID output for a series of equally spaced scans above the left leg of a typical subject.

THE MULTIPOLE EXPANSION

The mathematical model assumes that the current density does not vary in the longitudinal or y-direction but may depend arbitrarily on position $\underset{\sim}{r} = (x,z)$ in the transverse plane. Ampere's circuital law gives the following expression for the magnetic induction at a point $\underset{\sim}{R}$, also in the transverse plane, but outside the source current region;

$$\underset{\sim}{B}(\underset{\sim}{R}) = \frac{\mu_o}{2\pi} \int \frac{d^2r\, J_y(\underset{\sim}{r})\, \hat{\underset{\sim}{j}} \times (\underset{\sim}{R} - \underset{\sim}{r})}{|\underset{\sim}{R} - \underset{\sim}{r}|^2} \tag{1}$$

The integral is taken over the transverse area where the current density is non-zero. A vector potential giving rise to $\underset{\sim}{B}$ is

$$\underset{\sim}{A}(\underset{\sim}{R}) = -\frac{\mu_o}{2\pi}\hat{\underset{\sim}{j}} \int d^2r\, J_y(\underset{\sim}{r}) \log|\underset{\sim}{R} - \underset{\sim}{r}| \tag{2}$$

A multipole expansion of $\underset{\sim}{A}$, and therefore also of $\underset{\sim}{B}$, results from an expansion of the logarithm in a Taylor series about $\underset{\sim}{r} = 0$.

The monopole term follows directly from (2) by setting $\underset{\sim}{r} = 0$ in the logarithm. Since, physiologically, there is no net current through the transverse plane, this term vanishes for our application. The first non-vanishing term, and the one with which this paper is exclusively concerned, is the 'dipole' contribution

$$\underset{\sim}{A}_d(\underset{\sim}{R}) = \left(\frac{\mu_o}{2\pi}\right)\left(\frac{\underset{\sim}{m}.\underset{\sim}{R}}{R^2}\right)\hat{\underset{\sim}{j}} \tag{3}$$

where $\underset{\sim}{m}$, which we term the 'line-dipole moment', is

$$\underset{\sim}{m} = \int d^2r\, J_y(\underset{\sim}{r})\underset{\sim}{r} \tag{4}$$

Although the current $J_y(\underset{\sim}{r})$ can be very complex, the dipole field $\underset{\sim}{B}_d(\underset{\sim}{R}) = \text{curl}\, \underset{\sim}{A}_d(\underset{\sim}{R})$

is the same as would result from two oppositely directed line-currents lying parallel to the y-axis at negligible separation. In this case the magnitude m of $\underset{\sim}{m}$ is the product of current strength and separation and the direction of $\underset{\sim}{m}$ is from the negative to the positive line-current in the transverse plane.

In order to characterize A_d it is necessary to specify the line-dipole moment $\underset{\sim}{m}$ and its position. Thus, in a line-dipole fitting procedure, there are four adjustable parameters; the line-dipole strength m and its orientation ϕ, together with the coordinates x_o and z_o that define the origin of the multipole expansion in the laboratory coordinate system. We define the best-fit line-dipole as that which minimises I where

$$I = \int (S - S_d)^2 dx \tag{5}$$

Here S is the measured signal from a given trace and S_d is the four-parameter pure line-dipole signal. The integral is evaluated by Simpson's rule from the data in the trace and the calculation takes into account the geometry of the SQUID gradiometer.

The fitting procedure involves varying the four parameters to minimise I. Mathematical tricks reduce this effectively to a computer-based two parameter process (Grimes et al., 1984). For a perfect fit, when the signal is purely line-dipolar, the minimised value of I, denoted by I_{min}, vanishes.

TESTS ON THE MODEL

Experimental data from a real system such as a leg will contain noise and be neither axially-uniform, not truly line-dipolar. We have investigated the validity of our 2-d modelling in such circumstances using simulated data generated by computer and arrangements of current carrying wires. In these tests, the parameter $R = I_{min}/\int S^2{}_d dx$ describes the quality of the fit to the data. There are three types of test.

The first involves computer generated pure line-dipole signals. These are fitted essentially perfectly for all possible line-dipole parameters. The addition to the pure signal of reasonable ($\sim$ 10%) levels of white noise produces small but effectively negligible inaccuracies.

In the second type of test we generated data corresponding to a pair of oppositely oriented parallel line-currents of finite separation. Such a configuration is axially uniform but produces a field with contributions from higher moments than the dipole. The line-dipole term is calculated from equation 4. The fitting of the data to the line-dipole term decreases in accuracy with increasing separation. However, for separations up to $\sim$ 75% of the average depth of the line-currents below the lowest gradiometer coil, the discrepancies between the fitted and calculated parameters are less than Δm = 40%, $\Delta\phi = 15^o$, $\Delta x_o = \Delta z_o$ = 8 mm, the exact values depending on orientation. Such discrepancies are sufficiently small to allow us to distinguish between different anatomical sources and we use these values as arbitrary definitions of acceptability in the remainder of this paper. Computer and wire generated data gave identical results in this test, to within the uncertainties in the measurement of the line-dipole parameters for the wires (±3% in ϕ and ±2 mm in x_o and z_o).

The final type of test was undertaken to see how well the method could assign <u>local</u> values of the four line-dipole parameters to current distributions that vary longitudinally, ie., with y. This was done by means of the various configurations of current carrying wires summarised in Figure 2. These configurations, which might reflect actual current distributions in the leg, have local line-dipole moments (calculated from equation 4) that vary with y. The position of the local moment is just the mid-point between the two wires.

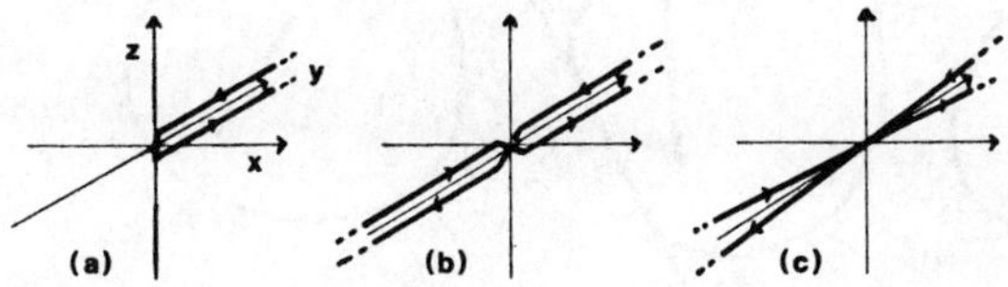

Figure 2 Three arrangements of wires; (a) a semi-infinite line-dipole, (b) parallel wires with an abrupt twist, and (c) uniformly diverging wires. In each case, the wire separation was small compared with z_o ($\sim$ 5 cm).

For each configuration the quality of the fitting depended on the orientation ϕ of the pair of wires, but, because this dependence is not central to our discussion, we report only the results for a somewhat unfavourable angle, $\phi = 70^o$. Figure 3 summarises the deviations of the fitted line-dipole parameters from their local values as a function of y > o. Indicated on the graphs are the arbitrary acceptable levels of deviation. The main conclusion from Figure 2 is that fitting to a local line-dipole does work, especially when changes in the current flow along the longitudinal direction are fairly gradual, as in Figure 2c, or when measurements are not being made too close to an abrupt change, as occurs at y = o for configurations 2a and 2b. In all the tests, loss in fitting accuracy is signalled by an increase in R.

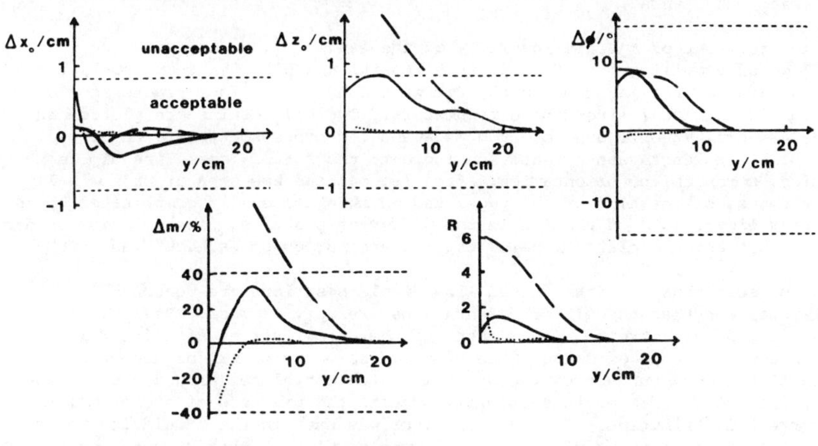

Figure 3 The results of tests on the axially non-uniform wire configurations of Figure 2a (solid line), 2b (dashed line) and 2c (dotted line). The discrepancies between the fitted and measured local line-dipole parameters are plotted against y.

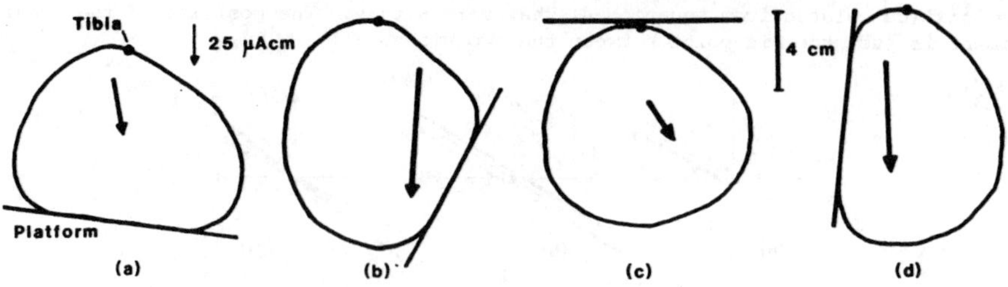

Figure 4 Four fitted local line-dipoles for one leg cross-section. In the four experiments, the (a) anterior, (b) lateral, (c) posterior and (d) medial leg surfaces were uppermost and, therefore, passed closest to the gradiometer.

LINE-DIPOLE ANALYSIS OF LEG DATA - CONCLUSIONS

The results of these tests suggest that the 2d line-dipole analysis is sufficiently robust to be usefully applied to the leg data. The fitted parameters should reflect the 'local' current flow below a given line of scan, and the parameter R will suggest the level of confidence to be placed in a particular fit. In Figure 1 the solid lines are the best fit line-dipole curves to the data of the associated traces. The quality of the fits are excellent, with most values of R less than 10%. This supports our assumption of a relatively simple, predominantly axial current distribution in the leg, particularly near cross-sections 3 and 7.

As a further test of the applicability of the 2-d model, we have scanned the left legs of a small group of subjects at position 3 with, in turn, each of the four surfaces of the leg uppermost. For each subject the four traces should give consistent local line-dipole parameters. Typical results are plotted in Figure 4 where, superimposed on each measured leg cross-section is an arrow whose position, length and orientation indicate the fitted local line-dipole position, strength and orientation. Each leg outline has been drawn nominally with the same orientation on the page, though this process is complicated by the differing distortions of the leg in the different postures. The anterior border of the tibia and the platform below the leg are marked on each outline.

It can be seen that the four fitted line-dipole positions are reasonably consistent, and that the fitted orientations are very similar. However, the fitted line-dipole strength (the length of the arrow) varies significantly between the four sets of data. Some of the discrepancies may be accounted for by leg distortion, but the variation in m, which correlates with the depth and orientation of the fitted line-dipole, suggests the presence of higher moments in the current distribution. Similar behaviour was seen in the simulation tests on parallel current carrying wires, where it was noted that overestimates of z_o led to significant overestimates of m. We may therefore have evidence for two regions of current flow in the leg of relatively large separation. However, overall the consistency of the results indicates that the fitted local line-dipoles reflect the local current distribution.

REFERENCES

Grimes D, Lennard R, and Swithenby S (1983), Il Nuovo Cimento, 2D, N2, 650-59.

Grimes D I F, Lennard R, Smith T B, and Swithenby S J (1984), To be published in IEEE Trans. Mag.

VIBRATIONS OF NATURAL AND ARTIFICIAL MIDDLE-EAR MEMBRANES MEASURED BY A SQUID-MAGNETOMETER

W.L.C. Rutten, C.A. v. Blitterswijk, C.J. Brenkman and J.J. Grote

Ear-Nose-Throat Dept., University Hospital Leiden, The Netherlands

ABSTRACT

A new SQUID magnetometer method for middle ear research is described. Results on natural and artificial membrane vibrations are presented, supplemented by electron micrograph results.

KEYWORDS

SQUID magnetometer, middle ear membranes, biomaterials.

INTRODUCTION

In developing a total artificial middle-ear (Grote et al., 1983) one of the goals is to achieve sound transmission properties which equal those of the natural human middle ear. In order to measure middle ear vibrations the method to be used must be sensitive enough to detect submicroscopic displacement amplitudes of about 10^{-9} meter, up to a frequency of about 10 kHz. Measurement in a closed middle ear cavity connected to an intact inner ear must be possible, in vitro as well as in vivo. Moreover, the possibility of testing artificial tympanic membranes is a requisite.
At our ENT department, after initial research in collaboration with Twente University of Technology, we developed such a method, based on the use of a SQUID magnetometer. The method itself and first results obtained on tympanic membrane vibrations in human temporal bones in vitro have been described earlier (Rutten et al., 1982). In this paper we present three applications of the method

1) vibration spectrum of a guinea pig tympanic membrane
2) vibration spectrum of a freely vibrating (i.e. not connected to middle ear ossicles) human tympanic membrane in vitro.
3) vibration spectrum of two polyetherurethane membranes and one silicone rubber membrane (Silastic^{R}).

Earlier investigations in our group have indicated that a subtotal middle ear prosthesis consisting of 1) hydroxyapatite canal wall 2) hydroxyapatite ossicular chain replacement connected to this wall by a hinge 3) a natural tympanic membrane, gives hearing improvement, but is not an equivalent of the normal middle ear transmission. After longer post-operative periods the hinge connection gets loose due to the biodegradation of hydroxyapatite (v. Blitterswijk et al., 1984). Therefore, a selection is made of porous and dense materials (Homsy et al., 1973) for our artificial tympanic membrane which will function as suspension of the artificial ossicular chain thereby avoiding a hinge construction. For this purpose we selected seven materials for artificial membranes and investigated vibratory and elastic properties (the latter not described in this paper) before and after sterilization. Results were compared with the vibrations of a natural tympanic membrane in a temporal bone in which the middle ear ossicles were removed, in order to test the free vibration.

In this paper we present vibration results of a selection of three out of the seven materials documented by detailed scanning electron micrographs.

METHODS AND MATERIALS

SQUID set-up

Mechanical vibrations are transformed into magnetic flux variations by glueing a tiny magnet to the vibrating point of interest. A commercial SHE-330 r.f. SQUID magnetometer set-up, supplemented by an adjustable second order gradiometer (built at Twente TU) measures the flux variations. Sound vibrations are generated by a two-way loudspeaker system in a semi-closed sound system. High frequencies above 2000 Hz are generated by a nearby positioned piëzo-ceramic loudspeaker, the low frequency part below 2000 Hz by a magneto dynamic speaker, which is heavily magnetically shielded and positioned 2 meter from the measuring compartment. Via a 25 cm diameter PVC tube, the sound is led to the compartment. The sound level is held automatically constant by a Bruel & Kjaer 4134 probe microphone in a feedback loop. The cryostat is mounted in a wooden man-size support. The vibrating structure under study can be positioned in heighth precisely (accuracy 0.1 mm) by a hydraulically adjustable PVC table. The tiny magnets used in this study are two Samarium Cobalt parallellopipeds with masses of 1.05 and 0.4 mgrams. A magnet is glued with zinc-phosphate cement to the center of the membrane. The SQUID output is high pass filtered (48 dB/oct) above 200 Hz and led to a Brookdeal 9505 two-phase lock-in amplifier. The probe microphone signal serves as reference input. Amplitude and phase spectra between 200 Hz and 10 kHz are written on an X-Y writer and off-line processed by a DEC PDP-11 computer system with Houston HIPAD digitizer and Houston HI-PLOT plotter. The latter system corrects for the not perfectly flat amplitude response of the probe microphone as well as for phase shifts in this microphone and in the high-pass filter.

Calibration is performed by measuring the SQUID voltage output in response to the calculated field of a small magnetic coil, acting as a magnetic dipole, positioned at several distances axially centered under the gradiometer. The calibration factor, thus obtained is 1.15×10^{-11} Wb/Volt.

Using earlier results (see Rutten et al., 1982, formulas 5 and 6) and inserting the calibration factor, one easily shows that the sensitivity in the axial direction of the gradiometer (z-direction)

$$\Delta z = 0.035\ \Delta V\ (R^2 + z^2)^{5/2} / M_m\ z$$

in which R is the radius of the gradiometer, Δz = vibration amplitude, ΔV = SQUID-voltage, z = distance between magnet and plane of lower gradiometer loop, M_m = magnetic moment of the SmCo magnet. As an example, with $M_m = 8 \times 10^{-5}$ Am^2, $z = 2.15 \times 10^{-2}$ meter, $R = 1.5 \times 10^{-2}$ meter one obtains a SQUID output of 0.4 μVolt/Å.

Because high sound levels induce stray fields from dynamic speaker to gradiometer and mechanically set the cryostat into vibration the maximum usable sound pressure is 90 dB SPL (re. 2×10^{-5} N/m^2).

Overall characteristics are: sound level range between 60 and 90 dB SPL, frequency range between 200 Hz and 10.000 Hz, minimal measurable vibration amplitude with the 1 mg magnet is 10^{-9} meter, i.e. 10 Å. The magnet should be chosen always such that its mass is small compared to the mass of the vibrating structure on which the magnet is glued. If not, resonances and antiresonances introduced by the too large masses will acomplish a fine structure on the frequency response. As the mass of natural as well as artificial membranes is typically 10-20 mg, a magnet mass of 0.4 to 1.05 mg seems acceptable, provided of course that the membrane material's damping factor is not too small.

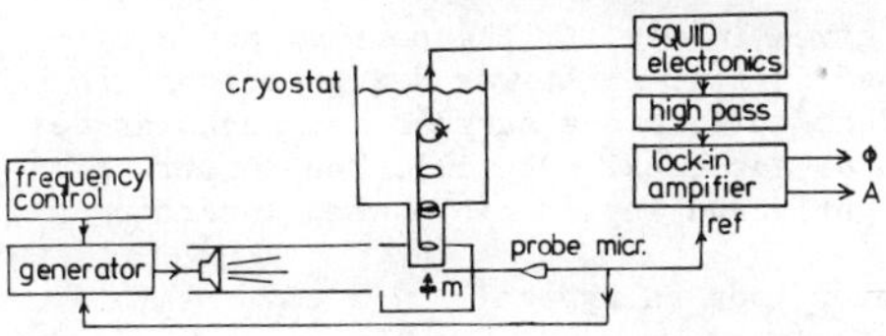

Fig. 1

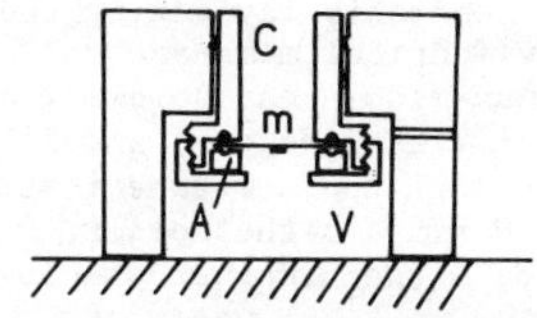

Fig. 2

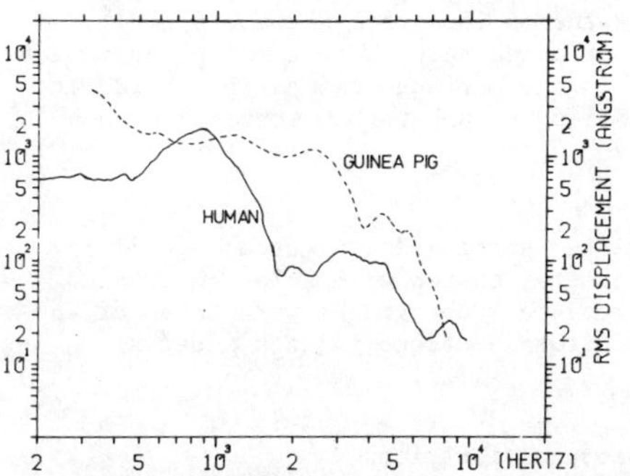

Fig. 3

Fig. 1. Set-up scheme. m is the vibrating magnet. ϕ and A are phase and Amplitude output. Frequency is swept continuously between 200 Hz and 10 kHz.

Fig. 2. Cross-section of the artificial membrane holder. The membrane is clamped between O-ring (•) and aluminium ring A. Dimensions of "middle-ear" volume V and "ear canal" C are given in the text. m is the magnet.

Fig. 3. Vibration spectrum of guinea pig tympanic membrane (the magnet is on the umbo) in vivo and human tympanic membrane (magnet also on the umbo) in vitro.
Stimulation sound level is 80 dB SPL. In the human ear, middle ear ossicles have been removed and the middle ear cavity closed.

Fig. 4, 5 and 6. Scanning electron micrographs. Cross sections through polyetherurethane 1 (fig. 4), polyetherurethane 2 (fig. 5) and Silastic[R] (fig. 6). Bar lengths: 30 µm. Note the macropores in the center in fig. 4, in the material periphery in fig. 5 and the absence of pores in fig. 6.

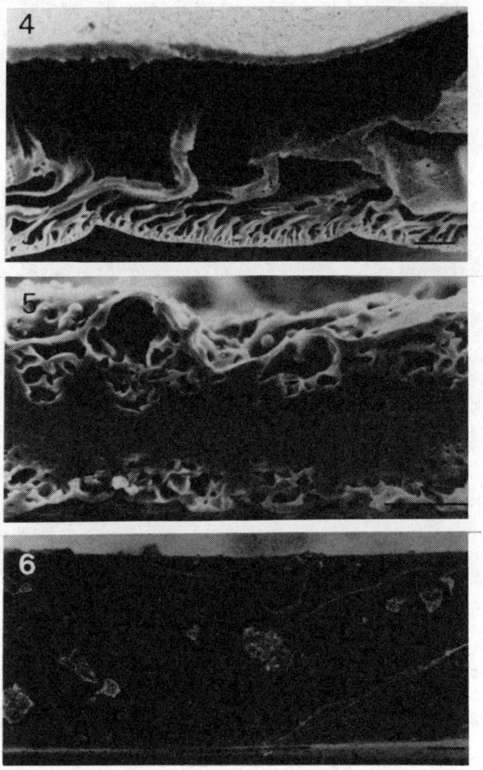

Figs. 4, 5 and 6

A very schematic lay-out of the set-up is given in fig. 1. The perspex and aluminium cylindrical membrane holder is sketched in fig. 2. It was designed to match the human middle ear volume (volume V is 2 cm^3), membrane surface (surface inside the O-ring is 0.86 cm^2) and diameter of outer ear canal (0.6 cm). The construction is such that the membrane is stretched slightly and reproducibly when inserting it on the O-ring in the holder.
Distance between magnet and lower gradiometer loop is typically 2-3 cm.

Membrane materials

Three materials were tested: two polyetherurethanes (Goodrich esthane 5714F1^R) of which one (to be called polyurethane 1) was prepared such that it had pores in its center and the other (polyurethane 2) pores in the surface region. The third material Silastic^R is a commercially available Silicone rubber (Dow Corning) with a dense structure.

Scanning electron microscopy

Samples of the three materials, before and after sterilization (steam sterilization during thirty minutes at 120°C), were sputter coated with gold and studied with a Cambridge S180 scanning electron microscope. Photographs were taken at varying magnifications of both surfaces and of cross sections in each material.

RESULTS

Human and guinea pig tympanic membrane vibrations

Figure 3 shows amplitudes of vibratory response of the tympanic membrane at 80 dB SPL in a living, sedated guinea pig and in a human temporal bone, 22 hours after death, in which the ossicles have been removed. The spectrum can be grossly characterized as that of a low-pass filter sloping down in guinea pig above 2000 Hz with about 18 dB/octave and in human above 1000 Hz with about 15 dB/octave.

Electron microscopy of the artificial membranes

Before sterilization:
Polyurethane 1 (fig. 4) has an approximate thickness of 120 µm, micropores at its surface (diameter ≈ 2 µm) and macropores (diameter ≈ 150 µm) in the center, sparsely interconnected with the surface.
Polyurethane 2 (fig. 5): In contrast to no. 1 the center is dense covered by a zone of macropores (diameter ≈ 15 µm) at either side. Only a small number of these pores are interconnected with the outer surface in which micropores (3 µm) could be seen. Thickness is ≈ 130 µm. Silastic^R (fig. 6) is a completely dense elastomere without any pores and a thickness of ≈ 130 µm.

After sterilization:
Polyurethane 1 is more smooth at its surface ensuing from what looked like a melting process. Micropores are less in number and smaller in diameter. Macropores show no apparent changes although the interpore walls become more compact.
Polyurethane 2: the outer surface of this material shows quite impressive changes. The number of micropores is decreased and a large part of the macropores has disappeared completely.
Silastic^R shows no visible changes as a consequence of sterilization.

Vibration amplitude spectra of the membranes

Figure 7 shows the spectrum of polyurethane 2 (P2) and Silastic^R (S), at 80 dB SPL before and after sterilization. Silastic is hardly influenced by the sterilization, while the P2 curve has become more smooth and resonance peaks have shifted to lower frequencies. In a somewhat lesser extent the same was observed in P1, results are not shown for clarity. Both aspects point at the fact of increased damping inside the material probably due to the heating process of sterilization. Figure 8 compares the three materials S, P1 and P2. It shows that P1 and P2 vibrate with greater amplitude than Silastic^R. High-frequency roll-off has about the same slope as in human and guinea pig.

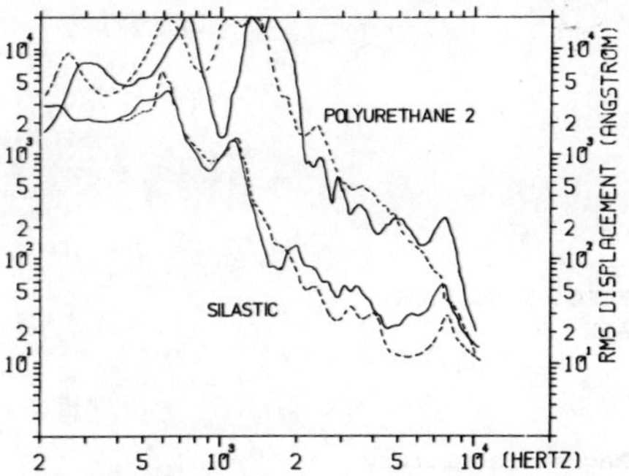

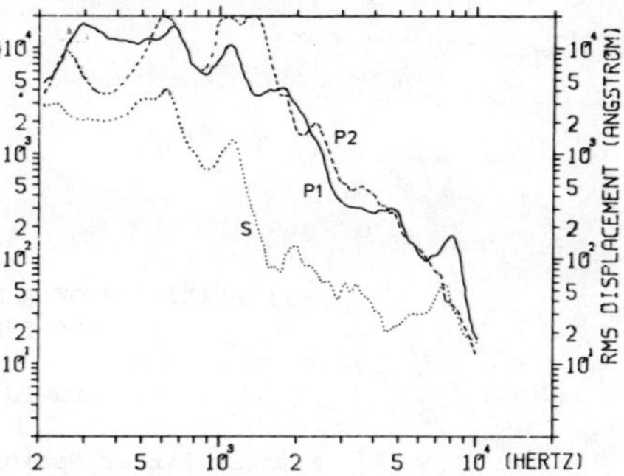

Fig. 7. Vibration spectra before (drawn curve) and after (dashed curve) sterilization of polyurethane 2 and Silastic.
Note that resonance fine structures become less pronounced and have shifted to lower frequencies in P2 after sterilization. Sound pressure level is 80 dB SPL.

Fig. 8. Vibration spectra of the three artificial membrane materials polyurethane 1 (P1), polyurethane 2 (P2) and Silastic^R (S). Sound pressure level is 80 dB SPL.

CONCLUSIONS

With this study the SQUID magnetometer method demonstrates its capacity to study small vibrations.
Comparison of figs. 3 (humans) and 7, 8 leads to the conclusion that on the average the Silastic membrane has comparable vibration amplitudes as the freely vibrating human membrane, while P1 and P2 vibrate even better.
Electron microscopy as well as vibration outcomes show that Silastic does withstand sterilization very well while P2 becomes more damped/viscous. According to these criteria Silastic is a candidate for an artificial tympanic membrane. Of course, aspects of pore size, macroscopic elasticity and biocompatability must be taken into account in a final stage of selection.

ACKNOWLEDGEMENTS

This study is supported by the Dutch Organization of the Advancement of Pure Research and by the Heinsius Houbolt Fund. We thank the dept. of Chemical Technology of Twente University of Technology for preparing the two polyurethanes.

REFERENCES

Blitterswijk, C.A. van, Kuypers, W., Blok-van Hoek, C.J.G., Daems, Th., Grote, J.J. (1984). Bioreactions at the tissue/hydroxyapatite interface. Biomaterials (submitted).

Grote, J.J., Kuijpers, W. (1983). Use of biomaterials in reconstructive middle ear surgery. In: Biomaterials in reconstructive surgery. Ed. L. Ruben, C.V. Mosby Company, St. Louis, 987-995.

Homsy, C.A., Kent, J.N., Hinds, E.C. (1973). Materials for oral implantation - biological and functional criteria. JADA, Vol. 86, 817-832.

Rutten, W.L.C., Peters, M.J., Brenkman, C.J., Mol, H., Grote, J.J. and van der Marel, L.C. (1982). The use of a SQUID magnetometer for middle ear Research. Cryogenics, 457-460.

FEASIBILITY OF A MAGNETIC STIMULATOR FOR THE BRAIN

David Cohen

Francis Bitter National Magnet Laboratory
Massachusetts Institute of Technology
Cambridge, Massachusetts 02139 USA

ABSTRACT

A design is presented for a non-invasive stimulator for a localized area of the cortex. This design overcomes several problems which would be present if attempting to stimulate by applying a current pulse directly through electrodes mounted on the scalp.

KEYWORDS

Non-invasive, magnetic stimulator, neural stimulation, induced current, current pulse.

INTRODUCTION

A non-invasive stimulator of the brain which is localized would have many clinical and research applications. For example, clinically it could be used to map functional sites before brain surgery, or to define a site of an epileptic focus by precipitating a discharge, where both procedures are now performed only on the exposed brain. In research it could be used to localize brain functions such as control of motor systems and motivational states, now only possible with implanted electrodes. Unfortunately, non-invasive stimulation of the cortex by passing a current pulse directly through scalp electrodes is not used on humans, because of three major problems. However, a magnetic stimulator, which is placed very near the scalp and induces a current pulse in the brain, should reduce some of these problems and could be successful. The purpose of this paper is to suggest a design of a magnetic stimulator, and theoretically show its feasibility. While a reduced experimental model has been used only for some electrical tests, a model which actually stimulates has not yet been built.

Two major problems in stimulating the cortex non-invasively with electrical current generally would be both the pain (electrical shock) and damage (mostly heat) in the tissues along the current path to the cortex; this is because in any method, using either scalp electrodes or a magnetic stimulator, the current density generally increases with approach from the cortex to the scalp surface. The special problem of damage to the target area itself has been largely solved by using an opposing pulse-pair (Lilly, 1954). In addition to the pain and heat problems, there is the third problem of current localization in the cortex; either with electrodes or magnetically, the current in the cortex cannot be "focussed" and the larger the cortical area of stimulation, the less useful is the stimulator. Because of these three problems, non-invasive stimulation via scalp electrodes has been studied almost always in animals (Gualtierotti and Paterson, 1954). However, there have recently been experiments on the human head where the application of an extremely short high-voltage pulse appears to have produced some cortical stimulation without uncomfortable shock (Merton and Morton, 1980), although the stimulation was not well localized.

A magnetic stimulator is proposed here which would reduce both the heat and the localization problems. The stimulator would consist of a one-turn coil of small diameter but thick wire placed on the scalp over the cortical area to be stimulated. A pulse of large current would be passed through the coils; this would induce a localized pulsed current through the cortex. The heat problem would be alleviated because there would be almost no current induced in the high-resistivity regions (skin, skull). As shown by lead field studies (Cuffin and Cohen, 1979), the current induced magnetically would be more localized than that produced by surface electrodes. Although there has been a previous report on magnetically induced currents in the brain (Rentsch, 1966), this was for the purpose of electrosleep where a large coil was used over the human head, hence there was no cortical localization or attempt at alleviating the heat or pain.

GENERAL DESIGN

The idea here is to magnetically induce enough current density in the cortex for stimulation, with the current as localized as possible. These are opposing requirements; a large coil on the scalp would be most efficient for the former, while a small coil (or several small loops in opposition) would be efficient for the latter. Calculations show that a single, small coil should do the job, and the general design is shown in Fig. 1. Flux compressors have been utilized

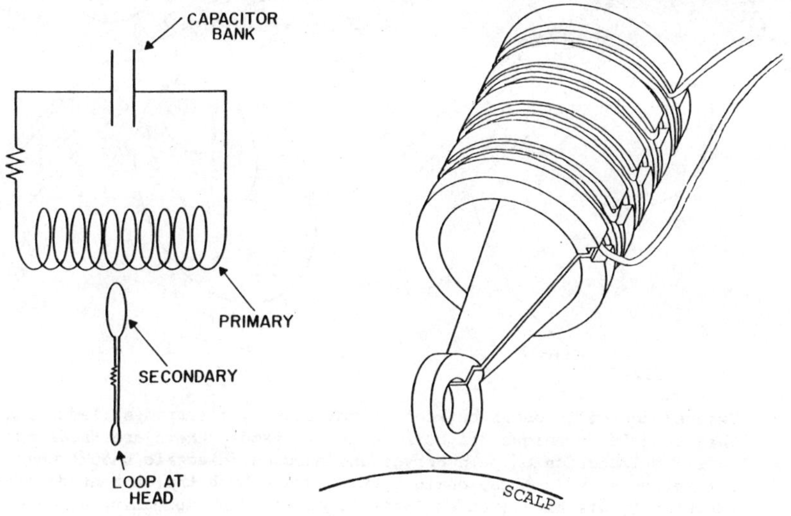

Fig. 1. Right: Sketch of the coil unit, called a flux-compressor. The entire unit is machined from one piece of metal. A low-current pulse in the primary, shown as 12 turns placed in four grooves, is efficiently converted into a high-current pulse in the one-turn loop (2 cm diameter) at the scalp. Left: Equivalent circuit of flux-compressor and power source, which is a capacitor bank.

previously at the Magnet Lab for high-field purposes, and their technology is well understood. The primary winding is on the outside, the secondary winding is the metal cylinder itself, and the load is the loop at the end. Therefore, this is a device for impedance matching the capacitor bank to the final loop, where only a low-current line need be used from the capacitor bank to the unit; this is important because the capacitor bank is bulky and hence will be located some distance from the subject. The way this design alleviates the heating problem, and produces localization, is shown in Fig. 2. For emphasis, the differences between electrode and magnetic current loops are here exaggerated. The two main points are that magnetically there is far less ohmic heating and there is far less spread of current in the cortex.

SOME DETAILED CALCULATIONS

It has been estimated (Fry, 1968) that the necessary charge density for cortical stimulation is $Q = 0.064$ coulombs/m^2; on several grounds this appears to be too conservative, and instead I chose $Q = 0.030$ coulombs/m^2. The charge transfer is assumed to take place during $t = 100$ µsec. Therefore, the current density $J = Q/t = 3 \times 10^2$ amps/m^2. Now the electric field amplitude $E = J\rho$, where the resistivity $\rho = 1.5$ ohm-meter, therefore for stimulation $E = 450$ volts/meter in the cortex. Assuming the stimulation takes place 1 cm below a loop of 2 cm radius,

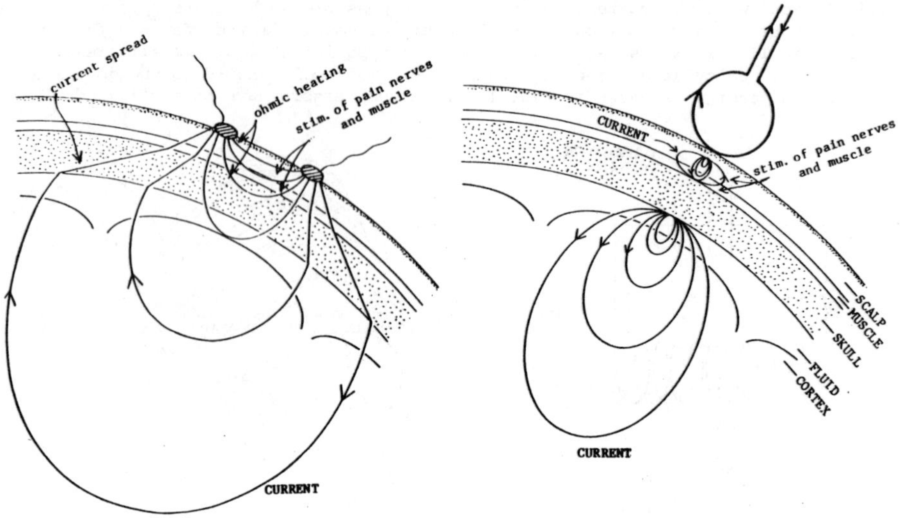

Fig. 2. Exaggerated differences between stimulation by electrodes (left) and by the magnetic technique (right). With electrodes there are three problems: stimulation of pain nerves and muscles (electric shock) where the current density is high, ohmic heating where both the current density and resistivity are high (shaded layers), and current spreading along the cortex due to the high resistivity of the skull. In the magnetic method, the pulsed magnetic field (not illustrated) induces a current in the low-resistivity areas. Because almost none of this current flows in the high resistivity layers (zero current is shown for emphasis), the heating problem is greatly reduced. Also, there is no current spreading because spherical shells do not magnetically affect each other.

this yields a rate of current change in the loop of $dI/dt = 2.8 \times 10^9$ amps/sec. Therefore the final current level after 100 μsec is 2.8×10^5 amps, and the magnetic field amplitude at the loop center is 9 Tesla. This current and field may appear to be high, but are certainly attainable when a capacitor bank is fired through a single small loop. This has routinely been done at the Magnet Lab in firing projectiles electromagnetically. However, the single-turn coil must be strong, hence designed and machined with care; Be-Cu is the alloy of choice. The entire design appears to be feasible.

Some verification of the flux-compressor design was obtained by performing some measurements with a "scrap" flux compressor, of reduced strength so that no actual stimulation was possible. Firing a capacitor bank through the coil and measuring the pulsed field in and around the loop showed that the electrical engineering principles were indeed sound; for example, essentially all the flux generated in the primary was transferred to pass through the loop.

SPATIAL LOCALIZATION

For our purposes here, we can define localization as the effective area of stimulation in the cortex, say bounded by the half-value of E. The improvement in localization of this magnetic design over electrodes can be seen by examining the lead fields in Cuffin and Cohen (1979). By the definition of lead fields, stated there, the lead field distribution is exactly the same as the distribution of the induced E, due to either electrodes or to coils which are parallel to the head. For the case here of a single erect (non-parallel) coil, the induced E can be estimated from the lead field of the closely related case of two opposing parallel coils. This should be compared to that of the single electrode; although two elecrodes are actually used in stimulation, they are far enough apart (to avoid local effects) so that the lead field more resembles that of the single electrode. In this estimate, the area of stimulation of the loop is considerably smaller on the cortex than is the area of one electrode, by a factor of at least four. This advantage is maintained after corrections for the proximity of the second electrode and the increased spacing of an actual coil center from the scalp.

FINAL REMARKS

It should be noted that the first or shock problem can perhaps be handled in the same way as is suggested with electrodes (Merton and Morton, 1980); that is, by a stronger but shorter magnetic pulse. If it turns out to be effective with electrodes, then it should also be effective magnetically. Therefore, this design of magnetic stimulator should handle the shock problem equally to electrode stimulation, in addition to significantly reducing the second and third problems. Physiologically, it should be noted that with any electrical stimulation there is the problem of "kindling," where repeated stimulation of cortical tissue appears to lower its stimulation threshold and produces predisposition to spontaneous firing. Initial and careful testing of this magnetic device should therefore be done in laboratory animals, with collaboration of electrophysiologists, not only to evaluate the improvement in the three stated problems, but also to assess neural phenomena such as kindling.

REFERENCES

Cuffin, B.N. and D. Cohen (1979). Comparison of the magnetoencephalogram and electroencephalogram. Electroenceph. clin. Neurophysiol., 47, 131-146.

Fry, W.M. (1968). Electrical stimulation of brain localized without probes--theoretical analysis of a proposed method. J. Acoust. Soc. Am., 44, 919-931.

Gualtierotti, T. and A.S. Paterson (1954). Electrical stimulation of the unexposed cerebral cortex. J. Physiol., 125, 278-291.

Lilly, J.C. and others (1954). Brief noninjurious electric waveform for stimulation of the brain. Science, 212, 468-469.

Merton, P.A. and H.B. Morton (1980). Stimulation of the cerebral cortex in the intact human subject. Nature, 285, 277.

Rentsch, W. Magneto-inductive transmission of stimuli to the brain, in Electro therapeutic Sleep and Electro-Anaesthesia, Proc. 1st Int. Symp. Graz, Austria, 12-17 September, 1966; Eds., F.M. Wageneder St. Schully, J. Nussmuller, J. Clutton-Brock, and B.M.Q. Weaver, Int. Cong. Series No. 136, Excerpta Medica Foundation, Amsterdam, pp 161-166.

FRONTIERS IN THE NEW SCIENCE OF BIOMAGNETISM*

S.J. Williamson and L. Kaufman

Neuromagnetism Laboratory
Departments of Physics and Psychology
New York University
New York, New York, 10003, U.S.A.

ABSTRACT

The notion that magnetism is associated with the human body was introduced in the 18th Century by charlatans. Despite this inauspicious beginning, with the advent of modern, high-sensitivity magnetic sensors it now has a firm scientific basis and is playing a significant role in fundamental research and clinical studies.

KEYWORDS

Biomagnetism; magnetic sensors; SQUID; liver; magnetic susceptibility; lungs; action potential; magnetocardiogram; His-Purkinje system; magnetoencephalogram; evoked fields; event-related brain activity; biological development

INTRODUCTION

Our purpose is to provide an overview of the exciting areas of biomagnetism that presently represent frontiers in the development of this science. To make this accessible to the widest readership our presentation will be at a largely non-technical level. Our choice of topics is necessarily subjective, yet we believe it remains comprehensive. By "frontiers" we mean those efforts that hold promise for exciting developments within the next few years. Research reports on nearly all of these frontiers are included in this book, and they explain the technical details first-hand. As these reports contain all of the relevant references to past work, we limit ourselves here to giving only a few of the most recent references.

The notion that magnetic fields are associated with the human body traces back at least two hundred years, though the first detection of such fields was achieved only twenty years ago (Baule and McFee, 1963). The first association between magnetism and humans to gain prominence was advanced by a charlatan, not a scientist, and it came at a time when society had become receptive to the ideas of invisible forces. The setting was Paris, just prior to the French revolution, when the wonders of Newton's gravity had been popularized by Voltaire, Franklin's electrical fluid had been given a practical application in the lightning rod, and the gas laws of Charles and others provided a dramatic display when man first rose into the sky under a hot air balloon. Science, in short, had become a fad.

Thus when a Viennese physician Franz Anton Mesmer arrived in Paris in February 1778 to proclaim an amazing "discovery" his ideas found a receptive audience (Darnton, 1970). Mesmer claimed that an invisible fluid permeates the universe, and in its continual motion it passes through all objects, including the human body. He went on to claim that the body contains poles not unlike those of a magnet, and if they become improperly aligned they impede the flow of this fluid: the result is illness. Capitalizing on this notion, Mesmer and his followers established salons in Paris and other large cities throughout France where they accepted patients for treatment. As illustrated in a contemporary drawing (Fig. 1), patients grasped iron bars, and to the accompaniment of wind instruments in the background, stroked their neighbors to help realign the poles within their bodies. One woman on the right is experiencing a crisis, which generally was taken as a sign of cure by the mesmerists. Hundreds of cures were claimed, though there were also many disbelievers. Indeed, the topic of "animal magnetism" was hotly debated in the newspapers of the time.

In response to scientific and medical criticism a royal commission was established in 1784 to investigate Mesmer's claims. Its members were drawn from the Faculty of Medicine of the University of Paris and from the Academy of Sciences. Among them were found Guillotin, Bailly, Lavoisier, and Benjamin Franklin. After extensive studies with the most sensitive electrical instruments and magnetic compasses, this commission issued its conclusion, stating that no evidence was found of a physical manifestation of magnetism to support Mesmer's claims. Nevertheless, despite this official indictment, Mesmer and his followers continued to exert an influence on french society for several years.

We know now, with the benefit of considerably more sensitive instruments, that magnetic fields are indeed associated with the human body - but their origin is of quite a different nature than supposed by Mesmer. Indeed, to identify one origin we need only travel a thousand kilometers south of Paris to observe Luigi Galvani in the city of Bologna beginning studies of a scientific nature of an effect that in 1791 became known as "animal electricity." Among his work is the well-known experiment showing that an electrical stimulus applied to the sciatic nerve in the leg of a frog causes the leg to twitch. This suggested quite rightly to him that electrical signals within the body control certain of its functions.

Fig. 1. Mesmerist seance, with its apparatus of iron bars protruding from an oak tub, around which the patients stand. (With permission of the Bibliotheque Nationale, Paris.)

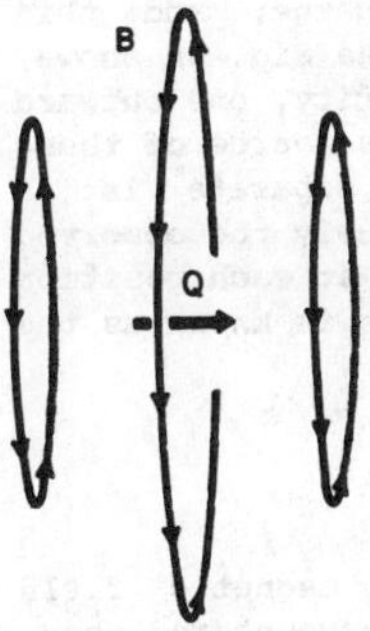

Fig. 2. Magnetic field (denoted by the symbol B) encircling a current flowing along a short, narrow path. The current dipole (denoted by Q) has a strength given by the product of the current and the length of its path.

CURRENTS AND FIELDS

The connection between Galvani's studies and the modern discipline of biomagnetism was provided by the work of two french scientists Jean Baptiste Biot and Felix Savart. Using compasses they found that whenever an electrical current flows, a magnetic field appears in the surrounding space. In 1820 they published an article showing how to predict its direction and strength. Figure 2 gives an example of the relationship between the field and a current in a particularly simple case: We imagine positively charged ions within the body (such as sodium and potassium) moving in the direction of the arrow along a short, narrow path confined by the membrane of a cell. Any net movement of charge from one place to another is an electrical current, and the magnetic field associated with this example encircles the current, becoming progressively weaker at greater distances away to the side and to the front and back of the arrow. The strongest field due to this short segment of current is therefore found within the region close to the current.

Such a segment of current that is driven by biochemical processes is given the name "current dipole." Why the name "dipole", implying two polarities? To answer this question we must look at the total current pattern associated with a current dipole situated within a conducting medium such as the human body. Movement of positive electrical charge in a current dipole produces a buildup of positive charge at the head of the arrow, and the outward electric field from this charge drives ions in the surrounding medium to form an outward spreading pattern of current (Fig. 3a). Similarly, the removal of positive charge from the

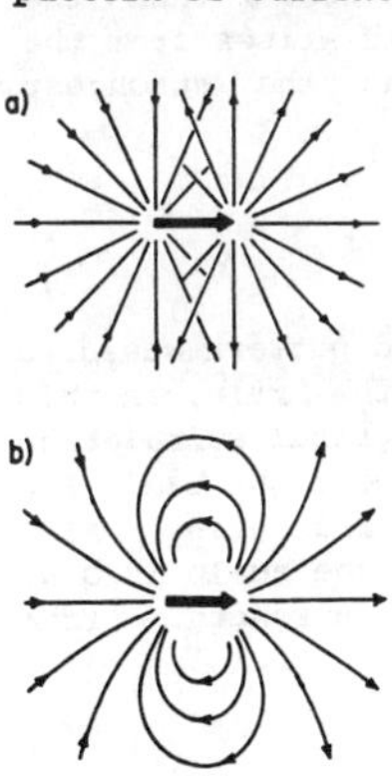

Fig. 3. (a) Outflow of current at the "head" of the current dipole of Fig. 2 and inflow at the "tail." (b) Superposition of the outward and inward flows produces a dipolar flow pattern from head to tail that completes a closed loop.

tail of the arrow leaves an increasing imbalance of negative charge, and this causes an inward flow of current from the surrounding medium. As Fig. 3a shows, the total effect produces two current patterns of opposite polarity, one outward and the other inward - hence the name "current dipole" for the source of these patterns. Of course considering the two patterns in Fig. 3a as separate is an artificial distinction; in reality, since both flow simultaneously the observed effect is the superposition of the two patterns. Adding the two at each position in space produces the smooth pattern shown in Fig. 3b. This is known as the "dipolar" pattern.

Biomagnetic Fields

This is all by way of introduction to the simplest pattern of magnetic field that is found to arise from currents in the human body. It is encountered when, say, a person touches a table with a finger. The earliest magnetic response in the brain is observed over the side of the head near the sensory center of the brain devoted to tactile response of that finger. Figure 4 gives an example that is more easily reproduced in the laboratory: applying an electrical stimulation to the little finger of the right hand. Less than a tenth of a second later there is one area of the scalp where magnetic field emerges from the head and an adjacent area where it enters. This is exactly what is expected if the electrical source can be represented as a current dipole located directly under the center of the field pattern. The field on entering the head continues around the dipole (qualitatively as in Fig. 2) to form a continuous loop. The direction of the dipole's arrow is at a right angle to the line joining the two positions of maximum field. By "maximum" we mean where the field perpendicular to the scalp is strongest. The pattern across the scalp even predicts the depth of the dipole beneath the scalp: its depth is related to the distance across the scalp between these two maxima, the greater the distance the greater the depth.

By detecting such fields, carefully measuring the field pattern, and computing the source locations, investigators using the biomagnetic technique for brain research are discovering significant information about various types of acitivty, such as how our brains deal with sensory information. Before giving examples of these and of other applications at the frontier of research, we shall turn briefly to consider how such weak fields are measured.

SENSING BIOMAGNETIC FIELDS

Biomagnetic fields are roughly one million to one billion times weaker than the earth's geomagnetic field. The strongest biomagnetic field arises from the heart, which should not be surprising in view of the fact that the strongest

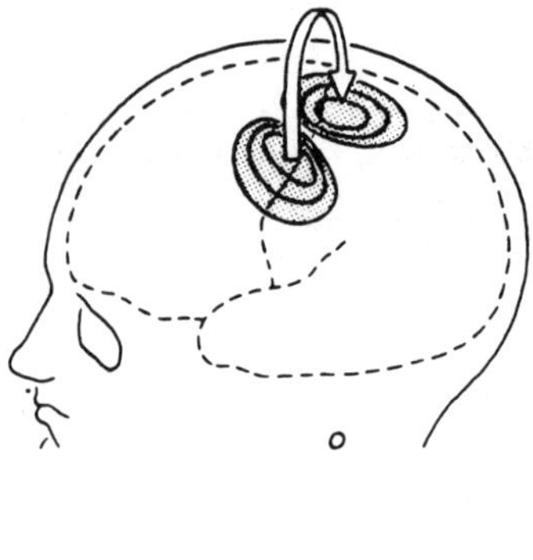

Fig. 4. Portrayal of the field pattern measured over the tactile center of the brain, in this case in response to rapid electrical stimulation of the little finger of the right hand. The curves are isofield contours indicating where the field perpendicular to the scalp is 0.9, 0.7, and 0.5 times the maximum strength. (from Brenner and colleagues, 1978)

electrical currents are found in heart muscle. Somewhat weaker are fields from the eye and fields from skeletal muscles. The weakest detected to date arise from neural activity in the brain. For discussions of these effects the reader is directed to a review article (Williamson and Kaufman, 1981), to the proceedings of a recent workshop (Romani and Williamson, 1983), and to a recent multi-authored textbook (Williamson and colleagues, 1983). There are two other classes of biomagnetic fields as well. One class consists of fields arising from the response of molecules in body tissue to the application of a magnetic field, an example being the field from the human liver arising from iron-bearing proteins. Another class consists of fields associated with magnetic particles in the body, as may be found in the lungs and stomach. In humans, these particles are commonly introduced by environmental exposure, particularly for workers in industries fabricating or processing iron or steel. There are several important applications of this kind of biomagnetic measurement, but we shall here just touch on a few of them. A general discussion of these various effects has been given by Cohen (1975), and more recent references may be found in the review articles just given.

The most sensitive instrument for measuring magnetic fields is based on a device called a SQUID, for superconducting quantum interference device. This sensor was developed in the laboratories of low temperature physicists almost twenty years ago (e.g., see Zimmerman and colleagues, 1970), and it relies on the unusual properties of superconducting materials. The superconductors of interest are metals that when cooled to sufficiently low temperatures (less than about -260° Celsius) exhibit zero electrical resistivity. In addition, their electrons that conduct current behave in a cooperative fashion, so that the wave-like properties of their quantum state plays an important role in how they respond to a magnetic field. One type of SQUID - called the "dc SQUID" - is designed so that when a constant ("dc") electrical current is sent into one end of the device, the current divides into two parts to take different routes to the other end. There, where they join, interference takes place between the two waves, similar to the effect when the wakes from two passing motorboats overlap. Depending on the relative lengths of the two paths taken by the two waves, their amplitudes may add constructively to produce an enhanced effect or destructively to diminish the overall effect. When a magnetic field is applied to the region separating the two paths, it gives the electrons in one path a little "boost" while tending to "retard" the electrons in the other path. This changes the conditions of interference, which can be detected by incorporating a "weak link" in each of the paths and a suitable electronic system that monitors the voltage across the parallel combination of weak links. This effect is very sensitive to the presence of a magnetic field, and it is for this reason that SQUIDs are useful as magnetic sensors. SQUID sensors must therefore be kept cold to operate, and this is presently done by immersing the device in liquid helium (at a temperature of about -269 Celsius). The bath is protected from excessive heating by a vacuum-insulated, container called a "dewar" (named for Sir James Dewar who invented it in about 1892), almost identical to a common glass Thermos bottle.

Biomagnetic fields are much weaker than typical magnetic fields in our environment, such as those produced by motor vehicles, subways, elevators, and even the earth's ionosphere. Consequently, one of the major challenges has been to reduce the effects of this environmental noise. One way has been to work inside a magnetically shielded room. Another way, which avoids the cost and confinement of such a room, is to use a sensing arrangement that discriminates against the sources of noise. Discrimination can be achieved by taking advantage of special geometries for the sensing coil that conventionally is used to provide the field information to the SQUID. As illustrated in Fig. 5, a small coil of superconducting wire of appropriate geometry is mounted within the dewar

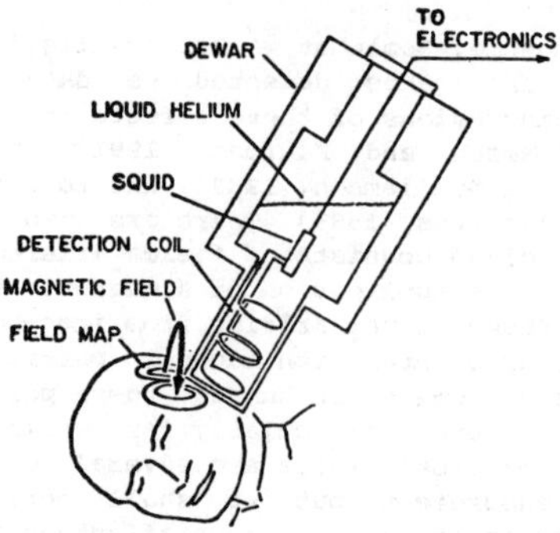

Fig. 5. Recording a biomagnetic field from the human brain, with a SQUID and a detection coil in the geometry of a second-order

some distance below the SQUID so it can be placed near the biomagnetic source, and the ends of this wire are connected to an input coil in close proximity to the SQUID. When the first coil, called the "detection coil", responds to the field a current is sent around the superconducting circuit, and the SQUID responds to the magnetic field applied to it by the second coil. The advantage here is that the detection coil's geometry can be chosen to reduce the pickup of magnetic noise in the environment, thereby improving the quality of the detected signal. For instance, if the detection coil consists of two identical loops of wire wound in opposite directions but separated by a few centimeters (called a "first-order gradiometer"), it will not respond to a field that is spatially uniform, because the positive signal imposed on one loop is just cancelled by a negative signal of identical magnitude imposed on the second loop. Since the field from a distant noise source is fairly uniform, this technique discriminates against such sources. Yet the detection coil retains sensitivity to a nearby source, such as a human subject. This is because the field strength decreases rapidly with distance close to a source, imposing a much stronger signal on the closer loop of the detection coil than the farther one. The most popular detection coil is a "second-order" gradiometer, which provides even stronger spatial discrimination because it consists of two gradiometers mounted back-to-back as illustrated in Fig. 5.

Even the weak magnetic fields of the brain can be detected in this way without recourse to shielding. Figure 6 shows an example of a laboratory where an unshielded sensor is being used to study neuromagnetic fields of the brain. The detection coil is mounted in the narrow tail of a fiberglass dewar positioned just above the subject's head, with the SQUID supported above this coil, at the bottom of the larger diameter reservoir for liquid helium. Analogous setups can be used to study biomagnetic fields of the heart, lungs, skeletal muscle, and limbs, as well as other organs. The further improvement of instruments such as this can be said to be a "frontier" in biomagnetism - a technical frontier. For instance, it is desirable to further improve the ability of such instruments to reject environmental noise, so that weak, low-frequency biomagnetic effects can be detected. Even more important is the development of multiple-sensor systems to measure the field at a number of positions simultaneously. This would greatly enhance the accuracy, speed, and capability of biomagnetic measurements. Systems with four sensors (Ilmoniemi and colleagues, 1984) and five sensors (Williamson and colleagues, 1984) have recently been developed, and even larger arrays are contemplated. Much progress has been made in the past two years, but we still need more improvements, especially for use of such instruments in noisy environments such as laboratories and hospitals.

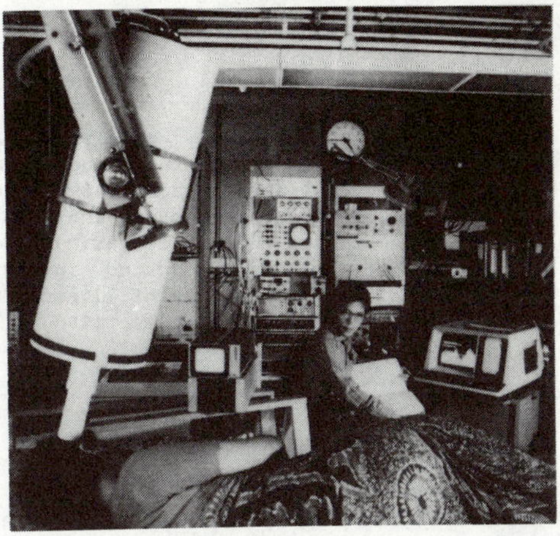

Fig. 6. The Neuromagnetism Laboratory at New York University, showing a subject under the magnetic field sensor, with an arrangement for studying the magnetic field of the brain when it responds to a pattern presented on an oscilloscope screen in the background.

AN OVERVIEW OF BIOMAGNETIC FIELDS

We shall now turn to a few select biomagnetic studies to provide a brief survey of the frontiers of biomagnetic applications. The references are not comprehensive but merely serve as an introduction to the literature.

Storage of Iron in the Liver

An important application of biomagnetism in clinical medicine is its ability to measure noninvasively the iron concentration in the liver when the body causes large amounts to be stored there. Previously this could be done with high sensitivity only by chemical analysis of tissue removed from the liver through biopsy, a procedure which is painful and not without risk. There are several diseases such as thalassemia major and hemochromatosis where the condition may lead to excessive storage of iron, either as a result of the conventional treatment or as a direct consequence of the disease, and it is highly desireable to follow the change in liver iron stores in response to various treatments.

The basic idea is to apply a magnetic field to the region of the liver and measure the responding field produced by the molecules in body tissue. The response of the tissue in comparison with the strength of the applied field is called the magnetic susceptibility of the tissue. Magnetic susceptibilties of most biologic materials are very small, the responding field being only one-millionth of the applied field. This is because the most abundant material in the body is water, which is only weakly affected by a magnetic field. When it responds, the field it produces is opposite to the applied field, and the susceptibility is said to be "diamagnetic". However iron is stored in the body in the protein ferritin, and to some extent as hemosiderin. These are strongly responding: their field lies in the same direction as the applied field, and such materials are said to be "paramagnetic". Therefore, with increasing iron concentration in the liver the net magnetic susceptibility first decreases from its weak diamagnetic value, passes through zero and then increases in the paramagnetic sense. To determine the iron concentration it is only necessary to measure the susceptibility of the liver.

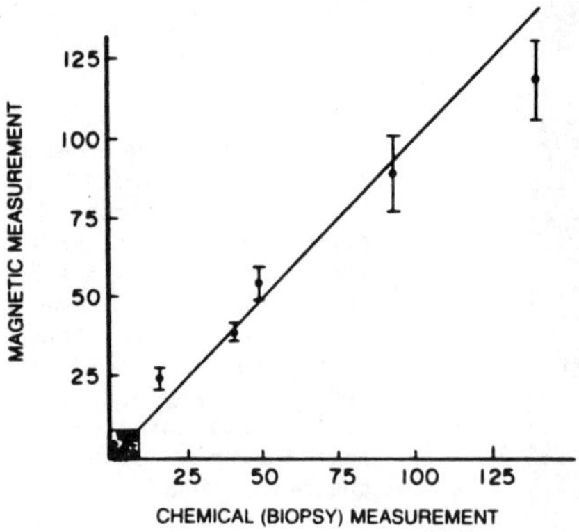

Fig. 7. Measurements of liver iron concentrations obtained magnetically (vertical axis) track very nicely those obtained by chemical analysis of biopsy tissue (horizontal axis). Concentrations are indicated in micromoles of iron per gram of liver. (from Brittenham and colleagues, 1982)

The method developed by Harris and his colleagues (1978), shares common elements with an earlier technique introduced by Wikswo et al. (Wikswo, 1980; Wikswo and colleagues, 1980) to measure the magnetic susceptibility of the heart. A magnet wound from superconducting wire is mounted within the tail section of a dewar, and nestled within its central space is the detection coil of a SQUID sensor. The sensor is placed directly above the liver of the patient, and then the patient is raised toward the sensor while it records the change in field produced by the patient's body. There is a clever trick to minimize complications due to effects of the shape of the torso: the space between the patient and sensor is occupied by water supported by a rubber bag, so that when the patient is raised the susceptibility of the water is replaced by the susceptibility of the patient. Thus any change in the sensor output is due largely to the iron within the tissue underneath and not to the normal body tissue itself.

The accuracy of this technique has been verified by comparing the magnetic estimate with that of the traditional biopsy analysis, as illustrated in Fig. 7. This shows data for five patients following reasonably well a straight line at a 45 degree angle, indicating that the two measures of iron concentration are similar. In the lower left corner is a shaded area indicating the upper limit for iron concentrations in normal individuals.

With the success of this technique now established for determining substantial iron overloads, the real frontier in liver studies has become further development of the procedure to improve its accuracy in determining iron concentrations just above, and perhaps even below, the normal range. There is substantial interest in this capability for widespread applications in diagnosis of a wide range of disorders which affect the iron balance of the body.

Particles in the Lungs

One of the first biomagnetic applications of practical importance is measurement of the magnetic field from magnetic particles in the lungs. The effect is easily detected in workers of certain occupations where the air they breathe contains high concentrations of iron or steel-bearing material. Iron foundries, arc welding shops, and machine shops are examples. To detect the presence of magnetic particles it is first necessary to apply a strong magnetic field over the region of the lungs, so that the particles become magnetized in that common direction. This common alignment of the particle's magnetization produces a magnetic field, which may be detected most easily when the applied field is removed so that a sensitive detector can be brought near the chest. Then a scan of the detector over the chest gives information about the amount and distribution of the magnetic material.

This technique, which was developed by Cohen (1973) in the early 1970s, has been applied to the study of the occupational health of humans (Kalliomaki and colleagues, 1983). Some of these dusts, such as asbestos dust to which miners are exposed, are considered harmful; whereas other materials such as finely divided magnetite (the lodestone of the Chinese!) are generally considered benign. Use of magnetite in experimental studies with humans and animals such as rabbits and hamsters is leading to some important findings about the mechanisms by which the lungs are cleansed of dust. This is illustrated by a remarkable feature that can be seen once the applied magnetizing field is removed: the magnetic field from the lungs declines in strength over a period of several minutes, first changing very rapidly and later much more slowly. An example is shown in Fig. 8, as observed only 20 minutes after breathing the dust, then 8 hours after, and finally 10 days after. This so-called "relaxation" of the magnetization within the lungs was interpreted in the early days as due to rotation of the particles owing to the buffeting they receive from respiratory movements. The ramdomizing of the particles' directions of magnetization means that the magnetic fields of the individual particles no longer reinforce, so the field external to the chest decreases in strength. Now, however, it is recognized that after a few hours of exposure particles in the lungs are no longer free, because they are engulfed by nature's garbage trucks, cells called "phagocytes" which move about on the alveolar surface. There is evidence that most of the relaxation seen a few hours after exposure is due to the motion of

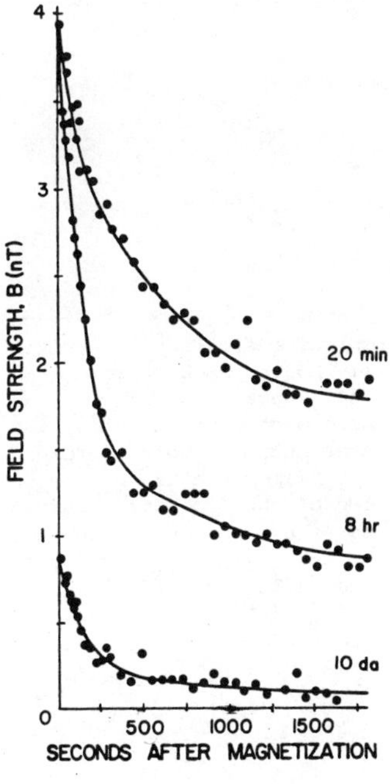

Fig. 8. Decline of the magnetic field strength outside the chest of a rabbit during the first 30 minutes following removal of the magnetizing field, recorded 20 minutes, 8 hours, and 10 days after inhalation of magnetite dust. The field strength is given in the units of nanotesla, which may be compared with the earth's steady field of about 50 microtesla, some 10,000 times stronger. (from Brain and colleagues, 1984).

the particles <u>within</u> the individual phagocytes (Gehr and colleagues, 1983). Thus, studies of relaxation provide the intriguing prospect of determining the condition of phygocytes without removing them from the lungs - a study that was not previously possible. The fact that most particles are taken up by phagocytes explains why relaxation shown in Fig. 8 is more important 8 hours or more after exposure than only 20 minutes after, because rotation of a particle within a phagocyte is less hindered than when the particle makes direct contact with the alveolar surface.

Figure 8 reveals another interesting feature when we examine the initial field right after the magnetizing field is removed (zero seconds after magnetization). This initial field diminishes over a period of days, decreasing from nearly 4 nanotesla to less than 1 nanotesla after 10 days. A reasonable explanation is that particles are being cleared from the lungs over this period, many being carried by phagocytes up the mucociliary escalator of the respiratory tract, eventually to be swallowed. Magnetically it is thus possible to follow the clearance of magnetic particles to learn how this ability differs among individuals and with changing conditions of the lungs.

One example illustrating how clearance measurements may be important is given in Fig. 9. The curves show how clearance of magnetite particles from smokers' lungs is greatly slowed in comparison with clearance from a group of nonsmokers. This indicates that smokers are subject for a longer time to the insult imposed by the particles, which may reasonably be viewed as a greater health risk. Similar studies on other populations exposed to harmful conditions are being carried out by numerous groups around the world, so this is a very active frontier of biomagnetic activity.

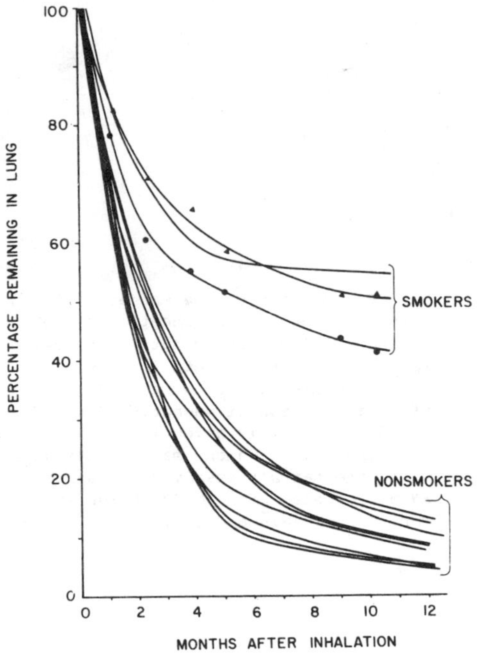

Fig. 9. Long-term clearance curves indicating that after 11 months smokers retain in the lungs about five times more magnetite dust than the nonsmokers. Dots represent the experimental points on two of the smokers' curves to show the typical scatter of the data. (from Cohen and colleagues, 1979)

Excitations in Cells

The nervous system carries communications over long distances by means of a sequence of pulses, or excitations, called action potentials. Similarly, contraction of muscle fibers such as those in the heart begins with the propagation of an action potential along the length of the individual cells. An action potential is launched within such active cells when the cell is appropriately stimulated by electrical or chemical means. Understanding the action potential is therefore basic to electrophysiology. Through technical developments by Wikswo and his colleagues (1980a) it is possible to detect and study the magnetic field associated with the current of an action potential of a single nerve axon, nerve bundle, or muscle fiber. To bring the sensing coil as close as possible to, say, an axon in order to detect its weak field, a small doughnut-shaped sensing coil at room temperature is used so that the axon may be threaded through it.

In the earliest experiments the action potential was initiated by an electrical pulse applied near one end of the axon. The leading portion of the action potential consists of a short length of current flowing within the axon in the direction of motion, with an accompanying backflow in the surrounding saline solution in which the axon is immersed. This current is associated with the "depolarization" front of the action potential where the voltage across the thin membrane of the axon quickly decreases from its resting value, passing through zero to momentarily reverse its polarity. Then follows a similar pattern with opposite flow associated with the "repolarization" portion of the action potential when the voltage across the membrane is re-established to its resting value. The current pattern within the axon can then be modeled by two current dipoles, one pointing forward in the direction of motion followed by one pointing backward. Consequently, the magnetic field of the action potential is found to first have one direction and then the opposite direction as the action potential moves through the sensing coil.

The rapid change in magnetic field as the action potential passes along the axon induces a voltage in the sensing coil. In the original experiments the current from this voltage was sent along a copper wire to a second coil wrapped around the tail of a dewar, where a detection coil and SQUID were located. Now, however, a high-sensitivity electronic amplifier can detect the induced voltage directly, without recourse to a SQUID system. By measuring how the field varies with time it is then possible to calculate the corresponding time variation of the voltage across the axon's membrane. Comparing this prediction with actual measurements of the transmembrane voltage provides an important test for our understanding of the source of the observed magnetic field. Figure 10 compares this prediction with the observed voltage for the crayfish axon. The predicted rise and fall of the voltage as the action potential passes a given position is in excellent agreement with the observed variation, aside from a difference in amplitude. This difference in fact is probably due to an inaccurate value for the resistivity of the material within the axon. Indeed, it may very well be that such a comparison provides the most accurate measurement for this important parameter, when the value is adjusted to bring the predicted and observed voltages into agreement.

This technique is useful not only for addressing fundamental questions about the action potential, and possibly other types of cellular response, but also may turn out to be a useful clinical tool. Wikswo and his collaborators have developed a new sensing coil with a segment removed from its doughnut shape so that it can be clamped around the middle of a nerve which remains intact at both ends in the body . The idea is that in cases of injury to a peripheral nerve in the body, a physician can free a short section of the damaged region and

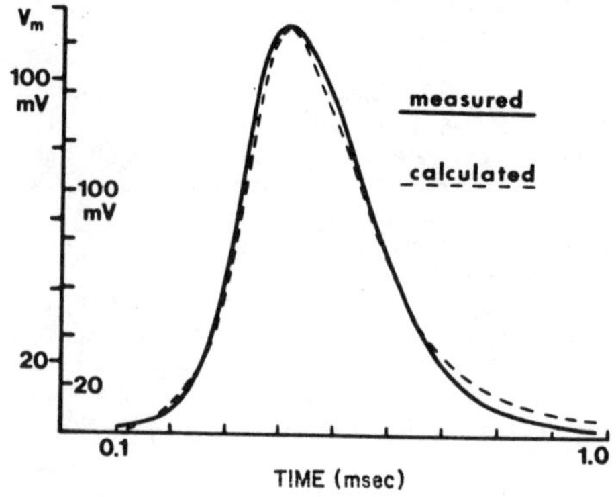

Fig. 10. Observed increase and subsequent decrease of the transmembrane voltage of the action potential of the crayfish axon. The electrically measured values refer to the scale on the outside of the vertical axis, and values calculated from magnetic measurements refer to the scale on the inside. (from Roth and colleagues, 1984b)

magnetically test how well it functions, without having to manipulate it as extensively as would be required with electrodes used with conventional electrical techniques. Clinical research on this approach is just beginning (Wikswo and colleagues, 1984).

Heart Activity

During the early development of biomagnetism more effort was devoted to studies of the heart than to any other organ. This was due in part to the fact that the currents in the heart are the strongest of any organ, and so the signal is comparatively easy to detect magnetically. Figure 11 shows the normal electrocardiogram through almost two heart cycles. When the corresponding magnetic signal is monitored, strengths and polarities of the individual features vary across the chest, indicating where the field perpendicular to the chest is strongest and whether it is directed outward or inward. The QRS peak is due to the depolarization front of the action potential as the leading edge of the current spreads out over the heart muscle of the ventricles, thereby initiating contraction. This pumps blood into the arteries. The QRS peak is followed by a smaller, broader peak called the T wave which accompanies the subsequent repolarization of the muscles. There is then a more or less horizontal portion of the waveform followed by a small wave denoted by P. The P wave is associated with the depolarization front of the action potential spreading over the muscle of the atria, causing them to contract and force blood into the ventricles. Shortly after, the atrial muscles repolarize, and then the process continues with another cycle of QRS and T waves.

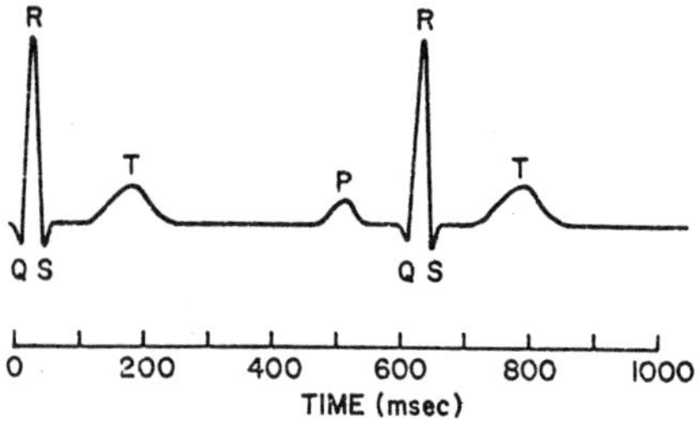

Fig. 11. Features of the normal electrocardiogram.

Although there appears to be considerable information in the pattern of waveforms of the magnetocardiogram, because they vary from one position to another across the chest, in fact we have not yet learned significantly more about normal heart function than can be gleaned from the electrocardiogram, at least as presently analyzed. Certain anomalies are emphasized differently in the MCG than ECG for abnormal heart function (Fenici, 1983; Saarinen and colleagues, 1984), but in those cases where the magnetic record provides unique information there are generally other diagnostic tools that provide equivalent information. Consequently, the MCG has so far been of little use in clinical applications. Indeed, if the overall heart function can be represented electrically by a current dipole changing in strength and orientation throughout the cycle, then in principle no new information is available from the MCG than is provided by the ECG (Katila and Karp, 1983; Wikswo, 1983). However, because the heart's size is comparatively large in comparison with its distance to a magnetic sensor placed just outside the chest, the current dipole alone is not an adequate model. The actual spatial distribution of the current pattern along the muscle fibers must be taken into account; in this case the MCG can at least in principle provide information not available in the ECG. Indeed, a degree of independence recently has been demonstrated (see for example MacAulay and colleagues, 1984), and it is of considerable interest to see how far this new information can be exploited.

One of the potential advantages of cardiomagnetism is only just being realized: the possibility of localizing the position of abnormal activity in the heart from the observed field pattern. By explicitly recognizing the heart is an extended organ, researchers may obtain new information by the magnetic technique. One example being explored is in determining the location of pathology in heart muscle (Erne and colleagues, 1984; Gonelli and colleagues, 1984) from the spatial pattern of the anomalous magnetic signal. Another potential advantage is in localizing disfunction in the heart's conduction system, which carries the action potential from the pacemaker at the atria downward to the left and right ventricles. This latter possibility was based on an observation by Farrell and co-workers (1980) of a weak signal appearing in the segment of the MCG between the end of the P wave and the beginning of the QRS wave. The variation of this ramp-like signal with position is shown in Fig. 12. Its polarity reverses across a line that runs diagonally across the heart.

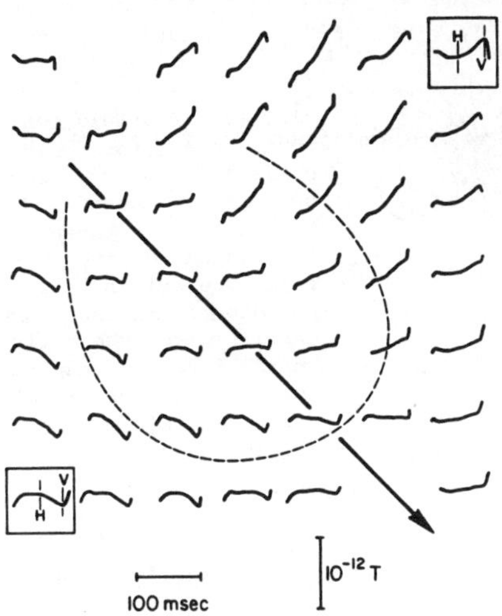

Fig. 12. Weak, ramp-like signal (between H and V labels in the boxed prototypical signals) observed in the P-R interval of the magnetocardiogram displays a reversal of polarity across the diagonal arrow. (from Farrell and colleagues, 1980)

This is approximately the direction taken by the His bundle and Purkinje fibers of the conduction system as they run downward from the artia along the septum dividing the ventricles, to then spread out over the ventricular surface. Therefore this feature in the magnetocardiogram was identified with the action potential passing along the conduction system from the atrial-ventricular node to the ventricles.

However a number of questions have arisen concerning this interpretation, because a portion or all of the signal may be due to repolarization of the atria which occurs at about the same time. Fenici and colleagues (1984) and Lorenzana and colleagues (1984) describe the most recent studies bearing on this problem. Interest in the conduction system is high, because there is a greater incidence of cardiac failure from arrythmias than from abnormal functioning of heart muscle, and many arrythmias involve the conduction system. To be able noninvasively to selectively monitor activity at various places between the pacemaker and Purkinje fibers, which terminate on the ventricles, would be a valuable clinical procedure. The MCG may turn out to serve in this capacity as a a means for diagnosis of precursors to arrythmias, which presently can be done only by insertion of a catheter through a blood vessel into the heart.

Brain Studies

Brain research is the most rapidly growing area of biomagnetism. The advantage of the magnetic approach is clear: in many situations neuromagnetic measurements can localize active regions of the brain where other approaches have failed. Therefore the magnetic technique provides advantages in basic research on brain function and in clinical applications for such diseases as focal epilepsy.

In Fig. 4 we already gave an example of localization as applied to the portion of the surface layer of the brain (the cortex) dealing with sensory inputs from the skin. The ability of neuromagnetism to discriminate between such primary sensory areas and other, perhaps deeper lying, regions has been also demonstrated for the visual and auditory areas as well. These studies have been recently reviewed by Okada (1983). One example showing the direction in which such research is going is typified by auditory studies (see Fig. 13). The

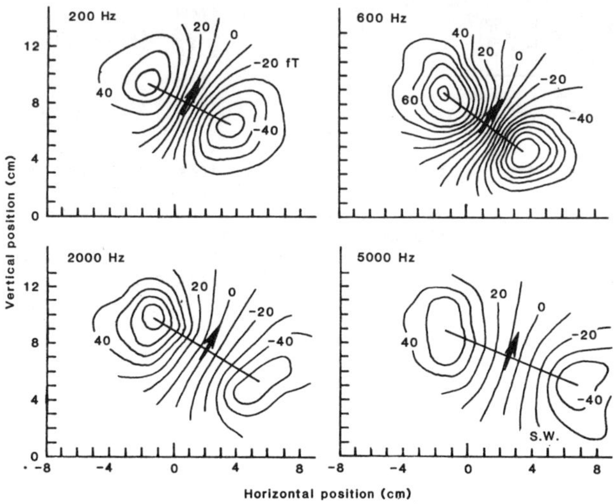

Fig. 13. Isofield contours over the right hemisphere for responses at the side of the head with tones of the indicated frequencies. The origin of the coordinates is the ear canal, with the horizontal axis running toward the corner of the eye. (from Romani and colleagues, 1982)

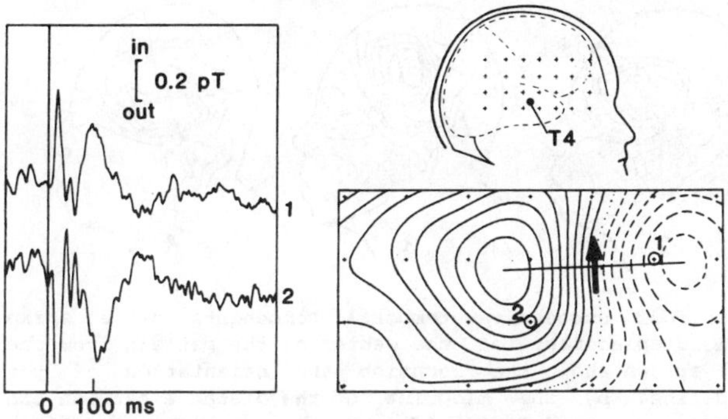

Fig. 14. Averaged magnetic responses to dental stimulation (left panel), measured at positions 1 and 2 on the isofield contour map for the component at 100 ms (right panel). The grid on the map corresponds to positions shown on the head profile. (from Hari and colleagues, 1984)

question being addressed is whether specific areas of human auditory cortex are devoted to identifiable features of an auditory stimulus, as has been found in several species of animals. The plots in this figure provide a definitive "yes", in that they show the source of the field pattern for a tone lies deeper within the Sylvian fissure for tones of higher frequency (Romani and colleagues, 1982). In this experiment the subject listened to a tone whose amplitude was slowly modulated, and a magnetic field outside the head was detected at the modulation frequency, corresponding to activity of a population of neurons that was slowly turned on and off at that frequency. The distance between the positions of maximum inward and maximum outward field increased systematically with tone frequency, indicating that the underlying current dipole modeling the source lies at greater depths beneath the scalp. Thus the auditory cortex has a "tonotopic" representation, with responses to tones of different frequencies appearing at various positions along a path of several centimeters length. The localization power of the neuromagnetic method was extended by Pelizzone and co-workers (1984) to show that other locations of the cortex can be identified with field sources evoked by an abrupt onset of a tone.

Recent studies show that the magnetic technique is capable of detecting a wide range of signals that have electrical analogs. These include the "contingent negative variation," a signal that anticipates by several hundred milliseconds the performance of a task (Weinberg and colleagues, 1984); a "readiness field" detected prior to voluntary movement of a part of the body (Deeke and colleagues, 1982, 1984); and the "P300," a signal detected 300 to 450 milliseconds following a stimulus that does not correspond with the subject's expectation (Okada and colleagues, 1982). The P300 study also showed that sources deep within the brain can be monitored magnetically. Space restrictions unfortunately preclude our discussing these interesting studies in detail.

Work by Hari and co-workers (1983, 1984) has opened a new frontier in pain research through their studies of the responses to electrical stimulation of a tooth. It could very well turn out that the strength of the magnetic response may provide an objective measure of a painful stimulus, which could then afford an important control in pain research. Figure 14 shows the observed magnetic response to a brief current pulse to an upper incisor. The polarity changed between the anterior and posterior regions over the anterior portion of the temporal area, thus allowing localization of the source to a position known as the temporal opper culum. This suggests that there is a region of the cortex anterior to the Rolandic fissure where tooth pulp is represented in humans. Control experiments indicated that the responses were transmitted to the brain by neural pathways that elicit pain, and consequently this area may have a specific function related to painful experiences. The relationship of the level

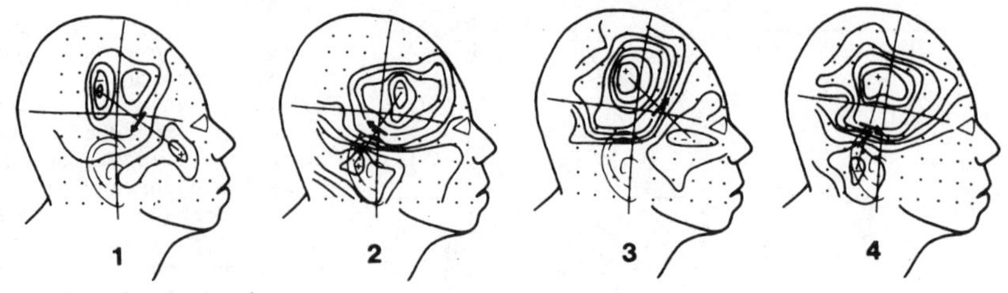

Fig. 15. Isofield maps for four successive temporal components of a spike complex, showing lateral displacement of the center of the pattern from one component to the next. The arrows show the location and orientation of the current dipole source defined by the midpoint of the distance between the positions of maximum outward and inward field. (from Barth and colleagues, 1984)

of activity in this region to the type and degree of perceived pain is now being explored.

Another rapidly advancing frontier is the application of magnetic studies to studies of localized epileptic activity. Here the hemispheric localization in three-dimensions comes into play to guide the neurosurgeon in those cases which are appropriate for surgical intervention to aliviate the condition. To obtain records of acceptable qualtiy, however, it is necessary to average together many repetitions of the activity to enhance the signal relative to the ambient field noise. One example from work at U.C.L.A. by Barth and colleagues (1984) is shown in Fig. 15 where the activity during a short-duration "spike" complex was

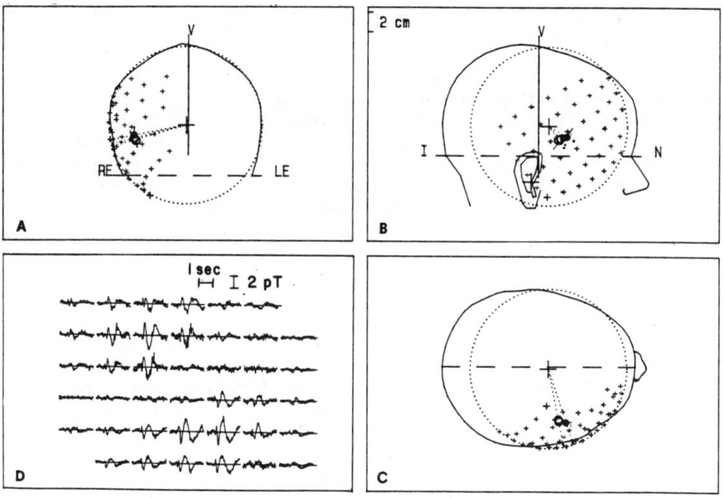

Fig. 16. Coronal (A), sagittal (B), and transverse (C) cross sections of the head, with a spherical model of the head (dotted outline) used to deduce the location of the source of the time-varying signals (D) measured at each of the grid points indicated by crosses. (from Ricci and colleagues, 1984)

averaged over 1000 spikes, using the simultaneous electrical spike signal as a trigger for the averaging computer. Isofield maps for four successive positive and negative components of the complex show that two separate locations displaced laterally are responsible, the more anterior one producing the first and third components and the more posterior and inferior one producing the second and fourth. These non-invasive measures provide a detailed characterization of discharge patterns between seizures. In some cases, where CT scans are available, the localization of activity has been found to correlated well with the presence of a tumor, thereby providing an independent measure of the location of pathology.

In Rome, Ricci and co-workers (1984) obtained direct surgical confirmation of the localizing ability of the magnetic technique. In some cases a novel averaging technique was used to improve the signal to noise ratio. Simply put, they first determined the degree to which components of the neuromagnetic signal at positions across the scalp are in step with components of the electrical voltage measured between two scalp electrodes. Since the ECG thus provides a time reference for comparison with the MCG, this measure of the "covariance" between the signals indicates where the field reverses sign across the scalp and the positions of strongest outward and inward field. The result of this process is illustrated in panel D of Fig. 16 for measurement positions indicated by the crosses in panels A, B, and C. Two regions of strong magnetic signals of opposite polarities are apparent in panel D, with the location of the best-fitting current dipole being indicated by the circles in panels A, B, and C. This location was found to correspond to the center of the region of pathological tissue, which was subsequently found in the temporal lobe during surgery. Ricci and colleagues conclude that the relative covariance method provides a significant degree of localization for sources oriented tangential to the overlying scalp.

For these reasons neuromagnetic studies have attracted much attention, and it seems that the frontiers will continue to expand as further progress is made in basic and clinical research.

The Egg: Where It All Begins?

Studies just described of individual cells, the liver, lungs, heart, and brain have evolved from work begun five or more years ago. These may be considered mature areas of research, although their cutting edges represent modern frontiers. Meanwhile other biomagnetic effects continue to be discovered. To illustrate one example, we shall mention the recent observation by Grimes and co-workers (1984) of a steady magnetic field near a developing chick egg. It is known that a number of developing organisms produce steady or slowly varying currents which can be measured by sensitive electrodes. Such voltages can be detected in the chick egg after about 15 hours of incubation, and the corresponding magnetic fields after about 40 hours. The underlying current source appears to lie parallel to the long axis of the egg. While the relationship of this field to the embryo's development remains to be established, this observation opens the possibility for many other applications of SQUID sensors to the non-invasive study of developmental currents in various organisms.

PERSPECTIVE

This brief appraisal of the frontiers of biomagnetism cannot do justice to the many fine studies that we did not have space to discuss. We have touched only some of the high points. In sum there are several exciting areas where contributions of considerable significance have been made and much more can be expected in the future. There are some areas, such as cardiac studies, where the value of the magnetic technique has been slower to appear. And finally, there are some more speculative areas, such as the investigation of developmental currents where work is just beginning but the prospects appear bright. In all, there are so many interesting topics that more researchers of inquisitive and innovative minds would be welcome. The rapid advance of technology, especially the introduction of multiple-sensor systems, is greatly enhancing the efficiency of magnetic measurements. This coupled with more sophisticated use of computerized signal processing techniques is opening opportunities for imaginative studies that could not have been dreamed just a few years ago. This book provides just a glimpse of what the future offers.

ACKNOWLEDGEMENTS

We thank D. Barth, Y. Okada, G. Stroink, and J.P. Wikswo, Jr., for informative discussions and helpful comments on this manuscript.

REFERENCES

* Supported by Air Force Office of Scientific Research Grant F49620-82-K-0014, National Institutes of Health Grant NS-19463-01, and Office of Naval Research Contract N00014-76-C-0568.

Barth, D.S., W. Sutherling, J. Engle Jr., and J. Beatty (1984). Neuromagnetic evidence of spatially distributed sources underlying epileptiform spikes in the human brain. Science, 223, 293-296.

Baule, G.M. and R. McFee (1963). Detection of the magnetic field of the heart. Am. Heart J., 66, 95-96.

Brain, J.D., S.B. Bloom, P.A. Valberg, and P. Gehr (1984). Experimental Lung Research, 6, 115-131.

Brittenham, G.M., D.E. Farrell, J.W. Harris, E.S. Feldman, E.H. Danish, W.A. Muir, J.H. Tripp, and E.M. Bellon (1982). Magnetic-susceptibility measurement of human iron stores. New England Journal of Medicine, 307, 1671-1675.

Brenner, D., J. Lipton, L. Kaufman, and S.J. Williamson (1978). Somatically evoked magnetic fields of the human brain. Science, 199, 81-83.

Cohen, D. (1973). Ferromagnetic contamination in the lungs and other organs of the human body. Science, 180, 745-748.

Cohen, D. (1975). Magnetic fields of the human body. Physics Today, 28, 34-43.

Cohen, D. (1976). Report of the low-field group: the magnetocardiogram. Publication MIT/FBNML-76/1, National Technical Information Service, Springfield, VA, 22161, U.S.A.

Cohen, D., S.F. Arai, and J.D. Brain (1979). Smoking impairs long-term dust clearance from the lung. Science, 204, 514-517.

Darnton, R. (1970). Mesmerism and the End of the Enlightenment in France, Schocken Books, New York.

Deeke, L., H. Weinberg, and P. Brickett (1982). Magnetic fields of the human brain accompanying voluntary movement: Bereitschaftsmagnetfeld. Exp. Brain Res. 48, 144-148.

Deeke, L., J. Boschert, P. Brickett, and H. Weinberg (1984). Evidence from the magnetoencephalogram for supplementary motor area participation in human voluntary movement. This volume.

Erne, S.N., H.-D. Hahlbohm, J. Korsukewitz, and H.P. Lehmann (1984). Disciminating late potentials using biomagnetic techniques. This volume.

Farrell, D.E., J.H. Tripp, and R. Norgren (1980). Magnetic study of the His-Purkinje system in man. IEEE Trans. Biomed. Engrng., BME-27, 345-350.

Fenici, R. (1983). Clinical assessment of the magnetocardiogram. In Williamson et al. (1983), pp. 285-298.

Fenici, R., G.L. Romani, S.N. Erne (1983). High-resolution magnetic measurements of human cardiac electrophysiological events. Il Nuovo Cimento, 2D, 231-247.

Fenici, R., M. Masselli, S.N. Erne, and H.-D. Hahlbohm (1984). Magnetocardiographic mapping of the P-R interval phenomena in an unshielded hospital laboratory. This volume.

Gehr, P., J.D. Brain, S.B. Bloom, and P.A. Valberg (1983). Magnetic particles in the liver: a probe for intracellular movement. Nature London, 302, 336-338.

Gonelli, R., P. Galeone, M. Sicuro, and A. Tartaglia (1984). Magnetocardiographic isofield mapping in the characterization of the infarcted area. This volume.

Grimes, D.I.F., B.S. Janday, R.F. Lennard, and S.J. Swithenby (1984). DC magnetic fields of developing chick embryos. This volume.

Hari, R., E. Kaukoranta, K. Reinikainen, T. Huopaniemi, and J. Mauno (1983). Neuromagnetic localization of cortical activity evoked by painful dental stimulation in man. Neurosci. Lett., 42, 77-82.

Hari, R., E. Kaukoranta, K. Reinikainen, and J. Mauno (1984). Neuromagnetic responses to painful dental stimulation. This volume.

Ilmoniemi, R., R. Hari, and K. Reinikainen (1984). A four-channel SQUID magnetometer for brain research. Electroenceph. Clin. Neurophysiol., in press.

Kalliomaki, P-L., et al. (1983). Magnetopneumography. In Williamson et al. (1983), pp. 533-568.

Katila, T., and P. Karp (1983). Magnetocardiography: morphology and multipole presentations. In Williamson et al. (1983), pp. 237-263.

Lorenzana, H.E., P.B. Pipes, M.P. Zaitlin, and D. James (1984). Structures in the P-R interval of the human magnetocardiogram. This volume.

MacAulay, C.E., G. Stroink, and B.M. Horacek (1984). Signal analysis of magnetocardiograms and electrocardiograms to test their independence. This volume.

Okada, Y., L. Kaufman, and S.J. Williamson (1982). Hippocampal formation as a source of endogeneous slow potentials. Electroenceph. Clin. Neurophysiol., 55, 417-426.

Okada, Y., (1983). Brain studies. In Williamson et al. (1983), pp. 399-468.

Ricci, G.B., R. Leoni, G.L. Romani, F. Campitelli, S. Buonomo, and I. Modena (1984). 3-D neuromagnetic localization of sources of interictal activity in cases of focal epilepsy. This volume.

Roth, B.J., J.E. Woosley, and J.P. Wikswo, Jr. (1984a). An experimental and theoretical analysis of the magnetic field of a single axon. This volume.

Roth, B.J. and J.P. Wikswo, Jr. (1984b). The magnetic field of a single nerve axon: a comparison of theory and experiment. Submitted for publication.

Siltanen, P., T.E. Katila, T. Poutanen, M. Seppanen, and T. Varpula (1983). Comparison between measured and simulated magnetocardiograms in a case of anterolateral myocardial infarction. Il Nuovo Cimento, 2D, 301-310.

Weinberg, H., P.A. Brickett, J. Vrba, A.A. Fife, and M.B. Burbank (1984). The use of a Squid third order spatial gradiometer to measure magnetic fields of the brain. Annals New York Acad. of Sci., 425, 743-752.

Wikswo, J.P. Jr.(1980). Noninvasive magnetic detection of cardiac mechanical activity: theory. Med. Phys., 7, 297-306.

Wikswo, J.P. Jr.(1983). Theoretical aspects of the ECG-MCG relationship. In Williamson et al. (1983), pp. 311-326.

Wikswo, J.P. Jr., J.P. Barach, and J.A. Freeman (1980a). Magnetic field of a nerve impules: first measurements. Science, 208, 53-55.

Wikswo, J.P. Jr., J.E. Opfer, and W.M. Fairbank (1980b). Noninvasive magnetic detection of cardiac mechanical activity: experiments. Med. Phys., 7, 307-314.

Wikswo, J.P. Jr., G.S. Abraham, and V.R. Hentz (1984). Magnetic assessment of regeneration across a nerve graft. This volume.

Williamson, S.J. and L. Kaufman (1981). Biomagnetism. J. Magn. Magn. Mat., 22, 129-201.

Williamson, S.J., G.L. Romani, L. Kaufman, and I. Modena, Eds. (1983). Biomagnetism: An Interdisciplinary Approach. Plenum Press, New York.

Williamson, S.J., M. Pelizzone, Y. Okada, L. Kaufman, D.B. Crum, and J.R. Marsden (1984). Five channel SQUID installation for unshielded neuromagnetic measurements. This volume.

Zimmerman, J.E. (1972). Josephson effect devices and low-frequency field sensing. Cryogenics, 12, 19-31.

AUTHOR INDEX

CONFERENCE PARTICIPANTS

Richard Abbou
115 Rue Du La Porte
75116 Paris
France

Y. Aizawa
Kitasato University
School Of Medicine
Kanaigawa, Japan

R. Bain
Sprsthelyde
Dept of Applied Medicine
Glasgow, Scotland

Bryan Barnard
G.E.C. Hirst Research Centre
East Lane
Wembley, Middlesex, England

Jackson Beatty
Department of Psychology
U.C.L.A. 405 Hilgard Avenue
Los Angeles, Cal. USA 9004

A. Blackburn
Navy Personnel Research & Dev.
Centr., San Diego
California, USA 92152

J.D. Brain
Harvard Sch. of Public Health
665 Huntington Ave.
Boston, Mass. USA 02115

A.C. Bruno
Pontif'icia Universidade
Cato'lica, Cx.P. 38071
Rio de Janeiro, RJ, Brazil

Jr. Cameron
Univ. of Wisconsin-Madison
Dept of Medical Physics
Madison, WI. USA 53706

Robert M. Chapman
Center for the Visual Sciences
University of Rochester
Rochester, N.Y. USA 14627

Antti I. Ahonen
Helsinki University of Tech.
02150 Espoo
Finland

Mario Baff
Brain Behaviour Laboratory
Simon Fraser University
Burnaby, B.C. Canada V5A 1S6

Janday Balwinder
The Open University
Walton Hall, Milton Keynes
MK 7 6AA, England

Daniel S. Barth
U.C.L.A., Dpt. Psychology
405 Hilgard Avenue
Los Angeles, Ca. USA 9004

William C. Black
S.H.E. Corporation
4174 Sorrento Valley Blvd.
San Diego, Cal. USA 92121

Jurgen Boschert
Army Hospital
Ulm,
West Germany

P. Brickett
Brain Behaviour Laboratory
Simon Fraser University
Burnaby, B.C. Canada, V5A 1S6

M. B. Burbank
C.T.F. Systems, Inc.
15 1750 McLean Ave.,
Port Coquitlamd, B.C. Canada

Paul J. Carroll
Sperry Corporation Mail-U1H14
Sperry Park, P.O. Box 64525
St. Paul Minn. USA 55164-0525

Keizo Chiyotani
Rosai Hospital for Silicosis
623 Fujowara Tochigi
821-25, Japan

Alice Chmielowska
Australia Medical Association
106 Hortoh Street, Leichardt
Sydney, NSW Australia

David Cohen
Mass. Inst. of Technology
170 Albany Street
Cambridge, MA 02139

F. Coolsma
Brain Behaviour Laboratory
Simon Fraser University
Burnaby, B.C. Canada, V5A 1S6

Duane Crum
S.H.E. Corporation
4174 Sorrento Valley Blvd.
San Diego, CA 92121

Neil Cuffin
Mass. Inst. of Technology
170 Albany Street
Cambridge, Ma 02139

C. Cuza
Desert Psychological Services
107 South 5th St., Suite 231
El Centro, Cal. USA 92243

L. Deecke
Univ. of Ulm, Dep. Neurology
Steinhovelstr. 9
D-7900 Ulm., West Germany

J.M.R. Delgado
Centro Ramon Y Cajal
CTRA De Colenar, Km.9
Madrid-34, Spain

P. Descouts
Institut De Physique
24 Quai E Ansermet
CH 1211 Geneve 4, Switzerland

Roy Douglas
Dept. of Mathematics
Univ. of British Columbia
B.C. Canada, V5A 2L2

Mark Dowding
C.T.F. Systems, Inc.
15 1750 McLean Ave.,
Port Coquitlam, B.C., Canada

Denis Duret
CEA/CEN-G-LETI/MEM
85X 38041 Grenoble Cedex
France

A. Edwards
B.C. Research,
3650 Westbrook Mall
Vancouver, B.C. Canada V6S 2L2

Sergio Erné
Physikalisch-Technische
Bundesanstalt-1B, Abbest. 2-12
1 Berlin 10, Germany

R.R. Fenici
Universita' Cattolica del S.
Facolta' di Medicina
e Chirurgia, 00168 Roma, Italy

Allan A. Fife
C.T.F. Systems, Inc.
15 1750 McClean Ave.
Port Coquitlam, B.C. Canada

E.R. Flynn
Los Almos Nat'l Lab
MS D443 Box 1663
Los Almos, N.M. USA 87545

Mary Forehend
Amonea Corporation
Suite 116-2150 New Market
Marietta, Georgia, USA 30067

Alan P. Freedman
Hahnemann Med. Coll. Hospital
230 North Broad Street
Philadelphia, PA USA 19202

Alfred Fregly
Department of the Air Force
AFSC, Bolling Air Force Base
D.C., USA 20332

Peter Gehr
Anatomisches Institut
Univ. Berne, Postfach 139
CH-3000 Bern 9, Switzerland

R. Gonnelli
Politechnico di Torino
c.so Duca degli Abruzzi 24
10129 Torino, Italy

E. Gordon
Westmead Center
Dept. of Psychiatry
Westmead NSW 2145, Australia

J. Guardian
7300 Ave. LA, Costa, N.E.
Albuquerque, New Mexico
USA

H.D. Hahlbohm
Physikalische-Technische
Bundesanstalt-1B, 1 Berlin 10
West Germany

A. Hardt
1052 Rhode Island Rd.
San Francisco, Ca
USA 94107

Charles P. Hatsel
U.S. Air Force Aerospace
Wright-Patterson AFB AFAMRL/CV
Ohio, USA 45433

Joachim F. Hohnsbein
Institut fur Arbeitsphysiol.
Ardeystr. 67 D-4600 Dortmund
West Germany

Y. Huang
Nicolet Instruments
Madison
Wisconsin, USA

Toshiaki Imada
Musashino Elec. Comm. Lab.
NTT 3-9-11 Midori-Cho
Musashino-Shi, Tokyo, Japan

B.S. Janday
23 Artemis Close
Gravesend, Kent
England DA12-2QW

K. Kalliomäki
Dept of Electrical Eng.
Oulu University, SF-90570
Pulu, Finland

Toivo E. Katila
Helsinki Univer. of Technology
Dept. of Technical Physics
02150 Espoo 15, Finland

Lloyd L. Kaufman
N.Y. University, Dpt. Psychol.
4 Washington Place,
New York City, N.Y. USA 10003

Axel Klaudius
Phys. Institute
Robert Mayerstr. 2-4
6 Frankfurt, West Germany

Philip Kubik
C.T.F. Systems, Inc
15 1750 McLean Ave.
Port Coquitlam, B.C., Canada

M.S. Hämäläinen
Low Temperature Laboratory
Helsinki Univrsity of Technol.
SF-02150 Espoo, Finland

Riitta Hari
Univer. Hospital of Helsinki
Dept. of Neurology
00270 Helsinki 27, Finland

H.E. Hoenig
Siemens Research Labs
Siemens' ZFE TPH 13
PO Box 3240 D-8520 y Germany

M. Hoke
Westf. Wilhelms Univ.
Kardinal-von-Galen-Ring 10
D-4400 Munster, Germany

Risto J. Ilmoniemi
Helsinki Univ. of Technology
Rakentajanaukio 2,
02150 Espoo 15, Finland

H. Jablonski
S.H.E. GMBH Kryotechnische
Gruner Weg 83,
D-5100 Aachen, Germany

F.W. Jones
Univ. of Alberta
Dept. of Physics
Edmonton, Alt., Canada T6G 2J1

M. Kamali
Zenith Instruments
Toronto Ontario
Canada

Mikio Kato
108 Miyazaki-Cho
Shimokamo, Sakyo-Ku
Kyoto 606, Japan

E. Kaukoranta
Helsinki University of Technol
Low Temp. Laboratory
SF 02150 Espoo, Finland

Makoto Kotani
TokyoDenki University
Kanda-Nishikicho, Chiyoda
101 Tokyo, Japan

J. Lekkala
Tampere University of Technol.
P.O. Box 527, SF-33101
Tampere 10, Finland

Rosemary Lennard
The Open University
Milton Keynes, Bucks
England MK6 3DP

Gregory Lewis
Navy Personnel Research Ctr.
8583 Hydra Lane
San Diego, Cal. 92126

Z. Litman
Danter Electronics
Milepartren 22
2740 Shoulunde, Denmark

H.E. Lorenzana
Dartmouth College
Dept. of Physics
Hanover, N.H. 03755

W. Ludwig
Dornier Systems GMBH
Postfach 1360/7990
Friedrichsha, West Germany

C. MacAulay
Dalhousie University
Dept. of Physics
Halifax N.S. Canada B3H 3J5

J Malmivuo
P.O. Box 527 SF-33101
Tampere 10
Finland

R. Maniewski
Helsinki University of Technol
Dept. of Applied Physics
SF-02150 Espoo 15, Finland

J. Marsden
S.H.E. Corporation
4174 Sorrento Valley Blvd.
San Diego, Ca. USA 92121

Jack McCubbin
C.T.F. Systems, Inc.
15 1750 McClean Ave.
Port Coquitlam, B.C. Canada

C.D. McGillem
School of Electrical Eng.
Purdue University
W. Layfette IND, USA 74907

Douglas McKenzie
C.T.F. Systems, Inc.
15 1750 McClean Ave.
Port Coquitlam, B.C., Canada

James McNeil
C.T.F. Systems, Inc.
15 1750 McLean Ave.
Port Coquitlam, B.C. Canada

H.J. Michalewski
Department of Neurology
University of California
Irvine, Cal. USA 92717

Ivo Modena
Univ. of Rome, Tor Vergata'
Via Orazio Raimouda
Rome, Italy

P. Morgan
Zenith Instruments
Toronto Ontario
Canada

J. H. Nagel
University of Erlangen
Biomed. Engineering Turustr. 5
D-8520 Erlangen, West Germany

I. Nemoto
Tokyo Denki University
Htoyama-machi, Hiki-gun
Saitama-ken, 350-03 Japan

Nicolas P.
Inite Inserm U. 84 (Neuropsyc)
47 BD. De L'Hospital
F. 75651 Paris Cedex 13 France

Deborah Nielsen
4671 Maple Ridge Drive
North Vancouver, B.C.
Canada

J. Nousiaines
Tampere University of Technol.
P.O. Box 527, SF-33101
Tampere 10, Finland

Dr. J. Nousianinen
Tampere University of Technol.
P.O. Box 527
Sf-33101, Tempere 10, Finland

Yoshio Okada
N. Y. Univ. Dpt. Psychology
#6 Washington Place
New York, N.Y. USA 10003

J.E. Orme
Alpha Engineering Ltd.
225 West First Street
North Vancouver, B.C. Canada

Douglas Paulson
S.H.E. Corporation
4174 Sorrento Valley Blvd.
San Diego, Cal. USA 92121

Gerald Perlman
Department of Psychology
D140 Manhattan Psych. Center
Wards Island, N.Y. USA 10035

Maria J. Peters
Twente University of Tech.
P.O.Box 217 - 7500 AE Enschede
The Netherlands

W. H. Pitlick
NIH/Federal Bldg. Room 812
7550 Wisconson Avenue
Bethesda, MD, USA 20205

J. Randall
Alpha Engineering Ltd.
225 W. 1st. Street,
N. VAncouver, B.C. Canada

J. Rechemberg
AMOENA Corporation
Suite 116 2150 New Market Park
Marietta, Georgia USA 30067

Martin Reite
Department of Psychiatry
Box C268 4200 E. Ninth Ave.
Denver, Colorado USA

Paulo C. Ribeiro
Depto Fisica Universidade
Catolica RT CX Postal 38006
22452 Rio de Janeiro, Brazil

G. B. Ricci
Instituto de Neurchirurgia
Val Delle University
30/A, I-Roma Italy

Gian Luca Romani
Instituo di Electronica dello
Stato Solido C.N.R.
V. Cineto Romano 42 Rome Italy

B. Rutt
Univ. of California SF
389 Oyster Point Blvd
S. Sanfrancisco, Ca.94080

M. Saarinen
Cardiovascular Laboratory
Helsink Univ. Central Hospital
SF-00290 Helsinki, Finland

M. Pelizzone
N.Y. University, Dpt. Physics
4 Washington Place
New York City, N.Y. USA 10003

B. Persson
Lund University
Lasarettet S-22185
Lund, Switzerland

P.B. Pipes
Dartmouth College Dpt. Physics
Hanover, New Hampshire
U.S.A. 03755

P Popovich
Lucliff Place
700 Bay St. Suite 408
Toronto Ont. Canada M5G 1Z6

C.S. Rebert
SRI International
333 Ravenswood Avenue
Menlo Park, Ca. USA 94025

H. Reichenberger
Siemens AG Medical Division
P.O. Box 3260
D-8520 Erlangen West Germany

G. Renato
Politecnico di Torino
Corso Duca Delgli Abruzzi
24-10129 Torino Italy

R. Riccardo
Catholoic University
L.G.A Gemelli 8
00168 Rome Italy

L.J. Rogers
Vancouver General Hospital
Dpt. of EEG 855 W. 12th St.
Vancouver, B.C. Canada

D.F. Rose
National Institutes of Health
Bathesda, Md
USA

W. Rutten
ENT dept. (KNO) Univ. Hospital
10 Rijnsbergerweg 2333 AA
Leiden, The Netherlands

M. Sams
Low Temperature Laboratory
Helsinki University of Technol
SF-02150 Espoo, Finland

S. Sato
Epilepsy Branch
National Institutes of Health
Bethesda, Maryland USA 20205

R. J. Schneider
Sperry Computer Systems
M.S. U2T20/P.O. Box 64525
St. Paul, Minn. USA 55164

Tadashi Shinohara
Hitachi Metals International
2279 De Paul Place
Santa Clara, Cal. USA 95051

G. L. Sloggett
CSIRO, Nat. Measurement Lab.
P.O. Box 218 N.S.W.
2070 Australia

Sorensen R.
Sperry Computing Sys. MSV2 P20
P.O. Box 43525, St. Paul
Minn. USA 55164-0525

Gerhard Stroink
Dalhousie University
Dept of Physics, Halifax
Nova Scotia, Canada B3H 3J5

M. Sullivan
Los Almos Nat'L Lab
MS D443 P-DO, Los Alamos
New Mexico, USA

Stephen Swithenby
The Open University
Walton Hall, Milton Keynes
Bucks, MK6 3DP, England

Norman Tepley
Oakland University
Dept. of Physics
Rochester, Mi, USA 48063

H. Toyotama
Maruzen Oil Research Center
Satte. Kitakatsushika-gun
Saitama 340-01, Japan

Zvonko Trontelj
Physics Department
University of E. Kardelj
61111 Ljubljana Yugoslavia

Shoogo Ueno
Kyushu University
Department of Electronics
Fukuka 812, Japan

Walter Schmid
Fuchsweg 9, D-7914
Pfaffenhoffen
West Germany

K. Schorne
HNO Univ. Klinik. Grosshardern
Morchivuiui Str. 15
8000 Muchen 70 West Germany

Pemtto Siltanen
Leppakertuntie 4C
02120 Espoo 12,
Finland

C. Sonrel
CEA-CENG
LETI-IPM 85X
38041 Grenoble Cedex, France

M. Street
Hahnemann University
Div. of Pulmonary Disease
Philadelphia P.A USA 19102

C. I. J. M. Stuart
Medical School
University of Alberta
Edmonton, Alb. Canada TG6 2J1

William Sutherling
UCLA School of Medicine
Reed Neurological Center
Los Angeles Ca. USA 90024

Pavlov Tekel
Tecknuky CEFV SAV
Drubravska Cesta 9
Bratislava, Czechoslovakia

Mark Tillotson
C.T.F. Systems, Inc.
15 1750 McClean Ave.
Port Coquitlam, B.C. Canada

Z. Trontelj
Univ. E. Kardelj of Ljubja
Yu-61111 Ljubljana P.O.B. 39
Yugoslvia

E. Tward
J. P. L. Cal. Inst. of Techn.
4800 Oak Grove Dr. MS183-901
Pasadena, Cal. USA 91109

P.A. Valberg
Harvard School of Public Healt
Dept of Env. Sci. & Physiology
Boston, Mass. USA 02115

J. Veith
Neuroglischen Klinik der Univ.
8520 Erlangen, Schwabachanlage
Erlangen, Germany

H. Weinberg
Simon Fraser University
Brain Behavior Laboratory
Burnaby, B.C. Canada V5A 1S6

John P. Wikswo Jr.
Vanderbilt University
Dept. of Physics And Astronomy
Nashville, Tenn, USA 37235

Samuel J. Williamson
N. Y. Univ. Dpt. of Physics
#4 Washington Place
New York, USA 10003

Christopher Wood
Veterns Admin. Hospital
W. Spring St.
Westhaven, Conn, USA

James Zimmerman
National Bureau of Standards
325 Broadway - Div. 724.03
Boulder, Colorado USA 80262

Jiri Vrba
C. T. F. Systems Inc.
15 - 1750 McLean Avenue
Port Coquitlam, B.C. Canada

J.J. Wevers-Henke
Th Twente, Dpt. Physics
Room 4254 Postbus 217
7500 AE Enschede, Netherlands

Kerri Wilks
6291 Broadway
Apt. 4A
Bronx, N.Y.C. USA 10471

Calvin Winter
Quantum Technology Corp.
6237-148 St.
Surrey, B.C., V3S 3C3

J.T. Zimmerman
University of Colorado
School of Medicine
Denver, Colorado, USA 80262

Anna Zotov
79 Rue De La Sante
75013 Paris
France

A NEW STATISTICAL METHOD FOR RESOLVING SUPERPOSED SIGNALS

L. J. Rogers* and R. R. Douglas**

*EEG Department, Vancouver General Hospital
**Mathematics Department, University of British Columbia
Vancouver, B.C., Canada

ABSTRACT

Magnetoencephalograms (and electroencephalograms) are known to be composed of several superposed signals and noise (in almost all paradigms). To date, no analytical tool has been developed to separate the superposed signals. The present paper provides a brief sketch of a statistical tool that can resolve superposed signals under certain very general assumptions that are likely to be met by the experimental system.

KEYWORDS

Magnetoencephalogram; event-related field; factor analysis; multivariate statistics; event-related potential.

INTRODUCTION

The purpose of the work described here is to discover and measure linearly independent "signals" produced by distinct generating processes in brains, under the hypothesis that the measured time series are linear combinations of these basic signals, together with additive noise. Thus the methods apply equally well to magnetically or electrically measured time series data, since both follow a superposition law. In fact, the methods apply to a wide variety of types of data in which several basic signals add linearly to produce the observed signal.

The mathematical model employed is (maximum likelihood) factor analysis (MLFA); however, the new methods to be described use factor analysis in a more strictly empirical context than has been achieved in previous applications of the powerful MLFA algorithm. It is well known that factor analysis, as it is usually practiced, has two major indeterminacies: number of generators and rotation. This paper sketches resolutions of these two indeterminacies. Suppose that X is a (pxN) mean zero dataset, with p variables and N samples. We have the model equation: X = LY + E. L is the (pxm) "factor loading matrix" whose columns represent the m basic "signals". Y is the (mxN) matrix of coefficients representing the "amplitude" of each signal for a given sample. E is the (pxN) matrix of uncorrelated noise.

These methods provide an empirical solution for the basic signal vectors which appear as the columns of L. A brief sketch of the three-step method, illustrated by examples of results from each step, will be presented here for one subject. Results from the last step for a second subject will be presented for comparison. Theory, simulations, and an EEG application of these methods have been published in a series of four papers (Douglas and Rogers, 1983, and preprint; Rogers and Douglas, 1984, 1984a).

METHODS

Visual evoked potentials (VEPs) were recorded from each of eight electrode placements (referenced to "linked ears") for data from two experimental subjects. The stimulus was a reversing full-field checkerboard, viewed binocularly; this is a standard stimulus used at the EEG Laboratory, Vancouver General Hospital (Morton Low, Director).

RESULTS

Step 1, Stable Subspace: The first step is to solve for what we term the "stable subspace" of the given dataset. The dataset is divided into (sufficiently large, random) sub-datasets, and the factor analyses of the sub-datasets are compared (pairwise) to find the largest dimension for which the factor analysis "results" are the same (for all sub-datasets). Realizing such comparisons effectively, depends critically on the observation that because of the factor analysis rotation indeterminacy, the actual "result" of a factor analysis computation is a subspace. In other words, each column of the factor loading matrix produced by standard factor analysis algorithms is well-known to be some linear combination of the columns of L (the basic signals) (Morrison, 1976). Thus the columns of this factor loading matrix and L both form bases of the same subspace of p-space (where p is the number of variables). Therefore, we required and defined a comparison measure for subspaces, which we have termed "ambient matrix coherence". Using this ambient coherence measure in cross-validation comparisons, yields an empirical technique for the determination of the number of independent generators in a given dataset, and the corresponding "stable subspace". Here, "stability" refers to the requirement that the factor analysis results are nearly the same for all sub-datasets.

The formula for computing the square of the (non-negative) "Ambient Coherence" (AC) of two (pxm) matrices, B_1 and B_2 (both of rank m), is:

$$[AC(B_1,B_2)]^2 = [\det(B_1^tQB_2)]^2/[\det(B_1^tQB_1)\det(B_2^tQB_2)]$$

where "det" is the (mxm) determinant function and Q is the (pxp) symmetric, positive definite matrix of the ambient inner product. For the remainder of this discussion, we simply set Q = I, the (pxp) unit matrix. Thus:

$$[AC(B_1,B_2)]^2 = [\det(B_1^tB_2)]^2/[\det(B_1^tB_1)\det(B_2^tB_2)] \quad .$$

Notice that AC satisfies precisely the correct (very strong) invariance property:

$$AC(B_1,B_2) = AC(B_1U,B_2V)$$

for arbitrary non-singular (mxm) real matrices, U and V. Thus, AC is invariant of all so-called "oblique rotations", separately in each of its two arguments. This allows resolution of one factor analysis indeterminacy (determination of the number of generators) in a way that is independent of the other factor analysis indeterminacy (rotation). Resolution of the second factor analysis indeterminacy can then follow, under certain very general assumptions.

Each lead's dataset was randomly sorted into 6 bins with equal numbers (870) of (unaveraged) VEP samples. Correlation matrices were then calculated for each bin, and each was analyzed using Maximum Likelihood Factor Analysis, for m = 1,2,...,10. Thus, eight leads provided eight datasets that were analyzed separately for each of two subjects.

Figure 1 contains 15 superimposed "stability graphs" of AC as functions of the number of factors (dimension) for each pair of factor analysis solutions. The 15 graphs represent comparisons of the factor analysis solutions for all pairs of the 6 bins of data. Each point on the graph represents a comparison of results for a particular pair of bins at the given dimension. Each line connects the points for a given bin-pair in consecutive dimensions. The average of the coherences at a given dimension was used to judge the stability of a given stable subspace solution. A criterion of 0.87 average coherence was used to decide that O2 had five stable factors, for this subject. Similarly, F3, C3, P3, and P4 had seven factors. O1 and C4 had five factors, and F4 had four factors.

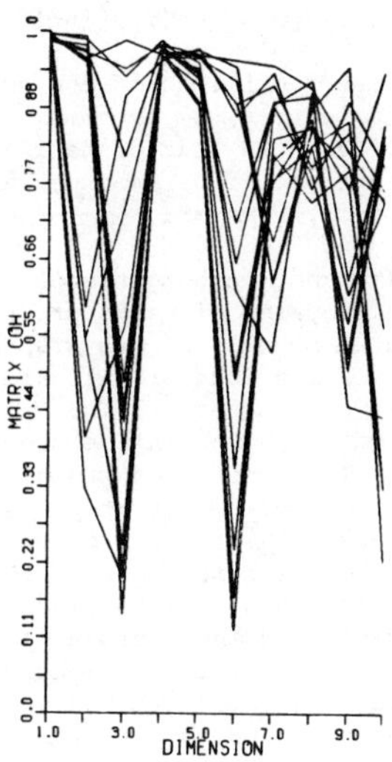

Fig. 1. O2 stability plot: 5 factors.

Step 2, Intersection: Having determined the stable subspaces (in Step 1), it is natural to ask whether the leads have any signals in common. For this determination, we calculate a statistical "intersection".

Suppose that B_1 and B_2 are both rank m (pxm) matrices, representing stable subspaces calculated from two leads' datasets. It is observed that $AC(B_1,B_2)$ is the product of (all m of) the canonical correlations between the subspaces, $\langle B_1 \rangle$ and $\langle B_2 \rangle$. If the datasets have k signals in common, then the product of the first k canonical correlations will be close to 1, and all canonical correlations beyond the first k should not be close to one.

One result of such intersection calculations showed that all of the leads had the same four basic signals in common (denoted I, II, III, and IV, in Fig. 2). Similarly, all the leads but O1 and F4 had an additional fifth signal (V), and F3, C3, P3, and P4 had a sixth signal (VI). Five leads (F3, C3, P3, O1 and P4) each had one signal that was not found at another lead. This accounts for all the signals indicated by the stability computations.

Step 3, Signal Detection: Finally, signal detection is used to estimate the signals within stable intersection subspaces. For the following we will make the simplifying assumption that the two datasets have the same m signals in common, and no more signals. Thus, $L_1 = L_2$. The general case of a lower dimensional intersection is treated in Douglas and Rogers (preprint).

Let B_1 and B_2 (both (pxm) matrices) represent the stable subspaces calculated in Step 1. $B_1 = L_1S_1U_1$ is a representation of the factor analysis rotation indeterminacy, assuming stochastically independent generators. U_1 is an arbitrary orthogonal matrix, and S_1 is a diagonal matrix. The columns of L_1 are unit length vectors. Let $Z_{12} = (B_1^tB_1)^{-1}B_1^tB_2$. Beginning with the assumption, $L_1 = L_2$, straightforward linear algebra can be used to show that

$$U_2^t[Z_{12}^t Z_{12}]U_2 = [S_2]^2[S_1]^{-2} \quad .$$

Since the right side of the equation is a diagonal matrix, U_2 is simply the matrix of eigenvectors that diagonalizes $Z_{12}^t Z_{12}$. This estimate of U_2 produces an estimate of L_2 (= L_1), and S_2.

Signal detection comparison of the subspaces of pairs of leads produced the estimated signal vectors graphed in Fig. 2. The first four signals are those in common for all leads. The signal labelled O1 is the unique signal observed at O1. The superposed graphs come from comparisons of different pairs of leads. The "constant" signal (CN) was calculated using the following observation. The dataset's average vector is a linear combination of the basic signal vectors derived here, plus any signals that are "constant" (i.e., with low variance) for the dataset. Using the stable subspace and the average, the constant signal can also be computed, and estimates of these signals (for all leads except F3, and both occipital leads) appear as the last set of superimposed graphs.

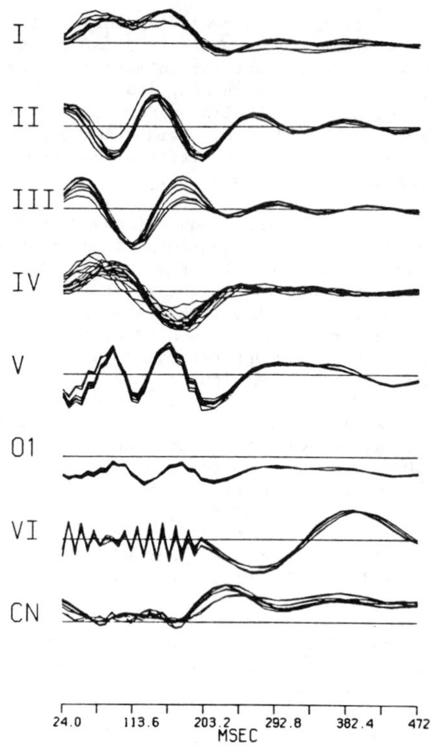

Fig. 2. Derived signals for first subject.

Figure 3 presents signal vectors derived by the same techniques, for a second subject. The first two signal vectors appear to be similar to vectors II and III of Fig. 2, because in each case the two vectors resemble damped sinusoids at the same frequency, but different phase. Furthermore, the constant vectors, which represent a kind of residual, are strikingly similar, both beginning at about 180 ms.

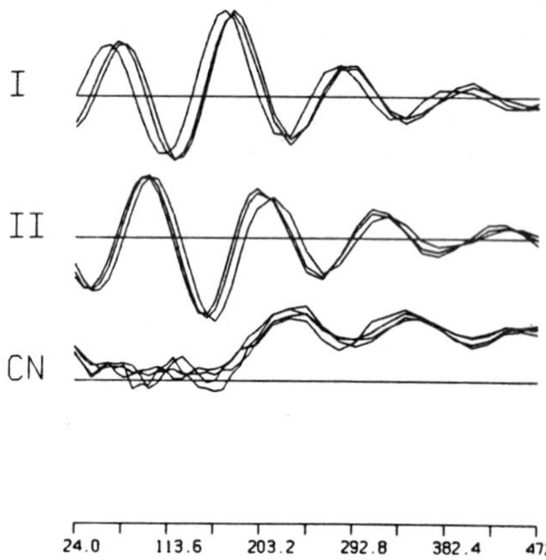

Fig. 3. Derived signals for second subject.

DISCUSSION

The signal vectors shown here are not derivable by any other multivariate technique (known to the authors), and it is likely that they have a physical interpretation. It should be emphasized that there was no mathematical restriction on the forms of the signal vectors. Therefore, it is encouraging that certain features of the observed signal vectors fall within physiological ranges that might be expected for visual system responses. For example, the two signal vectors that appear to be in common for both subjects (II and III, in Fig. 2,